T0181076

Advances in Intelligent Systems and Computing

Volume 554

Series editor

Janusz Kacprzyk, Polish Academy of Sciences, Warsaw, Poland
e-mail: kacprzyk@ibspan.waw.pl

About this Series

The series "Advances in Intelligent Systems and Computing" contains publications on theory, applications, and design methods of Intelligent Systems and Intelligent Computing. Virtually all disciplines such as engineering, natural sciences, computer and information science, ICT, economics, business, e-commerce, environment, healthcare, life science are covered. The list of topics spans all the areas of modern intelligent systems and computing.

The publications within "Advances in Intelligent Systems and Computing" are primarily textbooks and proceedings of important conferences, symposia and congresses. They cover significant recent developments in the field, both of a foundational and applicable character. An important characteristic feature of the series is the short publication time and world-wide distribution. This permits a rapid and broad dissemination of research results.

Advisory Board

Chairman

Nikhil R. Pal, Indian Statistical Institute, Kolkata, India
e-mail: nikhil@isical.ac.in

Members

Rafael Bello Perez, Universidad Central "Marta Abreu" de Las Villas, Santa Clara, Cuba
e-mail: rbellop@uclv.edu.cu

Emilio S. Corchado, University of Salamanca, Salamanca, Spain
e-mail: escorchado@usal.es

Hani Hagras, University of Essex, Colchester, UK
e-mail: hani@essex.ac.uk

László T. Kóczy, Széchenyi István University, Győr, Hungary
e-mail: koczy@sze.hu

Vladik Kreinovich, University of Texas at El Paso, El Paso, USA
e-mail: vladik@utep.edu

Chin-Teng Lin, National Chiao Tung University, Hsinchu, Taiwan
e-mail: ctlin@mail.nctu.edu.tw

Jie Lu, University of Technology, Sydney, Australia
e-mail: Jie.Lu@uts.edu.au

Patricia Melin, Tijuana Institute of Technology, Tijuana, Mexico
e-mail: epmelin@hafsamx.org

Nadia Nedjah, State University of Rio de Janeiro, Rio de Janeiro, Brazil
e-mail: nadia@eng.uerj.br

Ngoc Thanh Nguyen, Wroclaw University of Technology, Wroclaw, Poland
e-mail: Ngoc-Thanh.Nguyen@pwr.edu.pl

Jun Wang, The Chinese University of Hong Kong, Shatin, Hong Kong
e-mail: jwang@mae.cuhk.edu.hk

More information about this series at http://www.springer.com/series/11156

Sanjiv K. Bhatia · Krishn K. Mishra
Shailesh Tiwari · Vivek Kumar Singh
Editors

Advances in Computer and Computational Sciences

Proceedings of ICCCCS 2016, Volume 2

 Springer

Editors
Sanjiv K. Bhatia
Department of Mathematics and Computer
  Science
University of Missouri
St. Louis, MO
USA

Krishn K. Mishra
Department of Computer Science
  and Engineering
Motilal Nehru National Institute
  of Technology
Allahabad, Uttar Pradesh
India

Shailesh Tiwari
CSED
ABES Engineering College
Ghaziabad, Uttar Pradesh
India

Vivek Kumar Singh
Department of Computer Science
Banaras Hindu University
Varanasi, Uttar Pradesh
India

ISSN 2194-5357   ISSN 2194-5365 (electronic)
Advances in Intelligent Systems and Computing
ISBN 978-981-10-3772-6   ISBN 978-981-10-3773-3 (eBook)
https://doi.org/10.1007/978-981-10-3773-3

Library of Congress Control Number: 2017931526

Printed on acid-free paper

This Springer imprint is published by Springer Nature
The registered company is Springer Nature Singapore Pte Ltd.
The registered company address is: 152 Beach Road, #21-01/04 Gateway East, Singapore 189721, Singapore

Preface

The ICCCCS is a major multidisciplinary conference organized with the objective of bringing together researchers, developers and practitioners from academia and industry working in all areas of computer and computational sciences. It is organized specifically to help computer industry to derive the advances of next generation computer and communication technology. Researchers are invited to present the latest developments and technical solutions.

Technological developments all over the world are dependent upon globalization of various research activities. Exchange of information and innovative ideas are necessary to accelerate the development of technology. Keeping this ideology in preference, Aryabhatta College of Engineering & Research Center, Ajmer, India, has come up with an event—International Conference on Computer, Communication and Computational Sciences (ICCCCS-2016) during August 12–13, 2016.

Ajmer, situated in the heart of India, just over 130 km southwest of Jaipur, is a burgeoning town on the shore of the Ana Sagar Lake, flanked by barren hills. Ajmer has historical strategic importance and was ransacked by Mohammed Gauri on one of his periodic forays from Afghanistan. Later, it became a favorite residence of the mighty Mughals. The city was handed over to the British in 1818, becoming one of the few places in Rajasthan controlled directly by the British rather than being part of a princely state. The British chose Ajmer as the site for Mayo College, a prestigious school opened in 1875 exclusively for the Indian Princes, but today open to all those who can afford the fees. Ajmer is a perfect place that can be symbolized for demonstration of Indian culture and ethics and display of perfect blend of a plethora of diverse religions, communities, cultures, linguistics, etc., all coexisting and flourishing in peace and harmony. This city is known for the famous Dargah Sharif, Pushkar Lake, Brahma Temple, and many more evidences of history.

This is for the first time Aryabhatta College of Engineering & Research Center, Ajmer, India, is organizing International Conference on Computer, Communication and Computational Sciences (ICCCCS 2016), with a foreseen objective of enhancing the research activities at a large scale. Technical Program Committee and

Dr. Sushant Upadyaya, MNIT, Jaipur, India
Dr. Akshay Girdhar, GNDEC, Ludhiana, India

Publicity Co-Chair
Mr. Surendra Singh, ACERC, Ajmer, India

Tutorial Chairs
Prof. Lokesh Garg, Delhi College of Technology & Management, Haryana, India

Tutorial Co-Chair
Mr. Ankit Mutha, ACERC, Ajmer, India

Technical Program Committee

Prof. Ajay Gupta, Western Michigan University, USA
Prof. Babita Gupta, California State University, USA
Prof. Amit K.R. Chowdhury, University of California, USA
Prof. David M. Harvey, G.E.R.I., UK
Prof. Madjid Merabti, Liverpool John Moores University, UK
Dr. Nesimi Ertugrual, University of Adelaide, Australia
Prof. Ian L. Freeston, University of Sheffield, UK
Prof. Witold Kinsner, University of Manitova, Canada
Prof. Anup Kumar, M.I.N.D.S., University of Louisville, USA
Prof. Prabhat Kumar Mahanti, University of New Brunswick, Canada
Prof. Ashok De, Director, NIT Patna, India
Prof. Kuldip Singh, IIT Roorkee, India
Prof. A.K. Tiwari, IIT, BHU, Varanasi, India
Mr. Suryabhan, ACERC, Ajmer, India
Dr. Vivek Singh, BHU, India
Prof. Abdul Quaiyum Ansari, Jamia Millia Islamia, New Delhi, India
Prof. Aditya Trivedi, ABV-IIITM Gwalior, India
Prof. Ajay Kakkar, Thapar University, Patiala, India
Prof. Bharat Bhaskar, IIM Lucknow, India
Prof. Edward David Moreno, Federal University of Sergipe, Brazil
Prof. Evangelos Kranakis, Carleton University
Prof. Filipe Miguel Lopes Meneses, University of Minho, Portugal
Prof. Giovanni Manassero Junior, Universidade de São Paulo, Brazil
Prof. Gregorio Martinez, University of Murcia, Spain
Prof. Pabitra Mitra, Indian Institute of Technology Kharagpur, India
Prof. Joberto Martins, Salvador University-UNIFACS, Brazil
Prof. K. Mustafa, Jamia Millia Islamia, New Delhi, India
Prof. M.M. Sufyan Beg, Jamia Millia Islamia, New Delhi, India

Prof. Jitendra Agrawal, Rajiv Gandhi Proudyogiki Vishwavidyalaya, Bhopal, MP, India

Prof. Rajesh Baliram Ingle, PICT, University of Pune, India

Prof. Romulo Alexander Ellery de Alencar, University of Fortaliza, Brazil

Prof. Youssef Fakhri, Université Ibn Tofail, Faculté des Sciences, Brazil

Dr. Abanish Singh, Bioinformatics Scientist, USA

Dr. Abbas Cheddad, (UCMM), Umeå Universitet, Umeå, Sweden

Dr. Abraham T. Mathew, NIT, Calicut, Kerala, India

Dr. Adam Scmidit, Poznan University of Technology, Poland

Dr. Agostinho L.S. Castro, Federal University of Para, Brazil

Prof. Goo-Rak Kwon Chosun University, Republic of Korea

Dr. Alberto Yúfera, Instituto de Microelectrónica de Sevilla (IMSE), (CNM), Spain

Dr. Adam Scmidit, Poznan University of Technology, Poland

Prof. Nishant Doshi, S.V. National Institute of Technology, Surat, India

Prof. Gautam Sanyal, NIT Durgapur, India

Dr. Agostinho L.S. Castro, Federal University of Para, Brazil

Dr. Alok Chakrabarty, IIIT Bhubaneswar, India

Dr. Anastasios Tefas, Aristotle University of Thessaloniki

Dr. Anirban Sarkar, NIT-Durgapur, India

Dr. Anjali Sardana, IIIT Roorkee, Uttarakhand, India

Dr. Ariffin Abdul Mutalib, Universiti Utara Malaysia

Dr. Ashok Kumar Das, IIIT Hyderabad

Dr. Ashutosh Saxena, Infosys Technologies Ltd., India

Dr. Balasubramanian Raman, IIT Roorkee, India

Dr. Benahmed Khelifa, Liverpool John Moores University, UK

Dr. Björn Schuller, Technical University of Munich, Germany

Dr. Carole Bassil, Lebanese University, Lebanon

Dr. Chao Ma, Hong Kong Polytechnic University

Dr. Chi-Un Lei, University of Hong Kong

Dr. Ching-Hao Lai, Institute for Information Industry

Dr. Ching-Hao Mao, Institute for Information Industry, Taiwan

Dr. Chung-Hua Chu, National Taichung Institute of Technology, Taiwan

Dr. Chunye Gong, National University of Defense Technology

Dr. Cristina Olaverri Monreal, Instituto de Telecomunicacoes, Portugal

Dr. Chittaranjan Hota, BITS Hyderabad, India

Dr. D. Juan Carlos González Moreno, University of Vigo

Dr. Danda B. Rawat, Old Dominion University

Dr. Davide Ariu, University of Cagliari, Italy

Dr. Dimiter G. Velev, University of National and World Economy, Europe

Dr. D.S. Yadav, South Asian University, New Delhi

Dr. Darius M. Dziuda, Central Connecticut State University

Dr. Dimitrios Koukopoulos, University of Western Greece, Greece

Dr. Durga Prasad Mohapatra, NIT-Rourkela, India

Dr. Eric Renault, Institut Telecom, France

Dr. Felipe RudgeBarbosa, University of Campinas, Brasil

Dr. Fermín Galán Márquez, Telefónica I+D, Spain
Dr. Fernando Zacarias Flores, Autonomous University of Puebla
Dr. Fuu-Cheng Jiang, Tunghai University, Taiwan
Prof. Aniello Castiglione, University of Salerno, Italy
Dr. Geng Yang, NUPT, Nanjing, P.R. of China
Dr. Gadadhar Sahoo, BIT-Mesra, India
Prof. Ashokk Das, International Institute of Information Technology, Hyderabad, India
Dr. Gang Wang, Hefei University of Technology
Dr. Gerard Damm, Alcatel-Lucent
Prof. Liang Gu, Yale University, New Haven, CT, USA
Prof. K.K Pattanaik, ABV-Indian Institute of Information Technology and Management, Gwalior, India
Dr. Germano Lambert-Torres, Itajuba Federal University
Dr. Guang Jin, Intelligent Automation, Inc.
Dr. Hardi Hungar, Carl von Ossietzky University Oldenburg, Germany
Dr. Hongbo Zhou, Southern Illinois University Carbondale
Dr. Huei-Ru Tseng, Industrial Technology Research Institute, Taiwan
Dr. Hussein Attia, University of Waterloo, Canada
Prof. Hong-Jie Dai, Taipei Medical University, Taiwan
Prof. Edward David, UFS—Federal University of Sergipe, Brazil
Dr. Ivan Saraiva Silva, Federal University of Piauí, Brazil
Dr. Luigi Cerulo, University of Sannio, Italy
Dr. J. Emerson Raja, Engineering and Technology of Multimedia University, Malaysia
Dr. J. Satheesh Kumar, Bharathiar University, Coimbatore
Dr. Jacobijn Sandberg, University of Amsterdam
Dr. Jagannath V. Aghav, College of Engineering Pune, India
Dr. Jaume Mathieu, LIP6 UPMC, France
Dr. Jen-Jee Chen, National University of Tainan
Dr. Jitender Kumar Chhabra, NIT-Kurukshetra, India
Dr. John Karamitsos, Tokk Communications, Canada
Dr. Jose M. Alcaraz Calero, University of the West of Scotland, UK
Dr. K.K. Shukla, IT-BHU, India
Dr. K.R. Pardusani, Maulana Azad NIT, Bhopal, India
Dr. Kapil Kumar Gupta, Accenture
Dr. Kuan-Wei Lee, I-Shou University, Taiwan
Dr. Lalit Awasthi, NIT Hamirpur, India
Dr. Maninder Singh, Thapar University, Patiala, India
Dr. Mehul S. Raval, DA-IICT, Gujarat, India
Dr. Michael McGuire, University of Victoria, Canada
Dr. Mohamed Naouai, University Tunis El Manar and University of Strasbourg, Tunisia
Dr. Nasimuddin, Institute for Infocomm Research
Dr. Olga C. Santos, aDeNu Research Group, UNED, Spain

Dr. Pramod Kumar Singh, ABV-IIITM Gwalior, India
Dr. Prasanta K. Jana, IIT, Dhanbad, India
Dr. Preetam Ghosh, Virginia Commonwealth University, USA
Dr. Rabeb Mizouni, (KUSTAR), Abu Dhabi, UAE
Dr. Rahul Khanna, Intel Corporation, USA
Dr. Rajeev Srivastava, CSE, ITBHU, India
Dr. Rajesh Kumar, MNIT, Jaipur, India
Dr. Rajesh Bodade, Military College of Telecommunication, Mhow, India
Dr. Rajesh Kumar, MNIT, Jaipur, India
Dr. Ranjit Roy, SVNIT, Surat, Gujarat, India
Dr. Robert Koch, Bundeswehr University München, Germany
Dr. Ricardo J. Rodriguez, Nova Southeastern University, USA
Dr. Ruggero Donida Labati, Università degli Studi di Milano, Italy
Dr. Rustem Popa, University "Dunarea de Jos" in Galati, Romania
Dr. Shailesh Ramchandra Sathe, VNIT Nagpur, India
Dr. Sanjiv K. Bhatia, University of Missouri—St. Louis, USA
Dr. Sanjeev Gupta, DA-IICT, Gujarat, India
Dr. S. Selvakumar, National Institute of Technology, Tamil Nadu, India
Dr. Saurabh Chaudhury, NIT Silchar, Assam, India
Dr. Shijo M. Joseph, Kannur University, Kerala
Dr. Sim Hiew Moi, University Technology of Malaysia
Dr. Syed Mohammed Shamsul Islam, The University of Western Australia, Australia
Dr. Trapti Jain, IIT Mandi, India
Dr. Tilak Thakur, PED, Chandigarh, India
Dr. Vikram Goyal, IIIT Delhi, India
Dr. Vinaya Mahesh Sawant, D.J. Sanghvi College of Engineering, India
Dr. Vanitha Rani Rentapalli, VITS Andhra Pradesh, India
Dr. Victor Govindaswamy, Texas A&M University-Texarkana, USA
Dr. Victor Hinostroza, Universidad Autónoma de Ciudad Juárez
Dr. Vidyasagar Potdar, Curtin University of Technology, Australia
Dr. Vijaykumar Chakka, DAIICT, Gandhinagar, India
Dr. Yong Wang, School of IS & E, Central South University, China
Dr. Yu Yuan, Samsung Information Systems America—San Jose, CA
Eng. Angelos Lazaris, University of Southern California, USA
Mr. Hrvoje Belani, University of Zagreb, Croatia
Mr. Huan Song, SuperMicro Computer, Inc., San Jose, USA
Mr. K.K. Patnaik, IIITM, Gwalior, India
Dr. S.S. Sarangdevot, Vice Chancellor, JRN Rajasthan Vidyapeeth University, Udaipur
Dr. N.N. Jani, KSV University Gandhi Nagar, India
Dr. Ashok K. Patel, North Gujarat University, Patan, Gujarat, India
Dr. Awadhesh Gupta, IMS, Ghaziabad, India
Dr. Dilip Sharma, GLA University, Mathura, India
Dr. Li Jiyun, Donghua University, Shanghai, China

About the Editors

Dr. Sanjiv K. Bhatia received his Ph.D. in Computer Science from the University of Nebraska, Lincoln in 1991. He presently works as Professor and Graduate Director (Computer Science) in the University of Missouri, St. Louis. His primary areas of research include image databases, digital image processing, and computer vision. He has published over 40 articles in these areas. He has also been consulted extensively by industry for commercial and military applications of computer vision. He is an expert in system programming and has worked on real-time and embedded applications. He serves on the organizing committee of a number of conferences and on the editorial board of international journals. He has taught a broad range of courses in computer science and was the recipient of Chancellor's Award for Excellence in Teaching in 2015. He is a senior member of ACM.

Dr. Krishn K. Mishra is currently works as a Visiting Faculty, Department of Mathematics and Computer Science, University of Missouri, St. Louis, USA. He is an alumnus of Motilal Nehru National Institute of Technology Allahabad, India, which is also his base working institute. His primary areas of research include evolutionary algorithms, optimization techniques, and design and analysis of algorithms. He has published more than 50 publications in international journals and proceedings of international conferences of repute. He has served as a program committee member of several conferences and also edited Scopus and SCI-indexed journals. He has 15 years of teaching and research experience during which he made all his efforts to bridge the gaps between teaching and research.

Dr. Shailesh Tiwari works as Professor in Computer Science and Engineering Department, ABES Engineering College, Ghaziabad, India. He is also administratively heading the department. He is an alumnus of Motilal Nehru National Institute of Technology Allahabad, India. He has more than 15 years of experience in teaching, research, and academic administration. His primary areas of research include software testing, implementation of optimization algorithms, and machine learning techniques in software engineering. He has also published more than 40 publications in international journals and in proceedings of international conferences of repute. He has served as a program committee member of several

Approach for an Opinion Wrapping System–Using Focused Web Crawler

Gaurav Vats, Vishal Bhatnagar, Rajat Sharma, Ishan Setiya and Arushi Jain

Abstract Most of the search engine depends on web crawler to go through a large number of Webpages. Web crawler (i.e. web spider or scutter or bot) is used to fetch content and URL from the Webpages. It also indexes them so that browser can easily and quickly fetch pages related to the searched word. Tons of data is produced every day, 90% of data has been created in the last 2 years. This data contains opinions and thoughts of the people in unstructured form. Opinion Scutter goes through the content and fetch reviews and comments so that they can be grilled and processed to find useful information. While shopping online or searching any game to buy we are largely dependent on the reviews provided by people. If we can keep track of such reviews and opinion, it will be easy to track a good product and increase efficiency and efficacy of the search engine. The proposed system is a generic crawler which fetches all the reviews from a given site.

Keywords Web crawler · Opinion mining · World Wide Web · Reviews · Webpage parser · Product monitoring

G. Vats (✉) · V. Bhatnagar · R. Sharma · I. Setiya · A. Jain
Ambedkar Institute of Advanced Communication Technologies and Research,
New Delhi, India
e-mail: vats.gaurav101@gmail.com

V. Bhatnagar
e-mail: vishalbhatnagar@yahoo.com

R. Sharma
e-mail: rajvsrajat@gmail.com

I. Setiya
e-mail: setiya.ishan2781@gmail.com

A. Jain
e-mail: arushijain1391@gmail.com

© Springer Nature Singapore Pte Ltd. 2018
S.K. Bhatia et al. (eds.), *Advances in Computer and Computational Sciences*,
Advances in Intelligent Systems and Computing 554,
https://doi.org/10.1007/978-981-10-3773-3_1

Table 2 Example of knowledge table

Domain name	Class name
amazon.in	reviewText
Flipkart.com	rightCol
Tripadvisor.in	Entry

languages like JavaScript and Ajax makes it hard to track the comment or review on different pages. We observed that most website use CSS and the review section is contained and formatted using div tag with specific class, so to make a generalized System we used Knowledge Table (KT) in algorithm. KT is a table, which is formed with pair of domain name and class name of div containing review particular websites (see Table 2).

KT is implemented using hashing so searching time is O(1) which make it Generalized algorithm with better efficiency

```
Opinion_Hunter(){
    1 Providing A Knowledge_Table(KT) which contains
        Knowledge about ClassName of Div of Particular
        websites
    2 Find required ClassName Usong above KT
        String classNAme = compare(KT.URL, FilteredURL);
    3 Extract Reviews in String
        Elements divs = doc.select("div.className");
        for(Elements d : divs){
    4 Store d.text() in Opinion Buffer
    5 call Opinion Screener
    6 Store in Opinion Database
    }                  }
```

4.5 Opinion Screener

This block screens out reviews on the basis of redundancy, only one review from one customer will be entertained. There are people who post more than one reviews to increase the overall rating or to degrade the rival's product rating. Latest comment will be considered by each user. This will prevent our analysis from opinion warfare to some extent and the review will be genuine.

```
Opinion_Screener(){
    1 Check whether the customer have earlier commented
        on that product.
    2 If record exists for that particular customer
        product pair then update the record.
    3 Else enter a new record to Opinion Database.
    }
```

Our proposed model and algorithms will decrease the time taken by the wrapping crawler. As we can deduce from the algorithm:

For a generic crawler time complexity can be written as

$O(u^a)$; where u = total no. URLs & a = total anchor tags

For our crawler time complexity will be

$O(u^{a'} - K) = O(u^{a'})$; where u = total no. URLs & a' = a-total irrelevant tags; K = Keywords

Fig. 2 Module 1 (relevant URL retriever) execution on tripadvisor.in

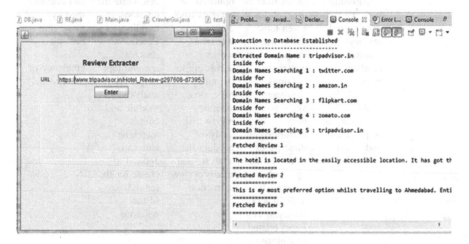

Fig. 3 Module 2 (review extractor) execution on tripadvisor.in

We tested Scalability and Effectiveness our proposed model and above mentioned algorithms in two separate modules on different websites such as Flipkart, TripAdvisor, Amazon and Twitter.

Module-1 (*Relevant URL Retriever*) with implementation of URL Retriever, URL Screener and URL Updater in JAVA (Fig. 2).

Module-2 (*Review Extractor*) with implementation of Opinion Hunter and Screener in JAVA (Fig. 3).

5 Comparison of Existing Framework with Our Proposed Framework

See Table 3.

Table 3 Comparison

	Title (Yahui Xi)	Title (Qinbao Song et al.)	Title (Honhoon kong et. al)	Title (Mfenyana et. al)	Proposed work
Year of publication	2013	2013	2009	2014	2016
Methods	Extracting product features from product reviews	FAST Algorithm for collecting high dimensional data	Modelling web crawler wrappers to collect user reviews on shopping mall	Facebook crawler architecture for opinion monitoring and trend analysis	Opinion wrapping system using focused web crawler
Methodology	Double propagation, i.e. there is somewhat relationship between opinionated text and features the product possesses which helps in the improvement of precision and recall	Focused crawling for collecting relevant pages from the web for a given topic and extracts the pages that contain opinions. Sentiment analysis is then performed on these pages to predict the opinion's polarity. It also finds out the latest web pages related to a given topic by making use of agents	Wrapper model designed to extract opinions or reviews from e-commerce websites works according to the analysis processes such as review list analysis, product list analysis and finally review extraction	Index searching technique is used to determine the opinions trending on Facebook about a specific topic frequency analysis module and context matching pattern is applied to check for the redundancy of the opinions	Focused web crawler which filters the URLs and opinions thereby increasing the efficacy and efficiency of the system
Precision or recall?	Yes	No	No	No	No
Incremental approach or not?	No	No	No	Yes	Yes
Opinion ranking	Yes	No	Opinions are extracted only, not ranked	No	No
URL filtering	No	No	No	No	Yes
Opinion filtering	No	No	No	No	Yes

6 Limitation and Future Work

The work proposed by us have some limitations that needs to be overcome.

We need manual input for URL filtering, as every website uses some keyword for the pages which contains product.

As complexity of websites are increasing day by day with the emergence of new languages it is very difficult to find the content. Languages like CSS, AJAX, Bootstrap and Angular.js are used for the styling and dynamic webpage creation, so basic knowledge of structure is needed for the tracing of user-generated data, we need a knowledge table that defines the naming convention of different site.

The growth of comments/reviews are not static, at times no comments are updated and sometime large number of reviews are updated, so this kind of system cannott be developed which updates the database when any comment is added.

So we have long road ahead in producing a perfect opinion wrapping system. This field has large application and a lot of improvement is needed.

- Our work have some limitations which is our prior goal is to eradicate them.
- To integrate the modules and implement the whole system.
- An automated system can be created to remove the human interference, which finds out the keyword and structure of website so that crawler can directly reach to reviews or comments.
- We will also increase the role of analysis, and restrict the entry of vague comment to the database and save it from opinion warfare.

7 Conclusion

We are successfully able to crawl through and extract reviews from websites like Flipkart, Amazon, and Zomato. The architecture proposed is efficient and can be backbone of any crawling system implementation. It increases the efficiency of the system. These types of system help in the better understanding of the current market and also the interest of the consumers or buyers. The demand can be better predicted and supplied. This architecture can be used for implementation of the blog opinion fetching for forecasting the people mindset regarding a particular topic/trend.

References

1. Qinbao Song, Jingjie Ni and Guangtao Wang "A Fast Clustering-Based FeatureSubset Selection Algorithm for High-Dimensional Data" IEEE, Vol. 25, No.1, January 2013.
2. Vladislav Shkapenyuk and Torsten Suel "Design and Implementation of High performance Distributed Crawler", 2002 IEEE.

3. Sanjeev Dhawan, Kulwinder Singh, Pratibha "Real-time Data Elicitation from Twitter: Evaluation and Depiction Strategies of Tweets Concerned to the Blazing Issues through Twitter Application", 2014 IEEE.
4. Hanhoon Kang, Seong Joon Yoo, Dongil Han "Modelling Web Crawler Wrappers to Collect User Reviews on Shopping Mall withVarious Hierarchical Tree Structure", 2009.
5. Sinesihle I. Mfenyana, Nyalleng Moorosi, Mamello Thinyane "Facebook Crawler Architecture for Opinion Monitoring and Trend Analysis Purposes", August 2014.
6. Mr. Dharav Samani, Mr. Girish Sahastrabuddhe, Mr. Rohit Darji "OPINION MINING BASED SPIDER BOT", IT-NCNHIT 2013.
7. Nikolaos Pappas, Georgios Katsimprasand Efstathios Stamatatos "A System for Up-to-date Opinion Retrieval and Mining in the Web".

Improved Environmental Adaption Method with Real Parameter Encoding for Solving Optimization Problems

Tribhuvan Singh, Ankita Shukla and K.K. Mishra

Abstract Environmental Adaption Method (EAM) was proposed by K.K. Mishra et al. in 2011. Further an improved version of EAM with binary encoding was proposed in 2012, known as Improved Environmental Adaption Method (IEAM) with some changes in adaption operator. IEAM uses adaption, alteration, and selection operators to generate new population. In this paper, we have implemented a real parameter version of IEAM. In IEAM, adaption window of variable bandwidth was used for evolution of solutions, due to this particles could not evolve properly in entire search space. Here, we have used adaption window of fixed bandwidth for proper evolution of solutions. Performance of Improved Environmental Adaption Method with real parameter encoding (IEAM-RP) is compared with other nature-inspired optimization algorithms on Black Box Optimization Test-bed at dimensions 2D, 3D, 5D, and 10D on a set of 24 benchmark functions. It is found that IEAM-RP performs better than other state-of-the-art algorithms.

Keywords Randomized algorithm · EAM · IEAM · Adaption factor · Mobility factor

1 Introduction

To solve optimization problems, randomized algorithms are always a good choice. Randomized algorithms are useful when direction of search is not known in the beginning. It starts search with random search space and finds optimal solution in

T. Singh (✉) · A. Shukla · K.K. Mishra
Computer Science and Engineering Department, MNNIT Allahabad,
Allahabad 211004, India
e-mail: tribhuvan.mnnit@gmail.com

A. Shukla
e-mail: shuklankita321@gmail.com

K.K. Mishra
e-mail: kkm@mnnit.ac.in

© Springer Nature Singapore Pte Ltd. 2018
S.K. Bhatia et al. (eds.), *Advances in Computer and Computational Sciences*,
Advances in Intelligent Systems and Computing 554,
https://doi.org/10.1007/978-981-10-3773-3_2

minimum time. There are various randomized algorithms that use nature inspired search technique such as PSO [1], EA, GA. Environmental Adaption Method (EAM), a randomized optimization algorithm is derived from natural phenomenon [2]. EAM is based on adaptive learning theory. As environment changes either a species can extinct or evolve. Key to survival is to adapt to environmental changes as earlier as possible. EAM has three operator namely adaption, alteration, and selection. After applying these three operator on initial population, next generation population will be generated. In EAM, diversity provided by alteration operator was not enough to solve multimodal problems. To improve diversity in solutions, a new version of EAM named as IEAM was proposed in 2012. New version of any algorithm can be created either by changing its operator or by parameter tuning. In IEAM, there are few changes in operators as well as there is fine-tuning of parameters. In IEAM, adaption operator has best particle which will explore other good regions in the search of optimal solution and other than best particles will use adaption window of variable bandwidth. Due to this, sometimes, solutions are not able to evolve properly. To overcome this problem, a new version of IEAM with real parameter encoding is proposed in this paper. In this paper, we mapped formulae of IEAM into real parameter version. Unlike IEAM, in IEAM-RP, an adaption window of fixed bandwidth is used, due to which solutions are able to properly evolve in the entire search space.

Remaining of this paper is organized as follows. Section 2 provides background details. Section 3 presents our proposed algorithm. Section 4 covers experimental setup. In Sect. 5, result analysis is done and in the last section conclusions are drawn.

2 Background Details

2.1 EAM

EAM uses the theory of adaptive learning [2]. EAM is a population-based algorithm. It uses binary encoding. It uses three operators namely adaption, alteration, and selection. It starts with randomly initialized population. All species will try to improve its phenotypic structure so as to adjust with new environmental conditions with adaption operator and those species which are not able to survive in new environment will be destructed. After this, alteration operator is applied to provide diversity to the solutions. When both of these operators are applied on old population, an intermediate population is generated. After that selection operator is applied. It combines old population and intermediate population and selects best individuals equal to initial population size on the basis of fitness. This process is repeated until we get the best solution or maximum number of iterations has been reached. There were following shortcomings of EAM.

- EAM works with binary encoding. So each time there is need of decimal to binary conversion, which is an extra overhead.
- In higher dimensions, convergent rate is not good enough and prone to Stagnation.

2.2 IEAM

To improve convergence rate of EAM and preventing it to converge on local optimal solution, Improved Environmental Adaption Method (IEAM) was proposed in 2012 by Mishra et al [3, 5]. IEAM uses basic framework of EAM except there is a change in adaption operator. In EAM, each solution updates its structure on the basis of environmental fitness and current fitness. There is no concept of best particles. Unlike EAM, IEAM uses concept used by PSO to update its phenotypic structure. As in PSO, particle updates its structure on the basis of its personal best fitness and global best fitness [1]. In the same way, in IEAM, direction of search is guided by best particle having optimal fitness so far and particle's own fitness. To promote diversity, best particle will explore whole search space and remaining particles will be guided by the best particle. Here, adaption operator is used for exploitation as well as exploration of search space. In IEAM, there is fine-tuning of parameters. With proper setting of parameters global optimal solution can be achieved in early generations.

3 Proposed Approach

Although IEAM has very high convergence rate but it is binary coded. The outcome of binary coding based optimization algorithm depends on how many solutions are considered. If very less number of solutions are taken then there may be a huge difference in obtained solution and desired solution. There is a need of binary to decimal conversion each time which is an extra overhead. There is a need of large number of bits to get accurate results in higher dimensions. To solve these problems, a real parameter version of IEAM, IEAM-RP is suggested. IEAM-RP uses basic framework of IEAM. Like PSO, IEAM uses concept of best particle. In IEAM, Best particle uses the following formula to generate new position

$$P_{i+1} = [\alpha * (P_i)^{F(X_i)/F_{avg}} + \beta]\%2^l, \tag{1}$$

where P_i is the position value of a particle that is updating its structure. α and β are tuning parameters and l is number of bits. $F(X_i)$ is fitness of ith particle and F_{avg} is current environmental fitness. Particles other than best update their positions with formula given below

$$P_{i+1} = [\alpha * (P_i)^{F(X_i)/F_{avg}} + \beta * [(G_b - P_i) + (P_b - P_i)]]\%2^l \tag{2}$$

where G_b is the position vector of best particle and P_b is personal best position vector of the particle that wants to change its position. Values of α and β are taken between 0 to 1.

In real coded version of IEAM, there is no need of binary to decimal conversion. Since we are dealing with real parameters directly, we do not need l (number of bits) any more. Like IEAM, in IEAM-RP clamping is done if the solutions move beyond the search space, but here we do not need modulus operator. Unlike IEAM, in IEAM-RP, we multiply old position by $F(X_i)/F_{avg}$ rather than putting it into the exponent, because if we put this term in exponent then it may result in a very large number or sometimes it may generate a complex number. So, in IEAM-RP best particle will use the following adaption operator

$$P_{i+1} = P_i * F(X_i)/F_{avg} + \beta, \tag{3}$$

where β is any random number between 0 to 1. Like IEAM, other than best particle will move in the direction of best particle to attain phenotypic structure of best particle. In binary IEAM, adaption window was different for each solution, as shown in Eq. 2. Due to variable bandwidth of adaption window, some solutions could not exploit the region properly. To resolve this issue, IEAM-RP uses adaption window of fixed bandwidth (difference between Best_Position and Worst_Position). With fixed bandwidth now solutions exploit properly in the region. Here, formula for adaption of other than best particles is mapped in the following manner

$$P_{i+1} P_i + \beta * (Best\_Position - Worst\_Position) \tag{4}$$

In proposed algorithm, alteration operator of IEAM is not used, because adaption operator here is powerful enough to produce diverse solutions. But alteration operator is not excluded from basic IEAM. In future, if any application needs more diversity than it is provided by IEAM-RP, alteration operator can be added to IEAM-RP. IEAM-RP uses selection operator in the same way as it was used in IEAM. Selection operator is used to select the best solutions equal to the number of initial population size from parent population and offspring. In this way, IEAM-RP ensures elitism. This process continues until stopping criteria is met.

3.1 Algorithms

Adaption Operator: Notations

P_i	Current Population
$Fitness$	Fitness of particle
P_{i+1}	Adapted Population
β	Randam Number between 0 to 1
F_{avg}	Environmental Fitness
AF	Adaption Factor
MF	Mobility Factor

Algorithm 1 Adaption(P_i,Fitness)

1: MF = best_position-worst_position
2: **for** each individual in P_i **do**
3:     $AF = Fitness/F_{avg}$
4: **end for**
5: **for** each individual in P_i **do**
6:     $P_{i+1} = AF * P_i + \beta$ ▷ Best particle will adapt with this formula
7:     $P_{i+1} = P_i + MF * \beta$ ▷ Other particles will adapt with this formula
8: **end for**
9: Clamp the position of particles if they move beyond the range
10: return P_{i+1}

Selection Operator:Notations

T_POP	Temporary Population
S_POP	Population after Sorting
P_{i+1}	Adapted Population
P_i	Current Population
ps	Population Size

Algorithm 2 Selection(P_i,P_{i+1},ps)

1: $T\_POP = \text{merge}(P_i, P_{i+1})$
2: $S\_POP = \text{sort}(T\_POP)$
3: P_i = select ps fittest individual from S_POP
4: return P_i

IEAM-RP: Notations

POP_i	Population at i^{th} generation
P_{i+1}	Adapted Population
MaxGen	Maximum number of generations
$IPOP_i$	Intermediate population at i^{th} generation

Algorithm 3 IEAM-RP

1: Initialize Population POP_1 randomly
2: **repeat**
3:     **for** i = 1 to MaxGen **do**
4:         Evaluate Fitness of each particle
5:         $P_{i+1} = \text{Adaption}(POP_i, Fitness_i)$
6:         $POP_{i+1} = \text{Selection}(POP_i, P_{i+1})$
7:     **end for**
8: **until** stopping criteria is not met or optimal solution is not found

---

3.2 Details of Algorithm

IEAM-RP uses the following steps to generate optimal solution

1. **Initialization Phase**: All solutions are randomly initialized in search space. This creates population for the first generation. Variables are also initialized in this phase
   ps = 100 * DIM;
   fbest = inf;
   xbound = 5;
   fun = FUN;
   Dim = DIM;
   oldpop = 2 * xbound * rand(ps,DIM) −xbound*ones(ps,Dim);
   fitness = feval(fun,oldpop');
   fitness = fitness';
   maxfunevals = min(1e5 * DIM, maxfunevals);
   maxiterations = maxfunevals;

2. **Next Generation Creation**: IEAM-RP uses two basic operators of IEAM that are adaption and selection. Functionality of each operator is explained below

   (a) **Adaption**: Adaption operator both explore and exploit the problem search space. Here, we use the term adaption factor (AF) to represent $F(X_i)/F_{avg}$ which is used for adaption of best particle, and difference of Best_Position and Worst_Position is called as Mobility Factor (MF). MF is responsible for adaption of other than best particles. Best Particle will move in whole search space with AF. AF depends on environmental fitness i.e., average fitness of all particles and their own fitness. Other than best particles will try to attain phenotypic structure of best particle with the help of mobility factor. Here, MF is responsible for providing fixed bandwidth. Function for adaption operator is given below
   $function[new\_pop, fitad] = adaption(oldpop, xbound, Dim, fitness, ps)$
   $xmin = -xbound * ones(1, Dim);$
   xmax = xbound * ones(1,Dim);
   fitad = fitness; favg = mean(fitad);
   favg = favg*ones(ps,Dim);
   fitad3 = repmat(fitad,1,Dim); c = fitad3./favg;
   mb = oldpop(1,:)−oldpop(ps,:);
   $new\_pop(1, :) = c(1, :). * oldpop(1, :) + rand(1, Dim);$
   for h = 2:ps
   $new\_pop(h, :) = oldpop(h, :) + mb. * rand(1, Dim);end$
   $s = new\_pop < repmat(xmin, ps, 1);$
   $new\_pop = (1 - s). * new_pop + s. * repmat(xmin, ps, 1);$
   $b = new\_pop > repmat(xmax, ps, 1);$
   $new\_pop = (1 - b). * new\_pop + b. * repmat(xmax, ps, 1);end$

   (b) **Selection**: This works in the same way as it was in binary encoded IEAM. In each generation, intermediate population and old population are merged, then best individuals equal to initial population size are chosen on the basis of fitness to generate new population. Function for selection operator is given below

$function[sel\_pop, fitsel] = selection(oldpop, new1\_pop, ps, fun, Dim, fitad1)$

$fin = cat(2, oldpop, fitad1);$

$x = feval(fun, new1\_pop');$

$fin1 = cat(2, new1\_pop, x');$

$marge = cat(1, fin, fin1);$

$final\_sort = sortrows(marge, Dim + 1);$

$sel\_pop = final\_sort(1 : ps, 1 : Dim + 1);$

$fitsel = sel\_pop(:, Dim + 1);$

$sel\_pop = sel\_pop(1 : ps, 1 : Dim);$

3. **Generation step and evolution**: In each generation, adaption and selection is applied on parent population to generate offspring. This process is repeated until either maximum number of generations are reached or we get desired optimal solution.

4 Experimental Setup

For experiments, Black Box Optimization Test-bed is used where search domain for all 24 benchmark functions is $[-5, 5]$. The algorithm is tested for dimensions 2-D, 3-D, 5-D, and 10-D with population size 100 * DIM without any restart mechanism.

5 Result Analysis

The performance of IEAM-RP is compared with other optimization algorithms like CMA-TPA, GPSO1, GP1-CMAES, IPOPCMAv3p61, BSrr, GP5-CMAES, BSqi, BSifeg, and BSif. The rank of IEAM-RP for separ, lcond, hcond, multi, multi2, and for all and for dimension 2, 3, 5 and 10 is given in Table 1. To check the efficiency of IEAM-RP, proposed algorithm has been applied to standard 24 COCO benchmark functions [4, 6, 7]. From Table 1 it is obvious that adaption operator of proposed algorithm provides solutions that are divesed enough and due to this it gives better results as compared to other algorithms.

Table 1 IEAM-RP rank in different dimensions

	separ	lcond	hcond	multi	multi2	all
2D	5	2	2	2	2	2
3D	5	2	2	2	1	1
5D	5	2	3	2	2	2
10D	5	2	5	2	2	2

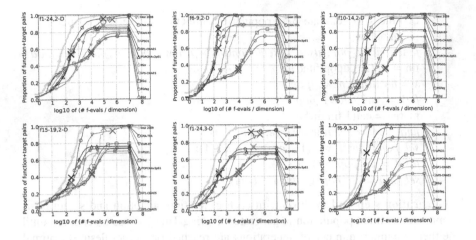

6   Conclusion

A real coded version of IEAM is proposed here. It is different from IEAM in two ways. First it works with real parameters. Second, in adaption operator, we have used adaption window of fixed bandwidth for evolution of other than best particle which makes adaption operator more powerful. Results show that IEAM-RP outperform other state-of-the art algorithms.

References

1. J. Kennedy, R. Eberhart, "Particle swarm optimization," Neural Networks, 1995, IEEE International Conference on Neural Networks, vol. 4, no., pp.1942–1948 vol.4, Nov/Dec 1995.
2. Mishra, K. K., Shailesh Tiwari, and A. K. Misra. "A bio inspired algorithm for solving optimization problems." In Computer and Communication Technology (ICCCT), 2011 2nd International Conference on, pp. 653–659. IEEE, 2011.
3. Mishra, K. K., Shailesh Tiwari, and A. K. Misra. "Improved Environmental Adaption Method for Solving Optimization Problems." In Computational Intelligence and Intelligent Systems, pp. 300–313. Springer Berlin Heidelberg, 2012.
4. http://coco.gforge.inria.fr/.
5. Mishra, K.K., Shailesh Tiwari and A.K. Misra. "Improved environmental adaption method and its application in test case generation" In Journal of Intelligent and Fuzzy Systems xx (20xx) x–xx DOI:10.3233/IFS-141195 IOS Press.
6. N. Hansen et al. Real-parameter black-box optimization benchmarking 2009: Noiseless functions definitions. Technical Report RR-6829, INRIA, 2009. Updated February 2010.
7. N. Hansen et al. Real-parameter black-box optimization benchmarking 2012: Experimental setup. Technical report, INRIA, 2012.

Grouping-Aware Data Placement in HDFS for Data-Intensive Applications Based on Graph Clustering

S. Vengadeswaran and S.R. Balasundaram

Abstract The time taken to execute a query and return the results, increase exponentially as the data size increases, leading to more waiting times of the user. Hadoop with its distributed processing capability can be considered as an efficient solution for processing such large data. Hadoop's default data placement strategy (HDDPS) places the data blocks randomly across the cluster of nodes without considering any of the execution parameters. Also, it is commonly observed that most of the data-intensive applications show grouping semantics. During any query execution only a part of the big data set is utilized. Since such grouping behavior is not considered, the default placement does not perform well, leading to increased execution time, query latency, etc. Hence an optimal data placement strategy based on grouping semantics is proposed. Initially by analyzing the user history log, the access pattern is identified and depicted as an execution graph. By applying Markov clustering algorithm, grouping pattern of the data is identified. Then optimal data placement algorithm based on statistical measures is proposed, which re-organizes the default data layouts in HDFS. This in turn increases parallel execution, resulting in improved data locality and reduced query execution time compared to HDDPS. The experimental results have strengthened the proposed algorithm and has proved to be more efficient for Big-Data sets to be processed in hetrogenous distributed environment.

Keywords Big data · Hadoop · Interest locality · Grouping semantics · Graph clustering · Data placement

S. Vengadeswaran (✉) · S.R. Balasundaram
National Institute of Technology, Tiruchirappalli 620015, Tamil Nadu, India
e-mail: meetvengadesh@gmail.com
URL: http://www.nitt.edu

S.R. Balasundaram
e-mail: blsundar@nitt.edu

© Springer Nature Singapore Pte Ltd. 2018
S.K. Bhatia et al. (eds.), *Advances in Computer and Computational Sciences*,
Advances in Intelligent Systems and Computing 554,
https://doi.org/10.1007/978-981-10-3773-3_3

21

1   Introduction

Large volume of data is being generated every day in a variety of domains such as Social networks, Health care, Finance, Telecom, Government sectors etc., The data which these domains generate are voluminous (GB, PB, and TB), varied (structured, semi-structured, or unstructured) and ever increasing at an unprecedented pace. Big data is thus the term applied to such large volume of data sets whose size is beyond the ability of the commonly used software tools to capture, manage, and process within a tolerable elapsed time [1]. This deluge of data has led to the use of Hadoop to analyze and gain insights from the data. The Apache Hadoop software library is a framework that allows for the distributed processing of large data sets across clusters of computers using simple programming models [2–4]. HDFS is a filesystem designed for storing very large files reliably and streaming data with high bandwidth [5]. By optimizing the storage and processing of HDFS, the queries can be solved earlier.

Hadoop follows master–slave architecture with one Name-Node and multiple Data-Nodes. Whenever a file is pushed into HDFS for storage, the file is split into number of blocks of desired size and placed randomly across the available Data-Nodes. While executing a query, meta information is obtained from the Name-Node about the location of the required blocks and then query is executed in the Data-Node where the required blocks are located. The most important feature of Hadoop is this movement of the computation to the data rather than the way around [1, 6]. Hence, the position of data across the data-nodes plays a significant role in exhibiting an efficient query processing. We focus on finding an innovative data placement strategy so that the queries are solved at the earliest possible time to enable quick decisions as well as to derive maximum utilization of resources. [7–9].

Also, it is commonly observed that most of the data-intensive applications are showing interest locality. It may be different for different domain analysts based on the Geographical location, time, person etc. That is, domain scientists are only interested in a subset of the whole dataset, and are likely to access one subset more frequently than others. For example, in the bioinformatics domain, X and Y chromosomes are related to the offsprings gender. Both chromosomes are often analyzed together in generic research rather than all the 24 human chromosomes [8]. These data blocks will then have the highest frequency to be accessed as a group during query executions. HDDPS does not consider the grouping semantics among the dataset and places the data blocks randomly across the cluster of nodes. Such random placement results in an uneven concentration of these grouped data blocks within few set of nodes. This will ultimately lead to reduced parallel execution thereby increasing query execution time, query latency. In order to overcome such lacunas, an ODPA is proposed based on grouping semantics.

2 Related Works

Several works have been done in the field of data placement for larger data in a distributed environment. Kumar et al. [10] proposed a workload-aware replica selection and placement algorithms that attempt to minimize the total resources consumed in a distributed environment. This strategy minimizes average query span and total energy consumed. But the drawback in this paper is increased query execution time. Also, keeping the available resources underutilized is not a viable solution since the cost of storage is at a declining stage especially after the establishment of cloud environment. Yuan et al. [7] give a matrix-based k-means clustering strategy for data placement in scientific cloud workflows. Accordingly the Grouping-Aware Data Placement in HDFS 3 existing datasets in k data centers are grouped during the workflow build-time stage, and dynamically clusters newly generated datasets to the most appropriate data centers—based on dependencies—during the runtime stage. This strategy guarantees balanced distribution of data and reduced data movements. Wang et al. [8] have proposed an ODPA based on grouping semantics. This proposed strategy reduces the query execution time and improves the data locality compared to default strategy. It improves parallel execution of data sets having interest locality. The drawback of these papers is the use of Bond Energy Algorithm to cluster the Dependency Matrix, since the time complexity of finding permutations of all rows every time for BEA is very high. Also for further execution of any new task the entire iterations of BEA has to be repeated. Schaeffer [11] has suggested different approaches in graph clustering that can be used for identification of the grouping pattern in the given dataset. In this paper, both global and local approaches are reviewed and the delicate issues of selecting an appropriate method for the task at hand, selecting good parameter values are discussed. However the applicability of these approaches on the data-intensive application to arrive at an ODP strategy is not discussed. Golab et al. [12] have presented practical algorithms for minimizing the data communication cost of evaluating a query workload in a distributed setting through graph partitioning. The algorithm is centered on the assumption that each query and its data requirement are independent. Accordingly, the data placement is done by taking into consideration that each query is executed separately within those nodes with least data movement. However, the queries being executed in an environment are seldom arbitrary and exhibit some grouping behavior. Hence these queries with interest locality, i.e., (*queries which are executed normally together or dependent on similar set of data*) will require further consideration. The data placement has to be modified based on the grouping semantics of the queries.

3   Optimal Data Placement Strategy—Proposed work

In this paper, an optimal data placement strategy-based grouping semantics is proposed. The entire workflow diagram is shown in Fig. 1. The various steps involved
in the work are explained subsequently.

Step 1: The input for this step will be the user history log and meta-informations.
These log files are typically large in size and are semi-structured. All Map-reduce
applications executed in cluster, save the task execution details as a log file, which
consists of two files (i) *Job Configuration file*: (ii) *Job Status file* for each job executed
in the machine. Using this as an input, the log files are processed to construct task frequency table containing the list of different tasks executed, frequency of occurrence,
and the required blocks for each task.

Step 2: In this step, the task frequency table obtained in the earlier step is depicted
as a task execution graph shown in Fig. 2. by using graph visualization tool. Task
execution graph is an unordered pair $GTex = (B, T)$ where B represents set of vertices
as blocks and T represents set of edges as tasks executed. GTex is undirected, and
may hold parallel edges since some set of blocks ($B' \subseteq B$) may require different task
executions Ti.

Step 3: The task execution graph ($GTex$) is then converted into clustered task
execution graph ($CGTex$) by applying graph clustering algorithm as shown in Fig. 2.
Graph clustering is the task of grouping the vertices of the graph into clusters taking
into consideration the edge structure of the graph in such a way that there should be
many edges within each cluster and relatively few between the clusters [11]. When

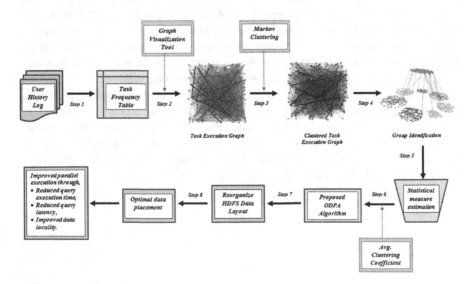

Fig. 1   Work flow diagram for the proposed work

| Blocks | 10 | B1, B2, B3, B4, B5, B6, B7, B8, B9, B10 |
| Nodes | 5 | DN1, DN2, DN3, DN4, DN5 |

Tasks	Blocks required	Task frequency
T1	1, 3, 5, 9, 10, 7	6
T2	3, 6, 8, 2, 4, 9	4
T3	7, 5, 8, 2, 10, 9, 4	3
T4	1, 5, 7, 8, 6	5

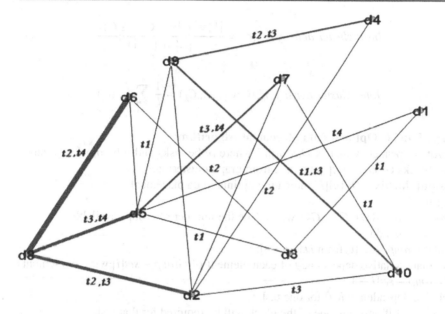

Fig. 2 Task frequency table and task execution graph for sample graph consisting of 10 blocks and 5 nodes

a clustering algorithm is applied to a uniformly distributed graph, the algorithm will arbitrarily group the graph into cluster-based similarity metric. The quality of cluster graph depends on the type of clustering algorithm applied over the input graph. *Markov Cluster [MCL] Algorithm* [13] is fast and scalable unsupervised cluster algorithm, which is applied for finding the natural groupings in the graph.

Step 4: In this step, the clusters obtained from clustered task execution graph (CGTex) are separated into various groups as shown in Fig. 3. A subset of vertices forms a good cluster if the induced subgraph is dense, but there are relatively few connections from the included vertices to vertices in the rest of the graph. Accordingly, each group in the cluster will have individual characteristics showing high Intra cluster density and low Inter cluster density shown in Eqs. (1) and (2).

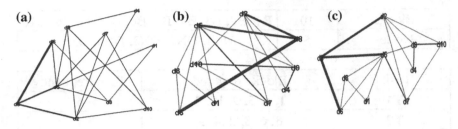

Fig. 3 Various stages of clustered graph by applying MCL algorithm

$$Intra\ cluster\ density\ \delta_{int}(c) = \frac{|\{\{v, u\} | v \in C, u \in C\}|}{|C|(|C| - 1)} \tag{1}$$

$$Inter\ cluster\ density\ \delta_{int}(G | C_1, ..., C_k) = \frac{1}{k} \sum_{i=1}^{k} \delta_{int}(c_i) \tag{2}$$

Steps 5 and 6: Optimal data placement Algorithm:

Input: Dependency matrix $DM[m][n]$ where m be tasks and n be blocks, grouped data blocks $Group - set[$]obtained from graph clustering.

Output: Matrix $ODP[n][k]$ indicating optimal data placement

Begin

$Group - Set = \{G1, G2, ..Gk\}$ where k be the number of groups available

For $i = 1$ to n do

Select $Group - set[i]$ from $DM[m][n]$;

Calculate Relative dependency of each element of $Group - set[i]$ with all elements of $Group - set[i + 1]$;

Relative dependency $R[i]$ for one task is

(a) $R[i] = 1$ if only any one of the block will be required for that task

(b) $R[i] = -1$ if both blocks are required for that task

(c) $R[i] = 0$ if none of the blocks are required for that task

Calculate $sum[i]$ for each $R[i]$;

$S = $ Choose the largest value from $sum[i]$;

$ODP[0][k] = GP[i]$;

$ODP[1][k] = S$;

Delete $GP[i]$ and S from $Group - set$;

Increment i;

Repeat upto $Group - set = \{empty\}$;

End For;

End;

Default-Random						Load Balancer				
Nodes	N1	N2	N3	N4	N5	Nodes	N1	N2	N3	N4	N5
Data placement	1	2	3	6		Data placement	1	2	3	6	7
	5	4	9	8			5	4	9	8	10
	7		10								

Grouping						Proposed				
Nodes	N1	N2	N3	N4	N5	Nodes	N1	N2	N3	N4	N5
Data placement	2	4	7	9	10	Data placement	2	4	7	9	10
	1	3	5	6	8		1	5	3	6	8

Default-Random						Load Balancer					
Tasks	Involved Nodes	#	U_t (%)	T_f	Avg. U_t (%)	Tasks	Involved Nodes	#	U_t (%)	T_f	Avg. U_t (%)
T1	N1,N3	2	40 %	6		T1	N1,N3,N5	3	60 %	6	
T2	N2,N3,N4	3	60 %	4	51.1 %	T2	N2,N3,N4	3	60 %	4	66.7 %
T3	N1,N2,N3,N4	4	80 %	3		T3	N1,N2,N3,N4,N5	5	100 %	3	
T4	N1,N4	2	40 %	5		T4	N1,N4,N5	3	60 %	5	

Grouping						Proposed					
Tasks	Involved Nodes	#	U_t (%)	T_f	Avg. U_t (%)	Tasks	Involved Nodes	#	U_t (%)	T_f	Avg. U_t (%)
T1	N1,N2,N3,N4,N5	5	100 %	6		T1	N1,N2,N3,N4,N5	5	100 %	6	
T2	N1,N2,N4,N5	4	80 %	4	90 %	T2	N1,N2,N3,N4,N5	5	100 %	4	100 %
T3	N1,N2,N3,N4,N5	5	100 %	3		T3	N1,N2,N3,N4,N5	5	100 %	3	
T4	N1,N3,N4,N5	4	80 %	5		T4	N1,N2,N3,N4,N5	5	100 %	5	

\# - Number of nodes utilized, U_t - Utilization percentage,
T_f - Task frequency, Avg. U_t - Average utilization

Fig. 4 Different data placement strategy and resource utilization

Steps 7 and 8: Then ODPA reorganizes default data layouts in HDFS to achieve the maximum parallel execution per group. A motivating example with a cluster of 5 nodes with 10 blocks has been considered for execution of 4 sample tasks. The results are tabulated in Fig. 4, which clearly shows an improved performance of 100% when placed as per ODPA after grouping wherein the performance of default (51.1%), load balancing (66.7%), and grouping (90%). It does not guarantee 100% local map task execution every time but it will always produce an improved result over the default data placement strategy which is tested with several miniature examples.

4 Experimental Results and Analysis

Our proposed strategy has been tested in a 15-node cluster in a single rack environment with Hadoop installed in every node. The cluster has been configured with one Name-Node and 15 Data-Nodes, the nodes are chosen in such a way to hold a heterogeneous environment. The program for execution of ODPA can be executed based on demand and the situation manually. The dataset used for our experiment was Amazon product data [14] which contains product reviews and metadata from Amazon, including 142.8 million reviews spanning May 1996–July 2014. This dataset includes reviews on various categories, viz., Books, Electronics, Movies, Music, Health and personal care, etc. The size of the dataset is 20 GB which is available in Stanford Network Analysis Project [15].

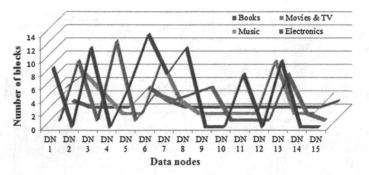

Fig. 5 Default data placement for Amazon product review data set across DN

when this dataset is uploaded in bulk, it is split into even-sized data blocks, and randomly placed across the available Data-Nodes in the cluster. This placement is done without any consideration to the nature of queries likely to be executed, however the data set consist of detail of various categories of reviews each with unique field of interest and hence the data blocks required for executing the queries may also be limited to part of the complete dataset based on grouping behavior. Since the data blocks are randomly placed there is a likelihood of group of required data blocks being concentrated within few Data-Nodes which in turn reduce parallel execution. Even when the uploading of data is done category wise instead of bulk uploading we found that this scenario has not shown any improvement.

Figure 5 shows the placement of data blocks in the cluster for four major categories (Books review −4.4 GB, Movies & TV review −2.8 GB, Music −2.1 GB, and Electronics 325 MB) of our dataset. It is observed that the data related to each of the different category are concentrated within few nodes for, e.g., Data blocks relating to books (71 blocks) is available for execution only in 8 of the 15 nodes (only 53% utilization), in this case we can consider books review as an interest locality since queries relating to books are to be executed only among this 71 blocks rather than the total of about 345 blocks, thereby establishing that Hadoop's default data placement does not consider any grouping semantics of dataset. But our proposed ODPA based on graph clustering identifies grouping semantics of data and are then optimally distributed across the Data-Nodes for improved parallel execution. Different tasks with interest locality are executed and performance of the proposed data placement strategy was studied in comparison with various existing data placement policies such as default, load balancer [16], groupings. The output shown in Table 1 gives an interesting result showing improved local map task and reduced execution time. The graphical representation of the results are shown in Figs. 6 and 7.

Table 1 Performance improvement in Local maps and exe. time—for various tasks

Dataset	Blocks	Interest domain	Task	Total maps	Default		Load balance		Groupings		Proposed	
					Local maps (%)	Exe.time (s)	Local maps (%)	Exe.time (s)	Local maps (%)	Exe.time (s)	Local maps (%)	Exe.time (s)
Amazon product review data (20 GB)	345	Music	T1	34	55.9	692	64.7	630	76.5	552	81.4	507
		Book	T2	71	62.7	1425	67.6	1359	78.9	1275	82.8	1126
		Electronics	T3	6	68	122	76	110	95	90	95	90
		Movies & TV	T4	45	52.5	827	57.8	798	71.1	667	77	560

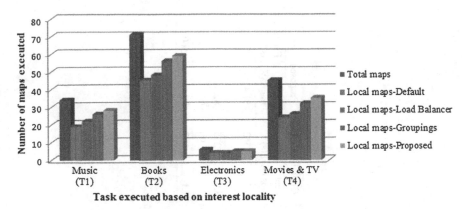

Fig. 6 Graph showing the improvement in local map tasks for proposed strategy

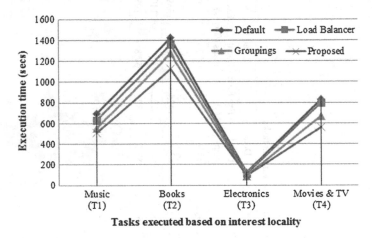

Fig. 7 Graph showing the improvement in execution time for proposed strategy

5 Conclusion and Future work

HDDPS places data randomly across the data-nodes without considering grouping semantics among the data set resulting in concentration of required data blocks
within a few data-nodes thereby reducing parallel execution. Hence, we proposed an
ODPS based on grouping semantics by applying graph clustering algorithm. This in
turn re-organizes the default data layouts in HDFS to achieve the maximum parallel
execution per group resulting in improved performance for Big Data sets in a distributed environment. Our proposed strategy is tested in 20-node cluster. The result has
strengthened proposed algorithm and has proved to be more efficient for massive
datasets by reducing query execution time by 26% and significantly improves the

data locality by 38% compared to HDDPS. Even though the results are optimistic, the behaviors of proposed work in a cross rack environment with replicas have to be further studied for improved performance.

Acknowledgements The research work reported in this paper is supported by Department of Electronics and Information Technology (DeitY), a division of Ministry of Communications and IT, Government of India., under Visvesvaraya PhD scheme for Electronics and IT.

References

1. White, Tom.:Hadoop: The definitive guide. OReilly Media, Inc., (2012).
2. Apache Hadoop, https://hadoop.apache.org/
3. Sammer, Eric.:Hadoop operations. O'Reilly Media, Inc., (2012).
4. Yahoo! Hadoop Tutorial, https://developer.yahoo.com/hadoop/tutorial/
5. Shvachko, K., Kuang, H., Radia, S., & Chansler, R.: The hadoop distributed file system. In: 26th IEEE Symposium on MSST, pp. 1–10, IEEE (2010).
6. Dean, Jeffrey, Sanjay Ghemawat.:MapReduce: simplified data processing on large clusters.Communications of the ACM 51.1, 107–113 (2008).
7. Yuan, D., Yang, Y., Liu, X., & Chen, J.: A data placement strategy in scientific cloud workflows. Future Generation Computer Systems. 26(8), 1200–1214 (2010).
8. Wang, Jun, Pengju Shang, & Jiangling Yin.: DRAW: a new data-grouping-aware data placement scheme for data intensive applications with interest locality. Cloud Computing for Data-Intensive Applications, Springer New York, 149–174 (2014).
9. Lee, C., Hsieh, K., Hsieh, S., & Hsiao,H.: A dynamic data placement strategy for hadoop in heterogeneous environments. Big Data Research, 1, 14–22 (2014).
10. Kumar, A., Deshpande, A., & Khuller, S.: Data placement and replica selection for improving co-location in distributed environments. arXiv:1302.4168 (2013).
11. Schaeffer, S. E.: Graph clustering.: Computer Science Review, 1(1), 27–64 (2007).
12. Golab, L., Hadjieleftheriou, M., Karloff, H. & Saha, B.: Distributed Data Placement via Graph Partitioning. arXiv preprint arXiv:1312.0285 (2013).
13. Van Dongen, Stijn Marinus.: Graph clustering by flow simulation. (2001).
14. McAuley, J., Pandey, R., & Leskovec, J.: Inferring networks of substitutable and complementary products. In: 21st ACM SIGKDD International Conference on Knowledge Discovery and Data Mining, pp. 785–794, ACM (2015).
15. Stanford Network Analysis, https://snap.stanford.edu/data/web-Amazon.html
16. Hadoop Load balancer, https://issues.apache.org/jira/browse/HADOOP-1652

Parameter Estimation for PID Controller Using Modified Gravitational Search Algorithm

Ankush Rathore and Manisha Bhandari

Abstract This paper focuses on the Gravitational Search Algorithm (GSA) which depends on the law of gravity and law of motion. GSA lacks the exploitation property. To improve the exploitation skill, efficiency, and accuracy of GSA, a modified GSA (MGSA) is proposed. The proposed algorithm keeps up a suitable stability between the exploitation and exploration skills of GSA by using an intelligence factor (IF). The MGSA is applied to speed control problem of an induction motor system. Simulations results showed that the modified GSA performs better than GSA and Adaptive Tabu Search (ATS) in terms of computational time and closed-loop response.

Keywords Gravitational search algorithm · Induction motor · Integral square error · Modified GSA

1 Introduction

The induction motor is mainly used for the constant speed control applications. The familiar approach Ziegler Nichols [1] furnishes a bare performance with a large settling time and wide load disturbances for design of PID controller.

Many approaches have been evolved to solve the speed control of an induction motor problem in the past decades. The major approaches append as fuzzy PD and PI controller [2], optimally tuning of the input weight factors by using adaptive tabu search (ATS) [3], MRAC technique [4], integral sliding mode variable structure [5] and robust QFT-based control of DTC [6] has been used to solve speed control of an induction motor problem. These methods cannot conveniently deal with distinct constraints and not to implement easily to find the desired solutions.

In recent years, a metaheuristic algorithm noted as gravitational search algorithm (GSA) proposed by EsmatRashedi et al. [7] and it is depends on the law of motion

A. Rathore (✉) · M. Bhandari
Department of Electronics, Rajasthan Technical University, Kota, India
e-mail: ankushrathore777@gmail.com

© Springer Nature Singapore Pte Ltd. 2018
S.K. Bhatia et al. (eds.), *Advances in Computer and Computational Sciences*,
Advances in Intelligent Systems and Computing 554,
https://doi.org/10.1007/978-981-10-3773-3_4

and Newton's law of gravity. The conventional metaheuristic algorithms do not provide an appropriate solution with a multi-dimensional search space [7] because the search space increases exponentially with respect to the population size.

To find out metaheuristic issues, GSA has been authenticated for eminent performance such as wastewater treatment Process [8], Forecasting future oil demand using GSA [9], optimal power flow in power system [10], Automatic Generation Control [11] and field-sensed magnetic suspension system [12]. Though, utilization of GSA in the metaheuristic problem is still limited.

In this paper a modified gravitational search algorithm is proposed to resolve the speed deviation issue of induction motor [13] and to overcome the problem associated with conventional techniques GSA and adaptive tabu search (ATS).

Rest of the paper is organized as follows: Basic GSA is explained in Sect. 2. The performance of the proposed strategy is analyzed in Sect. 3. Application of MGSA for induction motor system is explained in Sect. 4. Finally, Sect. 5, concludes summarizing the work with results obtained.

2 Gravitational Search Algorithm (GSA)

Gravitational Search Algorithm (GSA) is inspired by the law of gravity [7]: "Every particle in the universe attract every other particle with a force that is directly proportional to the product of their masses and inversely proportional to square of the distance between them." Gravitational search algorithm can be described as

Step-1 The position of an agent is described as

$$X_i = (x_i^1, \ldots, x_i^d, \ldots, x_i^n) \quad for \, i = 1, 2, \ldots\ldots N \tag{1}$$

Here, x_i^d acquaints the position of ith individual in the dth dimension.
Step-2 Evaluation of fitness function:
For minimization problems:

$$\text{best(t)} = \min fit_j(t), \quad j \in 1, \ldots\ldots\ldots, N \tag{2}$$

$$\text{worst(t)} = \max fit_j(t), \quad j \in 1, \ldots\ldots\ldots, N \tag{3}$$

Step-3 Computation of Gravity Constant (G):
The gravity constant $G(t)$ is calculated as

$$G(t) = G_0 e^{-\alpha t/T} \tag{4}$$

where G_0 and α are constant. T is the total number of iteration.

Step-4 Masses of the agent calculation:
Evaluation of masses of the agent are described as inertia mass (M_{ii}), passive gravitational mass (M_{pi}), and active gravitational mass (M_{ai}). This can be identified as the position of the mass.

$$(M_{ii}) = (M_{ai}) = (M_{pi}) = M_i$$

$$m_i(t) = \frac{fit_i(t) - worst(t)}{best(t) - worst(t)} \tag{5}$$

$$M_i(t) = \frac{m_i(t)}{\sum_{j=1}^{N} m_j(t)} \tag{6}$$

where $fit_i(t)$ is value of the agent i at time t.
Step-5 Accelerations of individuals calculation:
Acceleration of candidate solutions depends upon the ratio of force to the mass of the individual as follows:

$$a_i^d(t) = \frac{F_i^d(t)}{M_{ii}(t)} \tag{7}$$

$M_{ii}(t)$ is the mass of ith individual and $F_i^d(t)$ is the total force reacting on individual as procured as

$$F_i^d(t) = \sum_{j \in Kbest, j \neq 1} rand_j F_{ij}^d(t) \tag{8}$$

Force on ith individual by jth individuals mass during iteration t is computed using the following equation:

$$F_{ij}^d(t) = G(t) \frac{M_i(t) * M_j(t)}{R_{ij}(t) + \varepsilon} (x_j^d(t) - x_i^d(t)) \tag{9}$$

Here $G(t)$ is the gravity constant at time t, $M_j(t)$ and $M_i(t)$ are the masses of individual j and i respectively, ε is a small constant, and $R_{ij}(t)$ is the Euclidian distance between two candidate solutions i and j:

$$R_{ij}(t) = \|X_i(t) - X_j(t)\|_2 \tag{10}$$

Step-6 Update the velocity and position of each agent:
Moreover, the velocity of an individual is identified as a fraction of its acceleration value of the agent added to its current velocity. Therefore, new velocity and new position of the ith individual in dth dimension could be evaluated as

$$v_i^d(t+1) = rand_i * v_i^d(t) + a_i^d(t) \tag{11}$$

$$x_i^d(t+1) = x_i^d(t) + v_i^d(t+1) \tag{12}$$

where, $rand_i$ is a constant random variable in the interval [0, 1].

Step-7 Repeat steps 2–6 until their stopping criteria is reached.

3 Modification on GSA-PID Controller

3.1 Drawback of GSA

Two opposing forces are necessary for the metaheuristic algorithms as exploration and exploitation in the search space. Exploration is related to global search in the search space for convenient agents as well as exploitation is disclosed to local search with improve the solution and endeavor to keep away from big leaps in the search space (Fig. 1).

However, GSA lacks the exploitation skill in the last stage of iterations and at the same time the computational cost would elevate acutely. Although the agents cluster to a small domain, the agents cannot meet to the desired result due to the exploitation ability. As a result, the agents get snared in the search space. In the following part, we will incorporate MGSA into the process of GSA to achieve the suitable stability between exploitation and exploration of GSA.

3.2 Modified GSA (MGSA)

The modification part was done by examining the algorithm deportment and concluding that changing the value of new updated velocity. Modified GSA is based on the fact that velocity update equation includes number of iterations so as to increase the search space at the same time an intelligence factor is introduced to keep the search within boundary and balance the updated velocity. This intelligence factor may be system specific. In general, this strategy overcomes the lack of exploitation property of GSA. The update velocity equation for the best solution is given by

$$v_i^d(t+1) = IF * rand_i * v_i^d(t) * iteration + a_i^d(t) \tag{13}$$

where $v_i^d(t+1)$ and $v_i^d(t)$ represents the velocity of the ith solution during iteration $t+1$ and t respectively. $a_i^d(t)$ demonstrates the acceleration of the ith solution

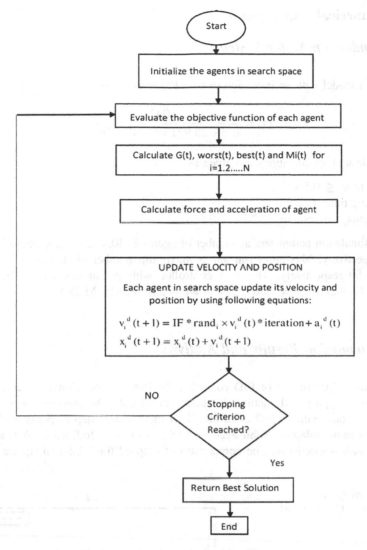

Fig. 1 Flowchart of modified gravitational search algorithm [7]

during iteration t. rand is a random number specified in the interval [0, 1]. The value of an intelligence factor (IF) is set to 0.12 empirically, because according to our experimental results this value of an intelligence factor provides better results for designing the PID controller for an induction motor. The effect of the intelligence factor will be discussed in the following section. This characteristic is important for implementing on a real-time application.

4 Numerical Experiments

4.1 Induction Motor System

The plant model with an induction motor [13] is given as

$$G_p(s) = \frac{168.0436}{s(s^2 + 25.921\,s + 168.0436)} \tag{14}$$

The desired specifications are given as

- Rise time ≤ 0.5 s
- Settling time ≤ 2 s
- Percentage overshoot ≤ 20%

The simulation parameters as number of agents is 30, α and G_0 is set to 20 and 100 respectively, Maximum run and a maximum number of iterations is set as 30 and 50 respectively. The PID controller with parameters $K_P = 10.1705$, $K_i = 1.1664$ and $K_d = 1.9631$ which are determined by MGSA.

4.2 Simulation Results and Analysis

To obtain the parameters of PID controller, MGSA is used. Convergence performance of the proposed method is shown in Fig. 2. The step response of an induction motor without PID controller, Z-N method and step response with PID controller using Adaptive Tabu Search (ATS), GSA and Modified GSA is shown in Fig. 3a, b respectively. The comparison of elapsed time, integral square error,

Fig. 2 Convergence performance of the proposed method

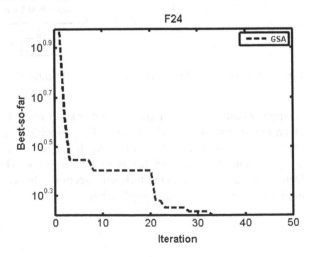

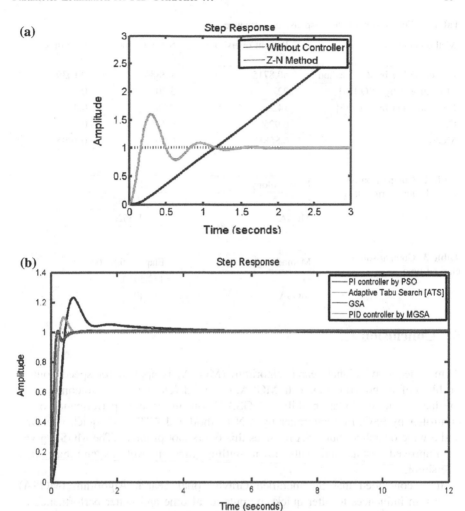

Fig. 3 a Step response without PID controller and Z-N method for an induction motor system.
b Step response with PID controller for an induction motor system

and evaluated responses for an induction motor are summarized in Tables 1, 2, and 3 respectively.

This demonstrates that the modified GSA exhibits better performance in both transient and steady-state response as compared to Adaptive Tabu Search and GSA. This is because MGSA can keep up a suitable stability between the exploitation and exploration proficiency of GSA by using an intelligence factor (IF).

Table 1 Comparison of step response performance

Methodology	Percent overshoot	Settling time (s)	Rise time (s)
PID controller by Z-N method	58.8715	1.3884	0.1039
PI controller by PSO [14]	22	3.20	0.64
PID controller by ATS [3]	14.23	0.58	0.20
GSA	0.9259	0.4958	0.1101
MGSA	0.8588	0.5046	0.1095

Table 2 Comparison of integral square error (ISE)

Methodology	ISE
GSA	2.0056
MGSA	1.6622

Table 3 Comparison of elapsed time

Methodology	Elapsed time (s)
GSA	152.5913
MGSA	84.4076

5 Conclusion

A modified gravitational search algorithm (MGSA) is applied for speed control problem of an induction motor. In MGSA, modified velocity is used to compensate the lack of the exploitation ability of GSA. From the results, performance of PI controller by PSO, PID controller by Z-N method and ATS has a quick rise time and a wide overshoot but GSA resolves this overshoot problem. The MGSA gives an improved results and halts down settling time as well as eliminates the overshoot.

It is concluded that the modified Gravitational Search Algorithm (MGSA) algorithm influences to offer quick computational time and better performance as compared with PI controller by PSO, Z-N method, GSA, and Adaptive Tabu Search (ATS) for design of PID controller for an induction motor system.

References

1. Katsuhiko Ogata. Modern control engineering. Prentice Hall PTR, 2001.
2. Saha-guHadan-dong. A realization fuzzy pi and fuzzy pd controller using a compensation fuzzy algorithms. 2002.
3. D Puangdownreong, T Kulworawanichpong, and S Sujitjorn. Input weight-ing optimization for pid controllers based on the adaptive tabu search. Proc. IEEE TENCON 2004, 4:451–454, 2004.
4. Khaled Halbaoui, DjamelBoukhetala, and FarèsBoudjema. Hybrid adaptive control for speed regulation of an induction motor drive. Archives of Control Sciences, 19(2), 2009.

5. Jianan Wang, Chunming Li, and Huiling Li. The application of integral sliding mode variable structure in induction motor vector control system. In 2010 2nd International Conference on Mechanical and Electronics Engineering, 2010.

6. Luis Ibarra, Pedro Ponce, and Arturo Molina. Robust qft-based control of dtc-speed loop of an induction motor under di erent load conditions. IFAC-Papers Online, 48(3):2429–2434, 2015.

7. EsmatRashedi, Hossein Nezamabadi-Pour, and SaeidSaryazdi. Gsa: a gravitational search algorithm. Information sciences, 179(13):2232–2248, 2009.

8. Chang KyooYoo, Dong Soon Kim, Ji-Hoon Cho, Sang Wook Choi, and In-Beum Lee. Process system engineering in wastewater treatment process. Korean Journal of Chemical Engineering, 18(4):408–421, 2001.

9. MA Behrang, E Assareh, M Ghalambaz, MR Assari, and AR Noghrehabadi. Forecasting future oil demand in iran using gsa (gravitational search algo-rithm). Energy, 36(9): 5649–5654, 2011.

10. SerhatDuman, UğurGüven,c, Yusuf Sönmez, and NuranYorükeren. Optimal power flow using gravitational search algorithm. Energy Conversionand Management, 59:86–95, 2012.

11. Rabindra Kumar Sahu, Sidhartha Panda, and SarojPadhan. Optimal gravitational search algorithm for automatic generation control of interconnected power systems. Ain Shams Engineering Journal, 5(3):721–733, 2014.

12. Jen-Hsing Li and Juing-ShianChiou. Gsa-tuning ipd control of a field-sensed magnetic suspension system. Sensors, 15(12):31781–31793, 2015.

13. Ghang-Ming Liaw and Faa-Jeng Lin. A robust speed controller for induction motor drives. Industrial Electronics, IEEE Transactions on, 41(3):308–315, 1994.

14. DeachaPuangdownreong and SupapornSuwannarongsri. Application of eitelberg's method for pi controllers via particle swarm optimization. In Control, Automation, Robotics and Vision, 2008. ICARCV 2008. 10th International Conference on, pages 566–570. IEEE, 2008.

Auto Improved-PSO with Better Convergence and Diversity

Ashok Kumar, Brajesh Kumar Singh and B.D.K. Patro

Abstract Particle swarm optimization [PSO] is one of the most accepted optimization algorithm and due to its simplicity it has been used in many applications. Although PSO converges very fast, it has stagnation and premature convergence problem. To improve its convergence rate and to remove stagnation problem, some changes in velocity vector are suggested. These changes motivate each particle of PSO in different directions so that full search space can be covered and better solutions can be captured. Moreover, autotuning of random parameters are done to remove stagnation problem and local optima. This auto-improved version is named as AI-PSO algorithm. The performance of the proposed version is compared with various state-of-the-art algorithms such as PSO-TVAC and basic PSO. Results show the superiority of the algorithm.

Keywords PSO · Stagnation · Local optima · Premature convergence

1 Introduction

There are many real-life problems that can be stated as optimization problems. An optimization problem's main objective is to find out minimum or maximum value of an objective function. Optimization problems are Np-hard problem therefore to

A. Kumar (✉)
Department of Computer Science & Engineering,
Dr. A.P.J. Abdul Kalam Technical Universty, Lucknow, India
e-mail: ash_chh@rediffmail.com

B.K. Singh · B.D.K. Patro
Department of CSE, R.B.S. Engineering Technical Campus,
Bichpuri, Agra 283105, India
e-mail: brajesh1678@gmail.com

B.D.K. Patro
e-mail: bdkpatro@rediffmail.com

© Springer Nature Singapore Pte Ltd. 2018   43
S.K. Bhatia et al. (eds.), *Advances in Computer and Computational Sciences*,
Advances in Intelligent Systems and Computing 554,
https://doi.org/10.1007/978-981-10-3773-3_5

solve such problems randomized algorithms like differential evolution and particle swarm optimization [1] can be used.

PSO [1, 2] is a popular population-based nature inspired algorithm which works on the theory of bird flocking and fish schooling. It was proposed in 1995 by James Kennedy and Russell Eberhart. PSO works with random particles and updates the positions of particles in upcoming generations to search the optimum solution. In a n-dimensional search space, the position of ith particle can be represented by an n-dimensional vector $X_i = (x_{i1}, x_{i2},..., x_{in})$ and velocity can be represented by vector $V_i = (v_{i1}, v_{i2},..., v_{in})$. During search, the fitness of each particle is evaluated and two best positions, i.e., previously visited best position by any particle also known as Pbesti $(p_{i1}, p_{i2},..., p_{in})$ and the best of whole swarm global known as Gbest position $G = (g_1, g_2,..., g_n)$ is identified. In each generation, following two equations are used to calculate the velocity and position of ith particle

$$V_i = \omega * V_i + c_1 * r_1 * (P_i - X_i) + c_2 * r_2 * (G - X_i), \tag{1}$$

$$X_i = X_i + V_i, \tag{2}$$

where some parameters like ω, r_1, r_2, c_1, and c_2 are used. ω is known as inertia weight and it is used to decide the impact of previous velocity of particle on its current one. Random variables r_1, r_2 are uniformly distributed random variables within range (0, 1). c_1, c_2 constant parameters are used for exploitation and exploration. In PSO, Eq. (1) is used for calculating the new velocity on the basis of its previous velocity and the distance of its current position from both its personal best position and the best position of the entire population obtained till yet. The value of each component in V can be clamped to the range $[-vmax, vmax]$ to control excessive roaming of particles outside the search space. Then the particle flies toward a new position according Eq. (2). This process is repeated until it reaches a user-defined stopping criterion.

There are two major drawbacks of PSO. The major drawback is its premature convergence, i.e., it could converge to the local minimum solution. Even though convergence rate of PSO is very high, it may converge on local optimum solution. The second drawback is related to its performance based on the problem. This dependency occurs by the way parameters are set, i.e., assigning different parameter settings to PSO will result in high performance variance.

The rest of the paper is organized as follows. In Sect. 2, popular variants of PSO are discussed. In Sect. 3, proposed approach is presented. Section 4 covers the experimental setup and benchmark functions. Section 5 presents comparison of the proposed algorithm AI-PSO with various PSO variants over 24 benchmark functions mentioned in Sect. 4. Finally Sect. 6 draws the conclusion.

2 Related Work

2.1 Variants of PSO

After PSO was developed, several modifications have been done to improve the performance of Basic PSO algorithm. Some of these considerations were focused on limiting the maximum velocity of the particles and selecting the appropriate values of acceleration constants and constriction factors, etc. Both the position and the velocity vectors of particles are updated in each generation t (t = 1, 2, 3 ...) using the Eqs. (3) and (4), respectively:

$$V_i = V_i + c_1 * r_1 * (P_i - X_i) + c_2 * r_2 * (G - X_i), \qquad (3)$$

$$X_i = X_i + V_i, \qquad (4)$$

Many new variants of PSO have been introduced in which either the operators were modified or parameter tuning have been made.

Research article [1, 2] provide detailed information about PSO. M. Senthil Arumugam et al. [3] proposed a modified version of PSO called GLBestPSO, which approaches the global best solution in different ways. Further modifications of same algorithm are done in 2009 by the same authors [4]. Niching concept was first introduced to PSO in [5] so as to increase its ability to solve more complex optimization problems that can search for multiple solutions at the same time. This same concept is used in many other variants of PSO, like ASNPSO (adaptive sequential niche particle swarm optimization) [6], PVPSO (parallel vector-based particle swarm optimizer), etc. Some enhancements were further done in niche based PSO algorithms such as enhancing the Niche PSO [7], adaptively choosing nicking parameters in a PSO [8]. Shi and Eberhart [9] who introduced the concept of using inertia weight ω into the velocity component. They updated the velocity update of Eq. (3) as given below:

$$V_i = \omega * V_i + c_1 * r_1 * (P_i - X_i) + c_2 * r_2 * (G - X_i), \qquad (5)$$

Further, time-dependent linearly varying inertia weight (ω = 0.9 to 0.4) PSO known as PSO-TVIW was introduced by same authors in [10]. Aging leader and challengers-particle swarm optimization (ALC-PSO) was proposed by Chen et al. [11], in which if leader is unable to lead other particles toward the optimal solution, new challengers will be generated that would replace older leader to come out from local optima. PSO-TVIW Ratnaweera et al. [12] designed another variation with time varying acceleration coefficients (TVAC). Moreover, two new variants, i.e., hierarchical PSO with TVAC (HPSO-TVAC) and mutation PSO with TVAC (MPSO-TVAC) [12] are proposed by same authors.

PSO-DR proved itself as a robust optimizer rather than a common PSO implementation on benchmark functions [13]. In [14] has introduced a new set-based particle swarm optimization (S-PSO) algorithm for solving combinatorial optimization problems (COPs) in discrete space [15]. S-PSO is able to characterize the discrete search space of COPs and candidate solution and velocity is defined as a crisp set. S-PSO replaces the updating rules for position and velocity in basic PSO by the procedures and operators defined on crisp. For searching in a discrete search space, S-PSO follows a similar approach to the basic PSO.

3 Auto Improved-PSO with Better Convergence and Diversity (AI-PSO)

It has been observed that the velocity operator of PSO need to be redefined because the operator which was used in PSO to update the velocity of particles is not effective for those particles whose value is equal to pbest and/or gbest. For all such particles whose values are equal to their pbest and/or gbest, the contribution of cognitive and/or social part for the solution will be zero. For example, the particle that is at gbest will be motivated by the velocity equation shown in Eq. (6). Due to this velocity vector, there will be very little change in the velocity of this particle. Hence, there will be a little change in the position of this particle. Similar situation will be faced by those particles in which at least one such component is zero. As these particles are already at the best position so exploitation will not work, they must be directed to search the entire search space and must be used for exploration.

$$V_{i(t+1)}^k = \omega * V_{i(t)}^k \tag{6}$$

For such particles that are already at pbest and gbest, we have updated the equation as follows.

$$V_i = \omega * V_i + c_1 * r_1 * (P_i - X_i) + c_2 * r_2 * (G - X_i) + r_3 * (G - P_i), \tag{7}$$

This additional factor will motivate these particles somewhere in the direction of gbest and pbest so that new good solutions can be identified. Moreover to prevent each particle to converge on local optimum, following values of w, c_1, and c_2 are taken.

$$w = |G - Pavg| \, / \, |G| \tag{8}$$

$$c_1 = |P_i - X_i| \, / \, |P_i| \tag{9}$$

$$c_2 = |(G - X_i)| \, / \, G \tag{10}$$

where Pavg is the average of fitness of all pbest particles. Symbol ‖ indicates the fitness of the mentioned particles not the position particles.

3.1 AI-PSO Procedure

The steps involved in AI-PSO are given as follows.

Symbol	Description
F	Objective function (minimization)
itr	Current iteration number
M	Number of particles (population size)
X_i	Position vector of ith particle, i = 1 to M
V_i	Velocity vector of ith particle, i = 1 to M
$pbest_i$	Personal best position of the ith particle, i = 1 to M (personal best)
gbest	Global best position (group best)

Begin
Step 1: Initialize the position and velocity vectors X and V randomly for each particle in swarm
Step 2: Generate initial values of random parameters and in new generation use Eqs. (8–10).
Step 3: $itr = 1$
Step 4: while $FEs < MaxFEs$
Step 5: Evaluate the fitness of each particle
Step 6: Update velocity vector V appropriately using Eqs. (5) and (7) as applicable
Step 7: Update position vector X using Eq. (4)
Step 8: $itr = itr + 1$
Step 9: end while
End

4 Test Functions and Experimental Setup

To check the efficiency of AI-PSO, this algorithm has been compared with other variants of PSO on standard benchmark functions using comparing continuous optimizers (COCO) software [11].

4.1 Benchmark Functions

Black-box optimization benchmarking (BBOB) functions are taken as benchmark functions. This benchmark function set includes 24 noise-free real-parameter single-objective benchmark functions. Among these 24 functions, five functions form group of separable functions (f1–f5), moderate conditioning functions are collected from (f6–f9) functions, set (f10–f15) contains unimodal functions, multimodal functions with adequate global structure are collected in (f16–f19) set, and group (f20–f24) is used for multimodal functions with weak global structure.

4.2 Experimental Setup

Each variable of these functions is bounded in domain [−5, 5]. These functions can be extended for any number of dimensions; however, in our experiment we have checked these functions for 2, 3, 5, 10, and 20 dimensions. Difference between f_{opt}–f_{target} is checked to terminate the optimization algorithm. The number of runs and maximum allowed function evaluations (FEs) that are allowed for each run are already fixed for BBOB benchmarking and calculated by Eq. (11). See [11] for more details. A size of 20*D is used for all dimensions

$$Maximum\_FEs = 1000*D*Population\_Size \tag{11}$$

where,

$$Population\_Size \sum = \begin{array}{lll} 20*Dimension & for & D = 2, 3 \\ 15*Dimension & for & D = 5 \\ 10*Dimension & for & D = 10, 20 \end{array} \tag{12}$$

Parameter settings of PSO Variants used in comparison

In order to measure the performance of proposed algorithm, this paper compares the AI-PSO with 6 existing PSO variants on all the 24 black-box optimization benchmark functions. Each algorithm is run for up to maximum number of function evaluations defined in Eq. (11). Eq. (12) is used to set the number of particles (population size) in each algorithm.

1. Global version PSO (PSO) [16]: Fix inertia weight $\omega = 0.792$ and fix acceleration coefficients $c_1 = 1.4944$, $c_2 = 1.4944$.
2. PSO with time dependent linearly varying inertia (PSO-TVIW) [11]: Varying inertia weight ($\omega = 0.9$ to 0.4) and fix acceleration coefficients $c_1 = 1.4944$, $c_2 = 1.4944$.
3. Hierarchical PSO with time varying acceleration coefficients (HPSO-TVAC) [17]: Inertia weight $\omega = 0$ and acceleration coefficients change as ($c_1 = 0.5$ to 2.5), ($c_2 = 2.5$ to 0.5).
4. Mutation PSO with time varying acceleration coefficients (MPSO-TVAC) [17]: Inertia weight ($\omega = 0.9$ to 0.4) and acceleration coefficients change as ($c_1 = 0.5$ to 2.5), ($c_2 = 2.5$ to 0.5).
5. Aging leader and challengers—particle swarm optimization (ALC-PSO) [14]: Fix inertia weight $\omega = 0.4$ and fix acceleration coefficients $c_1 = 2.0$, $c_2 = 2.0$. Lifespan of leader $\Theta_0 = 60$ and number of steps to evaluate leading power of challenger $T = 2$.

6. Particle swarm optimizer with adaptive bounds (PSO_Bounds): Dataset is available for noise-free BBOB testbed.
   (http://coco.lri.fr/BBOB2009/rawdata/PSO_Bounds_elabd_noiseless.tar.gz)
7. Auto improved-PSO with better convergence and diversity [Proposed algorithm]: inertia weight $w = |G-Pavg|/|G|$ and the values of $c_1 = |P_i-X_i|/|P_i|$ and $c_2 = |(G-X_i)|/G$

5 Experimental Results and Discussion

It is computed over all relevant trials as the number of function evaluations executed during each trial while the best function value did not reach ftarget, summed over all trials and divided by the number of trials that actually reached ftarget [17–20].

5.1 Comparison of Convergence Rate and Solution Accuracy

The performance of the proposed algorithm is compared with peer algorithms for convergence rate and solution accuracy on 2, 3, 5, 10 and 20 dimensions. Figure 1 shows that in 2-D, the overall performance of AI-PSO is better than other algorithms. For separable functions (f1–f5), ALC-PSO and AI-PSO are comparable to

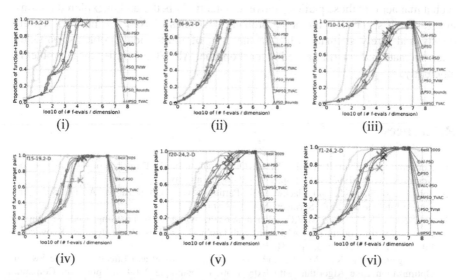

Fig. 1 Convergence graph for BBOB benchmark functions with 2-Dimension

the proposed algorithm. In low or moderate conditioning and unimodal with high conditioning function, AIS-PSO converges faster than the other algorithms. In terms of solution accuracy, AI-PSO achieves optimal solution for all 24 benchmark functions. Further, applying proposed algorithm on 3D, 5D, 10D, and 20D, following results were obtained.

1. Result on 3-D: In 3-D, the performance of AI-PSO is better than all the other algorithms used for comparison. AI-PSO demonstrates significant improvements especially in multimodal functions and it avoids the problem of getting stuck in local optima as well.
2. Result on 5-D: 5-D, includes the comparison of different algorithms. AI-PSO maintains its performance for the functions in 5-D also.
3. Result on 10-D: for 10-D problems overall performance of AI-PSO is best among all algorithms used in comparison.
4. Result on 20-D: AI-PSO performs significantly better than in multimodal problems (f15–f24). As finding global optimal solution for multimodal problems is more difficult, AI-PSO performs comparatively superior than other algorithms.

6   Conclusion

Proposed AI-PSO is able to achieve its aim, i.e., to avoid the stagnation and local optima problem of PSO. With updated value of velocities and random parameters, AI-PSO provides diversity to the particles that get stuck in local optimal solution in such a manner that those particles move to potentially better and unexplored regions in the search space. For unimodal problems particles of AI-PSO improve continuously and rarely stagnate, while in the case of multimodal problems, particles often stagnate and swarm trapped in local optima. AI-PSO identifies these problems efficiently.

References

1. Kennedy, J., Eberhart, R.: Particle Swarm Optimization. In: IEEE International Conference on Neural Networks 1995, vol. 4, pp. 1942–1948 Piscataway.
2. Poli, R., Kennedy, J., and Blackwell, T.: Particle Swarm Optimization: An Overview. In: Swarm Intelligence 2007, Springer, New York, pp. 33–57.
3. Arumugam, S.M., Rao, M.V.C., Chandramohan, A.: A New and Improved Version of Particle Swarm Optimization Algorithm with Global-Local Best Parameters. In: Knowledge Inf. Syst. 2008, vol. 16–3, pp. 331–357.
4. Arumugam, S. M., Rao, M.V.C., and Tan, A.W.C.: A Novel and Effective Particle Swarm Optimization Like Algorithm with Extrapolation Technique 2009. In: ppl. Soft Comput., vol. 9–1, pp. 308–320.

5. Riaan, B., Engelbrecht, A.P., and Van den Bergh F.: A niching particle swarm optimizer. In: Proceedings of the 4th Asia-Pacific conference on simulated evolution and learning 2002. Vol. 2. Singapore: Orchid Country Club.

6. Zhang, Jun, et al.: A novel adaptive sequential niche technique for multimodal function optimization. In: Neurocomputing 69.16 (2006): 2396–2401.

7. Liang, J.J., Qin, A.K., Suganthan, P.N., and Baskar, S.:Comprehensive learning particle swarm optimizer for global optimization of multimodal functions. In: Proceedings of IEEE Congress on Evolutionary Computation 2006, vol. 10, no. 3, pp. 281–295.

8. Stefan, B., and Li, X.: Adaptively choosing niching parameters in a PSO. In: Proceedings of the 8th annual conference on Genetic and evolutionary computation 2006, ACM.

9. Shi, Y., Eberhart, R.C.: A modified particle swarm optimizer. In: Proceedings of IEEE Congress on Evolutionary Computation 1998, pp. 69–73 USA.

10. Shi, Y., Eberhart, R.C.: Empirical study of particle swarm optimization. In: Proceedings of IEEE Congress Evolutionary Computation 1999, pp. 1945–1950.

11. Chen, Y.P., Peng, W.C., and. Jian, M.C.: Particle swarm optimization with recombination and dynamic linkage discovery. In: IEEE Trans. Syst. Man Cybern 2007, vol. 37, no. 6, pp. 1460–1470.

12. Mendes, R., Kennedy, J., and Neves, J.: The fully informed particle swarm: Simpler, maybe better, In: Proceedings of IEEE Congress on Evolutionary Computation 2004, vol. 8, no. 3, pp. 204–210.

13. Shinn-Ying, H., Lin, H.S., Liauh, W.H., and Ho, S.J.: OPSO: Orthogonal particle swarm optimization and its application to task assignment problems. In: Systems, Man and Cybernetics, Part A: Systems and Humans 2008, IEEE Transactions on vol. 38, no. 2 pp. 288–298.

14. Zhou, Di, Jun Sun, and WenboXu. "An advanced quantum-behaved particle swarm optimization algorithm utilizing cooperative strategy."Advanced Computational Intelligence (IWACI), 2010 Third International Workshop on.IEEE, 2010

15. Pena, J., Upegui, A., and Sanchez, E.: Particle swarm optimization with discrete recombination: an online optimizer for evolvable hardware. In: Proceedings of the 1st NASA/ESA Conference on Adaptive Hardware and Systems (AHS '06) 2006, pp. 163–170, Istanbul, Turkey.

16. Angeline, P.J.: Using selection to improve particle swarm optimization. In: Proceedings of IEEE Congress on Evolutionary Computation 1998, pp. 84–89.

17. Hansen, N., Auger, A., Finck, S., and Ros, R.: Real-parameter black-box optimization benchmarking 2009: Experimental setup. Technical Report RR-6828, INRIA, 2009.

18. Hansen, N., Auger, A., Finck, S., and Ros, R.: Real-Parameter Black-Box Optimization Benchmarking 2009. In: Noiseless Functions Definitions," INRIA Technical Report RR-6829, 2009.

19. Goldberg, D. E.: Genetic Algorithms in Search, Optimization, and Machine Learning. Reading MA 1989, Addison-Wesley.

20. Storm, R., and Price, K..V.: Minimizing the real functions of the ICEC 1996 contest by differential evolution. In: Proceedings of IEEE Congress Evolutionary Computation, 1996, pp. 842–844.

A Novel Hybrid PSO–WOA Algorithm for Global Numerical Functions Optimization

Indrajit N. Trivedi, Pradeep Jangir, Arvind Kumar, Narottam Jangir and Rahul Totlani

Abstract Recent trend of research is to hybridize two and more algorithms to obtain superior solution in the field of optimization problems. In this context, a new technique hybrid particle swarm optimization (PSO)–whale optimizer (WOA) is exercised on some unconstraint benchmark test functions. Hybrid PSO–WOA is a combination of PSO used for exploitation phase and WOA for exploration phase in uncertain environment. Analysis of competitive results obtained from PSO–WOA validates its effectiveness compared to standard PSO and WOA algorithm.

Keywords Particle swarm optimization · Whale optimization algorithm · HPSO–WOA

I.N. Trivedi (✉)
Electrical Engineering Department, G.E. College, Gandhinagar, Gujarat, India
e-mail: forumtrivedi@gmail.com

P. Jangir · N. Jangir
Electrical Engineering Department, LEC, Morbi, Gujarat, India
e-mail: pkjmtech@gmail.com

N. Jangir
e-mail: nkjmtech@gmail.com

A. Kumar
Electrical Engineering Department, S.S.E.C, Bhavnagar, Gujarat, India
e-mail: akbharia8@gmail.com3

R. Totlani
Electrical Engineering Department, JECRC, Jaipur, Rajasthan, India
e-mail: rhl.totlani@gmail.com

1 Introduction

HPSO–WOA comprises best characteristic of both particle swarm optimization [1] and whale optimizer algorithm [2]. HPSO–WOA result expresses that it has ability to converse faster with comparatively optimum solution for both unconstrained function.

Population-based algorithms based on randomization consists of two main phases for obtaining better results that are exploration (unknown search space) and exploitation (best solution).

In this HPSO–WOA, WOA is applied for exploration as it uses logarithmic spiral path so covers large uncertain search space with less computational time to explore possible solution or to converse particle toward optimum value. Most popular PSO algorithms have ability to attain near optimal solution avoiding local solution.

Contemporary works in hybridization are PBIL-KH [3] the population-based incremental learning (PBIL) with KH, a type of elitism is applied to memorize the krill with the best fitness when finding the best solution, KH-QPSO [4] is intended for enhancing the ability of the local search and increasing the individual diversity in the population, HS/FA [5] the exploration of HS and the exploitation of FA are fully exerted, CKH [6] the chaos theory into the KH optimization process with the aim of accelerating its global convergence speed, HS/BA [7], CSKH [8], DEKH [9], HS/CS [10], HSBBO [11] are used for the speeding up convergence, thus making the approach more feasible for a wider range of real-world applications.

The structure of the paper is given as follows: Introduction; description of participated algorithms; competitive results analysis of unconstraint test benchmark problem; and finally, conclusion based on results is drawn.

2 Particle Swarm Optimization

The particle swarm optimization algorithm (PSO) was discovered by James Kennedy and Russell C. Eberhart in 1995 [1]. This algorithm is inspired by simulation of social psychological expression of birds and fishes. PSO includes two terms P best and G best. Position and velocity are updated over the course of iteration from these mathematical equations:

$$v_{ij}^{t+1} = w v_{ij}^t + c_1 R_1 (Pbest^t - X^t) + c_2 R_2 (Gbest^t - X^t) \tag{1}$$

$$X^{t+1} = X^t + v^{t+1}, (i = 1, 2 \ldots NP) \ and \ (j = 1, 2 \ldots NG) \tag{2}$$

where,

$$w = w^{\max} - \frac{\left(w^{\max} - w^{\min}\right) * iteration}{\max iteration} \tag{3}$$

$w^{\max} = 0.4$ and $w^{\min} = 0.9$.

v_{ij}^{t}, v_{ij}^{t+1} is the velocity of jth member of ith particle at iteration number (t) and (t + 1). (Usually C1 = C2 = 2), r1 and r2 Random number (0, 1).

3 Whale Optimization Algorithm

In the meta-heuristic algorithm, a newly purposed optimization algorithm called whale optimization algorithm (WOA), which inspired from the bubble-net hunting strategy. Algorithm describes the special hunting behavior of humpback whales; the whales follows the typical bubbles causes the creation of circular or '9-shaped path' while encircling prey during hunting. Simply bubble-net feeding/hunting behavior could understand such that humpback whale went down in water approximate 10–15 m and then after the start to produce bubbles in a spiral shape encircles prey and then follows the bubbles and moves upward the surface. Mathematic model for whale optimization algorithm (WOA) is given as follows:

3.1 Encircling Prey Equation

Humpback whale encircles the prey (small fishes) then updates its position toward the optimum solution over the course of increasing number of iteration from start to a maximum number of iteration.

$$\vec{D} = \left| C \cdot \vec{X} * (t) - X(t) \right| \tag{4}$$

$$\vec{X}(t+1) = \vec{X} * (t) - \vec{A} \cdot \vec{D} \tag{5}$$

where $\vec{A}$, $\vec{D}$ are coefficient vectors, t is a current iteration, $\vec{X} * (t)$ is position vector of the optimum solution so far, and $X(t)$ is position vector.

Coefficient vectors $\vec{A}$, $\vec{D}$ are calculated as follows:

$$\vec{A} = 2\vec{a} * r - \vec{a} \tag{6}$$

$$\vec{C} = 2 * r \tag{7}$$

where $\vec{a}$ is a variable linearly decrease from 2 to 0 over the course of iteration and r is a random number [0, 1].

3.2 Bubble-Net Attacking Method

In order to mathematical equation for bubble-net behavior of humpback whales, two methods are modeled as

(a) **Shrinking encircling mechanism**

This technique is employed by decreasing linearly the value of $\vec{a}$ from 2 to 0. Random value for a vector $\vec{A}$ in a range between [−1, 1].

(b) **Spiral updating position**

Mathematical spiral equation for position update between humpback whale and prey that was helix-shaped movement given as follows:

$$\vec{X}(t+1) = \vec{D} * e^{bt} * \cos(2\pi l) + \vec{X} * (t) \tag{8}$$

where l is a random number [−1, 1], b is constant defines the logarithmic shape, $\vec{D'} = \left| \vec{X} * (t) - X(t) \right|$ expresses the distance between ith whale to the prey mean the best solution so far.

Note: We assume that there is 50–50% probability that whale either follow the shrinking encircling or logarithmic path during optimization. Mathematically we modeled as follows:

$$\vec{X}(t+1) = \begin{cases} \vec{X} * (t) - \vec{A}\vec{D} & if \quad p < 0.5 \\ \vec{D} * e^{bt} * \cos(2\pi l) + \vec{X} * (t) & if \quad p \geq 0.5 \end{cases} \tag{9}$$

where p expresses random number between [0, 1].

(c) **Search for prey**

The vector $\vec{A}$ can be used for exploration to search for prey; vector $\vec{A}$ also takes the values greater than one or less than −1. Exploration follows two conditions

$$\vec{D} = \left| \vec{C} \cdot \vec{X_{rand}} - \vec{X} \right| \tag{10}$$

$$\vec{X}(t+1) = \vec{X_{rand}} - \vec{A} \cdot \vec{D} \tag{11}$$

Finally follows these conditions:

- $\left|\overrightarrow{A}\right| > 1$ enforces exploration to WOA algorithm to find out global optimum avoids local optima
- $\left|\overrightarrow{A}\right| < 1$ For updating the position of current search agent/best solution is selected.

Flowchart of whale optimization algorithm is shown in Fig. 1.

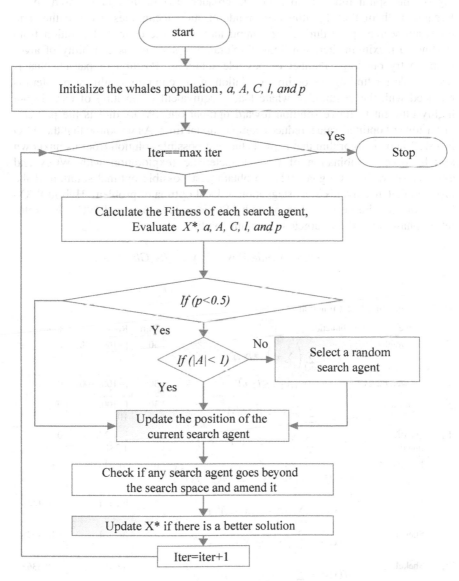

Fig. 1 Flowchart of WOA

4 The Hybrid PSO–WOA Algorithm

A set of hybrid PSO–WOA is a combination of separate PSO and WOA. The drawback of PSO is the limitation to cover small search space while solving higher order or complex design problem due to constant inertia weight. This problem can be tackled with Hybrid PSO–WOA as it extracts the quality characteristics of both PSO and WOA. Whale optimizer algorithm is used for exploration phase as it uses logarithmic spiral function so it covers broader area in uncertain search space. Because both of the algorithms are randomization techniques, we use the term uncertain search space during the computation over the course of iteration from starting to maximum iteration limit. Exploration phase means capability of algorithm to try out large number of possible solutions. Position of particle that is responsible for finding the optimum solution of the complex nonlinear problem is replaced with the position of whale that is equivalent to position of particle but highly efficient to move solution toward optimal one. WOA directs the particles faster toward optimal value, reduces computational time. As we know that that PSO is a well-known algorithm that exploits the best possible solution from its unknown search space, combination of best characteristic (exploration with WOA and exploitation with PSO) guarantees to obtain best possible optimal solution of the problem that also avoids local stagnation or local optima of problem. Hybrid PSO–WOA merges the best strength of both PSO in exploitation and WOA in exploration phase toward the targeted optimum solution.

$$v_{ij}^{t+1} = wv_{ij}^{t} + c_1 R_1 (Whale\_Pos^{t} - X^{t}) + c_2 R_2 (Gbest^{t} - X^{t}) \qquad (12)$$

Table 1 Benchmark test function

No.	Name	Function	Dim	Range	Fmn
F_1	Schwefel 1.2	$f(x) = \sum\limits_{i=1}^{n} \left(\sum\limits_{j-1}^{i} x_j \right)^2 * R(x)$	30	[−100, 100]	0
F_2	Schwefel 2.21	$f(x) = \max\limits_{i} \{|x_i|, 1 \le i \le n\}$	30	[−100, 100]	0
F_3	Step function	$f(x) = \sum\limits_{i=1}^{n} ([x_i + 0.5])^2 * R(x)$	30	[−100, 100]	0
F_4	Quartic function	$f(x) = \sum\limits_{i=1}^{n} i x_i^4 + random[0, 1) * R(x)$	30	[−1.28, 1.28]	0
F_5	Hartman 2	$f(x) = - \sum\limits_{i=1}^{4} c_i \exp\left(- \sum\limits_{j=1}^{6} a_{ij}(x_j - p_{ij})^2 \right)$	6	[0, 1]	−3.32
F_6	Shekel 1	$f(x) = - \sum\limits_{i=1}^{5} \left[(X - a_i)(X - a_i)^T + c_i \right]^{-1}$	4	[0, 10]	−10.1532
F_7	Shekel	$f(x) = - \sum\limits_{i=1}^{7} \left[(X - a_i)(X - a_i)^T + c_i \right]^{-1}$	4	[0, 10]	−10.4028
F_8	Shekel	$f(x) = - \sum\limits_{i=1}^{10} \left[(X - a_i)(X - a_i)^T + c_i \right]^{-1}$	4	[0, 10]	−10.5363

5 Simulation Results for Unconstraint Test Benchmark Function

Unconstraint test benchmark functions (Schwefel 1.2, Schwefel 2.21, Step Function, Quartic Function, Hartman 2, and Shekel's family) are given in Table 1. Search agent no is 30 used for all functions. Convergence curve of test function is shown in Fig. 2 proof that PSO–WOA algorithm has very good results compare to standard particle swarm optimization and recently proposed whale optimization algorithm.

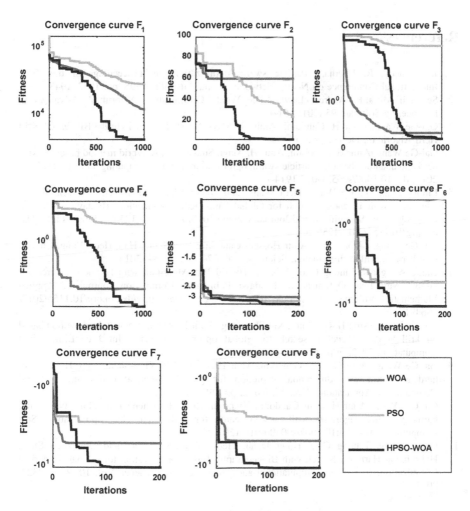

Fig. 2 Convergence characteristics of benchmark test functions

6 Conclusion

The drawback of PSO is the limitation to cover small search space while solving higher order or complex design problem due to constant inertia weight. This problem can be tackled with hybrid PSO–WOA as it extracts the quality characteristics of both PSO and WOA. WOA is used for exploration phase as it uses logarithmic spiral function so it covers broader area in uncertain search space. So WOA directs the particles faster toward optimal value, reduces computational time. HPSO–WOA is tested on some unconstrained problems. HPSO–WOA gives better results in most of the cases and in some cases results are inferior that demonstrate the enhanced performance with respect to original PSO and WOA.

References

1. J. Kennedy, R. Eberhart, Particle swarm optimization, in: Proceedings of the IEEE International Conference on Neural Networks, Perth, Australia, 1995, pp. 1942–1948.
2. Seyedali Mirjalili, Andrew Lewis "The Whale Optimization Algorithm" Advances in Engineering Software 95 (2016) 51–67.
3. Gai-Ge Wang, Amir H. Gandomi, Amir H. Alavi, Suash Deb, A hybrid PBIL-based Krill Herd Algorithm, December 2015.
4. Gai-Ge Wang, Amir H. Gandomi, Amir H. Alavi, Suash Deb, A hybrid method based on krill herd and quantum-behaved particle swarm optimization, Neural Computing and Applications, 2015, doi:10.1007/s00521-015-1914-z.
5. Lihong Guo, Gai-Ge Wang, Heqi Wang, and Dinan Wang, An Effective Hybrid Firefly Algorithm with Harmony Search for Global Numerical Optimization, Hindawi Publishing Corporation the Scientific World Journal Volume 2013, Article ID 125625, 9 pages http://dx.doi.org/10.1155/2013/125625.
6. Gai-Ge Wang, Lihong Guo, Amir Hossein Gandomi, Guo-Sheng Hao, Heqi Wang. Chaotic krill herd algorithm. Information Sciences, Vol. 274, pp. 17–34, 2014.
7. Gaige Wang and Lihong Guo, A Novel Hybrid Bat Algorithm with Harmony Search for Global Numerical Optimization, Hindawi Publishing Corporation Journal of Applied Mathematics Volume 2013, Article ID 696491, 21 pages http://dx.doi.org/10.1155/2013/696491.
8. Gai-Ge Wang, Amir H. Gandomi, Xin-She Yang, Amir H. Alavi, A new hybrid method based on krill herd and cuckoo search for global optimization tasks. Int J of Bio-Inspired Computation, 2012, in press.
9. Gai-Ge Wang, Amir Hossein Gandomi, Amir Hossein Alavi, Guo-Sheng Hao. Hybrid krill herd algorithm with differential evolution for global numerical optimization. Neural Computing & Applications, Vol. 25, No. 2, pp. 297–308, 2014.
10. Gai-Ge Wang, Amir Hossein Gandomi, Xiangjun Zhao, HaiCheng Eric Chu. Hybridizing harmony search algorithm with cuckoo search for global numerical optimization. Soft Computing, 2014. doi:10.1007/s00500-014-1502-7.
11. Gaige Wang, Lihong Guo, Hong Duan, Heqi Wang, Luo Liu, and Mingzhen Shao, Hybridizing Harmony Search with Biogeography Based Optimization for Global Numerical Optimization, Journal of Computational and Theoretical Nanoscience Vol. 10, 2312–2322, 2013.

Moth-Flame Optimizer Method for Solving Constrained Engineering Optimization Problems

R.H. Bhesdadiya, Indrajit N. Trivedi, Pradeep Jangir and Narottam Jangir

Abstract A recently proposed swarm inspired optimization algorithm based on the navigation approach of Moths in space entitle as Moth-Flame Optimization (MFO) algorithm, is used for solve equality and inequality constrained optimization, real challenging layout problems. The navigating strategy of moths in universe entitles transverse orientation, a well active mechanism for travel so far distance in the straight direction. In fact, artificial lights trick moths, so they follow a deadly spiral path. MFO algorithm gives the competitive results with both continuous and discrete control variables. Real Challenging Constrained Optimization is a way of optimising an objective function in presence of constraints on some control variables. MFO have an ability to solve both constraints that may be either hard constrained or soft constrained. A statical representation of MFO algorithm expresses the best objective function value with reference to accuracy and standard deviation over recently proposed and most popular optimization algorithms. Fourteen constrained benchmark function of real engineering problems have been calculated and gained solutions were compared with the solution obtained by various recognized algorithms. The results obtained through MFO algorithm represent better solutions in the field of engineering design problems among many recently developed algorithms.

R.H. Bhesdadiya (✉)
Electrical Engineering Department, School of Engineering, RK University,
Rajkot, Gujarat, India
e-mail: rhbhesdadiya@gmail.com

I.N. Trivedi
Electrical Engineering Department, G.E. College, Gandhinagar, Gujarat, India
e-mail: forumtrivedi@gmail.com

P. Jangir · N. Jangir
Electrical Engineering Department, LEC, Morbi, Gujarat, India
e-mail: pkjmtech@gmail.com

N. Jangir
e-mail: nkjmtech@gmail.com

© Springer Nature Singapore Pte Ltd. 2018
S.K. Bhatia et al. (eds.), *Advances in Computer and Computational Sciences*,
Advances in Intelligent Systems and Computing 554,
https://doi.org/10.1007/978-981-10-3773-3_7

Keywords Meta-heuristic · Moth-flame optimizer · Constraint optimization · Constrained handling · Global optimal · Design · Navigation

1 Introduction

A recently proposed swarm inspired optimization algorithm MFO [1] based on the navigation mechanism for flying in the straight direction in night called transverse position of Moths in universe. Transverse position for navigation apply a fixed angle by Moths with reference to Moon as Moon is far away from Moths so this mechanism guarantees for flying in linear direction. This transverse mechanism is applied for artificial lights nearer to Moths compared to Moon, then Moths are tricked as they do not keep constant angle with respect to flame or angle is reducing so fly over flames in scale of Logarithmic spiral way and after merge close to the flame. Spiral path represent best exploration area of an algorithm with respect of unknown search space.

In former years, different types of optimization methods introduce among that some optimization algorithm used derivative concept to solve different type of problems due to gradient that algorithms suffer from more than one best and peak values. Population-based MFO is a meta-heuristic optimization methods have an ability to jump local best value and get global optimal solution that make it appropriate for practical applications without structural modifications for solving different constrained Optimization problems. All numerical results obtained with different techniques shown in tables, are explicitly taken from [1–3, 4] directly without modification.

Remaining paper organize the following manner: The next section describes the Moth-flame optimizer algorithm and its equations are given in Sect. 2. Section 3 includes equality, inequality constraint optimization and classical well-known engineering layout problems including their structure, convergence curve and results compared with former algorithms. In Sect. 4 conclusions are drawn.

2 Moth-Flame Optimizer

Moth-Flame [1] optimization method first proposed by S. Mirjalili in 2015. It is population-depended meta-heuristic approach. The MFO method uses three rows for approximate, global solution for problems that describe as given steps:

$$\text{Moth-Flame Optimizer} = [Q, \ Z, \ X], \tag{1}$$

Q is the objective that yields an uncertain population of different moths with respect to fitness values. Considering Q, Z and X these points, we use a log scale Spiral for the MFO method below describe as:

$$s\left(m_p, f_q\right) = d_p * e^{bt} \cos 2\pi t + f_q \tag{2}$$

where d_p expresses the distance of the moth for the qth flame, b is a constant for expressing the shape of the log (logarithmic) spiral, and t is a random value in $[-1, 1]$.

$$d_p = \left| f_q - m_p \right|, \tag{3}$$

where m_i indicate the pth moth, f_q indicates the qth flame, and where d_i expresses the path length of the pth moth for the qth flame. The absolute value of moths represents minimum distance between two swarms.

The flame numbers are decreased over when number of iterations are increased until that have not equal to maximum number of iterations. We use a formula to calculate flame number as given below:

$$flame\ number = round\left(n - iter * \frac{n-1}{t} \right), \tag{4}$$

where $iter$ represent recent iteration, n represent max. number of flames, and t represent total number of iteration that required to solve problems.

3 Constrained Engineering Design Problem

Fourteen constrained engineering classical design problems are evaluated and they may have several inequality constraints. The process of MFO algorithm does not need to be changed during Optimization. For this purpose, we assign simplest constrained handling, penalty function, method where search agents are work at objective function to evaluate optimum result if they violate any constrains.

3.1 Car Side Impact Design

In the mechanical area impact design of car side [5] is to objective minimize weight using nine different constraints influence parametric mixers including materials of B-Pillar inner [5], floor side inner [6], barrier height [6], hitting position, thicknesses of B-Pillar inner [7], B-Pillar inner reinforcement, floor side inner, cross members [7], door beam, door beltline reinforcement and roof rail, the car side impact design problem is entitling by Gandomi et al. [8]. Figure 1 show iteration versus objective function value of car side impact design, Comparative statistical results from many optimization techniques such as PSO [5], FA [6] and DE [7] are compared with MFO results shown in Table 1.

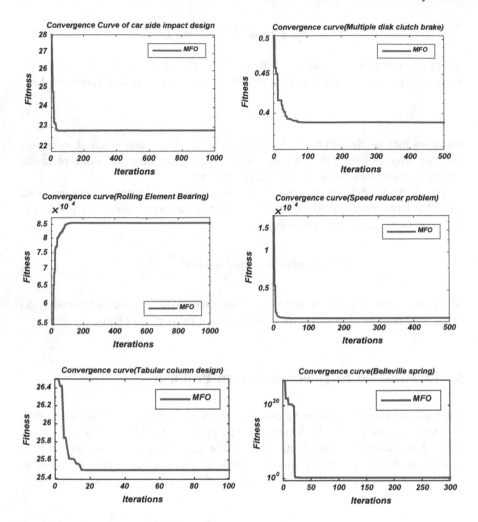

Fig. 1 Convergence curve of design problems

Table 1 Comparative
statistical results from other
former optimization
techniques car side impact
design problem

Method	Mean value	Best value	S.D.
PSO	22.89429	22.84474	0.15017
DE	23.22828	22.84298	0.34451
FA	23.22828	22.84298	0.16667
MFO	22.85979	22.84252	0.065966

Table 2 Comparative statistical results from other former optimisation techniques for multiple disk clutch brake design problem

D.V	NSGA-II	WCA	MFO
$X1$	70.0000	70.0000	70.009291
$X2$	90.0000	90.0000	90.000
$X3$	1.50000	1.00000	1.5000
$X4$	1000.0000	910.0000	1000.00
$X5$	3.00000	3.00000	2.311174
$f(X)$	0.4704	0.313656	0.389479670

3.2 Multiple Disk Clutch Brake Design Problem

Optimization design problem consists of discrete control variable; whose main aim is to obtain the minimum mass of multiple disk clutch brake. This Optimization problem is taken from [9]. The design problem is also analyzed by Deb and Srinivasan [10] with NSGA-II Optimization technique. Figure 1 shows the convergence curve for the multiple disk clutch brake layout problem. Competitive solution with NSGA-II [10], WCA [11] are shown in Table 2.

3.3 Rolling Element Bearing Design Problem

In the mechanical area rolling element bearing [12] layout is to maximize the dynamic load carrying capacity. The oversight criteria are inner and outer raceway curvature [13] coefficient (f_i, f_0), number of ball (Z), pitch diameter (D_m), ball diameter (D_b), K_{Dmax}, K_{Dmin}, ε, e and ζ. All control variables nature is continuous left of balls, a discrete variable, as it is rounded to the nearest integer value. Rolling element consist of manufacturing consideration [13], kinematic and according to application requirement. Rolling element exists in between two surfaces, that are moveable and provides a facility of less friction. Figure 1 show iteration versus objective function value of Rolling element bearing design problem. This constrained problem is analyzed by GA [12], ABC [13], MBA and results are compared in Table 3.

3.4 Speed Reducer Design Problem

In the mechanical area to minimize the weight of speed reducer in that constrained Optimization problem various parameters are involved such as bending stress of the gear teeth [14], transverse deflections of shafts, and stresses in shafts and surface stress [14]. This problem involved eleven [4] constraints, produces high complexity of problem reported in [15] solutions are also given but that are infeasible. Figure 1 shows iteration versus objective function value of speed reducer problem.

Table 3 Competitive solution obtained from various methods for the rolling element bearing design problem

D.V	MBA	GA	MFO
X1	125.7153	125.7171	125.7227
X2	21.423300	21.423	21.42330
X3	11	11	11
X4	0.515000	0.515	0.515000
X5	0.515000	0.515	0.515000
X6	0.488805	0.4159	0.486226
X7	0.627829	0.651	0.700000
X8	0.300149	0.300043	0.300000
X9	0.097305	0.0223	0.020000
X10	0.0646095	0.751	0.6000163
f(X)	85535.9611	81843.3000	85539.1929

Table 4 Competitive solution attained from many optimization methods for the speed reducer design problem

D.V	PSO-DE	MDE	MBA	MFO
X1	3.50000	3.500010	3.500000	3.49924
X2	0.70000	0.700000	0.700000	0.70000
X3	17.0000	17.00000	17.000000	17.0000
X4	7.30000	7.300156	7.300033	7.30000
X5	7.80000	7.800027	7.715772	7.71475
X6	3.35021	3.350221	3.350218	3.35016
X7	5.28668	5.286685	5.286654	5.28633
f(X)	2996.34	2996.3566	2994.48245	2994.20622

Competitive solutions with MFO are shown in Table 4 from many optimization techniques such as PSO-DE [16], MDE [4], MBA are compared with MFO results. In mechanical engineering layout area has highly challenging problems due to large numbers of complex constraints.

3.5 Tubular Column Design Problem

The function of tubular column design has carry a compressive load at minimum cost via a uniform column of tabular section and load is $P = 2500 kgf$. The force induced in column must be less than yield and buckling stress. The cost consists of material and suction cost. Figure 1 shows convergence curve of Tubular column design problem, Comparative statistical results from many optimization techniques such as Hsu and Liu [17], Rao [18] and cuckoo search (CS) [3] are compared with MFO results as shown in Table 5.

Table 5 Comparative statistical results from other former optimisation techniques for the tubular column design problem

Parameter	CS	Hsu and Liu	Rao	MFO
f min	25.5316	25.5316	25.5323	25.4933
D	5.45139	5.4507	5.44	5.5528
T	0.29196	0.292	0.293	0.25015

Table 6 Comparative statistical results from other former optimization techniques for the Belleville spring design problem

D.V	Gene AS II	Coello	Gene AS I	MBA	MFO
$X1$	0.210	0.208	0.205	0.204143	0.204143
$X2$	0.204	0.2	0.201	0.2	0.2
$X3$	9.268	8.751	9.534	10.030473	10.0305
$X4$	11.499	11.067	11.627	12.01	12.01
$f(X)$	2.16256	2.121964	2.01807	1.9796747	1.9796747

3.6 Belleville Spring Design Problem

In the mechanical area Belleville spring [13] to minimum weight and satisfying large complex constraints. The control variables are external diameter (D_e), thickness (t), internal diameter (D_i) and height (h) of the spring. Figure 1 show iteration versus objective function value of the Belleville spring design problem. The constrained problem was analyzed by Coello [19] and, with other techniques like ABC [13], Gene AS I, II [20], results have been presented in Table 6 for the comparisons of various algorithms for attains best solutions in reference of minimum value or accuracy.

4 Conclusion

In this paper, MFO is used to solve highly complex constrained, expensive, continuous and discrete control parameters benchmark problems. The various results obtained by MFO and its comparison with the solution achieved by other well-known heuristic algorithms prove the potential of MFO algorithm in solving challenging problems with unknown search space is good. In many design problems, MFO provides better result to other algorithms but in some cases, results are inferior. This Paper shows that applicability potential of MFO in solving real and complex engineering problems.

References

1. Seyedali Mirjalili, "Moth-flame Optimisation algorithm: A novel nature-inspired heuristic paradigm," Knowledge-Based System, vol. 89, pages 228–249, 2015.
2. Sadollah, A. Bahreininejad, H. Eskandar, and M. Hamdi, "Mine blast algorithm: A new population based algorithm for solving constrained engineering Optimisation problems," Applied Soft Computing, vol. 13, pp. 2592–2612, 2013.
3. A. H. Gandomi, X.-S. Yang, and A. H. Alavi, "Cuckoo search algorithm: a meta-heuristic approach to solve structural Optimisation problems," Engineering with Computers, vol. 29, pp. 17–35 2013.
4. E. Mezura-Montes, J. Velazquez-Reyes, C.A.C. Coello, Modified differential evolution for constrained Optimisation, in: Evol. Comput., CEC 2006, IEEE Congress, 2006, pp. 25–32.
5. J. Kennedy, R. Eberhart, Particle swarm Optimisation, in: Proceedings of the IEEE International Conference on Neural Networks, Perth, Australia, 1995, pp. 1942–1948.
6. X. S. Yang, "Firefly algorithm," Engineering Optimisation, pp. 221–230, 2010.
7. J. Lampinen, A constraint handling approach for the differential evolution algorithm, IEEE Transactions on Evolutionary Computation (2002) 1468–1473.
8. Gandomi AH, Yang XS, Alavi AH, Mixed variable structural Optimisation using firefly algorithm. Computers & Structures.
9. Osyczka A. Evolutionary algorithms for single and multi-criteria design Optimisation: studies in fuzzyness and soft computing. Heidelberg: Physica-Verlag; 2002.
10. Deb K, Srinivasan A. Innovization: innovative design principles through Optimisation, Kanpur genetic algorithms laboratory (KanGAL). Indian Institute of Technology Kanpur, KanGAL report number: 2005007; 2005.
11. Hadi Eskandar, Ali Sadollah, Ardeshir Bahreininejad, Mohd Hamdi "Water cycle algorithm-A novel metaheuristic Optimisation method for solving constrained engineering Optimisation problems", "Computers and Structures" 110–111 (2012) pages 151–166.
12. S. Gupta, R. Tiwari, B.N. Shivashankar, Multi-objective design Optimisation of rolling bearings using genetic algorithm, Mechanism and Machine Theory 42 (2007) 1418–1443.
13. B. Akay, D. Karaboga, Artificial bee colony algorithm for large-scale problems and engineering design Optimisation, Journal of Intelligent Manufacturing (2010).
14. E. Mezura-Montes, C.A.C. Coello, Useful infeasible solutions in engineering Optimisation with evolutionary algorithms, in: MICAI 2005: Lect. Notes Artif. Int., vol. 3789, 2005, pp. 652–662.
15. J.K. Kuang, S.S. Rao, L. Chen, Taguchi-aided search method for design Optimisation of engineering systems, Engineering Optimisation 30 (1998) 1–23.
16. H. Liu, Z. Cai, Y. Wang, Hybridizing particle swarm Optimisation with differential evolution for constrained numerical and engineering Optimisation, Applied Soft Computing 10 (2010) 629–640.
17. Hsu Y-L, Liu T-C (2007) Developing a fuzzy proportional derivative controller Optimisation engine for engineering design Optimisation problems. Eng Optmiz 39(6):679–700.
18. Rao SS (1996) Engineering Optimisation: theory and practice, 3rd edn. John Wiley & Sons, Chichester.
19. C.A.C. Coello, Treating constraints as objectives for single-objective evolutionary Optimisation, Engineering Optimisation 32 (3) (2000) 275–308.
20. K. Deb, M. Goyal, optimizing engineering designs using a combined genetic search, in: L.J. Eshelman (Ed.), Proceedings of the Sixth International Conference in Generic Algorithms, University of Pittsburgh, Morgan Kaufmann Publishers, San Mateo, CA, 1995, pp. 521–528.

Training Multilayer Perceptrons in Neural Network Using Interior Search Algorithm

R.H. Bhesdadiya, Indrajit N. Trivedi, Pradeep Jangir, Arvind Kumar, Narottam Jangir and Rahul Totlani

Abstract Multilayer perceptron (MLP) is the most popular neural network method and it has been widely used for many practical applications. In this paper, recently developed interior search algorithm (ISA) is proposed for training MLP. Five of most important standard classification datasets (balloon, XOR, Iris, heart, and breast cancer) are employed to evaluate the proposed algorithm performance. The obtained results from ISA-based are compared with five well-known algorithms including ant colony optimization (ACO), genetic algorithm (GA), particle swarm optimization (PSO), population-based incremental learning (PBIL), and evolution strategy (ES). The statistical results reflect that the performance of the proposed algorithm can train MLPs with a very high degree of accuracy and it is capable of outperforming the well-known algorithms. The results also show that the high convergence rate of the ISA and it is potential to avoid local minima.

R.H. Bhesdadiya (✉)
Electrical Engineering Department, School of Engineering,
RK University, Rajkot, Gujarat, India
e-mail: rhbhesdadiya@gmail.com

I.N. Trivedi
Electrical Engineering Department, G.E. College, Gandhinagar, Gujarat, India
e-mail: forumtrivedi@gmail.com

P. Jangir · N. Jangir
Electrical Engineering Department, LEC, Morbi, Gujarat, India
e-mail: pkjmtech@gmail.com

N. Jangir
e-mail: nkjmtech@gmail.com

A. Kumar
Electrical Engineering Department, S.S.E.C., Bhavnagar, Gujarat, India
e-mail: akbharia8@gmail.com

R. Totlani
Electrical Engineering Department, JECRC, Jaipur, Rajasthan, India
e-mail: rhl.totlani@gmail.com

© Springer Nature Singapore Pte Ltd. 2018
S.K. Bhatia et al. (eds.), *Advances in Computer and Computational Sciences*,
Advances in Intelligent Systems and Computing 554,
https://doi.org/10.1007/978-981-10-3773-3_8

Keywords Multilayer perceptron · Feedforward neural network · ISA · Learning neural network

1 Introduction

In the area of computational intelligence techniques, neural networks (NN) are one of the maximum discoveries [1]. Learning methods are categorized into supervised and unsupervised. In supervised method, the NN is providing feedbacks from the peripheral source. In unsupervised method, a NN adjusts automatically to inputs without any additional source. Trainer provides learning for a neural network [2]. A trainer has a capability to acquire the best result for new inputs. The trainer has supposed to be very significant part of any NNs. The neural network science concepts was first proposed in 1943 [1]. There are various types of neural networks proposed in the literature study: Kohonen self-organizing network, feedforward network, recurrent neural network, radial basis function (RBF) network, and spiking neural networks. In feedforward neural networks (FNN) and multilayer perceptron (MLP), the information is cascaded in one direction throughout the networks. As its name implies, however, recurrent neural networks share the information between the neurons in two directions. Finally, spiking neural networks activates neurons with spikes. This also motivates the attempts to investigate the efficiencies of my recently proposed algorithm called interior search algorithm (ISA) [3] in training neural networks. In fact, the main motivations of this work are one field: Interior search algorithm (ISA) shows high exploration and exploitation that may assist it to outperform other trainers in this field.

2 Feedforward Neural Network and Multilayer Perceptron

Neurons of feedforward neural networks are settled in different parallel layers [1]. The initial layer is known as input layer and final layer is known as the output layer. Hidden layer is placed between input layer and output layer.

The inputs, weights, and biases, the output of MLPs are found out using following steps:

1. The weighted sums of I/P is [4]:

$$s_j = \sum_{i=1}^{n} (W_{ij}{}^* X_i) - \theta_j, j = 1, 2, 3, \ldots, h \tag{1}$$

where n = No. of input nodes, W_{ij} = Connection weight from the ith node in the input layer to the jth node in the hidden layer, θ_j = Bias (threshold) of the jth hidden node, and X_i = ith input.

2. The I/P of each hidden node is

$$s_j = sigmoid(s_j) = \frac{1}{(1 + \exp(-s_j))}, \quad j = 1, 2, 3, \ldots, h \tag{2}$$

3. The last outputs based on the calculated outputs of the hidden layer neurons nodes are [5]:

$$O_k = \sum_{j=1}^{h} (W_{jk} * s_j) - \theta_k', \quad k = 1, 2, 3, \ldots, m \tag{3}$$

$$O_k = sigmoid(O_k) = \frac{1}{(1 + \exp(-O_k))} \tag{4}$$

where θ_k' = *Bias of the kth output node.*

3 Interior Search Algorithm

Recently proposed ISA is a combined optimization analysis divine to the creative work or art relevance to interior or internal designing [3] consists of two stages first one composition stage where number of solutions are shifted toward to get optimum fitness. Second stage is reflector or mirror inspection method where mirror is placed in the middle of every solution and best solution to yield a fancy view [3] to design, satisfying all control variables to constrained design problem.

3.1 Algorithm Description

1. However, the position of acquired solution should be in the limitation of maximum bound and minimum bounds, later estimate their fitness amount [3].
2. Evaluate the best value of solution, the fittest solution has maximum objective function whenever aim of optimization problem is minimization and vice versa is always true. Solution has universally best in jth run (iteration).
3. Remaining solutions are collected into two categories mirror and composition elements in respect to a control parameter α. Elements are categorized based on the value of random number (all used in this paper) ranging 0–1.

Whether $rand_1()$ is less than equal to α it moves to mirror category else moves toward composition category. For avoid problems α must be carefully tuned.

4. Being composition category elements, every element or solution is however transformed as described below in the limited uncertain search space.

$$x_i^j = lb^j + (ub^j - lb^j)*r_2 \tag{5}$$

where x_i^j represents $i-$th solution in $j-$th run, ub^j, lb^j upper and lower range in $j-$th run, whereas its maximum and minimum values for all elements exists $(j-1)$th run and $rand_2()$ in a range 0–1.

5. For $i-$th solution in $j-$th run spot of mirror is described [3]:

$$x_{m,i}^j = r_2 x_i^{j-1} + (1 - r_3)*x_{gb}^j \tag{6}$$

where $rand_3()$ ranging 0–1. Imaginary position of solutions is dependent on the spot where mirror is situated defined as:

$$x_i^j = 2x_{m,i}^j - x_i^{j-1} \tag{7}$$

6. It is auspicious for universally best to little movement in its position using uncertain walk defined:

$$x_{gb}^j = x_{gb}^{j-1} + r_n*\lambda \tag{8}$$

where r_n vector of distributed random numbers having same dimension of x, $\lambda = (0.01*(ub-lb))$ scale vector, dependable on search space size.

7. Evaluate fitness amount of new position of elements and for its virtual images. Whether its fitness value is enhanced then position should be updated for next design. For minimization optimization problem updating are as follows [3]:

$$x_i^j = \{ {}^{x_i^j \ldots f(x_i^j) < f(x_i^{j-1})}_{x_i^{j-1} \ldots\ldots\ldots Else} \tag{9}$$

8. If termination condition not fulfilled, again evaluate from second step.

3.2 Parameter Tuning

A curious component in algorithm [3] is α for unconstrained benchmark test function it is almost fixed 0.25, but requirement is to increase its value ranging 0–1 randomly as increment in maximum number of runs selected for particular problem. It requires to shift search emphasized from exploration stage to exploitation optimum solution toward termination of maximum iteration.

3.3 Nonlinear Constraint Manipulation

Nonlinear constraint manipulations have following rules:

- Both solutions ares possible, then consider one with best objective functional value.
- Both solutions are impossible, then consider one with less violation of constraints.

Evaluation of constraint violation:

$$V(x) = \sum_{k=1}^{nc} \frac{g_k(x)}{g_{\max k}} \tag{10}$$

where

nc = Number of constraints,

$g_k(x) = k$th constraint consisting problem,

$g_{\max k}$ = maximum violation in kth constraint till yet.

3.4 ISA-Based MLP Trainer

Multilayer perceptron are very vital variables in training biases and weights [6]. The interior search algorithm [3] accepts the variables in the form of a vector, the variables of an MLP (multilayer perceptron) for interior search algorithm is [1]:

$$\vec{V} = \{\vec{W}, \vec{\theta}\} = \{W_{1,1}, W_{1,2}, \ldots, W_{n,n}, h, \theta_1, \theta_2, \ldots, \theta_h\} \tag{11}$$

where n = No. of the input nodes and θ_j = Bias of the jth hidden node.

A common metric for the evaluation of a multilayer perceptron is the mean square error (MSE) [7]. In this performance metric, a specified bundle of training models is given to the MLP (multilayer perceptron). The difference between the desired output and the MLP value can be calculated by the equation given below [2]:

$$M.S.E. = \sum_{i=1}^{m} (o_i^k - d_i^k)^2 \tag{12}$$

where m = Number of outputs, o_i^k = Real output of the ith input unit, and d_i^k = Desired output of the ith input unit.

The average of MSE with every training sample is used to assess the MLP performance:

$$\overline{M.S.E.} = \sum_{k=1}^{s} \frac{\sum_{i=1}^{m} (o_i^k - d_i^k)^2}{S} \tag{13}$$

where s = Number of training samples.

The problem of training a multilayer perceptron can be formulated with variables and average mean square error for the interior search algorithm is

$$Minimize: F(\overrightarrow{V}) = \overline{MSE} \tag{14}$$

error of every training samples [5]. By varying the weights and biases iteratively, the interior search algorithm gives reduced average mean square error of each training samples.

4   Results and Discussion

In this part, five standard classification datasets like XOR, balloon, iris, breast cancer, and heart are used for benchmarked the ISA-based multilayer perceptron trainer [1].

A. **XOR dataset**

   (See Table 1 and Fig. 1).

B. **Balloon dataset**

   (See Table 2 and Fig. 1).

C. **Iris dataset**

   (See Table 3 and Fig. 1).

Table 1 Comparative results for the XOR dataset

Algorithms	Classification rate (%)	MSE (AVE ± STD)
ISA-M.L.P.	100	0.0001125 ± 0.00031
Particle swarm optimization-M.L.P. [8]	37.50	0.084050 ± 0.035945
Genetic algorithm-M.L.P. [9]	100	0.000181 ± 0.000413
Ant colony optimization-M.L.P. [10]	62.50	0.180328 ± 0.025268
Evolution strategy-M.L.P. [11]	62.50	0.118739 ± 0.011574
Population-based incremental learning-M.L.P. [12]	62.50	0.030228 ± 0.039668

D. **Breast Cancer dataset**

(See Table 4 and Fig. 1).

E. **Heart dataset**

(See Table 5 and Fig. 1).

Table 2 Comparative results for the balloon dataset

Algorithms	Classification rate (%)	MSE (AVE ± STD)
ISA-M.L.P.	100	3.28e-24 ± 1.04e-23
Particle swarm optimization-M.L.P. [8]	100	0.000585 ± 0.000749
Genetic algorithm-M.L.P. [9]	100	5.08e-24 ± 1.06e-23
Ant colony optimization-M.L.P. [10]	100	0.004854 ± 0.007760
Evolution strategy-M.L.P. [11]	100	0.019055 ± 0.170260
Population-based incremental learning-M.L.P. [12]	100	2.49e-05 ± 5.27e-05

Table 3 Comperative results for the Iris dataset

Algorithms	Classification rate (%)	MSE (AVE ± STD)
ISA-M.L.P.	92.65	0.054732 ± 0.026515
Particle swarm optimization-M.L.P. [8]	37.33	0.228680 ± 0.057235
Genetic algorithm-M.L.P. [9]	89.33	0.089912 ± 0.123638
Ant colony optimization-M.L.P. [10]	32.66	0.405979 ± 0.053775
Evolution strategy-M.L.P. [11]	46.66	0.314340 ± 0.052142
Population-based incremental learning-M.L.P. [12]	86.66	0.116067 ± 0.036355

Table 4 Comperative results for the breast cancer dataset

Algorithms	Classification rate (%)	MSE (AVE ± STD)
ISA-M.L.P.	99	0.0023452 ± 0.00142
Particle swarm optimization-M.L.P. [8]	11	0.034881 ± 0.002472
Genetic algorithm-M.L.P. [9]	98	0.003026 ± 0.001500
Ant colony optimization-M.L.P. [10]	40	0.013510 ± 0.002137
Evolution strategy-M.L.P. [11]	06	0.040320 ± 0.002470
Population-based incremental learning-M.L.P. [12]	07	0.032009 ± 0.003065

Table 5 Comperative results for the heart dataset

Algorithms	Classification rate (%)	MSE (AVE ± STD)
ISA-M.L.P.	82.70	0.083589 ± 0.003871
Particle swarm optimization-M.L.P. [8]	68.75	0.188568 ± 0.008939
Genetic algorithm-M.L.P. [9]	58.75	0.093047 ± 0.022460
Ant colony optimization-M.L.P. [10]	00.00	0.228430 ± 0.004979
Evolution strategy-M.L.P. [11]	71.25	0.192473 ± 0.015174
Population-based incremental learning-M.L.P. [12]	45.00	0.154096 ± 0.018204

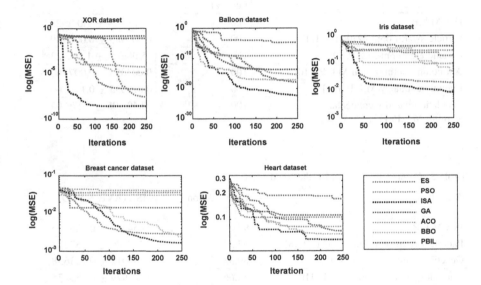

Fig. 1 Convergence curve of ISA-MLP algorithm

5 Conclusion

In this work, the newly developed ISA is introduced for very first time as a multilayer perceptron trainer. The multilayer perceptron training problem is used with ISA. ISA used to find the optimal fitness values for weights and biases. The ISA-based trainer is employed to five standard classification datasets (XOR, balloon, Iris, breast cancer, and heart). For confirmation, the results obtained with the ISA–MLP algorithm are compared with five different stochastic optimization trainers: PSO, ES, GA, ACO, and PBIL. The results proved that the ISA have ability to be very effective and accurate in neurons training MLPs. ISA–MLP algorithm can be able to best level of local optima avoidance, which increases the possibility of searching more accurate approximations for optimal weights and biases for MLPs. Furthermore, the efficiency of the optimal values for weights and biases obtained by suggested method is quite more, which is due to the high exploitation of the ISA-MLP trainer.

For future scope, it is interesting to see the efficiency and effectiveness of the proposed interior search algorithm (ISA) for training multilayer perceptron with radial basis function network, Kohonen network, and recurrent network as well as structures and parameters of MLPs such as number of hidden nodes and hidden layers. In addition, methods for improving the exploitation of ISA algorithm are worth studying.

References

1. Seyedali Mirjalili, "How effective is the Grey Wolf optimizer in training multi-layer perceptrons", Springer-january-2015.
2. Reed RD, Marks RJ, "Neural smithing: supervised learning in feedforward artificial neural networks", Mit Press-1998.
3. Gandomi A.H. "Interior Search Algorithm (ISA): A Novel Approach for Global Optimization" ISA Transaction, Elsevier, 53(4), 1168–1183, 2014.
4. Mirjalili S, Sadiq AS (2011), "Magnetic optimization algorithm for training multi-layer perceptron", In: Communication Software and Networks (ICCSN), 2011 IEEE 3rd International Conference, IEEE, pp 42–46.
5. Moallem P, Razmjooy N, "A multi layer perceptron neural network trained by invasive weed optimization for potato color image segmentation", Trends Appl Sci Res-2012-7:445–455.
6. Bebis G, Georgiopoulos M, "Feed-forward neural networks", Potentials, IEEE-1994.
7. Mirjalili S, Mirjalili SM, Lewis A, "Let a biogeography based optimizer train your multi-layer perceptron", Inf Sci-2014..
8. Mendes R, Cortez P, Rocha M, Neves J, "Particle swarms for feedforward neural network training, learning", 2002- vol. 6.
9. Whitley D, Starkweather T, Bogart C, "Genetic algorithms and neural networks: Optimizing connections and connectivity", Parallel comput-1990, 14:347–361.
10. Blum C, Socha K, "Training feed-forward neural networks with ant colony optimization: an application to pattern classification," In: 5th international conference on, Hybrid Intelligent Systems, 2005. HIS'05, p 6.
11. Yao X, Liu Y, "Fast evolution strategies", In: evolutionary programming VI-1997, pp 149–161.
12. Baluja S (1994), "Population-based incremental learning. A method for integrating genetic search based function optimization and competitive learning", DTIC Document-1997.

Sequence Generation of Test Case Using Pairwise Approach Methodology

Deepa Gupta, Ajay Rana and Sanjay Tyagi

Abstract There are various instruction specifications in the software system that interacts with each other. While designing software system, software engineers have to face a tedious task of ensuring the validation of every available software use case. If the instruction specifications are huge, the required number of use cases to validate becomes unmanageable within reasonable time budget and therefore designers have to choose one of the options of either delaying the projects or delivering without validating all available test case scenarios. Even for moderately sized software systems, comprehensive testing sometimes becomes impossible due to limited time availability for validation phase. Empirical studies have indicated that most of the faults in software systems often precipitate by interaction between smaller numbers of instruction specifications. If designers can cover all available interactions between all pairs of instructions specifications (instead of comprehensive testing), it can give reasonable confidence to designers about the software attribute within the limited time. This paper presents one such approach which uses the sequence origination approach for pairwise test case origination. This approach makes certain to propagate the required intent of trial run cases which cover all available interactions between all instructions pairs at least once. Trial run selection specification is this approach based on combinatorial testing. The paper also discusses the results with this new approach on one of the candidate software system to demonstrate its efficacy over already existing validation approaches.

Keywords Software testing · Trial run · Combinatorial Testing · Pairwise testing

D. Gupta (✉)
AIIT, Amity University, Noida, India
e-mail: deepa19july@gmail.com

A. Rana
Amity University, Noida, India
e-mail: ajay_rana@amity.edu

S. Tyagi
Kurukshetra University, Kurukshetra, India
e-mail: tyagikuk@gmail.com

© Springer Nature Singapore Pte Ltd. 2018   79
S.K. Bhatia et al. (eds.), *Advances in Computer and Computational Sciences*,
Advances in Intelligent Systems and Computing 554,
https://doi.org/10.1007/978-981-10-3773-3_9

1 Introduction

Combinative testing (CT) is an exemplification approach which develop trial run that focuses on the comportment of interaction of orderliness components with their collaborators [1]. Inclination is established on the surveillance that utmost software flaws precipitate through orderliness of solely two aspects analogous instruction expenses. Its search group that is much smaller than that of comprehensive trial earlier was efficacious in findings flaws [2]. Pairwise testing is an efficacious, combinative testing approach that, for each amalgamation of instruction attribution to a software arrangement, tests all available amalgamation of this attribution [3].

In software engineering, software testing and removing flaws is still very labor-comprehensive and expensive. Roughly software testing is a paramount of project. Therefore, the aim is to search for a mechanized worth-efficacious software testing and unscramble scheme to check high attribute software release [4]. Forthwith analysis and inquisition on testing the software that targets on the test coverage criterion design, obvious error and test localization difficulty. Among these difficulties, test case origination difficulty is a valuable one and has come forward in producing error-free programs. Explain this, pairwise strategy, known for its productive test case reduction scheme and adaptability to notice from superlative per hundred of the faults, can be used. It is necessary to state that an efficacious way of finding any flawless solution is not established and the time required for finding a smallest instance of trial run grows very fast when the number of attribution and available expenses rises.

A genetic iterative is an approach that resembles the familiar rule of progression' [5]. The following algorithm was discovered for handling exploration and augmentation related disputes and is known to be efficacious for determining results to the disputes with the very massive search space and complex disputes. In genital algorithm, we tend to find a better answer from community of possible choice called entity to a complication which tends to find much better solutions.

In this paper, metaheuristic algorithms are applied and also the concept of genital algorithm concepts have been applied for generating pairwise test sets as a search dispute and this paper presents more details of sequence test origination using a pairwise approach. Here we have provided with the sequence origination method which combines the feature of NP—Complete dispute [3]; as well as sequence origination approach which reduces the number of trial run.

2 Metaheuristic Algorithms

A metaheuristic iterative process can be explained as a continual origination process that guides a dependent heuristic by combining smartly various conceptions to analyze and minimize the search space. To find nearly flawless solutions, information should be framed using informative ways [6]. The subsequent part explains the sequence generation algorithm:

2.1 N-IPO (Novel-IPO)

- An adjustment with many instruction particularization, the IPO hereditary develops an amalgamation test intent for the early dual attributes, stretches the developed test intent to develop a pairwise test intent being the early trio attributes and go on to do so for each descriptive attributes [3]. The N-IPO yields into attention the illustrative expenses of attribution authority instead of all the available expenses. During spanning the preceding test intent for over and above attribution dual steps supervenes:
- In the early step the current trial run are stretched inconsistently by adding the expense of recent attribution. The following is known as inconsistent extension.
- In the dual stepping the test intent is enhanced by summing more trial run, which may be mandatory due to inclusion of new attribution. This impression is the perpendicular extension.
  Antecedent partitions must be decided in the instruction domains and take out the illustrative expenses from each domain. To escape any kind of viewpoint we endorse the illustrative expense from each separation on irregular support.

2.2 Pairwise Testing

The above-mentioned approach uses the combinative testing approach in which one by one dual set of instruction attribution of software is being tried [1, 7]. It is observed as a feasible adjustment amidst combinative testing method. It can be implemented much rapidly than comprehensive testing that test an amalgamation for all instruction attribution is more efficacious than less comprehensive plans that fail to act on all available pairs of instruction attribution. The interpretation after this type of testing is the larger part of the software flaws which precipitate by the single instruction attribution. Pairwise testing thus depends upon that one by one dual instruction attribution conscience be occupied at slightly one time.

3 Combination Sequence Generation Algorithm (Proposed Algorithm)

To begin with first we need to know how this optimized test sequence can be developed. For that we will use combination sequence origination algorithm. Then we will use this sequence generator in our test case origination algorithm to get the final outputs.

Combination Sequence Generation Algorithm

1. Let C be the array of elements of which Combinations are to be made.
2. Let M be the count of elements in C.
   Create a Multi-Way Tree with root node containing elements of C.
   Create all possible combination groups, when M − 1 elements are chosen from C.
   Make them the child nodes.
   Do steps 3–4 recursively in such a way that each parent has all possible combinations of child nodes with length one less than to itself, until each child node contains only one element.
3. Now Traverse this tree with depth first approach and generate the sequence.
4. Remove the duplicates from bottom to top, from the sequence.
5. Display sequence.

Example: C = [a, b, c, d].

M = 4   Tree Origination:

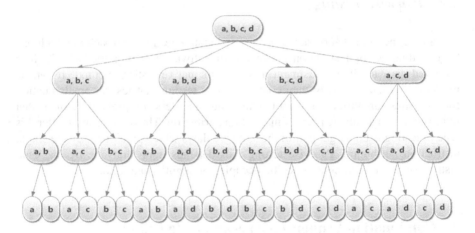

Traverse this tree with **depth first** approach and remove the duplicates from bottom to top.

3.1   Test Case Origination Algorithm

- This algorithm will develop the output containing test cases and their test results.
- The data structures to be maintained for this process:

• Name	• Description
• X	• Set of parameters to be tested
• V	• Set of parameter's value domain for each item
• TV	• Set of tested values
• xCount	• Number of items to be tested
• S	• Stack of parameter groups
• N	• Counter
• Output	• A table with three columns Group, Values, Result • Group = parameter group that has been tested • Values = values used for the test • Result = test result, Pass or Fail

Test Case Origination Algorithm

1. Input **X** with parameter names to be tested
2. Input **Y** with parameter value domains to be tested
3. **xCount** = count items in **X**
4. n = 0
5. n = n + 1
6. Select **Group** as first **n** parameters from **X**
7. Use **Combination Sequence Origination Algorithm** to generate combinations sequence of **Group** parameters and push in **S**, where combination is not present in **Output** [Group].
8. Select **testGroup** as pop(**S**).
9. Select **testParameterGroup** as **n** random values from **V** domain of each parameter in **testGroup**. If **V** has empty domains for any parameter then fetch random value from **TV** for that parameter.
10. Test **testGroup** with **testParameterGroup** and store in **Output**.
11. Add **testParameterGroup** to **TV**
12. Remove **testParameterGroup** values from V
13. If **S** is not empty go to step 8.
14. If **n** < **xCount** go to step 5.
15. Display Output.

4 Experimental Outcomes

A sequence origination algorithm which is based on this hereditary has been replaced in fraternization with intranet portal Nico minds Inc. [8]. The algorithm was used to propagate the pairwise intent of trial run during different development phases. The trial run was delivered to validation team which ran through trial run in parallel with comprehensive mode trial run to check their efficaciousness. The results indicated that pairwise trial run developed with this hereditary were able to figure paramount of the bugs found by comprehensive intent of trial run whereas the validation effort put in for validating pairwise trial run was much less as compared to comprehensive testing. The general findings were that error occurred due to inter-action between **a** and **b** attributions. The proposed sequence propagation hereditary was quite successful in downsizing the number of developed trial run Here is a sample table which depicts the results for one of the small module of the software:

Group	Values	Result
A	False	Pass
a, b	**True, 4**	**Fail**
B	2	Pass
a, b, c	**True, 1, 0**	**Fail**
a, c	False, 5	Pass
C	8	Pass
b, c	3, 5	Pass

5 Discussion and Conclusion

The paper presented a sequence propagation hereditary for generating the pairwise trial run based upon the pairwise testing of the software arrangements having the instruction attributions considering enormous spheres. In summary, sequence origination algorithm

- embraces gains of IPO hereditary, i.e., uniform and perpendicular extension;
- embraces (Novel-IPO) boundary value analysis;
- embraces benediction of Fibonacci series, i.e., resembling ascending–descend-ing path;
- embraces the benediction of tree traversal.

The presented approach was evaluated on a software project of intranet portal and results indicated that paramount faults were discovered with the intent of pairwise trial run developed in this case and number of trial run developed and tried were much less than comprehensive intent of trial run. This approach obtains the advantage of segregating all the spheres with reasonable budget. Also the dispute was formulated using depth first searching traversing approach and applied the test case propagation algorithm to find pairwise test sets which helped the trial run intent to cover the most of the instructions attributions interactions. With the approach of testing, the tradeoff for the carefulness of trial run coverage can be optimized by a tester whereas we have finite basis of time and liability that are available.

Going forward, the proposed need to be evaluated on multiple trial run and need to be enhanced and customized for domain specific disputes. A multilayered model is also available where this approach is first used to develop the top level of trial run to develop very few set of trial run to check the overall sanity of software. When these small number of trial run are validated successfully and all the bugs which are found are fixed, this approach is again used to cover more instruction attributions and develop more trial run to given better coverage.

References

1. Cohen, D. M., Dalal, S. R., Fredman, M. L., and Patton, G. C., The AETG system: An Approach to Testing Based on Combinatorial Design, IEEE Transaction on Software Engineering, in 1997, 23 (7), 437–443.
2. Tai, K. C., Lei, Y.: A Test Generation Strategy for Pairwise Testing. IEEE Trans. On Software Engineering, in Jan 2002, 28(1) 3.
3. Lei, Y and Tai, K. C., In-Parameter-Order: A Test Generating Strategy for Pairwise Testing, IEEE Trans. on Software Engineering, in 2002, 28 (1), 1–3.
4. Dr. Mohammed Abdul Kashem, Mohammad Naderuzzaman: On An Enhanced Pairwise Search Approach for Generating Optimum Number of Test Data and Reduce Execution Time. Computer Engineering and Intelligent Systems http://www.iiste.org ISSN 2222–1719 (Paper) ISSN 2222–2863 (Online) Vol. 4, No.1, 2013.
5. M. Mitchell, On an Introduction to Hereditary Algorithm. The MIT Press, in 1999.
6. I. H. Osman and G. Laporte, On "Metaheuristic: A bibliography", Ann. Oper. Res. vol. 63, (1996), pp. 513–623.
7. J. Czerwonka. Pairwise Testing, combinatorial test case generation in (2010, Dec.) [online]. Available: http://www.pairwise.org.
8. http://www.nicominds.com.

A Rule Extraction for Outsourced Software Project Risk Classification

Zhen-hua Zhang, Yong Hu, Kuixi Xiao, Shenguo Yuan
and Zhao Chen

Abstract A rule extraction algorithm based on K-means clustering with interval-valued intuitionistic fuzzy sets (IVIFS) information, which is the combination of K-means clustering and IVIFS, is proposed in this paper. First, we introduce IVIFS and its distance. Second, we introduce IVIFS method and its application to the classification of software project. Finally, we present a rule extraction model according to IVIFS fuzziness and its K-means clustering, and apply them to pattern classification of outsourced software project risk to demonstrate the advantages of this model. The experimental results show that the rules from IVIFS model are better than that from the conventional K-means clustering model in rule extraction, and the prediction effect from the former is more effective than that from the latter. According to this combinational rule extraction method, based on database from a special and professional investigation for Chinese small and medium software-outsourced enterprise, we obtain some valuable, realistic and available project development risks decision rules.

2016 International Conference on Computer, Communication & Computational Sciences [IC4S 2016]

Z. Zhang (✉) · K. Xiao · S. Yuan · Z. Chen
School of Economics and Trade, Guangdong University of Foreign Studies,
Guangzhou 510006, China
e-mail: zhangzhenhua@gdufs.edu.cn

K. Xiao
e-mail: xiaokuixi@gdufs.edu.cn

S. Yuan
e-mail: yuanshenguo@gdufs.edu.cn

Z. Chen
e-mail: chenzhao@gdufs.edu.cn

Y. Hu
Institute of Big Data and Decision, Jinan University, Guangzhou 510632, China
e-mail: huyonghenry@163.com

© Springer Nature Singapore Pte Ltd. 2018
S.K. Bhatia et al. (eds.), *Advances in Computer and Computational Sciences*,
Advances in Intelligent Systems and Computing 554,
https://doi.org/10.1007/978-981-10-3773-3_10

Keywords Interval-valued intuitionistic fuzzy sets · K-means clustering · Rule extraction · Outsourced software project risk

1 Introduction

By introducing membership and nonmembership degree, Zadeh launched fuzzy sets (FS, [1]) in 1965. In 1980s, Atanassov presented hesitant degree and proposed intuitionistic fuzzy sets (IFS, [2]) and IVIFS [3]. Hence, many studies on IFS and IVIFS had been carried out, such as pattern recognition [4–8], static and dynamic decision-making [9–12] etc. However, most studies of intuitionistic fuzzy system focused on IFS theory, decision-making algorithm, and design and application of distance, similarity, entropy, etc. At present, few researches on how to extract the rules based on IFS system have been done. More importantly, few traditional decision analysis models on IFS and IVIFS are applied to the risk discovery in outsourced software project. This paper proposes a rule extraction model of IVIFS, and then the presented model is utilized to knowledge discovery of outsourced software project risk.

In the research of software project risk, early scholars majorly focused on principles [13] and project risk index system [13–18]. Wallace, Keil, and Nidumolu et al. studied the constitute of output and decision attributes for software project risk, and analyzed the effect of condition attributes to decision attributes qualitatively [14, 19, 20]. Hu, Zhang, and Nagi et al. utilized Bayesian networks to construct causal analysis framework for software project risk [21, 22], and predicted the risk by supervised combinational model [23, 24]. In fact, for the research of unsupervised model applied to software project risk prediction, only conventional clustering method and fuzzy neural network are presented [19, 25].

According to survey data of software project risk, we conclude that most of the research data use Likert scale (five level or seven level scale), and the border between any two adjacent levels is vague. Thus the traditional rule algorithms, such as decision tree and rough set rule extraction etc., do not reflect the fuzziness and vagueness.

In summary, considering the shortage of exiting studies, we introduce a rule extraction method with IVIFS information based on k-means clustering algorithm. First, we present the definition of IVIFS and its distance measures. And then, the IVIFS model along with its distance measures are utilized to pattern recognition of outsourced software project risk. Finally, we present a rule extraction method with IVIFS information. The simulation results show that these IVIFS rules are more suitable for the study of outsourced software project risk than the rules derived from the conventional K-means model.

2 IVIFS and Its Distance

Definition 1 An IVIFS A over X is

$$A = \{ <x, M_A(x), N_A(x), H_A(x) > | x \in X \}. \tag{1}$$

$M_A(x) = [t_A^-(x), t_A^+(x)] \subseteq [0, 1]$, $N_A(x) = [f_A^-(x), f_A^+(x)] \subseteq [0, 1]$, $H_A(x) = [\pi_A^-(x),$ $\pi_A^+(x)]$, with the condition $\pi_A^-(x) = 1 - t_A^+(x) - f_A^+(x) \in [0, 1]$, $\pi_A^+(x) = 1 - t_A^-(x) - f_A^-(x) \in [0, 1]$, $t_A^+(x) + f_A^+(x) \le 1$ for x $\in$ X. And M_Ax, N_A(x), and H_A(x) represent the interval of membership degree, nonmembership degree, and hesitant degree, respectively.

Definition 2 According to the membership function and the nonmembership function of IVIFS, weighted standardized Minkowski distance can be defined as follows, where $w_{AB}(x) \ge 0$, $\sum\limits_{x \in X} w_{AB}(x) = 1$. When p = 2, Eq. (2) is Euclidean distance.

$$d_{IVIFS}(A^*, B^*)$$
$$= \sqrt[p]{\sum_{x \in X} w_{AB}(x) \left[|t_A^-(x) - t_B^-(x)|^p + |t_A^+(x) - t_B^+(x)|^p + |f_A^-(x) - f_B^-(x)|^p + |f_A^+(x) - f_B^+(x)|^p \right]}. \tag{2}$$

3 Application Example

We utilize IVIFS model above to a pattern recognition example about the classification of outsourced software project risk [5–7].

Example 1 Suppose that we have four types of outsourced software projects, which are expressed by four IVIFSs $A_i = \{ <x, M_{A_i}(x), N_{A_i}(x) > | x \in X \}(i = 1, 2, 3, 4)$ over the feature space $X = \{x_1, x_2,..., x_{12}\}$ with weight vector w = {0.1, 0.05, 0.08, 0.06, 0.03, 0.07, 0.09, 0.12, 0.15, 0.07, 0.13, 0.05}^{T}. And B is an unknown outsourced software project. We focus on determining which class the unknown pattern B belongs to. All the attributes we use are shown in Table 1, which have been proved to be important attributes between customers and contractors when outsourced software is under development [21–25].

$A_1 = \{ <x_1, [0.1, 0.2], [0.5, 0.6]>, <x_2, [0.1, 0.2], [0.7, 0.8]>, <x_3, [0.5, 0.6], [0.3, 0.4]>,$
$\quad <x_4, [0.8, 0.9], [0, 0.1]>, <x_5, [0.4, 0.5], [0.3, 0.4]>, <x_6, [0, 0.1], [0.8, 0.9]>,$
$\quad <x_7, [0.3, 0.4], [0.5, 0.6]>, <x_8, [1.0, 1.0], [0, 0]>, <x_9, [0.2, 0.3], [0.6, 0.7]>,$
$\quad <x_{10}, [0.4, 0.5], [0.4, 0.5]>, <x_{11}, [0.7, 0.8], [0.1, 0.2]>, <x_{12}, [0.4, 0.5], [0.4, 0.5]> \}.$
$A_2 = \{ <x_1, [0.5, 0.6], [0.3, 0.4]>, <x_2, [0.6, 0.7], [0.1, 0.2]>, <x_3, [1.0, 1.0], [0, 0]>,$
$\quad <x_4, [0.1, 0.2], [0.6, 0.7]>, <x_5, [0, 0.1], [0.8, 0.9]>, <x_6, [0.7, 0.8], [0.1, 0.2]>,$
$\quad <x_7, [0.5, 0.6], [0.3, 0.4]>, <x_8, [0.6, 0.7], [0.2, 0.3]>, <x_9, [1.0, 1.0], [0, 0]>,$
$\quad <x_{10}, [0.1, 0.2], [0.7, 0.8]>, <x_{11}, [0, 0.1], [0.8, 0.9]>, <x_{12}, [0.7, 0.8], [0.1, 0.2]> \}.$
$A_3 = \{ <x_1, [0.4, 0.5], [0.3, 0.4]>, <x_2, [0.6, 0.7], [0.2, 0.3]>, <x_3, [0.9, 1.0], [0, 0]>,$
$\quad <x_4, [0, 0.1], [0.8, 0.9]>, <x_5, [0, 0.1], [0.8, 0.9]>, <x_6, [0.6, 0.7], [0.2, 0.3]>,$
$\quad <x_7, [0.1, 0.2][0.7, 0.8]>, <x_8, [0.2, 0.3], [0.6, 0.7]>, <x_9, [0.5, 0.6], [0.2, 0.4]>,$
$\quad <x_{10}, [1.0, 1.0][0, 0]>, <x_{11}, [0.3, 0.4], [0.4, 0.5]>, <x_{12}, [0, 0.1], [0.8, 0.9]> \}.$
$A_4 = \{ <x_1, [1.0, 1.0], [0, 0]>, <x_2, [1.0, 1.0], [0, 0]>, <x_3, [0.8, 0.9], [0, 0.1]>,$
$\quad <x_4, [0.7, 0.8], [0.1, 0.2]>, <x_5, [0, 0.1], [0.7, 0.9]>, <x_6, [0, 0.1], [0.8, 0.9]>,$
$\quad <x_7, [0.1, 0.2], [0.7, 0.8]>, <x_8, [0.1, 0.2], [0.7, 0.8]>, <x_9, [0.4, 0.5], [0.3, 0.4]>,$
$\quad <x_{10}, [1.0, 1.0], [0, 0]>, <x_{11}, [0.3, 0.4], [0.4, 0.5]>, <x_{12}, [0, 0.1], [0.8, 0.9]> \}.$
$B = \{ <x_1, [0.9, 1.0], [0, 0]>, <x_2, [0.9, 1.0], [0, 0]>, <x_3, [0.7, 0.8], [0.1, 0.2]>,$
$\quad <x_4, [0.6, 0.7], [0.1, 0.2]>, <x_5, [0, 0.1], [0.8, 0.9]>, <x_6, [0.1, 0.2], [0.7, 0.8]>,$
$\quad <x_7, [0.1, 0.2], [0.7, 0.8]>, <x_8, [0.1, 0.2], [0.7, 0.8]>, <x_9, [0.4, 0.5], [0.3, 0.4]>,$
$\quad <x_{10}, [1.0, 1.0], [0, 0]>, <x_{11}, [0.3, 0.4], [0.4, 0.5]>, <x_{12}, [0, 0.1], [0.7, 0.9]> \}.$

From the data above, considering the realistic meaning of all the attributes for x_i (i = 1, 2, …, 12), we use x_1 (Project Manager) and x_2 (Development Team) to characterize Development Experience (y_1); x_3 (Plan and Control), x_4 (Development and Test) and x_5 (Engineering Support and Milestone Management) to describe Project Management Level (y_2); x_6 (Client Team Collaboration) and x_9 (Client Development Experience) to express Client Development Experience (y_3); x_7

Table 1 Outsourced software project risk factors

Customer risks (Support and collaboration risks)	References	Contractor risks	References
1 Client Team Collaboration (x_6)	[14–16]	1 Project Manager (x_1)	[14]
2 Top Management Support (x_7)	[14–16]	2 Development Team (x_2)	[13–15]
3 Client Department Support (x_8)	[14–16]	3 Plan and Control (x_3)	[13, 14]
4 Client Development Experiment (x_9)	[15, 16]	4 Development and Test (x_4)	[18]
5 Business Environment (x_{10})	[15]	5 Milestone Management (x_5)	[14, 17]
6 Level of IT Application (x_{11})	[14]		
7 Business Process (x_{12})	[15]		

(Top Management Support) and x_8 (Client Department Development) to explain the Client Support Level (y_4); x_{10} (Client Business Environment), x_{11} (Level of IT Application for Client) and x_{12} (The Level of Client Business Process standardization) to show the Level of Client Business Environment (y_5). According to IVIFS information, using weights w for each attribute x_i, we obtain

$A_1 = \{ <y_1, [0.1, 0.2], [0.57, 0.67] >, <y_2, [0.59, 0.69], [0.19, 0.29] >, <y_3, [0.14, 0.24], [0.66, 0.76] >,$
$\quad <y_4, [0.7, 0.74], [0.21, 0.26] >, <y_5, [0.56, 0.66], [0.24, 0.34] > \}.$

$A_2 = \{ <y_1, [0.53, 0.63], [0.23, 0.33] >, <y_2, [0.51, 0.56], [0.35, 0.41 >, <y_3, [0.90, 0.94], [0.03, 0.06] >,$
$\quad <y_4, [0.56, 0.66], [0.24, 0.34] >, <y_5, [0.17, 0.27], [0.63, 0.73] > \}.$

$A_3 = \{ <y_1, [0.47, 0.67], [0.27, 0.37] >, <y_2, [0.42, 0.52], [0.42, 0.48] >, <y_3, [0.53, 0.63], [0.2, 0.37] >,$
$\quad <y_4, [0.16, 0.26], [0.64, 0.74] >, <y_5, [0.44, 0.51], [0.37, 0.44] > \}.$

$A_4 = \{ <y_1, [1, 1], [0, 0] >, <y_2, [0.62, 0.72], [0.16, 0.28] >, <y_3, [0.27, 0.37], [0.46, 0.56] >,$
$\quad <y_4, [0.04, 0.14], [0.7, 0.86] >, <y_5, [0.44, 0.51], [0.37, 0.44] > \}.$

$B = \{ <y_1, [0.9, 1], [0, 0] >, <y_2, [0.54, 0.64], [0.22, 0.32] >, <y_3, [0.3, 0.4], [0.43, 0.53] >$
$\quad y_4, [0.1, 0.2], [0.7, 0.8] >, <y_5, [0.44, 0.51], [0.37, 0.44] > \}.$

From the result above, for each y_k ($k = 1, 2, 3, 4, 5$) we have

If $\frac{t_A^-(x)+t_A^+(x)}{2} \geq 0.8 \& \frac{f_A^-(x)+f_A^+(x)}{2} \leq 0.2$, then it is Good;

If $(t_A^-(x)+t_A^+(x)) - (f_A^-(x)+f_A^+(x)) \geq 0$, then it is Common;

If $(t_A^-(x)+t_A^+(x)) - (f_A^-(x)+f_A^+(x)) < 0$, then it is Bad;

According to the data above, we define three sets initially: Good, Bad, and Common. The rules are shown in Table 2.

According to Table 2, A_1 means project development and management is standardized, but their experience is poor. As for A_2, their developing experience is enough, while the project management is not standardized for Contractor. The performance of A_3 is quite common for the contractees and the contractors, and its Client support is terrible. For A_4, the contractors perform perfectly, but the contractees are extremely bad. Obviously, B is more close to A_4, which is the same with the result in [6, 7].

Table 2 Preliminary determination of outsourced software project risk

Pattern	Development experience x_1, x_2	Project management x_3, x_4, x_5	Client team experience x_6, x_9	Client support x_7, x_8	Client business environment x_{10}, x_{11}, x_{12}
A_1	Bad	Common	Bad	Common	Common
A_2	Common	Common	Good	Common	Bad
A_3	Common	Common	Common	Bad	Common
A_4	Good	Common	Bad	Bad	Common
B	Good	Common	Bad	Bad	Common

Xu introduced four similarity formulas in [6], and got the following results:

$s_1(A_1, B) = 0.597$, $s_1(A_2, B) = 0.561$, $s_1(A_3, B) = 0.833$, $s_1(A_4, B) = 0.976$;

$s_2(A_1, B) = 0.530$, $s_2(A_2, B) = 0.529$, $s_2(A_3, B) = 0.734$, $s_2(A_4, B) = 0.951$;

$s_3(A_1, B) = 0.545$, $s_3(A_2, B) = 0.503$, $s_3(A_3, B) = 0.810$, $s_3(A_4, B) = 0.956$;

$s_4(A_1, B) = 0.473$, $s_4(A_2, B) = 0.473$, $s_4(A_3, B) = 0.712$, $s_4(A_4, B) = 0.934$.

In [6], considering that the similarity degree between A_4 and B is largest, Xu concluded that the unknown pattern B should belong to the pattern A_4. In 2011, Wei et al. also obtain the same results as Xu using another similarity formula in [7].

4 Algorithm Steps

In order to comprehensively find out the risk of outsourced software project in its development process for small and medium enterprises (SMEs) of China, and extract useful and explicit decision rules in management process, we made a detailed survey. We collect 293 sample data from some SMEs of China undertaking outsourced software project from the USA, Japan, and Southeast Asia. And most of these companies come from the eastern coastal developed city of China. There are 260 complete and 33 incomplete individuals. 69 of complete individuals are failure and 191 successful. The decision attribute named Target attribute, including 8 output attributes: Function, Performance, Information Quality, Maintainability, Satisfaction of Customer and User, Company Profits, Completion Degree in Time, Completion Degree in Budget [19, 20]. Either Yes or No are their options. A software project is a successful project when all 8 output options are Yes.

Algorithm steps are shown below:

Step 1: Fuzzification of outsourced software project and its conditional attributes using fuzzy relationship matrix.

We use five level scales $\{1, 2, 3, 4, 5\}$ to denote the different level of attribute. Thus, for each "Bad" project set A and conditional attribute $x \in X$, we define

$$t_A^-(x) = 1 - 0.2x, t_A^+(x) = 1.2 - 0.2x, f_A^-(x) = 0.2x - 0.2, f_A^+(x) = 0.2x.$$

For each "Common" project set A and $x \in X$, if $x \leq 3$, we define

$$M_A(x) = [t_A^-(x), t_A^+(x)] = [\min\{|0.5x - 0.5|, |0.5x - 0.7|\}, \max\{|0.5x - 0.5|, |0.5x - 0.7|\}],$$
$$N_A(x) = [f_A^-(x), f_A^+(x)] = [\min\{1.5 - 0.5x, 0.5x - 0.2\}, \max\{1.5 - 0.5x, 0.5x - 0.2\}].$$

If $x > 3$, we define

$$M_A(x) = [t_A^-(x), t_A^+(x)] = [\min\{|2.5 - 0.5x|, |2.3 - 0.5x|\}, \max\{|2.5 - 0.5x|, |2.3 - 0.5x|\}],$$
$$N_A(x) = [f_A^-(x), f_A^+(x)] = [\min\{0.5x - 1.5, 0.5x - 1.3\}, \max\{0.5x - 1.5, 0.5x - 1.3\}].$$

For each "Good" project set A and $x \in X$, we define

$$t_A^-(x) = 0.2x - 0.2, t_A^+(x) = 0.2x, f_A^-(x) = 1 - 0.2x, f_A^+(x) = 1.2 - 0.2x.$$

We will obtain 3 fuzzy relationship matrixes $R_{A \to X}$. Obviously, each $R_{A \to X}$ is a 260 × 12 matrix. Where each $r_{i,j}$ have four values $[t_A^-(x), t_A^+(x), f_A^-(x), f_A^+(x)]$.

Step 2: Process of K-means clustering algorithm with IVIFS information.

On the basis of the result of fuzzification, we adopt 10-means to 20-means clustering algorithm according to membership and nonmembership interval with the same weights. Taking all the classifications into account, we merge several categories with individual sample into their most close categories, and we obtain 9 category centers:

$A_1 = \{ <x_1, [0.9, 1], [0, 0.1]>, <x_2, [0.7, 0.9], [0.1, 0.3]>, <x_3, [0.7, 0.9], [0.1, 0.3]>, <x_4,$
  $[0.7, 0.9], [0.1, 0.3]>, <x_5, [0.7, 0.9], [0.1, 0.3]>, <x_6, [0.9, 1], [0, 0.1]>, <x_7, [0.9, 1],$
  $[0, 0.1]>, <x_8, [0.7, 0.9], [0.1, 0.3]>, <x_9, [0.7, 0.9], [0.1, 0.3]>, <x_{10}, [0.9, 1],$
  $[0, 0.1]>, <x_{11}, [0.7, 0.9], [0.1, 0.3]>, <x_{12}, [0.7, 0.9], [0.1, 0.3]> \}.$

$A_2 = \{ <x_1, [0, 0.2], [0.8, 1]>, <x_2, [0.4, 0.6], [0.4, 0.6]>, <x_3, [0.9, 1], [0, 0.1]>, <x_4,$
  $[0.9, 1], [0, 0.1]>, <x_5, [0.9, 1], [0, 0.1]>, <x_6, [0.4, 0.6], [0.4, 0.6]>, <x_7, [0.7, 0.9],$
  $[0.1, 0.3]>, <x_8, [0.4, 0.6], [0.4, 0.6]>, <x_9, [0.7, 0.9], [0.1, 0.3]>, <x_{10}, [0.4, 0.6],$
  $[0.4, 0.6]>, <x_{11}, [0.9, 1], [0, 0.1]>, <x_{12}, [0.1, 0.3], [0.7, 0.9]> \}.$

$A_3 = \{ <x_1, [0.9, 1], [0, 0.1]>, <x_2, [0.9, 1], [0, 0.1]>, <x_3, [0.7, 0.9], [0.1, 0.3]>,$
  $<x_4, [0.9, 1], [0, 0.1]>, <x_5, [0.9, 1], [0, 0.1]>, <x_6, [0.7, 0.9], [0.1, 0.3]>,$
  $<x_7, [0.7, 0.9], [0.1, 0.3]>, <x_8, [0.7, 0.9], [0.1, 0.3]>, <x_9, [0.7, 0.9], [0.1, 0.3]>,$
  $x_{10}, [0, 0.1], [0.9, 1]>, <x_{11}, [0.4, 0.6], [0.4, 0.6]>, <x_{12}, [0.1, 0.3], [0.7, 0.9]> \}.$

$A_4 = \{ <x_1, [0.7, 0.9], [0.1, 0.3]>, <x_2, [0.7, 0.9], [0.1, 0.3]>, <x_3, [0.7, 0.9], [0.1, 0.3]>,$
  $<x_4, [0.7, 0.9], [0.1, 0.3]>, <x_5, [0.4, 0.6], [0.4, 0.6]>, <x_6, [0.7, 0.9], [0.1, 0.3]>, <x_7,$
  $[0.4, 0.6], [0.4, 0.6]>, <x_8, [0.7, 0.9], [0.1, 0.3]>, <x_9, [0.4, 0.6], [0.4, 0.6]>, <x_{10}, [0.7, 0.9],$
  $[0.1, 0.3]>, <x_{11}, [0.2, 0.4], [0.6, 0.8]>, <x_{12}, [0.4, 0.6], [0.4, 0.6]> \}.$

$A_5 = \{ <x_1, [0.7, 0.9], [0.1, 0.3]>, <x_2, [0.7, 0.9], [0.1, 0.3]>, <x_3, [0.7, 0.9], [0.1, 0.3]>,$
$<x_4, [0.7, 0.9], [0.1, 0.3]>, <x_5, [0.4, 0.6], [0.4, 0.6]>, <x_6, [0.7, 0.9], [0.1, 0.3]>, <x_7,$
$[0.4, 0.6], [0.4, 0.6]>, <x_8, [0.4, 0.6], [0.4, 0.6]>, <x_9, [0.4, 0.6], [0.4, 0.6]>, <x_{10}, [0.7, 0.9],$
$[0.1, 0.3]>, <x_{11}, [0.9, 1], [0, 0.1]>, <x_{12}, [0.4, 0.6], [0.4, 0.6]> \}$

$A_6 = \{ <x_1, [0.4, 0.6], [0.4, 0.6]>, <x_2, [0.4, 0.6], [0.4, 0.6]>, <x_3, [0.4, 0.6], [0.4, 0.6]>,$
$<x_4, [0.4, 0.6], [0.4, 0.6]>, <x_5, [0.4, 0.6], [0.4, 0.6]>, <x_6, [0.2, 0.4], [0.4, 0.6]>, <x_7,$
$[0.4, 0.6], [0.4, 0.6]>, <x_8, [0.2, 0.4], [0.4, 0.6]>, <x_9, [0.4, 0.6], [0.4, 0.6]>, <x_{10}, [0.2, 0.4],$
$[0.4, 0.6]>, <x_{11}, [0.2, 0.4], [0.4, 0.6]>, <x_{12}, [0.2, 0.4], [0.4, 0.6]> \}.$

$A_7 = \{ <x_1, [0.7, 0.9], [0.1, 0.3]>, <x_2, [0.7, 0.9], [0.1, 0.3]>, <x_3, [0.4, 0.6], [0.4, 0.6]>,$
$<x_4, [0.4, 0.6], [0.4, 0.6]>, <x_5, [0.1, 0.3], [0.7, 0.9]>, <x_6, [0.7, 0.9], [0.1, 0.3]>, <x_7,$
$[0.7, 0.9], [0.1, 0.3]>, <x_8, [0.7, 0.9], [0.1, 0.3]>, <x_9, [0.7, 0.9], [0.1, 0.3]>, <x_{10}, [0.9, 1.0],$
$[0, 0.1]>, <x_{11}, [0.7, 0.9], [0.1, 0.3]>, <x_{12}, [0.4, 0.6], [0.4, 0.6]> \}.$

$A_8 = \{ <x_1, [0.4, 0.6], [0.4, 0.6]>, <x_2, [0.4, 0.6], [0.4, 0.6]>, <x_3, [0.4, 0.6], [0.4, 0.6]>,$
$<x_4, [0.2, 0.4], [0.4, 0.6]>, <x_5, [0.2, 0.4], [0.4, 0.6]>, <x_6, [0.4, 0.6], [0.4, 0.6]>, <x_7,$
$[0.4, 0.6], [0.4, 0.6]>, <x_8, [0.7, 0.9], [0.1, 0.3]>, <x_9, [0.4, 0.6], [0.4, 0.6]>, <x_{10}, [0.7, 0.9],$
$[0.1, 0.3]>, <x_{11}, [0.4, 0.6], [0.4, 0.6]>, <x_{12}, [0.4, 0.6], [0.4, 0.6]> \}.$

$A_9 = \{ <x_1, [0.2, 0.4], [0.4, 0.6]>, <x_2, [0.2, 0.4], [0.4, 0.6]>, <x_3, [0.3, 0.5], [0.5, 0.7]>,$
$<x_4, [0.1, 0.3], [0.7, 0.9]>, <x_5, [0.1, 0.3], [0.7, 0.9]>, <x_6, [0.2, 0.4], [0.4, 0.6]>, <x_7,$
$[0.2, 0.4], [0.4, 0.6]>, <x_8, [0.3, 0.5], [0.5, 0.7]>, <x_9, [0.3, 0.5], [0.5, 0.7]>, <x_{10}, [0.3, 0.5],$
$[0.5, 0.7]>, <x_{11}, [0.3, 0.5], [0.5, 0.7]>, <x_{12}, [0.2, 0.4], [0.4, 0.6]> \}.$

Step 3: Rule extraction.

According to the results of K-means clustering, we obtain 9 clustering centers, and then we determine all classification rules. Considering the realistic meaning of all the attributes for x_i ($i = 1, 2, \ldots, 12$) and y_j ($j = 1, 2, \ldots, 5$) in Example 1, we obtain

$A_1 = \{ <y_1, [0.8, 0.95], [0.05, 0.2]>, <y_2, [0.7, 0.9], [0.1, 0.3]>, <y_3, [0.8, 0.95],$
$[0.05, 0.2]>, <y_4, [0.8, 0.95], [0.05, 0.2]>, <y_5, [0.77, 0.93], [0.07, 0.23]> \}.$

$A_2 = \{ <y_1, [0.2, 0.4], [0.6, 0.8]>, <y_2, [0.9, 1.0], [0, 0.1]>, <y_3, [0.55, 0.75], [0.25, 0.45]>,$
$<y_4, [0.55, 0.75], [0.25, 0.45]>, <y_5, [0.47, 0.63], [0.37, 0.53]> \}.$

$A_3 = \{ <y_1, [0.9, 1], [0, 0.1]>, <y_2, [0.83, 0.97], [0.03, 0.17]>, <y_3, [0.7, 0.9], [0.1, 0.3]>,$
$<y_4, [0.7, 0.9], [0.1, 0.3]>, <y_5, [0.17, 0.33], [0.67, 0.83]> \}.$

$A_4 = \{ <y_1, [0.7, 0.9], [0.1, 0.3] >, <y_2, [0.6, 0.8], [0.2, 0.4] >, <y_3, [0.55, 0.75],$
$[0.25, 0.45] >, <y_4, [0.55, 0.75], [0.25, 0.45] >, <y_5, [0.43, 0.63], [0.37, 0.57] > \}.$

$A_5 = \{ <y_1, [0.7, 0.9], [0.1, 0.3] >, <y_2, [0.6, 0.8], [0.2, 0.4] >, <y_3, [0.55, 0.75],$
$[0.25, 0.45] >, <y_4, [0.4, 0.6], [0.4, 0.6] >, <y_5, [0.67, 0.83], [0.17, 0.33] > \}.$

$A_6 = \{ <y_1, [0.4, 0.6], [0.4, 0.6] >, <y_2, [0.4, 0.6], [0.4, 0.6] >, <y_3, [0.3, 0.5], [0.4, 0.6] >,$
$<y_4, [0.3, 0.5], [0.4, 0.6] >, <y_5, [0.2, 0.4], [0.4, 0.6] > \}.$

$A_7 = \{ <y_1, [0.7, 0.9], [0.1, 0.3] >, <y_2, [0.3, 0.5], [0.5, 0.7] >, <y_3, [0.7, 0.9], [0.1, 0.3] >,$
$<y_4, [0.7, 0.9], [0.1, 0.3] >, <y_5, [0.67, 0.83], [0.17, 0.33] > \}.$

$A_8 = \{ <y_1, [0.4, 0.6], [0.4, 0.6] >, <y_2, [0.27, 0.47], [0.4, 0.6] >, <y_3, [0.4, 0.6],$
$[0.4, 0.6] >, <y_4, [0.55, 0.75], [0.25, 0.45] >, <y_5, [0.5, 0.7], [0.3, 0.5] > \}.$

$A_9 = \{ <y_1, [0.2, 0.4], [0.4, 0.6] >, <y_2, [0.17, 0.37], [0.63, 0.83] >, <y_3, [0.25, 0.45],$
$[0.45, 0.65] >, <y_4, [0.25, 0.45], [0.45, 0.65] >, <y_5, [0.27, 0.47], [0.47, 0.67] > \}.$

From the result above, for each y_k ($k = 1, 2, 3, 4, 5$) we obtain

If $\frac{t_A^-(x) + t_A^+(x)}{2} \geq 0.8$ and $\frac{f_A^-(x) + f_A^+(x)}{2} \leq 0.2$, then it is Good;

If $(t_A^-(x) + t_A^+(x)) - (f_A^-(x) + f_A^+(x)) \geq 0$ then it is Common;

If $(t_A^-(x) + t_A^+(x)) - (f_A^-(x) + f_A^+(x)) < 0$ then it is Bad;

If $\frac{t_A^-(x) + t_A^+(x)}{2} \geq 0.95 \& \frac{f_A^-(x) + f_A^+(x)}{2} \leq 0.1$, then it is Best.

5 Experimental Results Analysis

In this research, the data set is divided into two parts: testing set (with 200 individuals) and training set (with 60 individuals). The results shown in Table 4 are average accuracy of 10 sampling tests. Using these rules to distinguish 33 data with missing attribute, we gain 88% prediction accuracy. From Table 3, we obtain the following decision rule results:

1. A_1, A_2 and A_3 will lead to a successful project, which means that if not all the attributes are good in a software project development, we can also greatly improve the successful rate of outsourced software project. Our simulation results in Table 3 show that when the project management and the development process are very normal, most of the projects being developed will be successful (to 67.35%).

Table 3 Preliminary determination of outsourced software project risk

Pattern	Development experience x_1, x_2	Project management x_3, x_4, x_5	Client team experience x_6, x_9	Client support x_7, x_8	Client business environment x_{10}, x_{11}, x_{12}	Successful rate for development (%)	Prediction result	Account
A_1	Good	Good	Good	Good	Good	68.2	Success	44
A_2	Bad	Best	Common	Common	Common	100	Success	2
A_3	Best	Good	Good	Good	Bad	66.7	Success	3
A_4	Good	Common	Common	Common	Common	29.5	Failure	44
A_5	Good	Common	Common	Common	Common	18.9	Failure	37
A_6	Common	Common	Bad	Bad	Bad	8.3	Failure	24
A_7	Good	Bad	Good	Good	Common	20.0	Failure	40
A_8	Common	Bad	Common	Common	Common	11.5	Failure	52
A_9	Bad	Bad	Bad	Bad	Bad	0.0	Failure	14
Total						26.5		260

Table 4 Prediction accuracy of IVIFS and conventional methods

Method\Accuracy	Failure software	Success software	Total software
K-means with IVIFS information	0.916	0.478	0.800
K-means algorithm	0.911	0.420	0.781
Decision tree	0.859	0.638	0.800
FNN	0.780	0.710	0.762
BPNN	0.963	0.188	0.758

2. A_4, A_5, and A_6 tell us that when many attributes of project management for developer are common, the failure rate will be 79.09%. And A_4 and A_5 illustrate that even when the business condition of customers are common or even good, the failure rate of project will be also 75.31%. A_6 illustrates that if the business condition of customer is worst, then the failure rate of project will be 91.7%.

3. A_7, A_8, and A_9 reveal that if project management of developers is rather unnormal, then the successful rate for software project will be only 13.21%. And from A_9 we conclude that if there are not any outstanding performance for a project group including customers and developers, then the project failure rate will be 100%.

From the results above, we infer that if we are looking forward to the success of a software, we must try our best to do outstanding performance in some areas as we can, especially improve our project management normalization and standardization level, such as standardized development process, normal requirements development, timely follow-up development progress, a milestone standardization management and positive communication with each other, etc. Table 4 shows the accuracy of all methods.

Obviously, the results of rule discovery are similar to that from some companies (Manifesto 2011; CHAOS Summary 2009–2010, Standish Co. Ltd; Software Project Survival Guide, S. MCConnell, Microsoft Corporation, 1997).

6 Conclusion

We propose a rule extraction method based on K-means algorithm with IVIFS information, which not only involves membership interval information, but also involve nonmembership interval information. Therefore, it is more comprehensive and flexible than conventional K-means model. And we apply them to the rule extraction on outsourced software project risk and achieve some significant results. The presented method is a supplement for the prediction of outsourced software risk. Being a simple method, it is suitable for the practice of risk prediction for SMEs of China. We can improve it as follows: increasing samples, finding better algorithms, etc.

Acknowledgements This research is supported by National Natural Science Foundation of China (No. 71271061), Natural Science Projects (No. 2014A030313575, 2016A030313688) and Soft Science funds (No. 2015A070704051, 2015A070703019) and Social Sciences project (No. GD12XGL14) and Education Department Project (No. 2013KJCX0072) of Guangdong Province, Social Sciences funds of Guangzhou (No. 14G41), Team Fund (No. TD1605) and Innovation Project (No. 15T21) and Advanced Education Project (No. 16Z04, GYJYZDA14002) of GDUFS.

References

1. Zadeh L. A: Fuzzy sets. Information and Control, 8, 338–353 (1965)
2. Atanassov K.: Intuitionistic fuzzy sets. Fuzzy Sets and Systems, 20, 87–96 (1986)
3. Atanassov K.: Interval valued intuitionistic fuzzy sets. Fuzzy Sets and Systems, 31, 343–349 (1989)
4. Li D.F., Cheng C.T: New similarity measures of intuitionistic fuzzy sets and applications to pattern recognitions. Pattern Recognition Letters, 23, 221–225 (2002)
5. Wang W.Q., Xin X.L: Distances measure between intuitionistic fuzzy sets. Pattern Recognition Letters, 26, 2063–2069 (2005)
6. Xu Z.S: On similarity measures of interval-valued intuitionistic fuzzy sets and their application to pattern recognitions. Journal of Southeast University (English Edition), 23, 139–143 (2007)
7. Wei C.P., Wang P., Zhang Y.Z: Entropy, similarity measure of interval-valued intuitionistic fuzzy sets and their applications. Information Sciences, 181, 4273–4286 (2011)
8. Wan S.P: Applying interval-value vague set for multi-sensor target recognition. International Journal of Innovative Computing, Information and Control, 7, 955–963 (2011)
9. Xu Z.S: A method based on distance measure for interval-valued intuitionistic fuzzy group decision making. Information Sciences, 180, 181–190 (2010)
10. Xu Z.S.: Dynamic intuitionistic fuzzy multiple attribute decision making. International Journal of Approximate Reasoning, 28, 246–262 (2008)
11. Wei G.W: Some geometric aggregation functions and their application to dynamic attribute decision making in intuitionistic fuzzy setting. International Journal of Uncertainty, Fuzziness and Knowledge-Based Systems, 17, 251–267 (2009)
12. Su Z.X., Chen M.Y., Xia G.P., Wang L: An interactive method for dynamic intuitionistic fuzzy multi-attribute group decision making. Expert Systems with Applications, 38, 15286–15295 (2011)
13. Boehm, B.W.: Software risk management: principles and practices. IEEE Software, 8(1), 32–41 (1991)
14. Wallace L., Keil M.: Software project risks and their effect on outcomes. Communications of the ACM, 47(4), 68–73 (2004)
15. Xia W.D., Lee G.: Complexity of information systems development projects: Conceptualization and measurement development. Journal of Management Information Systems, 22(1), 45–84 (2005)
16. Schmidt R., Lyytinen K., Keil M.: Identifying software project risks: An international Delphi study. Journal of Management Information Systems, 17(4), 5–36 (2001)
17. Karolak D.W.: Software Engineering Risk Management. Los Alamitos: IEEE Computer Society Press (1996)
18. Jiang J.J., Klein G.: An exploration of the relationship between software development process maturity and project performance. Information & Management, 41(3), 279–288 (2004)
19. Wallace L., Keil M., Rai A.: Understanding software project risk: A cluster analysis. Information & Management, 42(1), 115–125 (2004)

20. Nidumolu S.: The effect of coordination and uncertainty on software project performance: Residual performance risk as an intervening variable. Information Systems Research, 6(3), 191–219 (1995)

21. Hu Y., Mo X.Z., Zhang X.Z., Zeng Y.R., Du J.F. and Xie K.: Intelligent analysis model for outsourced software project risk using constraint-based Bayesian network. Journal of Software, 7(2), 440–449 (2012)

22. Hu Y., Zhang X.Z., Nagi E.W.T., Cai R.C., Liu M.: Software project risk analysis using Bayesian networks with causality constraints. Decision Support Systems, 56(12), 439–449 (2013)

23. Hu Y., Du J.F., Zhang X.Z., Hao X.L., Nagi E.W.T., Fan M., Liu M.: An integrative framework for intelligent software project risk planning. Decision Support Systems, 55(11), 927–937 (2013)

24. Hu Y., Feng B., Mo X.Z., Zhang X.Z., Nagi E.W.T., Fan M., Liu M.: Cost-Sensitive and ensemble-based prediction model for outsourced software project risk prediction. Decision Support Systems, 72(2), 11–23 (2015)

25. Ouyang Z.Z., Zhang Z.H., Chen J.Z., Hu Y.: An unsupervised model on outsourcing software projects risks based on T-S fuzzy neural network. Advances in Engineering Research, 13(6), 411–414 (2015)

Prediction of Market Movement of Gold, Silver and Crude Oil Using Sentiment Analysis

Kunal Keshwani, Piyush Agarwal, Divya Kumar and Ranvijay

Abstract Prediction of stock movements and share market has always remained an area of great curiosity and concern for investors. It has already been established that the movement of market shares a big correlation with the sentiments about it. In this paper, we have applied sentiments analysis techniques and machine learning principles to foretell the stock market trends of three major commodities, Gold, Silver and Crude oil. We have used the SentiWordNet library to quantify the emotions expressed in the text. Further neural network has been trained over the calculated readings. Thereafter, the trained neural network is used to forecast the future values. The efficacy of the proposed model is measured on the basis of mean absolute percentage error. The results clearly reflect that there in fact lies a strong correlation between public mood and stock market variations.

Keywords Sentiment analysis · SentiWordNet · Neural network · Stock market · Microblogging data

1 Introduction

Emotions have intense effect on the decision making [1], similarly the variations in stock values are greatly affected by reaction of people towards a particular stock or commodity. These variations in turn greatly affect the investors and have a significant impact on the organizations as well. So, is it possible that public emotions can predict

K. Keshwani · P. Agarwal · D. Kumar (✉) · Ranvijay
Computer Science & Engineering Department, MNNIT Allahabad, Allahabad, UP, India
e-mail: divyak@mnnit.ac.in

K. Keshwani
e-mail: kunalkeshwani6@gmail.com

P. Agarwal
e-mail: piyushagarwal.mnnit@gmail.com

Ranvijay
e-mail: ranvijay@mnnit.ac.in

© Springer Nature Singapore Pte Ltd. 2018   101
S.K. Bhatia et al. (eds.), *Advances in Computer and Computational Sciences*,
Advances in Intelligent Systems and Computing 554,
https://doi.org/10.1007/978-981-10-3773-3_11

stock movements? Yes, stock market prediction has attracted much attention from academia as well as business. Early efforts to predict the stock market was based on the EMH (Efficient Market Hypothesis) theory [2] and random walk pattern [3] which states that stock market can be predicted with the help of new information. In respond to this, several researchers have attempted to develop computerized models to predict the stock prices and market movements [4–6].

Many sentiment analysis techniques have been discovered in the past years to extract indicators of public mood from social media content such as reviews [7, 8], blogs [9], and other content such as twitter [10, 11]. Mishne et al. in [12] used blogs sentiments to predict movie sales. Chevalier and Mayzlin in [13] used reviews of people from E-commerce sites to predict the book sales. Tetlock in [14] investigated the relations between the financial news media and movements in stock market activity. Asur and Huberman in [15] provided a correlation between the social media (Twitter) and the box office receipts.

It is true that emotions do affect decision making and behavioural finance thus provide another proof that financial decisions are certainly driven by public mood [16]. This can be observed from the sample data set as portrayed in Fig. 1. From this figure, the correlation can be easily observed between the positive SentiWordNet [17] score and stock price of Silver thus directly showing the effect of public mood in the stock movement. The dotted line represents the scores calculated from sentiments and its direct effect can be seen on the stock value movement. Understanding it with the help of an example, as the positive SentiWordNet score rise (line AB), its effect can be seen in the stock value (line DE). Similarly as the score decreases (line BC), the value also decreases (line EF). Although at many places value do not vary according to scores, still with the help of this correlation we can say that public

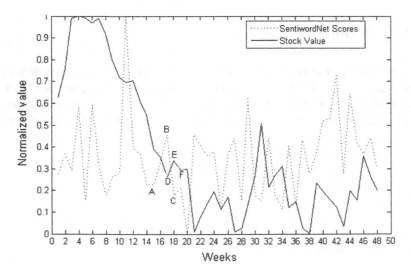

Fig. 1 Correlation between positive SentiWordNet scores and stock prices of silver

sentiments can be helpful in predicting the stock market movement. In this paper, we have used behavioural economics to predict the stock movement. We have tried to find a correlation between "Public Mood" and "Stock Movement" using artificial neural network [18]. To fulfill the object, the paper is organized as: we begin with the explanation of problem statement and then propose a solution methodology. Finally the results are presented which validate our approach.

2 Problem Statement and Proposed Methodology

For some n weeks time frame $[1, \ldots, n]$, the given is the publically expressed sentiments in sentiment files $S\{s_1, \ldots, s_n\}$, about a commodity CY in the form of n text files. The stock prices $X\{x_1, \ldots, x_n\}$ of the same commodity CY for the same duration is also provided. The task is to quantify the text files and then to obtain a correlation between S and X using neural network for future price prediction of CY.

The sentiment files $S\{s_1, \ldots, s_n\}$ when treated with SentiWordNet will result in four types of score, respectively $Pos\,score(A)$, $Neg\,score(B)$, $Pos\,tre\,score(C)$ and $Neg\,tre\,score(D)$ on weekly basis. The obtained scores A, B, C, D can be written as: $A = \{a_1, \ldots, a_n\}$, $B = \{b_1, \ldots, b_n\}$, $C = \{c_1, \ldots, c_n\}$ and $D = \{d_1, \ldots, d_n\}$ for n weeks. To produce these scores SentiWordNet passes through a series of steps: (1) Tokenization and Stemming (2) Speech Tagging (3) Sentiword interpretation. These steps are briefed in the Sect. 2.1 with the help of an example also.

The resultant scores when written against the stock prices can be viewed as shown in Table 1. As a part of next step neural network is trained to establish the correlation between A, B, C, D and X. Since the variation of stock values is nonlinear, we have used neural network that predicts the stock values of current week, on the basis of the data of past two weeks. The trained neural network adheres with the following specifications:

1. Number of inputs: to predict the price of ith week the input contains data of previous two weeks, i.e. $(i-1)$th and $(i-2)$th. Therefore the input is the set of ten attributes, i.e. $\{a_{i-1}, b_{i-1}, c_{i-1}, d_{i-1}, x_{i-1}, a_{i-2}, b_{i-2}, c_{i-2}, d_{i-2}, x_{i-2}\}$.

Table 1 SentiWordNet scores along with stock prices of Silver

Week	A	B	C	D	X
w1	5	3.25	3.62	2.37	18.99
w2	5.87	6.12	5.12	4.62	19.72
w3	5.12	5	3.25	3.37	20.95
w4	8	5	5.75	3.25	21.01
...	...	...	...	...	...
w47	6.62	4.25	4.5	3.75	17.03
w48	5.25	3.87	4.75	2.37	16.68

2. Number of hidden layers: two, each with five neurons.

3. Number of outputs: one, i.e. the predicted value of the stock y_i at ith week.

We have used neural networks because it has previously been used to decode the nonlinear time series which describes the characteristics of stock market [19]. Earlier work also shows that neural networks have been extensively used in predicting the stock movements [20, 21].

The obtained data set is divided into two parts, 90% for training (say k weeks) and 10% for testing. The output of the neural network can be represented as $Y = \{y_{k+1}, \ldots, y_n\}$. For accuracy validation, the Error E_i for week $i \in \{k + 1, \ldots, n\}$, the Mean Error μ and the Mean Absolute Percentage Error (*MAPE*) are used. The formulas of these errors are defined in Eqs. (1), (2) and (3). The trained and verified neural network can be used at any point of time to predict the future market prices of commodity *CY*. The goal is to minimize *MAPE* and the pictorial representation of proposed methodology is shown in Fig. 2.

$$E_i = \mid x_i - y_i \mid \tag{1}$$

$$\mu = \frac{1}{n-k} \sum_{i=k+1}^{n} E_i \tag{2}$$

$$MAPE = \frac{1}{n-k} \sum_{i=k+1}^{n} E_i / x_i \tag{3}$$

2.1 Generating Public Mood Time Series

For generating the public mood time series on a weekly basis we have used two tools namely NLTK (Natural language processing toolkit) [22] and SentiWordNet [17]. NLTK is publicly available toolkit for used for pre-processing. It is widely used toolkit for implementing tokenization, stemming, speech tagging, parsing etc. SentiWordNet is publicly available software package for sentiment analysis that can be used to classify the polarity of a sentence along with the positive, negative and objective scores for that particular sentence [17, 23]. It has been extensively used to analyse the emotional content of reviews [24, 25]. SentiWordNet is collection of words which works on the basis of WordNet. Each word in the SentiWordNet is given a positive, negative and objective score. We have used the lexicon of the SentiWordNet for analysing the polarity of each word occurring in the sentence. The different phases of SentiWordNet are as follows:

- **Tokenization and Stemming**—Each sentiment is first tokenized into words. After that stemming is performed in each word i.e. from each word affix is removed ending with only the stem. For example, feeling, feels will be converted to feel.

Fig. 2 Proposed
methodology

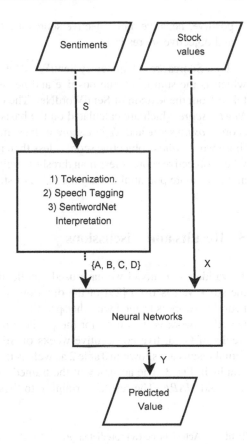

Consider a sentence "Price of Gold will rise again." After Tokenization and Stemming the sentence will be converted into "Price", "of", "Gold", "will", "rise", "again".

- **Speech Tagging**—After tokenization and stemming, speech tagging is performed on each word. Each word is attached with its corresponding tag in the sentence. i.e. whether the word is noun, pronoun, adverb, adjective etc. and the "word, tag" pair is given as input to SentiWordNet. After speech tagging the sentence will become ('Price', 'NN'), ('of', 'IN'), ('Gold', 'NN'), ('will', 'VB'), ('rise', 'VB'), ('again', 'RB').

- **SentiWordNet Interpretation**—Taking the word-tag pair as input, SentiWord-Net searches the word in its lexicon according to a maximum matching strategy. A word basically can have several meanings, so to rectify this problem SentiWordNet has divided the words into groups of its synonyms. Each group is known as Word-net. Each Wordnet has a brief explanation of the word. SentiWordNet searches in all Wordnets and do a maximum matching on the sentence and the explanation of that Wordnet. The Wordnet with maximum matching is selected and the positive, negative and objective scores are taken. The sum of positive, negative and objective scores is 1.

Four Scores are calculated by SentiWordNet, the first two are *Pos score*, *Neg score* which is the sum of all the positive and negative scores of the words in sentence taken from the lexicon of SentiWordNet. The other two scores are *Pos tre score* and *Neg tre score* which are calculated on the basis of another score, which is objective score. *Pos tre score* and *Neg tre score* is the sum of positive and negative scores of all those words whose objective score is less than the certain threshold. All those words whose objective score is less than threshold will have more positive or negative score and have more potential to vary the value of stock.

3   Results and Discussions

To validate our model we have used a collection of sentiments from [26–28] and the stock values from [29] of the duration April 2014 to May 2015. An artificial neural network is trained with the specifications as described in Sect. 2. The trained neural network is then used for the prediction of stock prices for four consecutive weeks of Gold, five consecutive weeks of Silver and Crude oil. The predicted vs actual results are shown in Table 2 as well as these results are elaborated through bar graphs in Fig. 3. The accuracy of the trained neural network is gauged on the basis of μ and *MAPE*. The results pertaining to these errors are also shown in the same

Table 2   Actual price (X), predicted price (Y), absolute error (E), mean error μ and MAPE of Commodities: Gold, Silver and Crude oil

Commodity	X_i	Y_i	E_i	μ	*MAPE* (in %)
Gold	1183.10	1199.00	15.90	28.64	2.38
	1203.10	1267.81	64.71		
	1201.50	1179.05	22.45		
	1186.90	1175.38	11.52		
Silver	16.67	17.23	0.56	0.37	2.22
	16.44	16.33	0.11		
	17.54	16.94	0.60		
	17.03	17.19	0.16		
	16.68	17.14	0.46		
Crude oil	59.96	62.48	2.52	1.91	3.23
	59.61	60.96	1.35		
	59.63	61.74	2.11		
	56.93	55.26	1.67		

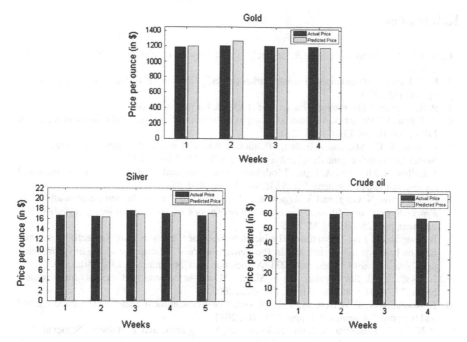

Fig. 3 Actual versus predicted stock prices of gold, silver and crude oil

table, Table 2. A mean error of 28.64, 0.37 and 1.91 and mean absolute percentage error of 2.38, 2.22 and 3.23 is observed for Gold, Silver and Crude oil respectively which is quite remarkable.

4 Conclusion and Future Work

This paper exhibits the process of churning out meaningful information from text corpus in terms of sentiments to quantify public mood. To experiment with sentiment analysis techniques we created a text corpus containing publicly expressed views about three universal commodities which are: Gold, Silver and Crude oil, of the duration of twelve months, on weekly basis. We have used SentiWordNet libraries to calculate positive and negative scores of the text corpus for each week. Then a neural network is trained to learn the correlation between the SentiWordNet scores and stock market behaviour of these commodities. Afterwards, the effectiveness of this model has been validated on the basis of mean absolute percentage error. The results establish the fact that publicly expressed sentiments indeed can predict the stock market behaviour. As a part of future work, more sophisticated methods can be applied for neural network training and the predicted market trends can be deployed for portfolio management.

References

1. R. J. Dolan, "Emotion, cognition, and behavior," *Science*, vol. 298, no. 5596, pp. 1191–1194, 2002.
2. E. F. Fama, "The behavior of stock-market prices," *The journal of Business*, vol. 38, no. 1, pp. 34–105, 1965.
3. P. H. Cootner, "The random character of stock market prices," 1964.
4. A. Pak and P. Paroubek, "Twitter as a corpus for sentiment analysis and opinion mining.," in *LREc*, vol. 10, pp. 1320–1326, 2010.
5. H. Mao, S. Counts, and J. Bollen, "Predicting financial markets: Comparing survey, news, twitter and search engine data," *arXiv preprint* arXiv:1112.1051, 2011.
6. J. Bollen, H. Mao, and A. Pepe, "Modeling public mood and emotion: Twitter sentiment and socio-economic phenomena.," *ICWSM*, vol. 11, pp. 450–453, 2011.
7. C. Whitelaw, N. Garg, and S. Argamon, "Using appraisal groups for sentiment analysis," in *Proceedings of the 14th ACM international conference on Information and knowledge management*, pp. 625–631, ACM, 2005.
8. B. Pang and L. Lee, "A sentimental education: Sentiment analysis using subjectivity summarization based on minimum cuts," in *Proceedings of the 42nd annual meeting on Association for Computational Linguistics*, p. 271, Association for Computational Linguistics, 2004.
9. N. Godbole, M. Srinivasaiah, and S. Skiena, "Large-scale sentiment analysis for news and blogs.," *ICWSM*, vol. 7, no. 21, pp. 219–222, 2007.
10. E. Kouloumpis, T. Wilson, and J. D. Moore, "Twitter sentiment analysis: The good the bad and the omg!," *Icwsm*, vol. 11, pp. 538–541, 2011.
11. P. Nakov, Z. Kozareva, A. Ritter, S. Rosenthal, V. Stoyanov, and T. Wilson, "Semeval-2013 task 2: Sentiment analysis in twitter," 2013.
12. G. Mishne, N. S. Glance, *et al.*, "Predicting movie sales from blogger sentiment.," in *AAAI Spring Symposium: Computational Approaches to Analyzing Weblogs*, pp. 155–158, 2006.
13. J. A. Chevalier and D. Mayzlin, "The effect of word of mouth on sales: Online book reviews," *Journal of marketing research*, vol. 43, no. 3, pp. 345–354, 2006.
14. P. C. Tetlock, "Giving content to investor sentiment: The role of media in the stock market," *The Journal of Finance*, vol. 62, no. 3, pp. 1139–1168, 2007.
15. S. Asur and B. A. Huberman, "Predicting the future with social media," in *Web Intelligence and Intelligent Agent Technology (WI-IAT), 2010 IEEE/WIC/ACM International Conference on*, vol. 1, pp. 492–499, IEEE, 2010.
16. J. R. Nofsinger, "Social mood and financial economics," *The Journal of Behavioral Finance*, vol. 6, no. 3, pp. 144–160, 2005.
17. S. Baccianella, A. Esuli, and F. Sebastiani, "Sentiwordnet 3.0: An enhanced lexical resource for sentiment analysis and opinion mining.," in *LREC*, vol. 10, pp. 2200–2204, 2010.
18. M. T. Hagan and M. B. Menhaj, "Training feedforward networks with the marquardt algorithm," *IEEE transactions on Neural Networks*, vol. 5, no. 6, pp. 989–993, 1994.
19. A. Lapedes and R. Farber, "Nonlinear signal processing using neural networks: Prediction and system modelling," tech. rep., 1987.
20. T. Kimoto, K. Asakawa, M. Yoda, and M. Takeoka, "Stock market prediction system with modular neural networks," in *Neural Networks, 1990., 1990 IJCNN International Joint Conference on*, pp. 1–6, IEEE, 1990.
21. X. Zhu, H. Wang, L. Xu, and H. Li, "Predicting stock index increments by neural networks: The role of trading volume under different horizons," *Expert Systems with Applications*, vol. 34, no. 4, pp. 3043–3054, 2008.
22. E. Loper and S. Bird, "Nltk: The natural language toolkit," in *Proceedings of the ACL-02 Workshop on Effective tools and methodologies for teaching natural language processing and computational linguistics-Volume 1*, pp. 63–70, Association for Computational Linguistics, 2002.
23. A. Esuli and F. Sebastiani, "Sentiwordnet: A publicly available lexical resource for opinion mining," in *Proceedings of LREC*, vol. 6, pp. 417–422, Citeseer, 2006.

24. K. Denecke, "Using sentiwordnet for multilingual sentiment analysis," in *Data Engineering Workshop, 2008. ICDEW 2008. IEEE 24th International Conference on*, pp. 507–512, IEEE, 2008.
25. B. Ohana and B. Tierney, "Sentiment classification of reviews using sentiwordnet," in *9th. IT & T Conference*, p. 13, 2009.
26. Online website, "Silver phoenix 500 (silver-phoenix500.com)," 2016.
27. Online website, "Gold eagle, empowering investors since, 1997 (gold-eagle.com)," 2016.
28. Online website, "Stock twits (stocktwits.com)," 2016.
29. Online website, "New york stock exchange (nyse.com)," 2016.

Social Influence and Learning Pattern Analysis: Case Studies in Stackoverflow

Sankha Subhra Paul, Ashish Tripathi and R.R. Tewari

Abstract Stackoverflow is one of the most popular question and answer (Q&A) websites. One of the reasons for the popularity is the use of badges and reputation score. The reputation score is cumulative of answers accepted by the community users on their posted questions on various topics. This paper tries to find out the most influential top users topic (tag) specific over a group of topics by forming social network based on the question owner and accepted user owner for a particular topic (tag) using SNAP (Stanford Network Analytics Platform). After formation of topic (tag)-specific social network graph, top 10 experts and learners are obtained using HITS algorithm. A simple scoring methodology is used to find top-n most influential experts and learners in the group of topics (tags). Three case studies are undertaken for three groups of topics (tags). We find that there are some common users appearing in the top 10 experts list in a group of topics (tags). It is also observed that there are no common users in the top 10 learners list for first two case studies. However, some users appear common in top 10 learners for more than 1 topic (tag) in the third case study which is more domain-specific. The latest Stackoverflow dataset has been used.

Keywords SNAP · HITS · Social influence · Experts · Learners

S.S. Paul (✉) · R.R. Tewari
University of Allahabad, Allahabad, India
e-mail: sankha.paul@gmail.com

R.R. Tewari
e-mail: tewari.rr@gmail.com

A. Tripathi
SPMIT, Allahabad, India
e-mail: ashish.mnnit44@gmail.com

© Springer Nature Singapore Pte Ltd. 2018
S.K. Bhatia et al. (eds.), *Advances in Computer and Computational Sciences*,
Advances in Intelligent Systems and Computing 554,
https://doi.org/10.1007/978-981-10-3773-3_12

1   Introduction

Finding effective people in societies has been a key curiosity for politicians, marketers, security analysts, social researchers, engineers, and computer scientists. Since any society can be modeled as a network, network analysis has provided significant insight in this area. Social and economic networks have been studied for decades. The motive has been to mine useful information in order to organize these networks in a way that maximum efficiency is accomplished. This has immensely helped the researchers to understand the role of nodes or groups in a network [1, 2]. The work here focuses on finding important nodes in a network based on their behavior as well as the structure of the network by forming social network based on the tags in "Posts.xml" dataset of Q&A site "Stackoverflow".

Stackoverflow is a website which is part of the Stack Exchange Network. It was created in 2008 by Jeff Atwood and Joel Spolsky as a more open alternative to earlier forums such as Experts Exchange. The name for the website was chosen by voting in April 2008 by readers of Coding Horror, Atwood's popular programming blog. It features questions and answers on a wide range of topics in computer programming.

Though the Stackoverflow site has provisions for reputation score of users interacting with other members of the Q&A community, there is no way to find out say "top 10 python experts." Also, we have no direct answer to know say "top 10 python learners in Stackoverflow Q&A community." Since this is not directly possible, therefore, some preprocessing is needed to be done on "Posts.xml" dataset of Q&A site "Stackoverflow." Our motive in this work is to study the most influential experts and learners in a group of topics (tags).

The motive to find most influential experts and learners in a group of topics (tags) can answer many interesting questions. Some of the questions the proposed work tries to answer are as follows:

- Can there be some members who appear in the top 10 expert list of more than one topic?
- Can there be overlap of users in top 10 learners lists of more than one topic?
- Can there be a user who appears in both top 10 learners and top 10 experts of the same topic?

The rest of the paper is organized as follows: Section 2 summarizes related work, Section 3 gives a summary of the HITS algorithm, Section 4 discusses the methodology, Section 5 gives the results and Section 6 summarizes the conclusion from the results obtained.

2   Related Works

Dana Movshovitz-Attias et al. [3] investigate data from Android applications and Stackoverflow together. Their motive is to find out what it is that programmers want to know and why. Amiangshu Bosu et al. [4] try to find out some guidance principles that may help new stackoverflow users to earn reputation scores quickly. They enumerate following do's for new user in their work: contributing to diverse areas, being active during off peak hours, being the first one to answer a question, answering questions promptly, and answering questions related to tags with lower expertise density. An ACT-R inspired Bayesian probabilistic model was developed by Clayton Stanley et al. [5] to predict the hashtags used by the author of the post. Benjamin V. Hanrahan et al. [6] aim to understand various Hybrid Intelligence Systems through the investigation of Stackoverflow. The long-term goal of their study is to find out how complex problems are handled and dispatched across multiple experts. The findings of B. Vasilescu et al. [7] confirm that men constitute vast majority of contributors to Stackoverflow.

Luca Ponzanelli et al. [8] present Prompter, a plug-in for the Eclipse IDE to automatically search and identify relevant Stackoverflow discussions. It also evaluates the relevance of the discussions given the code context in the IDE, and informs the developer if and only if a user-defined confidence threshold is surpassed. Luca Ponzanelli et al. [9] present an approach to improve the automated system to identify low quality posts in use at Stackoverflow. N. Novielli et al. [10] show that the probability of promptly obtaining a satisfying answer is affected by the emotional style of posting question. Alexander Halavais et al. [11] show that the numerous "tag" badges provide for more socially determined differentiation, while the more general badges are closely related to tenure on the site. Megan Squire et al. [12] make two contributions in their paper. The first contribution of this paper is to foster a more detailed understanding of whether the presence of "source code" (and how much) actually will produce the "best" Stackoverflow questions or answers. The second contribution of this paper is to determine how the non-code portions of the text might also contribute the "best" Stackoverflow postings. Nicole Novielli et al. [13] demonstrate that a state-of-the-art sentiment analysis tool, if applied, can be suitable for detecting affective expressions in Stackoverflow. Fabio Calefato et al. [14] investigate how Stackoverflow users can increase their chances of successful acceptance of their answers. Their research gave evidence that factors related to information presentation, time and affect all have an impact on the success of answers. Christoffer Rosen et al. [15] try to find out what mobile developers are talking about in Stackoverflow. They use Latent Dirichlet allocation-based topic models to help summarize the mobile-related questions.

3 HITS Algorithm

HITS stands for Hypertext Induced Topic Search. HITS is search query dependent. It was devised by Klienberg [16] of Cornell University around 1998–1999.

When the user issues a search query,

- HITS first expands the list of relevant pages returned by a search engine and
- then produces two rankings of the expanded set of pages, authority ranking and hub ranking.

Fig. 1 Illustration for hubs
and authorities in HITS
algorithm

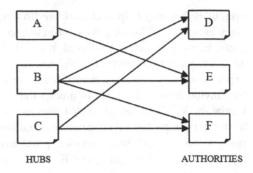

In this algorithm, a web page is named as authority if the web page is pointed by many hyper links and a web page is named as HUB if the page point to various hyperlinks (Fig. 1).

Given a broad search query, q, HITS collects a set of pages as follows:

- It sends the query q to a search engine.
- It then collects t (t = 200 is used in the HITS paper) highest ranked pages. This set is called the root set W.
- It then grows W by including any page pointed to by a page in W and any page that points to a page in W. This gives a larger set S, base set.

HITS works on the pages in S, and assigns every page in S an authority score and a hub score. Thus, these features of HITS has been used for centrality measure in the topic-wise social network digraph formed.

4 Methodology

We have investigated the latest dataset of Stackoverflow provided by Stack Exchange as shown below highlighted with reddish oval:

The dataset when compressed is of size 6.5 GB as shown in Fig. 2. The uncompressed dataset is of size 32.9 GB. This latest dataset contains 8,978,719 question posts and 15,074,572 answer posts. The experiment was done on Ubuntu 14.04 LTS operating system. The python libraries of SNAP for Unix was used. The system used had 4 GB RAM.

The first task is to preprocess the dataset. The dataset is Posts.xml. In this, PostTypeId = 1 signifies that the post is a question post. Similarly, PostTypeId = 2 means the post is an answer post. Each question has an AcceptedAnswerId showing which user's answer was accepted. The preprocessing steps are as follows:

stackexchange directory li ✕		
← → C 🔒 https://archive.org/download/stackexchange/		
spanish.stackexchange.com.7z	09-Apr-2015 19:27	5.4M
sports.stackexchange.com.7z	09-Apr-2015 19:27	5.1M
sqa.stackexchange.com.7z	09-Apr-2015 19:27	7.0M
stackapps.com.7z	09-Apr-2015 19:27	6.0M
stackexchange_archive.torrent	13-Mar-2015 19:17	317.9K
stackexchange_files.xml	02-Jul-2015 05:48	80.5K
stackexchange_meta.sqlite	10-Apr-2015 05:00	252.0K
stackexchange_meta.xml	16-Mar-2015 21:35	3.4K
stackexchange_reviews.xml	02-Jul-2015 05:48	4.1K
stackoverflow.com-Badges.7z	09-Apr-2015 19:28	90.9M
stackoverflow.com-Comments.7z	09-Apr-2015 19:42	2.0G
stackoverflow.com-PostHistory.7z	09-Apr-2015 21:06	10.7G
stackoverflow.com-PostLinks.7z	09-Apr-2015 21:06	30.8M
stackoverflow.com-Posts.7z	09-Apr-2015 21:57	6.5G
stackoverflow.com-Tags.7z	09-Apr-2015 21:57	538.2K
stackoverflow.com-Users.7z	09-Apr-2015 21:58	126.7M
stackoverflow.com-Votes.7z	09-Apr-2015 22:01	440.2M

Fig. 2 Regarding stackoverflow dataset

Fig. 3 Illustration of how social network digraph is formed

- (i) Extract questions posts
- (ii) Extract answer posts
- (iii) Extract questions with particular tag
- (iv) Join AcceptedAnswerId with Ids in answer posts

This helps to identify the user who posted questions in given tag as well as the user whose answer was accepted. Then a directed graph is created with the help of SNAP where the edges point from user who asked question to the user whose answer has been accepted (Fig. 3).

In this social network digraph, HITS algorithm is applied to list out top 10 experts and top 10 learners for each topic. Three case studies are done for a group of topics to find out whether common user appears in the group of topics in the top 10 list of more than one topic. A simple scoring methodology has been used where a score of ten is assigned to the topmost first user in the list, a score of nine to second topper in the list and so on, i.e., the tenth user gets a score of 1. When in more than one topic a common user appears in the list, then their scores are added up. The list of users is sorted according to this combined score.

The topics (tags) in the three case studies are as follows:

Case I: Programming languages group—C++, C#, Java, Python
Case II: Open Source unix lover group—unix, linux, redhat, ubuntu
Case III: Web Development Cluster group—ajax, css, dotnet, html, javascript

The common users' scores are added and top-n scorers are tabulated in Section 5.

5 Result

The top-n scorers are tabulated below. Colored id field represents overlap of users, i.e., users common in more than one topic and according to combined score sorting appear as top-n scorers. The tags field shows the topics in which the user was found in the corresponding list of top 10 experts or learners.

Case I:

The red-colored user with user id 22656 ranked as top expert in both C++ and Java as shown in Table 1. The tags field shows the topic in which the user appeared in top 10. The overlap of user with user id 22656 in this Case I group answers one of our initial question, i.e., whether in a group of topics there can be one user who

Table 1 Top 5 scorers in expert

id	combined score (sort)	tag(s)
22656	20	java/C++
100297	10	python
204847	10	c++
589924	10	perl
23354	9	c#
190597	9	python
571407	9	java
596781	9	c++
622310	9	perl
17034	8	c#
57695	8	java
168657	8	perl
179910	8	c++
2225682	8	python
29407	7	c#
139985	7	java
151292	7	c++
908494	7	python
1521179	7	perl

Table 2 Top 4 scorers in learners

id	combined score (sort)	tag(s)
146780	10	C++
651174	10	Python
859154	10	C#
892029	10	Java
4653	9	C#
610569	9	Python
882932	9	C++
1194415	9	Java
252000	8	C++
359862	8	Java
875317	8	C#
902885	8	Python
541686	7	C++
565968	7	C#
648138	7	Java
1391441	7	Python

appears in top 10 expert list of more than one topic (tags) in the group of topics (tags).

In Table 2, we represent our result for top 4 scorers in the group of topics in Case I.

It clearly shows that there has been no overlap of users in the group of topics in Case I. Therefore all the 4 top 10 learners with score 10 take the top position in Table 2 and so on.

Case II:

In Table 3, we find more number of users overlapping in group of topics in Case II. Since the topics were more correlated to open source programming, therefore, we find more users overlapping in top 5 scorers list.

Table 3 Top 5 scorers in experts

id	combined score(sort)	tag(s)
15168	21	unix/linux/redhat
20862	20	linux/ubuntu/redhat
548225	14	unix/linux
4249	10	redhat
841108	10	linux
874188	10	ubuntu
7552	9	unix/ubuntu
126769	9	redhat

Table 4 Top 4 scorers in learners

id	combined score(sort)	tag(s)
134713	10	unix
310139	10	ubuntu
575281	10	linux
757750	10	redhat
67405	9	unix
353829	9	redhat
760807	9	linux
837208	9	ubuntu
63051	8	redhat
540009	8	ubuntu
779111	8	linux
2630193	8	unix
247542	7	ubuntu
1003575	7	linux
2092392	7	redhat
2888846	7	unix

The users in colored background show the overlapping users in top 5 scorers in experts' field. Thus, users with user id 15168, 20862, 548225, and 7552 are overlapping when taken into account top 10 experts in each topic of Case II.

However, there is no common user appear in the top 10 learners as shown in Table 4.

This means that a user learning say ubutu platform is more dedicated to post questions regarding ubuntu and not say redhat at the same duration of learning and hence appear in top 10 learner list for one topic (tag) only.

Case III:

Here the topics (tags) are more concerned with Web Development cluster. Here, we find that a lot more number of users overlap for more than one topic. Table 5 is self-explanatory.

In Table 6, we have very interesting result. Here, overlap of users occurs in more than one topic. For example, user with user id 4653 appears in the top 10 learners list for topics (tags)-ajax, html, css.

Table 5 Top 10 scorers in expert

id	combined score(sort)	tag(s)
19068	28	ajax/javascript/html
106224	18	html/css
157247	18	ajax/javascript/html
29407	17	ajax/dotnet
405015	15	html/css
1084437	15	html/css
114251	12	javascript/html
816620	11	ajax/javascript
22656	10	dotnet
34397	10	ajax/javascript/dotnet
1542290	10	html/css
17034	9	dotnet
13249	8	ajax
23354	8	dotnet
182668	8	javascript
616443	8	html/css
19068	7	css

Table 6 Top 10 scorers in learners

id	combined score(sort)	tags
4653	29	ajax/html/css
84201	19	html/css
859154	15	javascript/html/dotnet
179736	14	javascript/html
247243	14	html/css
207381	11	javascript/css
51816	10	dotnet
48465	9	dotnet
533941	9	javascript/css
766532	9	ajax
84539	8	dotnet
383759	8	ajax
467875	8	css
565968	8	javascript
1738522	8	html
34537	7	dotnet
364312	7	ajax
172319	6	ajax
188962	6	html
536768	6	javascript
1444475	6	css

6 Conclusion

In Case I, we find there are no common users in the learners list. This means when user learns about one object-oriented language, he/she is concerned with only one language and hence posts question regarding that language. However, in case of experts, with due duration of years, one can learn another object-oriented language much easily and appear in top 10 expert list. In Case II, similar thing is seen in case of learners as happened with learners of Case I. However, more users appear in more than one topic (tag) in top 10 experts list. It shows experts become acquainted with more topics in Open source Unix lovers group. The topics in Case III are more prone to the effect of cluster or domain. The cluster is of .NET web developers. The topics are more interrelated and hence both experts and learners interact in more than 1 topic. This is because during webpage development, the developers can post questions about their requirement in css or html or javascript, etc. We extend our acknowledgments to Prof. Jure Leskovec for providing nice tutorials to learn SNAP [17]. The final conclusion that can be drawn from these 3 cases is that the design principle and reputation scores, etc., of Stackoverflow site avoid the unnecessary question posting and the users are serious members of this Q&A site.

References

1. M. Jackson. Social and economic networks. Princeton University Press (2008).
2. L. Freeman. The development of social network analysis. Empirical Press (2004).
3. Dana Movshovitz-Attias, Yair Movshovitz-Attias, Peter Steenkiste and Christos Faloutsos, 2013. Analysis of the reputation system and user contributions on a question answering website: Stackoverflow In *Proceedings of the 2013 IEEE/ACM International Conference on Advances in Social Networks Analysis and Mining* ACM, pages 886–893.
4. Amiangshu Bosu, Christopher S. Corley, Dustin Heaton, DebarshiChatterji, Jeffrey C. Carver, Nicholas A. Kraft, 2013. Building reputation in Stackoverflow: an empirical investigation In *Proceedings of the 10th Working Conference on Mining Software Repositories*. IEEE, pages 89–92.
5. Clayton Stanley, Michael D. Byrne, Predicting Tags for Stackoverflow Posts, Proceedings of ICCM, 2013.
6. Benjamin V. Hanrahan, Gregorio Convertino, Les Nelson, Modeling problem difficulty and expertise in Stackoverflow, Proceedings of the ACM 2012 conference on Computer Supported Cooperative Work Companion, pg. 91–94.
7. B. Vasilescu, A. Capiluppi, A. Serebrenik, Gender, Representation and Online Participation: A Quantitative Study of Stackoverflow, 2012 International Conference on Social Informatics (SocialInformatics), pg. 332–338.
8. Luca Ponzanelli, Gabriele Bavota, Massimiliano Di Penta, Rocco Oliveto, Michele Lanza. Prompter: A Self-confident Recommender System. In *30th International Conference on Software Maintenance and Evolution (ICSME'14)*. IEEE.
9. Luca Ponzanelli, Andrea Mocci, Alberto Bacchelli, Michele Lanza. 2014. Improving Low Quality Stackoverflow Post Detection. In *30th International Conference on Software Maintenance and Evolution (ICSME'14)*. IEEE.

10. N. Novielli, F. Calefato, and F. Lanubile. 2014. Towards Discovering the Role of Emotions in Stackoverflow In Proc. of the *6th International Workshop on Social Software Engineering* (colocated with FSE 2014) Hong Kong, nov. 2014 ACM.
11. Alexander Halavais, K Hazel Kwon, Shannon Havener and Jason Striker. 2014. Badges of Friendship: Social Influence and Badge Acquisition on Stackoverflow. In *47th Hawaii International International Conference on Systems Science (HICSS-47 2014)*. IEEE.
12. Megan Squire and Christian Funkhouser. 2014. "A Bit of Code": How the Stackoverflow Community Creates Quality Postings In *47th Hawaii International International Conference on Systems Science (HICSS-47 2014)*. IEEE.
13. Nicole Novielli, Fabio Calefato, FilippoLanubile. The Challenges of Sentiment Detection in the Social Programmer Ecosystem. In Proceedings of SSE '15, The 7th International Workshop on Social Software Engineering.
14. Fabio Calefato, FilippoLanubile, Maria ConcettaMarasciulo, Nicole Novielli. Mining Successful Answers in Stackoverflow. In Proceedings of MSR 2015, The 12th Working Conference on Mining Software Repositories.
15. Christoffer Rosen, EmadShihab, 2015. What are mobile developers asking about? A large scale study using Stackoverflow. In *Empirical Software Engineering, April 2015*.Springer Science+Business Media.
16. Jon M. Kleinberg, Authoritative sources in a hyperlinked environment, Proceedings of the ninth annual ACM-SIAM symposium on Discrete algorithms, p. 668–677, January 25–27, 1998, San Francisco, California, USA.
17. https://snap.stanford.cdu/snap/quick.html.

Classification Approach to Extract Strongly Liked and Disliked Features Through Online User Opinions

Juveria Fatima and Deepak Arora

Abstract In recent years, with the advent of emergence and growth of various web technologies and paradigms, an exponential increase can be seen regarding its usage and applicability. This growth has impacted tremendously the way of managing and analysis of data generated on web and how it is being utilized for further planning of any large business organization by exploring different hidden patterns and associated knowledge. Nowadays, internet is being popularized among its users with different dimensions of way of expressing their opinions. As a result, there are various sources of information in form of large repositories is being generated all around, such as social networking sites, e-commerce sites (Amazon, Flipkart, etc.) and forums, etc., which is beneficial to the customers as well as the manufacturers. Feature-based opinion mining aims to produce a feature-based summary of reviews and classifying it as positive, negative, and neutral. In this paper, a method has been proposed to extract the strongly liked and disliked features of product based on customers' online reviews. The Stanford POS tagger has been used to tag the sentences to extract information to identify the required features of any product. This work is implemented on Eclipse using JAVA.

Keywords Feature-based opinion mining · Implicit features · Opinion mining

1 Introduction

Web 2.0 has emerged very rapidly. Hence, the content available on the web has increased tremendously. The number of e-commerce sites, social networking sites, blogs, and forums is increasing constantly. These sites do not only display their respective content but they also offer the provision of giving suggestions and

J. Fatima (✉) · D. Arora
Department of Computer Science & Engineering, Amity University, Lucknow, India
e-mail: juveriafatima2@gmail.com

D. Arora
e-mail: deepakarorainbox@gmail.com

© Springer Nature Singapore Pte Ltd. 2018
S.K. Bhatia et al. (eds.), *Advances in Computer and Computational Sciences*,
Advances in Intelligent Systems and Computing 554,
https://doi.org/10.1007/978-981-10-3773-3_13

123

feedbacks. These suggestions are useful to the customers and manufacturers. Consider a situation where we have to buy something. We normally prefer taking opinions from the individuals. Opinion mining is a discipline that aims at mining reviews from the information available on the web sources. Opinions are basically the sentiments, attitude, views, and perspective of a person on any subjective matter. However, it is not an easy task as there are number of resources available on the web. It becomes difficult to extract the reviews from the web and then classify it as positive, negative or neutral based on the orientation of the reviews. Opinion mining mines the opinions from the web. These opinions are available in the form of reviews on the web like e-commerce sites, social networking sites, forums, etc. Mining the relevant opinions can help the customers as well as the organizations. Customers find it beneficial to consult reviews before buying anything whenever they are confused amongst several things and the manufacturers can take its help to analyze if their product is being liked by their customers or not. It will help the manufacturers to make better marketing plans and thereby improving their productivity. It is a very challenging task as there is no particular structure followed in reviews. Automated extraction of reviews from those available online require natural language processing, machine learning, information retrieval, data mining, etc.

Opinion mining is done at three different levels—sentence level, document level, and feature level. In the sentence level opinion mining, an assumption is made that each sentence contains opinion. It involves two basic tasks—(i) Identifying if the sentence has opinion or not and (ii) Identifying the polarity of sentence. In the document level opinion mining, two assumptions are made—(i) Each document contains the opinions on a single entity and (ii) Each document contains the opinions from a single opinion holder. It includes a single task, i.e., identifying the overall polarity of the document. In the feature-based opinion mining, the opinions corresponding to every feature are summarized after the features have been identified. It determines which feature of the product is liked or disliked by the customers. There are various products available on the e-commerce sites and each product has various features on which various customers express their reviews. It is of interest to the customers and the manufacturers because some customers value only certain features of the products and like to know reviews of those specific features only and the manufacturers are interested to know which feature of their product needs improvement. It becomes difficult for the customers to analyze the reviews available online. It is a very challenging task. Feature-based opinion mining aims to solve this problem by producing a summary of opinions for each respective feature. However, the customers do not generally use any fixed format or structure for writing reviews. So one cannot use a single format to extract features and opinions from the web. Another problem is identifying features from the reviews. It is a difficult task as some reviews may explicitly contain features and the others may have it implicitly. For example, "The camera of this phone is very nice." This review has explicitly mentioned the feature "camera." Consider another review —"This phone does not give good pictures." This review has mentioned the feature 'camera' implicitly which is difficult to be identified. Not much work has been done to identify the implicit features. For the process of mining reviews from various platforms like social blogs, e-commerce sites, etc., the reviews are first downloaded from the required sources and then the preprocessing of the downloaded reviews is done. It includes removing HTML tags and stop words. The downloaded reviews are preprocessed until the unnecessary information has not been removed. After this step, features and opinion words are identified, for which many approaches have been proposed. The objective of this paper is to identify the features of the products that customers strongly like or dislike. Knowing about the features that customers strongly like, the manufacturers can focus on the ways to draw attention of the customers toward those features and knowing about the features that customers strongly dislike, the manufacturers can focus on the ways to improve those features. The rest of the paper is organized as follows: Sect. 2 discusses about the related work that has been done with respect to the proposed topic, Sect. 3 discusses about the proposed technique, Sect. 4 discusses about the experiment and the results produced after implementing the proposed technique, Sect. 5 discusses about various challenges encountered during the system's execution, and Sect. 5 concludes the paper.

2 Background

Much of the work has been done to extract the features and their corresponding opinion words. Some have used the dictionary based approach; others have used syntactical, semantic or machine learning approach. David D. Lewis [1], Pedro Domingos and Michael J. Pazzani [2] used the supervised machine learning techniques to extract the features and the opinions. It used the Naive–Bayes method for extracting the same. It showed great results in classifying the reviews. Bing et al. [3] used the unsupervised method for extracting the features. It used double propagation method for performing the required function. In this method, first the features are extracted and then their corresponding opinion words are extracted. Once it is done, the extracted opinion words are used again to extract the infrequent features. It continues until every feature has not been extracted. It works well for medium sized domain but yields unsatisfactory results for small and large domains. Peter Turney [4] and Peter Turney and Michael L. Littman [5] used point-wise mutual information (PMI). First, the ratio of frequency of two words that occur together to the individual frequencies of two words is calculated in this method and then semantic orientation is evaluated. Semantic orientation is then determined by evaluating the difference of PMI values computed against negative words from PMI values computed against positive words. Hu and Liu [6–8] used WordNet, which is the standard dictionary capable of defining the sentiment orientation of words. It first prepares a seed list and then expands it iteratively using the synonyms and antonyms obtained from WordNet. It did not deal with the words that are context dependent. M. Taylor et al. [9–11] used the association rule mining to extract the features and the opinion words. It used the domain of reviews of restaurant and hotels.

M. Eirnaki et al. 2011 [12] used High Adjective Count for identifying the potential features from reviews. Features on which most of the reviews are available are considered to be potential and those are extracted using this algorithm. Once the potential features have been identified, the sentiment classification is to be done. For which, Max Opinion Score algorithm was used. For identifying the polarity of reviews, features extracted from title are considered separately.

Therefore, we rarely see the work been done to extract the best and the worst features of a product. In this paper, a method has been proposed to extract such features.

3 Experiment Design and Methodology

Details of the proposed system are presented below:

3.1 Crawl Reviews

The required data sources for the reviews can be the e-commerce sites, social networking sites, blogs or forums, etc. Any of these sources can be used for collecting the reviews. There are a lot of reviews available on these sources. Reviews from these sources can be downloaded using the web crawlers or they can be collected manually. The downloaded reviews are then stored in a file.

3.2 Preprocessing

After the reviews have been downloaded, the preprocessing of the downloaded reviews is done. It is done to retain the relevant information. It includes stop word removal, stemming, tokenization, etc. After this step, the downloaded reviews are free from stop words, HTML tags and all such irrelevant data and the collected data is then ready for further processing. The layout of proposed work has been depicted in Fig. 1.

3.3 POS Tagging

This is a very essential step as it is supposed to label every word in a sentence with its corresponding part of speech tag. For example,

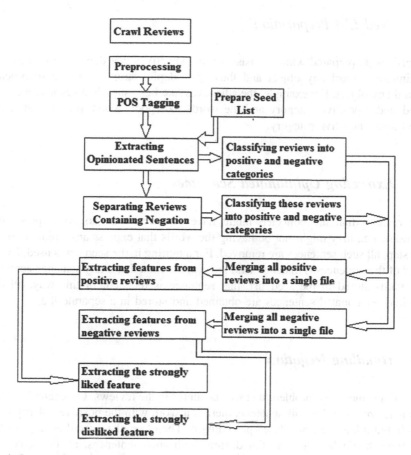

Fig. 1 Layout of the proposed work

The sentence

Its battery is super awesome.

is tagged as:

its_PRP$ battery_NN super_JJ awesome_JJ._.

In the above example, the word '*battery*' is tagged as '*NN*', which indicates that this word is a Noun. It is a general fact that the features are mostly nouns. So after the sentence has been tagged, it gets easier to extract the features. Also, the words that express opinions are generally adjectives which in the above example have been tagged as '*JJ*' (*super* and *awesome*). So, one can easily extract such words too.

3.4 Seed List Preparation

A seed list is prepared which consists of the adjectives that depict strong positive sentiments toward any object and those that depict strong negative sentiments toward any object. For example, the adjectives like best, superb, awesome, etc., are listed under positive category and the words like poor, worst, pathetic, etc., are listed under negative category.

3.5 Extracting Opinionated Sentences

The reviews that are collected might have sentences that do not express any opinions, i.e., they might not be having the words that express any opinion. So in this step, all such sentences are removed. For attaining it, the seed list is used. Every word of the sentences is matched with those in the seed list, if any sentence contains the words stored in the seed list, that sentence is retained. In this way, all the required opinionated sentences are obtained and stored in a separate file.

3.6 Handling Negation

Negation is the major problem that we encounter in the reviews. For example, *"The phone is not good."* In this sentence, there is a word with the positive polarity (i.e., *good*), but it has the word *"not"* preceding it. Therefore, the overall polarity of the sentence should be negative. For dealing with such sentences, all the sentences containing *"not"* are stored in a separate file and all the further processing is done on them separately.

3.7 Identifying the Polarity of Reviews

At this step, there are two files:

- The file containing opinionated sentences excluding those which have negation.
- The file that contains sentences having negation.

From the first file, every sentence is compared word by word with the seed list and if a match is found, then that sentence is assigned the polarity similar to that of the matched word. In the second file (containing sentences having *'not'*), every sentence is compared word by word with the seed list, if a match is found then that sentence is assigned polarity opposite to that of the matched word. In this way, the sentences from each of the two files are classified into positive and negative categories.

3.8 Merging All of the Reviews into a Single File

At this stage, there are four files, which contain:

- Positive reviews from opinionated sentences file.
- Negative reviews from opinionated sentences file.
- Positive reviews from the file containing sentences having negation.
- Negative reviews from the file containing sentences having negation.

So at this step, all of the positive reviews are merged into a single file and all of the negative reviews are merged into another file.

3.9 Extracting the Features

At this stage, there are two files:

- File containing positive reviews.
- File containing negative reviews.

From each file, words tagged as *"NN"* by the POS tagger are extracted. Since it is generally assumed that features are usually the nouns, therefore, extracting nouns from each sentences might give the required features.

3.10 Extracting the Strongly Liked/Disliked Feature

To extract the best and worst feature of a product, there is a need to look at the frequency of each feature in the positive and negative reviews. If a feature occurs more frequently in the file of positive reviews, then that feature is the best or strongly liked feature of the product and if it occurs more frequently in the file of negative reviews then that feature is the worst or strongly disliked feature of the product.

4 Experiment Results and Discussion

To see how effective this proposed strategy is, it is evaluated and discussed in this section. For evaluating this technique, first the reviews are crawled from the required sources. First, the reviews of a mobile phone are collected manually from http://www.amazon.in. After collecting the reviews, preprocessing is done to remove any irrelevant information like stop words, HTML tags, etc. After this step, the entire relevant information ready to be processed further is received. The proposed system is implemented using eclipse as the IDE and java as the coding language used.

There is a need to tag every word of the sentence with its corresponding part of speech tag for which we used POS tagger [13]. The output generated by POS tagger is shown in Fig. 2.

A seed list is prepared which consists of positive and the negative orientation categories each including their corresponding adjectives. For example, words like poor, pathetic, disappointing, etc., are stored under negative category and words like amazing, awesome, extraordinary, etc., are stored under positive category. This list is used to find out the opinionated sentences from the reviews that are collected and stored in a separate file. After separating the opinionated sentences, sentences containing negation are to be handled. In the reviews that possess negation (i.e., 'not'), the overall polarity of the sentence is to be inverted. For example,

```
C:\Windows\system32\cmd.exe

camera_NN phone_NN very_RB good_JJ ._.
produces_VBZ very_RB clear_JJ pictures_NNS ._.
awesome_JJ phone_NN ._.
its_PRP$ battery_NN super_JJ awesome_JJ ._.
did_VBD not_RB like_VB phone_NN ._.
its_PRP$ battery_NN backup_NN superb_JJ ._.
its_PRP$ display_NN not_RB very_RB good_JJ ._.
heats_VBZ up_RP ._.
awesome_JJ ._.
its_PRP$ display_NN poor_JJ ._.
camera_NN pathetic_JJ ._.
poor_JJ battery_NN life_NN ._.
mobile_JJ heating_NN issues_NNS ._.
very_RB bad_JJ performance_NN not_RB satisfied_JJ .
keyboard_NN touch_NN not_RB working_VBG properly_RB
camera_NN yields_NNS awesome_JJ results_NNS ._.
dispay_NN superb_NN ._.
display_NN very_RB bad_JJ ._.
memory_NN amazing_JJ ._.
sound_JJ wonderful_JJ ._.
sound_JJ great_JJ ._.
battery_NN great_JJ ._.
touch_RB great_JJ ._.
touch_NN awesome_JJ ._.
touch_RB very_RB poor_JJ ._.
```

Fig. 2 Output of the POS tagger

"The picture quality of this phone is not good."In this example, even when the sentence has a word with positive orientation, i.e., *good*, it is preceded by the word *'not'* and therefore the polarity of the sentence has to be inverted. So for handling sentences possessing negation, such sentences are separated in another file (as shown in Fig. 3) and then classified on the basis of the comparisons of the file made with the seed list and assigning the opposite polarity to the sentence for every match found.

Same procedure is repeated for sentences that do not possess negation but in this case, same polarity as that of the matched word is assigned to the sentence rather than the opposite one. After this, files containing the positive reviews are merged together and same is done with the negative reviews too. Figure 4 depicts the reviews classified as positive.

From the file containing positive reviews, features are to be extracted. It is generally assumed that the features of any product are nouns. So, from the file words tagged as nouns are extracted. The code and output of which are illustrated in Figs. 5 and 6, respectively. However, there are some implicit features too that one

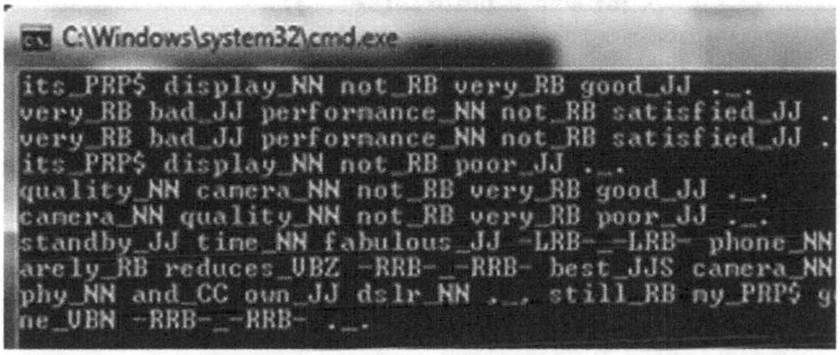

Fig. 3 Output of the reviews with negation

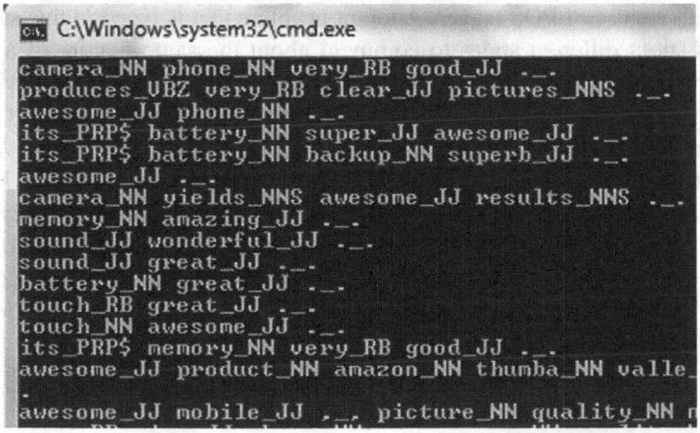

Fig. 4 Output of the reviews classified as positive

```
package feature;
import java.util.*;
    public class extractNN
    {
        public static void main(String args[]) throws IOException
    ClassNotFoundException
        {
        File file =new File("mergedpositive.txt");

            Scanner in = null;

        try {
            in = new Scanner(file);
            while(in.hasNext())
            {String str=in.nextLine();

            PrintWriter printWriter=new PrintWriter
            (new FileWriter("stronglyliked.txt",true));
            for(String s:str.split("\\s+"))
            {
                if(s.endsWith("_NN"))
                printWriter.println(s+" ");

            }
            printWriter.close();

        }} catch (FileNotFoundException e) {
            // TODO Auto-generated catch block
            e.printStackTrace();
        }
```

Fig. 5 Code for extracting the features

cannot obtain from POS tagging. For example, '*It is too large to fit in the pocket easily*'. In this sentence, the implicit feature is *size*, which can be detected manually but not with the use of POS tagger. Another problem that is faced in this step is that customers adopt different styles to comment about the same feature of a product. For example, they might write *battery, battery backup* and *battery life* for describing their reviews about the *battery* feature of a mobile. While the POS tagger reads every word independently and tags it. It will not tag *battery life* as *NN* (noun) and therefore, these two words cannot be extracted as a single feature. However, using the logic of extracting features as the words tagged as noun in the reviews has another problem too that there are some irrelevant words extracted too that are tagged as nouns in the reviews. After extracting the features from the file containing the positive reviews, the features from the file containing negative reviews are extracted following the same procedure.

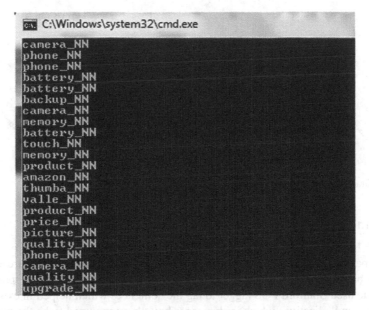

Fig. 6 Output of the extracted features

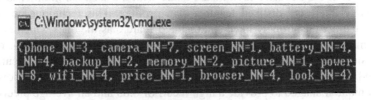

Fig. 7 Output of the frequency of disliked features

Then there is a final step of determining the strongly liked and the strongly disliked features. For extracting the strongly liked feature, the feature that is repeated the maximum number of times in file that is generated using the positive reviews is the strongly liked feature and the same procedure is adopted for extracting the strongly disliked feature using the extracted feature list that is generated from file containing negative reviews. The output of the frequency of the disliked features is shown in Fig. 7.

The problem with this system was that it classifies a statement twice if it contains two opinion words (or two adjectives in the approach followed by this system). The overall recall and precision obtained by the proposed technique is shown in Fig. 8.

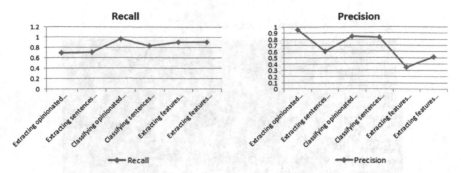

Fig. 8 Graph showing the recall and precision measures

5 Conclusion

In this paper, a method for extracting strongly liked and disliked feature of any product based on user's opinion found on web has been proposed. The proposed technique used Stanford POS tagger to tag the sentences which helped in extracting the nouns and adjectives which in this paper have been assumed to be the potential features and opinions, respectively. After extracting the nouns and adjectives, the comparisons of the extracted reviews are made with the seed list in order to extract the strongly liked and disliked features of a product. This technique could successfully extract the strongly liked and disliked feature as the ones which occurred for the maximum number of times in the positive and negative reviews, respectively. Since, negation is also handled in the system; the overall accuracy is improved as the misclassifying errors that occurred due to negation are removed. So this technique managed to provide a high recall of 0.83 and an average precision of 0.68. The reason for comparatively lower value of precision is due to the phrases used to describe a certain feature, while this technique extracts words as features and thereby this step yields a precision of 0.50 lowering the overall average precision. On the other hand, the step in which negation is handled has managed to yield a recall of 0.80 and a precision of 0.83 which has increased the overall efficiency quite effectively.

References

1. David D. Lewis.: The Independence Assumption in Information Retrieval. In: Proc. of the European Conference on Machine Learning (ECML), pp. 4–15. (1998).
2. Pedro Domingos, Michael J. Pazzani.: On the Optimality of the Simple Bayesian Classifier Under Zero-One Loss. Machine Learning, 29(2–3): 103–130 (1997).
3. Qiu, gang., Bing, Liu., Jiajun Bu, Chun Chen: Expanding Domain Sentiment Lexicon Through Double Propagation. In: Proceedings of IJCAI (2009).

4. Turney, Peter D.: Thumbs up or thumbs down? Semantic orientation applied to unsupervised classification of reviews. In: Proceedings of the 40th annual meeting on association for computational linguistics, Association for Computational Linguistics, pp. 417–424 (2002).
5. Turney, Peter D., Michael L. Littman.: Measuring praise and criticism: Inference of semantic orientation from association. In: ACM Transactions on Information Systems (TOIS), vol. 21, pp. 315–346 (2003).
6. Hu, Minqing, Liu Bing.: Mining and summarizing customer reviews. In: Proceedings of the tenth ACM SIGKDD international conference on Knowledge discovery and data mining, ACM, pp. 168–177 (2004).
7. Hu, Minqing, Liu Bing.: Mining opinion features in customer reviews. In: AAAI, vol. 4, pp. 755–760 (2004).
8. Liu, Bing, Minqing Hu, Junsheng Cheng.: Opinion observer: analyzing and comparing opinions on the web. In: Proceedings of the 14th international conference on World Wide Web, ACM, pp. 342–351 (2005).
9. Marrese-Taylor, Edison, Juan D. Velásquez, Felipe Bravo-Marquez, Yutaka Matsuo: Identifying customer preferences about tourism products using an aspect-based opinion mining approach. In: Procedia Computer Science, vol. 22, pp. 182–191(2013).
10. Marrese-Taylor, Edison, Juan D. Velásquez, Felipe Bravo-Marquez: Opinion Zoom: A Modular Tool to Explore Tourism Opinions on the Web. In: Web Intelligence (WI) and Intelligent Agent Technologies (IAT), IEEE/WIC/ACM International Joint Conferences, vol. 3, IEEE, pp. 261–264 (2013).
11. M. Taylor, Edison, Juan D. Velásquez, Felipe Bravo-Marquez.: A novel deterministic approach for aspect-based opinion mining in tourism products reviews. In: Expert Systems with Applications, vol. 41, pp. 7764–7775 (2014).
12. Eirinaki Magdalini, Shamita Pisal, Japinder Singh.: Feature-based opinion mining and ranking. In: Journal of Computer and System Sciences, vol. 78, pp. 1175–1184 (2012).
13. Stanford log-linear part-of-speech tagger http://nlp.stanford.edu/software/tagger.html (2010).

Part II
Internet of Things

A Multicriteria Decision-Making Method for Cloud Service Selection and Ranking

Rakesh Ranjan Kumar and Chiranjeev Kumar

Abstract Recent advancement in the field of Cloud computing shows that Cloud services play a crucial role in terms of on-demand services on a subscription basis. A number of available enterprises offer a large number of Cloud services to their customers. Subsequently, increased number of available Cloud services has drawn enormous attention of the customers as a challenging decision-making problem. Due to vast availability of Cloud services, it is very difficult for the users to decide about the best service and reason for its selection. In this paper, we have proposed a multicriteria decision-making method (TOPSIS) to help customers to select a better Cloud service among the set of existing Cloud services with the satisfaction of their requirements. As a result, we found that the proposed method is practical for solving Cloud service selection problem. Moreover, we have compared the simulated results with the existing approaches as analytical hierarchy priority (AHP) and fuzzy AHP method which shows proposed method is flexible and achieve better accuracy.

Keywords Cloud computing · Cloud services selection · SLA · MCDM · TOPSIS · Quality of service · Service measurement

1 Introduction

Over the last decade, Cloud computing has emerged as a distributed computing paradigm which provides an on-demand services and resources to the user on the basis of usages under "pay-as-you-go" pricing model. Cloud computing [1]

R.R. Kumar (✉) · C. Kumar
Department of Computer Science and Engineering, Indian School of Mines,
Dhanbad, India
e-mail: rakeshranjan.cdac@gmail.com

C. Kumar
e-mail: k_chiranjeev@yahoo.co.uk

© Springer Nature Singapore Pte Ltd. 2018   139
S.K. Bhatia et al. (eds.), *Advances in Computer and Computational Sciences*,
Advances in Intelligent Systems and Computing 554,
https://doi.org/10.1007/978-981-10-3773-3_14

provides three main services model such as SaaS, PaaS, and IaaS according to the need of the user. Based on the deployment model the Cloud has been classified as private, public, community, and hybrid Cloud [2, 3]. Due to Cloud's agility and flexibility, Cloud provides an opportunity for companies like IBM, Microsoft, Amazon, and Google to make their new business applications on the Cloud but also to migrate their existing business services onto the Cloud. With a huge choice of Cloud service providers offering the services, the clients often have to choose best Cloud service providers. To identify which service is the best for a service user, quality of service (QoS) [4] is usually exploited. QoS represents a set of non-functional attributes of service such as response time, throughput, reliability, and security. Due to the vast diversity of Cloud services, selection of the most suitable Cloud service becomes an important challenge for the customers. Such Cloud service, selection problem has recently attracted considerable attention from the industry and academia. A Cloud service, selection problem is considered as multicriteria decision-making (MCDM) problem [5]. Multicriteria decision-making (MCDM) is a well-known area of operational research and also proves its efficiency in different complex real-world decision-making problems. The main challenge in Cloud customer is to discover which Cloud service has satisfied their requirements and how to rank the functionally equivalent Cloud services. QoS plays an important role for service selection and ranking of services according to their priority. Therefore, an efficient and accurate decision-making approach is highly desirable to guarantee that the chosen services work well in all conditions. The cloud service, selection mechanism is broadly classified into two types, i.e., Cloud service selection for the single task and Cloud service selection for the different candidate task on a composite task. In this paper, we focus on Cloud service selection for the single task and used TOPSIS (technique for order preference by similarity to ideal solution) [6] method as MCDM approach for ranking of Cloud services. The remainder of this paper is organized as follows. In the next section, we have described related work. Section 3 describes the service measurement index (SMI) [7] framework and its attributes. Section 4 describes the service selection approach. Section 5 presents a case study and comparison of result with another MCDM method. Finally, conclusion and brief summative analysis are given in Sect. 6 with future work.

2 Related Work

With the increasing popularity of Cloud computing, many researchers studied the performance of Clouds for different type of applications such as web applications, scientific computing and e-commerce. For the past few years, decision-making theory has successfully applied to choose appropriate Cloud services. There is a wide range of mathematical methods for evaluation Cloud services based on QoS. Cloud Service Selection methods are generally an extension of the MCDM method for Web services [8, 9, 10]. Godse and Mulik [11] described the Cloud service selection problem as MCDM problem and presented a case study of a sales force automation service for an understanding of the importance and significance of quantitative method to solve SaaS selection and proposed a service recommender framework with dissimilar parameters for Cloud selection according to users requirement. Alhamad et al. [12] proposed a fuzzy-based model for selection of Cloud service with a set of criteria such as usability, security, scalability, and availability. Hussain and Hussain [13] described multicriteria decision-making methods for IaaS Cloud service selection in a case study which contains five basic performance measurements of thirteen Cloud services by a third party monitoring service. In this paper, ELECTRE method proposed for Cloud service selection based on user specification. Garg et al. [14] proposed a trust framework using AHP method for ranking of Cloud services with measurement of attributes weight. The author has measure Cloud Service Measurement Initiative Consortium (CSMIC) [7] QoS criteria and compares Cloud service in the term of Key Performance Indicators (KPIs). Nevertheless, the selection and ranking considered only quantifiable criteria of CSMIC and does not consider trustworthiness nonquantifiable QoS criteria for selection of CSPs. Based on literature survey, it shows that Cloud service selection procedure is an important MCDM problem among a set of QoS properties. In this paper, we present an idea that uses the SMI values and describes a mathematical model which is useful for ranking and selection of Cloud service based on user's requirement.

3 Service Measurement Index (SMI) of Cloud

The Service Measurement Index (SMI) [7] is a standard model designed for measuring the end-to-end customer experience for any number Cloud service provider. The SMI Model is a set of business-relevant Key Performance Indicators (KPI's) that provide a standardized method for measuring and comparing a business services. The SMI model provides a hierarchical framework for selecting Cloud services with primary area: Cost, Performance, Security and Privacy, Agility, Accountability, Assurance, and Financial. Each category is further divided by two and more attributes. In current scenario, different Cloud service provider offers similar services with similar functionality with different criteria (Fig. 1). From consumer point of view, the SMI criteria are important decision-making framework for Cloud service provider selection.

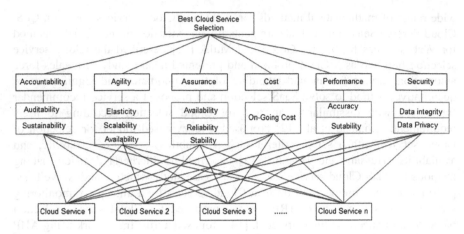

Fig. 1 A hierarchy model for cloud service selection

4 Cloud Service Selection

In this section, we propose a hybrid technique for ranking Cloud service which incorporates seven steps. In proposed techniques, we use the SMI model, which is based on a set of business-relevant KPIs. We used the Technique for Order Preference by Similarity to ideal Solution (TOPSIS) method is one of the most classical methods for solving MCDM problem. The TOPSIS method is based on the principle of determining a solution with the shortest distance to the ideal solution and the greatest distance from the negative ideal solution. The basic procedure of our proposed service ranking method is given below.

Step 1: Computation of the relative weights of each QoS and Define Decision matrix—Computation of the relative weights of each QoS is achieved by pairwise comparison matrix [14]. A scale of one–nine suggested in the paper [15] is used to quantify the strength of two alternatives with respect to a given attribute. Finally, construct a decision table for a given dataset.

Step 2: Construct normalize decision matrix—This step involves the development of matrix formats. To transform the various attribute dimensions in nondimensional attributes, which allows comparison across the attributes. The normalization is usually made through

$$r_{ij} = \frac{S_{ij}}{\sqrt{\sum_{i=1}^{m} S_{ij}^2}} \tag{1}$$

where i = 1, 2... m; j = 1, 2,...., n and S_{ij} is a crisp value.

Step 3   Construct the weighted normalized decision matrix—Multiply each column of the normalized decision matrix by associating weight of QoS parameter

$$v_{ij} = w_j * r_{ij} \tag{2}$$

where $i = 1, 2\ldots$ m; $j = 1, 2,\ldots,$ n and wj represent the weight of jth attribute.

Step 4:   Define the positive ideal (best) solution called v_j^+ and the negative ideal (worst) solution called v_j^- for each criteria.

$v_j^+ = \max\{v_{1j}\ldots v_{mj}\}$ and $v_j^- = \min\{v_{1j}\ldots v_{mj}\}$ for beneficial attribute

$v_j^+ = \min\{v_{1j}\ldots v_{mj}\}$ and $v_j^- = \max\{v_{1j}\ldots v_{mj}\}$ for non $-$ beneficial attribute

Step 5:   Calculate the separation measure for each attributes

$$s_j^* = \sqrt{\sum_{j=1}^n (v_j^+ - v_{ij})^2} \tag{3}$$

$$s_j^- = \sqrt{\sum_{j=1}^n (v_j^- - v_{ij})^2} \tag{4}$$

where $i = 1, 2\ldots$ m and $j = 1, 2,\ldots,$ n.

Step 6:   Calculate the closeness coefficient to the ideal solution for each alternative

$$cc_i = \frac{s_i^-}{s_i^- + s_i^+} \tag{5}$$

where $i = 1\ldots$ m.

Step 7:   Rank the alternative according to CC_i. The bigger value of CC_i is the better Cloud alternative.

5   Case Study: Using TOPSIS Method to Ranking Cloud Service Based on QoS Requirement

To clarify, the new proposed procedure and required calculations have been coded using MATLAB 13 on Intel core [TM] i7-4770 platform running Windows 8. Kivait graph (Fig. 2) illustrates the information regarding 3 different Cloud Service

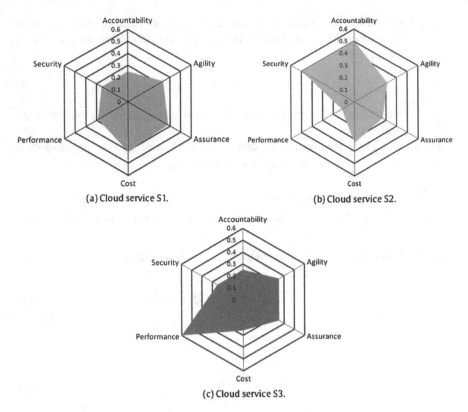

(a) Cloud service S1.

(b) Cloud service S2.

(c) Cloud service S3.

Fig. 2 Comparison of cloud services for different attributes

providers [14] with best values for their KPIs. From these three alternatives, the best Cloud services has to be selected. We have considered top six high level QoS attribute: **Accountability, Agility, Assurance, Cost, Performance, and Security**. Here we explain the process of proposed method step by step in real example.

Step 1: Construct a pairwise comparison matrix between Cloud Service for each QoS attributes. For each attributes we also construct pairwise comparison. Figure 2 shows values for each Cloud Services respect of top QoS attributes. Finally, we get a decision matrix [14] of given dataset.

Step 2: Using Eq. (1), the normalized decision matrix is calculated and shown in Table 1 with different attributes.

Step 3: The weighted normalized decision matrix is constructed using Eq. (2) and respective result is showing in Table 2.

Step 4: Positive ideal solution (PIS) and Negative ideal solution (NIS) are defined as

Table 1 Normalized decision matrix

QoS attribute with weight	Service (S1)	Service (S2)	Service (S3)
Accountability (0.05)	0.408	0.816	0.408
Agility (0.1)	0.576	0.550	0.603
Assurance (0.3)	0.654	0.457	0.603
Cost (0.3)	0.694	0.568	0.441
Performance (0.3)	0.422	0.175	0.889
Security (0.05)	0.408	0.816	0.408

Table 2 Normalized weighted decision matrix

QoS attributes	Service (S1)	Service (S2)	Service (S3)
Accountability	0.0204	0.0408	0.0204
Agility	0.0577	0.0551	0.0603
Assurance	0.1306	0.0916	0.1206
Cost	0.2082	0.1707	0.1324
Performance	0.1268	0.0527	0.2677
Security	0.0204	0.0408	0.0204

$$S^* = [0.0204, 0.0603, 0.1306, 0.1325, 0.2660, 0.0408]$$
$$S^- = [0.0408, 0.0551, 0.0916, 0.2084, 0.0527, 0.0204]$$

Step 5: The distance of each candidate from PIS and NIS are calculated, respectively, using Eqs. (3) and (4).

Step 6: The closeness coefficient is calculated for each candidate using Eq. (5). The result is

$$CC_1 = 0.3805; \quad CC_2 = 0.1634; \quad CC_3 = 0.9101$$

Step 7: Therefore, the relative ranking of all the Cloud services can be decided based on the resultant closeness coefficient (0.3805, 0.1634, 0.9101). Based on the user requirement, the Cloud services are ranked as follows.

$$S3 > S1 > S2.$$

From the above ranking, it is clear that S3 is the best Cloud services provider, whereas S2 is considered as the worst service provider.

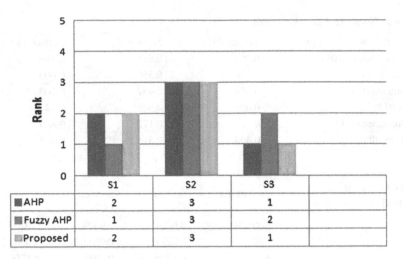

Fig. 3 Rank of cloud service providers using different methods

6 Conclusion and Future Work

With the growth of Cloud Computing paradigm, there are many Cloud providers who offer different Cloud services with different attributes. We have seen that the choice of the best Cloud Service satisfying the majority of available conditions and alternatives is an MCDM problem. The main objective of the paper is to provide a method to help the evaluation and selection Cloud service provider from the user's prospective. In this paper, we proposed TOPSIS approach for Cloud service selection mechanism which is used for ranking Cloud service based on QoS attributes proposed by CSMIC.

It is obvious from the experimental results that the use of nonfunctional QoS parameters during Cloud Service discovery process improves the probability of selecting highly appropriate Cloud service. Figure 3 compares the AHP approach and Fuzzy AHP approach with our proposed method which shows rank of different cloud service providers. Our proposed approach avoids the rank reversal problem, which is the disadvantage of the popular AHP based approach. For future work, we plan to examine a fuzzy extension of the proposed method in order to cope with variation in QoS attributes such as performance. We are also planning to extend the quality model to nonquantifiable QoS attributes.

References

1. Buyya, R, Yeo, C. S., Brandic, I.: Cloud computing and emerging IT platforms: Vision, hype, and reality for delivering computing as the 5th utility. Future Generation computer systems 25(6), 599–616 (2009).
2. Mell, P., Grance, T.: The NIST definition of cloud computing (2011).
3. Lecznar, M., Patig, S.: Cloud computing providers: Characteristics and recommendations. In E-Technologies: Transformation in a Connected World Springer Berlin Heidelberg, pp. 32–45 (2011).
4. Ardagna, D., Casale, G., Ciavotta, M., Prez, J. F., Wang, W.: Quality-of-service in cloud computing: modeling techniques and their applications. Journal of Internet Services and Applications, 5(1), 1–17 (2014).
5. Chen, C. T., Hung, W. Z., Zhang, W. Y.: Using interval valued fuzzy VIKOR for cloud service provider evalution and selection. In Proceedings of the International Conference on Business and Information, (2013).
6. Hwang, C. L., Yoon, K.: Multiple attribute decision making: methods and applications a state-of-the-art survey, Springer Science Business Media, (2012).
7. Cloud Service Measurement Index Consortium (CSMIC), SMI framework. URL:https://slate.adobe.com/a/PN39b/.
8. Tran, V. X., Tsuji, H., Masuda, R.: A new QoS ontology and its QoS-based ranking algorithm for Web services. Simulation Modeling Practice and Theory. 17 (8),1378–1398 (2009).
9. Lin, S. Y., Lai, C. H., Wu, C. H., Lo, C.: A trustworthy QoS-based collaborative filtering approach for web service discovery. Journal of Systems and Software. 93, 217–228 (2014).
10. Wang, P.: QoS-aware web services selection with intuitionist fuzzy set under consumers vague perception. Expert Systems with Applications. 36(3), 4460–4466 (2009).
11. Godse, M., Mulik, S.: An approach for selecting software-as-a-service (SaaS) product. IEEE International Conference, pp. 155–158 (2009).
12. Alhamad, M., Dillon, T., Chang, E.: A trust-evaluation metric for cloud applications. International Journal of Machine Learning and Computing. 1(4), (2011).
13. Hussain, O. K., Hussain, F. K.: Iaas cloud selection using MCDM methods. In eBusiness Engineering (ICEBE), IEEE Ninth International Conference, pp. 246–251 (2012).
14. Garg, S. K., Versteeg, S., Buyya, R.: A framework for ranking of cloud computing services. Future Generation Computer Systems. 29(4), 1012–1023 (2013).
15. Saaty, T. L.: How to make a decision-the analytic hierarchy process. European journal of operational rescarch. 48(1), 9–26 (1990).

Development and Analysis of IoT Framework for Healthcare Application

Anil Yadav, Nitin Rakesh, Sujata Pandey and Rajat K. Singh

Abstract In this paper, major building blocks of IoTized ecosystem is discussed. We have considered IoT application scenario for a simple healthcare system and have shown the interaction process between different applications. Proposed work in this paper is based on the Iotivity software framework. A simple use case of heartbeat sensor is considered as an IoT application. To distinguish the services offered to a normal user and a privileged user an algorithm is formulated along with the state machine, class diagram. Sequence of operations is also depicted to showcase the events and their subsequent reaction to simulate a real time heartbeat sensor.

Keywords IoT · App · Security · Privacy · Sensor · Wireless

A. Yadav (✉)
Smart TV Platform Security, Delhi, India
e-mail: anil.yadav@samsung.com

N. Rakesh
CSE Department, Noida, India
e-mail: nitin.rakesh@gmail.com

S. Pandey
ECE Department, Noida, India
e-mail: spandey@amity.edu

R.K. Singh
ECE Department, Allahabad, India
e-mail: rajatsingh@iiita.ac.in

A. Yadav
Samsung Research Institute, Delhi, India

N. Rakesh · S. Pandey
Amity University, Noida, Uttar Pradesh, India

R.K. Singh
IIIT Allahabad, Allahabad, India

© Springer Nature Singapore Pte Ltd. 2018   149
S.K. Bhatia et al. (eds.), *Advances in Computer and Computational Sciences*,
Advances in Intelligent Systems and Computing 554,
https://doi.org/10.1007/978-981-10-3773-3_15

1 Introduction

Advancement in wireless technology and availability of new applications has led to the development of new area of interest "Internet of Things" [1–10]. It is evident that growing usage of Wi-Fi and wireless internet access, which gave rise to the increasing use of smart devices that continuously, interact with each other for transfer of information.

Internet now can be looked upon as a distributed network system which dynamically respond to the physical environment and translate them into information data which can be used for triggering further action. The most important part of the network/IoT ecosystem is the smart objects with dedicated roles in the entire ecosystem. Various smart objects can be interconnected to create a smart environment. The IoT framework can be used for a wide range of applications including healthcare, transportation, agriculture, consumer electronics, etc.

Recently some research work has been done in the area of IoT. Stankovic [11] gave some research directions in IoT where vision of smart world is discussed along with the problems associated with massive scaling, architecture and dependencies, creating knowledge and big data, robustness, openness, security, privacy, and human in the loop. Zanella et al. [12] discussed IoT for smart cities, Palma et al. [13] described an example of IoT taking the example of classroom access control and Li et al. [14] showed the use of IoT in agriculture.

Collection of data from different sensor nodes, storing and processing them from the cloud becomes an important task. Security requirements are an important issue which must be taken care of. Also, management of such varied network of smart objects becomes complex and various use case scenario's need to be considered for designing the framework.

2 State of Art Techniques

An IoTized healthcare app [1] may consist of different modules to monitor blood pressure, ECG, blood glucose meter, calorie count meter, activity monitor, etc. The various sensor nodes used for recording above mentioned data must be able to communicate with each other and provide necessary inputs to the doctor/patient so that corrective actions can be taken. An IoTized Home app may be able to communicate with various home appliances like TV, washing machine, air conditioner, smart phone dish washer, etc. Sensor nodes should be capable of reading as well as generating data for each and every device in the IoT app for a smart home. Also it should provide remote monitoring facility.

Based on the use cases and the services required a common software platform is desired for these IoTized apps. Platform includes abstraction of state of the art hardware technologies along with software stacks. Physical infrastructure like RFID

[15], Zigbee [2], ANT/ANT+RFID [15], ZigBee [2], etc. are the popular standards. Also several communication protocols like CoAP: Constrained Application Protocol, which is a light-weight version of HTTP [8], MQTT [16], TLS&DTLS [9, 10], 6LowPAN [3, 5], Z-wave [17], BLE [18], DASH7 [19], Wavenis [20], RuBee [21], EuOcean [22] Wireless HART [23] are designed and developed to build a robust IoT framework.

The protocols described above can be collectively arranged to form paradigms for connectivity and various applications running across the devices. Several IoT frameworks have come up with various sensors supported in their frameworks. Some of the popular frameworks along with their features are enlisted below:

Brillo OS is based on a stripped down version of Linux operating system. The memory barrier for sensor devices is kept at as low as 32–64 MB [24] in this framework. HomeOS supports Z-wave-based sensor devices for various services over Arduino hardware and .NET framework [25]. HomeKit framework primarily supports MFI enabled devices without popular protocols like Z-wave, Zigbee. Furthermore, many more restrictions and customization from operational and control perspective require with HomeKit that increases additional hardware interfaces to bridge the in-compatibilities [26].

AllSeen alliance is created for connecting numerous devices based on rich open source platform. This platform supports major OS's like Android, iOS, Windows, Linux with language support of C, C++, Java and Objective-C [27]. Iotivity is an open source framework primarily targeted for Linux-based operating systems like Tizen, Ubuntu and Android. It is supported by C and C++ languages, several hardware specific access protocols like BLE, NFC, Bluetooth, etc. [28]. Moreover, Iotivity is well supported by Open Interconnect Consortium (OIC). There are more than 200 organizations including universities are contributing towards the completion of IoT specifications and their adaption [29].

In this paper we have considered IoT application scenario for a simple healthcare system and have shown the interaction process between different applications. Proposed work in this paper is based on the Iotivity software framework. A simple use case of heartbeat sensor is considered as an IoT application. To distinguish the services offered to a normal user and a privileged user an algorithm is formulated along with the state machine, class diagram. Sequence of operations is also depicted to showcase the events and their subsequent reaction to simulate a real time heartbeat sensor.

3 Proposed Work

This section is divided into sub-sections—Architecture, Use case and Proposed Design. Proposed Design sub-section is further decomposed into Algorithm, State Machine, Class Structure and Sequence Diagram.

3.1 Architecture

Iotivity-based software architecture is depicted in Fig. 1 for a heartbeat sensor application. Iotivity provides the software building blocks to create such an application. The building blocks include device management, storage management, discovery and notification, communication and security management through its rich set of device abstraction layer, network communication protocols, data capturing mechanisms and cryptographic libraries to ensure the security and privacy of the system and user data.

3.2 Use Case

For a heartbeat sensor; a minimal set of features are considered to showcase the capabilities of the device and its services offered to the user. These capabilities include authentication and authorization to ensure restricted access to the device to avail the services provided. Two different modes are created to distinguish between a regular user with limited capabilities and an admin user with full capabilities to perform administrative operations on the device. Additionally, to demonstrate the user's health parameter to concern people like friends, family members, doctor while using the sensor device over a regular period performance calculator capability is also included. The heartbeat sensor device is connected to a smart thing server to regularly record the events and subsequent operations carried out. All these events and operations alter the user data along with device's data on regular

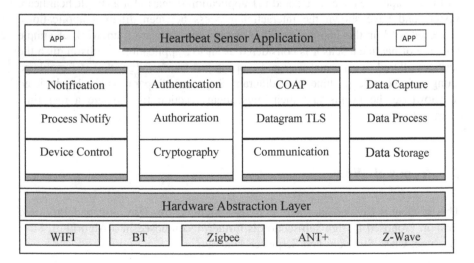

Fig. 1 Iotivity-based architecture for heartbeat sensor application

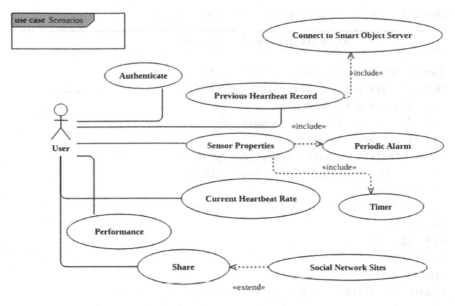

Fig. 2 Use case of the ecosystem

basis whenever the device provides its services to the user. These use cases are depicted in Fig. 2 by a use case diagram.

3.3 Proposed Design

Algorithm: Proposed algorithm for the heartbeat sensor application is described in Table 1. The algorithm starts with a registration process to the sensor device for a user. User provides its credentials like name, family name, date of birth, etc. as input to the sensor device. Device checks its user specific table for that user. If the user is already registered with the device it notifies same to the user otherwise the device creates public and private key-pair pertaining to the user. Additionally, the algorithm also differentiates between a normal user and an admin user by provision of **privilege** option to the admin user. This feature is disabled for the normal user. While updating the public and private key tables, algorithm also sets **privilege** option as true for admin user whereas false for normal user. Based on this demarcation admin user can avail privileged services offered by the device. Finally, the algorithm creates COAP connection to send and receive the user and device specific data to smart things server.

State Machine: The state machine of the device is shown in Fig. 3. The figure depicts the possible states in which the device operates, events that trigger the device to switch from one mode to another. Primarily a powered-on heartbeat sensor device operates in two states—Idle and Service. On power-on the device

Table 1 Describes proposed registration algorithm of application

Algorithm: HBS Registration

Input: Registration request from heartbeat sensor user.
Output: User pub_key entry into smart things server and update
priv_key_table for new user.
for all user registration request do
create pub key & prv_key;
prvkey_table ← update_prvkey(user_prvkey);
if user_data contains "privilege" then
privilege ← 1;
for all "services" request do
services ← 1;
else if
privilege ← 0;
end else if
update "privilege" in prvkey_table_entry;
update "services" in services_table_entry;
end if
user_id ← get_user_id(pubkey_table);
user_pubkey ← get_pubkey_from_pubkey_table (user_id);
connect ← create a things_server_conn req over COAP;
connect_id ← get_connection_id_coap(req);
send requests to things_server_over_coap (connect_id, user_data);
notify_user;
end

Fig. 3 State machine of heart beat sensor device

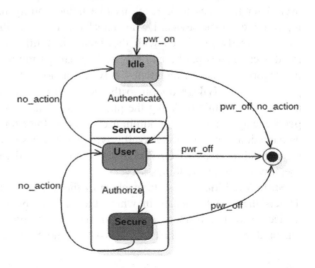

attains its default state, i.e. Idle state. In Idle state device listens to receive authentication information from its user and switches to Service state on successful completion of authentication. Service state is a composite state in which it operates either into User state or Secure state. A normal user only requires User state where it does not perform any secure service operations whereas secure state requires further authorization to perform admin operations on the device. While operating in

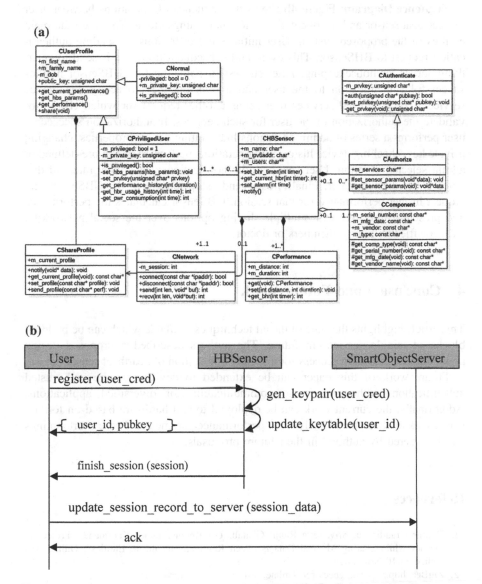

Fig. 4 **a** Class diagram of heart beat sensor device, **b** sequence diagram of heartbeat sensor application

Secure state if there is no action for certain pre-defined duration; device automatically switches back to User state and subsequently falls back to Idle state if no action detected further at User state. From Idle state device automatically switches-off after a pre-defined silence. From all the states Idle, User and Secure device goes back to power-off mode if user deliberately powered it off.

Class Structure: Figure 4a includes all the classes along with their member variables and functions.

Sequence Diagram: Figure 4b shows the sequence of operations between user to heartbeat sensor and heartbeat sensor to smart things cloud. These are the main entities of the proposed system. User initiates the operations by sending authentication request to BHSensor. This event and subsequent user data is send to smart thing server for housekeeping. After successful authentication HBSensor notifies successful authentication to the user and allows the user to access the device services. For privileged services user sends another request for which HBSensor validate the authorization of the user for such services. If authorization succeeded, user performs a series of admin operations like, setting device properties, changing private key, checking device history and operations performed over a pre-defined or admin selected duration. User sets periodic notification of heartbeat rate and distance covered by him/her. Finally, user sends stop request to the HBSensor and request his/her performance for that session. HBSensor calculates the performance and present to the user with multiple sharing options over the social networking sites for friends, family members or doctor.

4   Conclusion and Future Work

This article highlights the state of the art techniques available which can be building blocks of thing's services in future. The authors described a proposed design approach of registration process for a sample application of heartbeat sensor device.

Future work of this paper can be extended to devising secure and trusted authentication, authorization and communication for diversified applications. Additionally, the current work can be deployed to real hardware based on Iotivity framework to observe and analyze the performance parameters. These shortcomings will be catered by authors' in their future proposals.

References

1. Tyrone Grandisona, Srivatsava Ranjit Gantab, Uri Braunc, James Kaufmana, "Protecting Privacy while Sharing Medical Data Between Regional Healthcare Entities", Online, last visited on 10 Nov, 2015.
2. ZigBee. http://www.zigbee.org/. Online, last visited 20. April 2015.

3. E. Kim, D. Kasper, N. Chevrollier, and JP. Vasseur. Design and Application Spaces for 6LoWPANs draft-ietf-6lowpan-usecases-09, January 2013.
4. Tobias Heer, Oscar Garcia-Morchon, René Hummen, Sye Loong Keoh, Sandeep S. Kumar, Klaus Wehrle, "Security Challenges in the IP-based Internet of Things 2014", Wireless Personal Communications: An International Journal, Volume 61 Issue 3, December 2011 Pages 527–542 Kluwer Academic Publishers Hingham, MA, USA.
5. IETF 6LoWPAN Working Group. http://tools.ietf.org/wg/6lowpan/. Online, last visited 19. April 2015.
6. G. Montenegro, N. Kushalnagar, J. Hui, and D. Culler. Transmission of IPv6 Packets over IEEE 802.15.4 Networks. RFC 4944, September 2007.
7. IETF Constrained RESTful Environment (CoRE) Working Group. https://datatracker.ietf.org/ wg/core/charter/. Online, last visited 18. April 2015.
8. Z. Shelby, K. Hartke, C. Borman, and B. Frank. Constrained Application Protocol (CoAP). Draft-ietf-core-coap-04 (Internet Draft), January 2013.
9. T. Dierks and E. Rescorla. The Transport Layer Security (TLS) Protocol Version 1.2. RFC 5246, August 2008. Updated by RFC 5746, 5878.
10. T. Phelan. Datagram Transport Layer Security (DTLS) over the Datagram Congestion Control Protocol (DCCP). RFC 5238, May 2008.
11. John A. Stankovic, "Research Directions for the Internet of Things", IEEE, 2014. http://dx. doi.org/10.1109/JIOT.2014.2312291.
12. Andrea Zanella, Nicola Bui, Angelo Castellani, Lorenzo Vangelista, Michele Zorzi, "Internet of Things for Smart Cities", IEEE INTERNET OF THINGS JOURNAL, VOL. 1, NO. 1, FEBRUARY 2014, pp. 22–32.
13. Daniel Palma, Juan Enrique Agudo, Héctor Sánchez and Miguel Macías Macías, "An Internet of Things Example: Classrooms Access Control over Near Field Communication", Sensors 2014, 14, pp. 6998–7012.
14. Guohong Li, Wenjing Zhang, Yi Zhang, "A Design of the IOT Gateway for Agricultural Greenhouse", Sensors & Transducers, Vol. 172, Issue 6, June 2014, pp. 75–80.
15. https://en.wikipedia.org/wiki/Radio-frequency_identification, Radio-frequency identification, Online, last seen on 10 Dec 2015.
16. http://docs.oasis-open.org/mqtt/mqtt/v3.1.1/os/mqtt-v3.1.1-os.html#_Figure_2.1_-, "MQTT Version 3.1.1." Specification URIs. Online, last visited on 29. Sep 2015.
17. Jesus J. Martínez, Teresa García-Valverde, Francisco Campuzano, Pablo Campillo Sanchez, Alberto García-Sola, Juan A. Botía, "Multi-Agent Based Social Simulation Applied to Validation of Location Services", doi:10.3233/978-1-61499-050-5-91 In book: Ambient Intelligence and Smart Environments Volume 12: Agents and Ambient Intelligence, Publisher: IOS Press Books, Editors: Tibor Bosse, pp. 91–118.
18. https://en.wikipedia.org/wiki/Bluetooth_low_energy, Bluetooth low energy, Online, last seen on 10 Dec 2015.
19. https://en.wikipedia.org/wiki/DASH7, DASH7, Online, last seen on 10 Dec 2015.
20. http://www.powershow.com/view/79956-NzBkY/WAVENIS_powerpoint_ppt_presentation, Ultra Low Power WSNs, Hype or ripe? Online, last seen on 10 Dec 2015.
21. https://en.wikipedia.org/wiki/RuBee, RuBee, Online, last seen on 10 Dec 2015.
22. https://en.wikipedia.org/wiki/EnOcean, EnOcean, Online, last seen on 10 Dec 2015.
23. https://en.wikipedia.org/wiki/WirelessHART, WirelessHART, Online, last seen on 10 Dec 2015.
24. http://arstechnica.com/gadgets/2015/05/google-developing-brillo-internet-of-things-os-based-on-android/, "Brillo" Internet of Things OS based on Android, Online, last visited on 18 Feb 2016.
25. https://labofthings.codeplex.com/wikipage?title=Lab%20of%20Things%20Devices, Lab of Things Devices, Online, last visited on 18 Feb 2016.

26. http://www.macworld.com/article/2874486/home-tech/building-an-apple-powered-smart-home-heres-how-non-homekit-devices-could-work.html, Building an Apple-powered smart home? Here's how non-HomeKit devices could work, online, last visited on 18 Feb 2016.
27. https://allseenalliance.org/framework/documentation/supported-platforms, Allseenalliance Framework, online, last visited 19 Feb 2016.
28. https://www.iotivity.org/, IoTivity, online, last visited on 20 Feb 2016.
29. http://openconnectivity.org/, Open Connectivity Foundation, online, visited on 20 Feb 2016.

An Effective and Empirical Review on Internet of Things and Real-Time Applications

Surbhi Gill, Paras Chawla, Pooja Sahni and Sukhdeep Kaur

Abstract Internet of Things or IoT is one of the common and widely used technology used for the formation of smart environments based on sensor-based scenarios. At the current implementation, IoT is being implemented for assorted scenarios including environmental monitoring, production, manufacturing, healthcare, home automations, transportations, and many others. The key idea behind IoT is the machine to machine communication in which each and every device is connected using sensor technology. This paradigm is used for the development and deployment of smart cities and sensor-based infrastructures. Nowadays, IoT is being implemented in many of the transportation services and integrated in cars as well as freight vehicles which make it IoV (Internet of Vehicles). Using this approach, the remote IoT control servers can monitor any of the moving vehicles in real time and tracking can be done on multiple parameters. In the classical implementation, the remote monitoring and tracking is one using towers or base stations and satellites which is not very effective because of assorted quality of service factors. For the enhanced scenario integration, the global positioning system is used but number of security vulnerabilities can be there and cracking can be done. This paper underlines and highlights the scope of IoT in assorted applications with the association to different case scenarios.

S. Gill (✉) · P. Chawla · P. Sahni · S. Kaur
Department of Electronics and Communication Engineering,
Chandigarh Engineering College, Landran, Mohali, Punjab, India
e-mail: gillsurbhi@gmail.com

P. Chawla
e-mail: dr.paraschawla@cgc.edu.in

P. Sahni
e-mail: pooja.ece@cgc.edu.in

S. Kaur
e-mail: sukhdeep.ece@cgc.edu.in

© Springer Nature Singapore Pte Ltd. 2018   159
S.K. Bhatia et al. (eds.), *Advances in Computer and Computational Sciences*,
Advances in Intelligent Systems and Computing 554,
https://doi.org/10.1007/978-981-10-3773-3_16

Keywords Internet of things · Internet of vehicles · IoT · IoV · Smart
cities · Sensor-based communications

1 Introduction

Internet of Things that is classically known as IoT [1] refers to the high performance and all time connectivity-based domain that makes use of sensor-based
communication for machine-based communication. It is the recent technologies in
current era focusing on the interconnection of every real world object using smart
devices integrating wireless sensor connecting to satellite. We can imagine the real
life objects with embedded computing devices and communicating with each other.
By this technology, we can track everything from remote location using Internet
infrastructure.

Using IoT, the inter connection in every system, device, machine, human being,
home equipments, office products can be established using existing network
resources. As an example or case of IoT, we can track any train by using the
messaging service of Indian Railways. As per the instructions, we can send the
message SPOT <TrainNumber> to 139. After this message, we get the exact
location and upcoming station of that train. In the same way, many taxi or cab
services are trying to utilize and implement IoT. Currently, many taxi operators are
connected with GPS and we can track the location of that car on mobile phone,
tablet, or any network connected device. Smart Cities, Smart Home are implemented using IoT in which everything is connected and searchable.

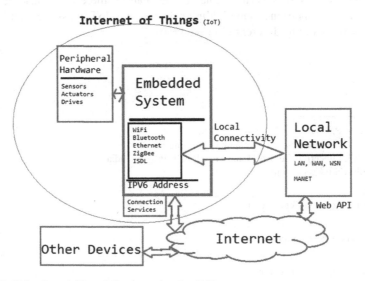

Fig. 1 Traditional model-based diagram related to IoT

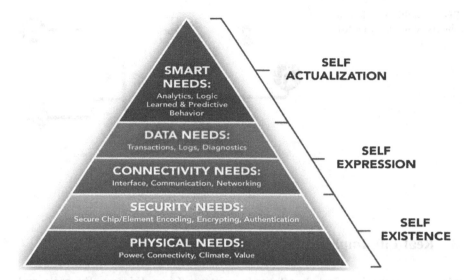

Fig. 2 Hierarchy of needs and objects in IoT

At the technology dimension, IoT makes use of sensors and embedded chips which are inserted in the system that we want to monitor and track. RFID (Radio Frequency Identification)-based devices are classically used for IoT implementation (Figs. 1, 2).

2 Usage of Sensor Data in Cloud Environment

In IoT, there is regular and huge amount of data that is generated and processed. As in IoT, millions of objects of interconnected with each other to share the data, there is need to devise and implement a cloud-based solution so that the fault tolerant and secured applications can be developed. Using this approach, the IoT generate data can be preserved and stored without any issue of intentional or unintentional damage. To manage and control all these aspects, cloud is required [2]. For this, the effective implementation of cloud infrastructure is required. In the upcoming years, the Internet of Things (IoT) will be transformed to the Cloud of Things (CoT) [3] because it will be very difficult to manage huge data or BigData [4] without cloud integration (Fig. 3).

Fig. 3 Multiple devices and objects connected with IoT

3 Real Life Implementations and Applications of IoT

Many corporate giants including Texas Instruments, Cisco, Ericsson, Freescale, and GE are working in the development as well as deployment of IoT scenarios. The companies are making and developing the apps easier using hardware, software, and sustain in getting all the things connected using IoT environment. A set of key markets exists for the IoT with potential for exponential growth.

- Medical and healthcare systems
- Transportation
- Wearable devices
- Health care
- Building automation
- Smart cities
- Smart manufacturing
- Predictive maintenance
- Employee safety
- Remote and enhanced monitoring
- Medical telemetry
- Tracking of drugs
- Asset tracking for medical care
- Access and authenticated control
- Automobiles/Automotive.

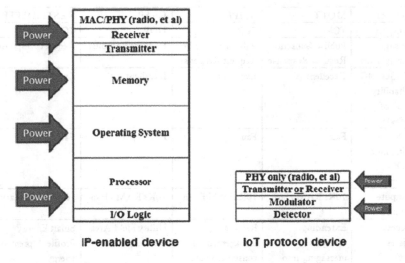

Fig. 4 IoT protocol-based components

4 Protocols Associated with Internet of Things

- Bluctooth
- Cellular
- 6LowPAN
- LoRaWAN
- Thread
- NFC
- Neul
- Sigfox
- WiFi
- Zigbee
- Z-Wave (Fig. 4).

5 Literature Review

Atzori et al. [5] addressed IoT as the powerful system for managing different types of devices and even vehicles. Using gadgets-based infrastructure and remote monitoring the virtual environment can be created and effectively managed.

Protocol	MQTT	XMPP	CoAP	RESTful HTTP
Transport	TCP	TCP	UDP	TCP
Message Transmission	Publish/Subscribe Request/Response	Publish/Subscribe Request/Response	Request/Response	Request/Response
2G / 3G / 4G Suitability (1000 of nodes)	Excellent	Excellent	Excellent	Excellent
LLN Suitability (1000s nodes)	Fair	Fair	Excellent	Fair
Compute Resources	10Ks RAM/Flash	10Ks RAM/Flash	10Ks RAM/Flash	10Ks RAM/Flash
Success Stories	Extending enterprise messaging into IoT applications	Remote management of consumer white goods	Utility Field Area Networks	Smart Energy Profile 2 (premise energy management, home services)

Fig. 5 Protocols with IoT scenarios

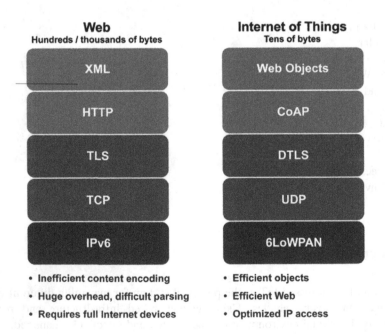

Fig. 6 Comparison of web and IoT protocol stacks

Ben et al. [6] proposed the application of IoT in the domain of medical sciences and tracking purpose. Using this approach, the patients can be monitored using IoT or software defined networking-based scenarios.

David et al. [7] introduced the concept of TinySec in which the completely integrated link layer integrity and security-based architecture for the wireless sensor networks can be there. In this approach and proposed design, the security and overall performance of the system can be improved.

Kopetz [1] proposed the integration and implementation of RFID-based infrastructure that ultimately forms a unique and high performance IoT system. In this scenario, the devices or vehicles can be connected with the unique address-based RFID chips and then monitoring can be done (Figs. 5, 6).

Following is the summarized comparative analysis of various IoT platforms based on the relative features and technologies (Fig. 7; Table 1).

Protocol	Transport	Messaging	2G,3G,4G (1000's)	LowPower and Lossy (1000's)	Compute Resources	Security	Success Stories	Arch
CoAP	UDP	Rqst/Rspnse	Excellent	Excellent	10Ks/RAM Flash	Medium - Optional	Utility field area ntwks	Tree
Continua HDP	UDP	Pub/Subsrb Rqst/Rspnse	Fair	Fair	10Ks/RAM Flash	None	Medical	Star
DDS	UDP	Pub/Subsrb Rqst/Rspnse	Fair	Poor	100Ks/RAM Flash +++	High- Optional	Military	Bus
DPWS	TCP		Good	Fair	100Ks/RAM Flash ++	High- Optional	Web Servers	Client Server
HTTP/ REST	TCP	Rqst/Rspnse	Excellent	Fair	10Ks/RAM Flash	Low- Optional	Smart Energy Phase 2	Client Server
MQTT	TCP	Pub/Subsrb Rqst/Rspnse	Excellent	Good	10Ks/RAM Flash	Medium - Optional	IoT Msging	Tree
SNMP	UDP	Rqst/Response	Excellent	Fair	10Ks/RAM Flash	High- Optional	Network Monitoring	Client- Server
UPnP		Pub/Subscrb Rqst/Rspnse	Excellent	Good	10Ks/RAM Flash	None	Consumer	P2P Client Server
XMPP	TCP	Pub/Subsrb Rqst/Rspnse	Excellent	Fair	10Ks/RAM Flash	High- Manditory	Rmt Mgmt White Gds	Client Server
ZeroMQ	UDP	Pub/Subscrb Rqst/Rspnse	Fair	Fair	10Ks/RAM Flash	High- Optional	CERN	P2P

Fig. 7 Comparative analysis of IoT protocols

Table 1 Comparative analysis of IoT platforms

Platform	Device management	Protocols (data collection)	Analytics	Visualization
Telemetry	Present	CoAP, MQTT, M3DA, STOMP	Storm Technology	Present
Appcelerator	Not present	HTTP as well as MQTT	Titanium	Present using Titanium UI Dashboard
Amazon Web Services	Present	MQTT, HTTP1.1	Using Amazon Kinesis AWS Lambda	Dashboard
Bosch	Present	MQTT, CoAP, AMQP, STOMP	Undefined	GUI
Ericsson Device Connection Platform	Present	SSL with Secured and High integrity-based TSL	Undefined	Not present
Everything	No	LIGHT WEIGHT MESSAGING PROTOCOL, CoAP, WebSockets	Real-time analytics (Rules Engine)	Present (EVRYTHNG IoT Dashboard)
ThingWorx	Present	LIGHT WEIGHT MESSAGING PROTOCOL, AMQP, XMPP, CoAP, DDS,	Machine Learning (ParStream DB)	Squeal
ParStream	No	LIGHT WEIGHT MESSAGING PROTOCOL	Real-time analytics, Batch analytics (ParStream DB)	Present (ParStream Management Console)
PLAT.ONE	Present	LIGHT WEIGHT MESSAGING PROTOCOL, SNMP	Deep	Present (Management Console for application enablement, data management)
IBM	Present	LIGHT WEIGHT MESSAGING PROTOCOL, HTTPS	Real-time analytics (IBM IoT Real-Time Insights)	Present (Web portal)
Xively	Not present	Secured HTTP and HTTPS	Enormous	Present

6 Conclusion

The current implementations of IoT make use of road side units which regularly collects the information and then transmit. There are number of attacks including vampire attacks, denial of service attack, and many others which can be fired on these wireless devices. Each and every car or vehicle in built in with a global positioning and tracking unit so that the actual location can be found. In such attacks, the location or any car or vehicle can be spoofed and damaged. It can be dangerous on road and that is why there is need to implement a novel and effective approach that will keep track of the actual location based on the dynamic key generated with the integration of actual timestamp and longitude with latitude. There is need to devise and implement a unique approach for higher security and integrity in such scenarios. Using this approach, the overall security and integrity can be enhanced by which way and performance of system can be improved.

References

1. Kopetz, H. (2011). Internet of things. In Real-time systems (pp. 307–323). Springer US.
2. Bojanova, I., Hurlburt, G., & Voas, J. (2014). Imagineering an Internet of Anything. Computer, (6), 72–77.
3. Aazam, M., Khan, I., Alsaffar, A. A., & Huh, E. N. (2014, January). Cloud of Things: Integrating Internet of Things and cloud computing and the issues involved. In Applied Sciences and Technology (IBCAST), 2014 11th International Bhurban Conference on (pp. 414–419). IEEE.
4. McAfee, A., Brynjolfsson, E., Davenport, T. H., Patil, D. J., & Barton, D. (2012). Big data. The management revolution. Harvard Bus Rev, 90(10), 61–67.
5. Atzori, L., Iera, A., & Morabito, G. (2010). The internet of things: A survey. Computer networks, 54(15), 2787–2805.
6. Ben Othman, S., Bahattab, A. A., Trad, A., & Youssef, H. (2014, May). Secure data transmission protocol for medical wireless sensor networks. In Advanced Information Networking and Applications (AINA), 2014 IEEE 28th International Conference on (pp. 649–656). IEEE.
7. David W, Karlof, C., Sastry, N., (2004, November). TinySec: a link layer security architecture for wireless sensor networks. In Proceedings of the 2nd international conference on Embedded networked sensor systems (pp. 162–175). ACM.

Operations on Cloud Data (Classification and Data Redundancy)

Sandeep Khanna, Nitin Rakesh and Kamal Nayan Chaturvedi

Abstract Cloud computing is a turning in the field of information technology as it provides resources over network. Besides the features, cloud services are widely available for all. Content duplicacy increases the data redundancy problem in cloud. Files on cloud need an effective classification method so that the problem of cloud server efficiency may be optimized. In this paper, we have proposed two algorithms: Checker's algorithm (to remove data redundancy from cloud) and Pronto-Key algorithm (to classify the files and enhance the performance of cloud) which overall increase the efficiency of the cloud.

Keywords Cloud computing · Redundancy · Classification · Searching

1 Introduction

Most of the devices nowadays are using the cloud platform to store their data. Many of the applications are running on the cloud. Cloud contains a lot of similar data in it which becomes a problem when client searches for a file in the cloud. The files are not classified in the cloud such that it becomes difficult for the user to search for a desired file. This affects the efficiency of the cloud services. The network cost is also high while accessing the cloud as the data is not transmitted from nearby location; it is from miles of distance from a server on which it resides. Miles of

S. Khanna (✉) · N. Rakesh
Amity School of Engineering and Technology, New Delhi, India
e-mail: sandyy.2511@gmail.com

N. Rakesh
e-mail: nitin.rakesh@gmail.com

K.N. Chaturvedi
Amity Institute of Telecom Engineering and Management, Noida, India
e-mail: kamalnc12@gmail.com

S. Khanna · N. Rakesh · K.N. Chaturvedi
Amity University, Uttar Pradesh, Noida, India

© Springer Nature Singapore Pte Ltd. 2018
S.K. Bhatia et al. (eds.), *Advances in Computer and Computational Sciences*,
Advances in Intelligent Systems and Computing 554,
https://doi.org/10.1007/978-981-10-3773-3_17

distance may also affect the efficiency in accessing the cloud service. There are many more errors in accessing the data from a distance server [1]. These are few issues in today's cloud service that is focusing in the paper.

We are considering 'TEXT' type of data to make the cloud to do smarter operations on the server side and make the services more efficient. For that we use to process the file using proposed algorithms in the paper. When the client wants to upload a file on the cloud, before uploading, the server processes the content of the file and compare the content data with the uploaded content data on cloud, if there is more than $X\%$ (Admin decided percent) of content data is similar to previously content data of uploaded file, the algorithm cancels the uploading of file. It depends on the server admin that what type of restrictions it creates for the client to upload the file. If the document content is not similar with the previous uploads, then document is uploaded on the cloud. This process is used to prevent the duplicacy of content in the cloud. The second operation is to classify the files in such a manner that client finds the desired file on cloud. For that we classify the file in a ratio of their upload days' count and access count. This helps the client to get the most popular and the most trending file at present. The classification of the files depends on both the scenarios, 1. How old is the file? 2. How many clients have accessed that file? The file with the highest score is on the top of the search. This operation gives the client a better visibility of files while searching in the cloud. The third operation is to enhance the accessing of file by the client from the cloud. This operation uses the location of client and obtains the nearest server from the client's location and gives an option to the client to access the document from the nearest server location. This can enhance the accuracy and time to access the file thereby helps the cloud to become smarter and enhance the performance. Further, there are algorithms that describe how to proceed these operations of cloud and makes the smart cloud.

2 Related Work

There are various techniques and algorithms present to enhancing the performance of the cloud. Enhancing the security of cloud data by applying 3D security; in this there is a framework which is divided in two phases. First phase contributes in doing encryption of data and classifying it before storing. It can be classified into three parameters namely confidentiality, integrity, and availability. The parameters measures the critical rating and protection ring is given to data on the bases of critical rating. In the second phase, the users who want to access the data need to authenticate [2]. Private clouds file management system which is able to do automatic clustering of files and enhance the document management in cloud and proposed an algorithm and k-means. The top-k frequent itemsets of the SHDC algorithms found the utilization parTFI algorithm. Documentation set, which incorporates all parts of a common set of terms, been observed as a cluster candidates and their mean value as the initial k-means cluster seed polymerization similar file the clustering by the k-means on the basis of set of term of the top-k frequent to accomplishing initial cluster returns [3].

Solving the problem on big data classification of network intrusion traffic by combining various effective learning techniques and big data technologies is suggested by [4] using combination of modern cloud technologies, distributed file systems techniques in a combination of big data classification to predict network intrusions. Classify encrypted data by k-NN classifier and this protocol protects the internal confidentiality information of the user [4]. Enhancing text clustering by genetic algorithm. The strategy comprises of Phase 1 and Phase 2 where the Phase 1 is clustering the sentences based on modified repeated spectral bisection. Here an initial clustering is done based on the sign of elements of the eigen vectors of the similarity matrix. Phase 2 is clustering sentences based on the genetic algorithm, where chromosomes are made [5]. Enhancing the security of files in cloud by classify them into sensitive and non-sensitive data with the help of k-NN classifier and then using RSA encryption on sensitive data stored them in separately irrespective of non-sensitive data. This classification is based on the confidentiality [6].

3 Problem Statement

The cloud services are openly available for all kinds of organizations and the users. Content duplicacy is a complicated problem and considered one of the biggest problems for increasing data redundancy in cloud. Content duplicacy increased with the extensive use of the Internet on which a huge amount of digital data is available. Plagiarism is one of the problems of content duplicacy in text documents [7]. The first problem is that cloud contains a lot of files so there is a duplicacy problem happens in cloud data. Because of number of files, the content of file can be similar as the file already present on cloud and when client uploads the file with different name, the problem of redundancy on cloud is occurred.

As we noticed in Table 1 that 'Networking.doc' and 'Advance_networking.pdf' content is same and 'Pin_codes.pdf' and 'Codes.doc' content is also same but both duplicate content files are uploaded to cloud with different names. $Total\,Size = 253 + 1208 + 452 + 1208 + 2183 + 452 + 412 = 6,168\,KB's$; When we analysis the cloud data, there are a many of the files detect which has the similar content that's the problem. Now, there are a lot of files in the cloud and the clients which uses the cloud service got confused to search the file he wants to access. We search for a word and documents related to the word are listed as follows:

Table 1 List of files on cloud

No.	Name	Size (KB)
1	Security.pdf	253
2	Networking.doc	1,208
3	Pin_codes.pdf	452
4	Advance_networking.pdf	1,208
5	Design_issues.doc	2,183
6	Codes.doc	452
7	Information storage and management.txt	915

Table 2 Files contains user searched word

ID	File	Download	Days
1	3D security	40	6
2	Document clustering algorithm	69	8
3	Big data classification problems and challenges	52	7
4	Classification of big point cloud data using cloud computing	78	5
5	Classification of cloud data using Bayesian	130	9
6	Classification of gene expression data on public clouds	89	8
7	Cloud classification for weather information	150	13
8	Data classification based on confidentiality in virtual cloud environment	452	20
9	Implementation of text clustering using genetic	230	45
10	k-nearest neighbour classification	521	54
11	Performance enhancement of classifiers using integration of clustering and classification techniques	632	45
12	Plagiarism detection based on SCAM algorithm	745	98
13	Web document clustering approaches using K-means algorithm	456	75

Table 3 Distance in terms of latitude and longitude from current location (Noida, India); It also shows the distance in km from Noida City

ID	Server	Latitude	Longitude	Distance from Noida (km)
1	Australia	−25.837	130.2848	8439.54
2	Algeria	28.1731	2.5908	8311.058
3	Germany	52.1786	10.2694	7906.55
4	Brazil	−9.019	−53.7216	15159.13
5	China	35.3789	104.3335	3096.395
6	Canada	59.5369	−112.2685	21361.85
7	U.S	41.5811	−100.3329	19808.08
8	Mexico	24.5375	−102.364	19985.81
9	Russia	66.1844	105.4268	5221.908
10	Korea	35.2743	127.6527	5644.406

In Table 2, download specifies number of files downloaded by clients and days specifies for how many days the file would be on cloud. The documents in the list are in the unclassified manner and it is difficult for the client to search the desire document he wants which is again a problem for the client. The third problem is how efficient the cloud services are, while client accessing files.

The file is available on these servers, but cloud service randomly gives any server to client to access. When it gives Canada server to access the file in place of any nearby server available, the efficiency of network decreases, which is again a big problem. Resolving these problems can enhance the performance of cloud services. Several other approached exists in literature which are working on optimization [8–10] (Table 3 and Fig. 1).

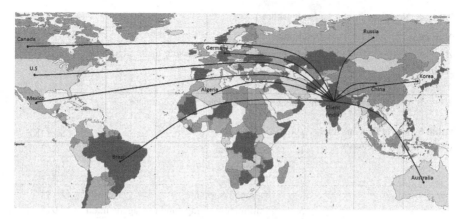

Fig. 1 Server-client locations on world map

4 Proposed Approach

Content Duplicacy Checker using text Summarization: In this paper, we have proposed two algorithms (Checker's algorithm, and Pronto algorithm) which deals with the duplicacy of the document content with the rest of the documents in the cloud. The algorithm deals with the process of text summarization of document with the help of following steps:

1. **Tokenization:** It is the process of removing the white spaces from the document. The whitespaces characters such as spaces, punctuations, line break. The process can removes these tokens and results document further goes for removal of stop word.
2. **Removal of stop word:** There are various stop words in the document which we cannot compare, so we have to remove those also to make a genuine content compare document. The most common stop words are such as **the, is, at, which, on** and so on. After removing these, the remaining data is saved to the database for comparison purpose (Table 4).

Checker's Algorithm:

In this paper we proposed an algorithm (checker's algorithm) which helps to restrict the duplicacy on cloud. This algorithm checks the content data of the file before uploading it on cloud. If it found $> X\%$ of similar content data with the existing content data in files, it rejects the file to upload else the file is uploaded on cloud. Notations used in the checker's algorithm:

Table 4 Notations in checker's algorithm

Variables	Description
d	The document which is processing under algorithm
Source	Content of source
Target	Content of target
N	File fetched from the database to compare

Table 5 List of files on cloud after removing duplicacy

No.	Name	Size (KB)
1	Security.pdf	253
2	Networking.doc	1,208
3	Pin_codes.pdf	452
4	Design_issues.doc	2,183
5	Information storage and management.txt	915

Steps:

1. Using functional API's in .Net framework, it provides the common pre-processing steps of document including tokenization and removal of stop words.

2. The data of file N is fetched out from the database and temporary saved to the *Target*.

3. Now when client is uploading a new document 'd', the document goes in a process state. In the process state:

   Firstly, API's analysis the whole content in the document and do apply the above steps of tokenizing, removing of stop word then saved the data to *Source*. Levenshtein algorithm is applied to the source and target that gives a percentage of identical content of source with target [11].

4. Call Function of *Levenshtein* algorithm: *Percentage = Levenshtein (Source, Target)*;

5. When the percentage is more than $X\%$, the document is marked as duplicate and pop up a message of duplicacy to the client.

The documents listed in Table 1 is processed through Checker's algorithm, it would not allow duplicate documents and data would look like as following as Table 5. Checker's algorithm will not allow identical content on the cloud.

Enhancing the searching and accessing of documents from cloud: In this paper, we also proposed an algorithm which deals with the classification of the cloud data. The classification in this algorithm is done on the basis of the date of file uploaded and the how many users would access the file. The algorithm used the ratio of both vectors (date and access times) and calculates a score which rates the document for the process of classification.

Pronto-key algorithm: These algorithms classify the data with respect to the ratio of number of downloads and number of days of file. Notations used in the Pronto-key algorithm:

Steps:

1. Declare a variable *SCORE*.

2. Initialize the variable *DAYS* and returns the value of number of days the file on cloud.

3. Initialize the variable *CLICK* and returns the value of number of clicks or number of times the file is accessed.

4. Let us say *N*th document to calculate the score.

5. *CALCULATE_SCORE*(*DAYS*, *CLICK*) which is used to calculate the score by using this formula: $SCORE(N) = \frac{CLICK(N)}{DAYS(N)}$;

6. The above function will run until and calculates the score of *N*th document and saves it to database.

7. Now, the files are classified in order of their score using bubble sort algorithm so that the high scored documents are on the top of the list and returns the list contains the top-*k* documents for the client.

   Now the clients have to select a document to proceed for the further steps of the algorithm. After the client selected a document.

8. The algorithm lists the number of servers on which the document is there.

9. Then it request client to give permission to detect the client's location.

10. After getting access to client's location, it detects your location coordinates and server coordinates and compares client's geo-distance with the lists of servers on which the file resides. $DISTANCE = \frac{Mylocation.DistanceTo(Server)}{1000}$;

11. The servers are arranged according to the nearest server from your location using bubble sorting algorithm.

    The servers are sorted with their least distance from the client $(28.5472°, 77.3366°)$

12. When the client click the server to access the document, the client may not fully access the document, which will return '**0**' and there is no change in the *CLICK*(*N*).

13. When the client click the server to access the document and fully access the document, which will return '**1**' and there is change in the *CLICK*(*N*) such that $CLICK(N) = CLICK(N) + 1$;

14. If returns '**1**', then goto **Step 5** to update Score in the database.

15. Else exit.

5 Results and Analysis

In this paper, a dataset of various text files is considered for analysing the redundancy in content, as in Table 1. The checker's algorithm detects the content duplicacy and rejects the file on the basis of their duplicacy result. The dataset contains various fields like downloaded number and how many days that file resides in the dataset. When client uploads the file, the data content is processed through the algorithm. The algorithm then compares the content of file with the uploaded files and calculates the percentage of similar content and if the calculated percentage is $>X\%$, the algorithm rejects the file to be uploaded on the cloud, but if the calculated percentage is $<X\%$, the file is accepted and uploaded on the cloud. Before applying the algorithm to the dataset of Table 1, the sizes of files are $Total Size = 253 + 1208 + 452 + 1208 + 2183 + 452 + 412 = 6, 168 \, KB's$; After applying the checker's algorithm to dataset, the size is: $Total Size = 253 + 1208 + 452 + 2183 + 412 = 4, 508 \, KB's$; $Difference = 6, 168 - 4, 508 = 1660 \, KBs$.

The algorithm saves 1,660 KBs of space on cloud which helps in reducing the cost of storage disks. While analysing the pronto-key algorithm, the first part of algorithm manages the data and provides a proper visibility of list of files by sorting the files on the basis of their scores as the client searches file from a list of related files from the cloud. The scores calculate on the basis of popularity of the file and helps to classify the files on the basis of their score (Table 6). The second part of algorithm classified the list of servers to access a particular file, such that, the server nearest to client location is on the top of the list and others are arranged in the order with the minimum distance with respect to client location (Table 7).

Comparing server efficiency on the basis of distance: The efficiency is calculated by taking:

Let's take frame length (L) of 4800 bits with a transfer speed (R) of 9600 bps. The velocity of propagation (V) is $20, 0000 \, Km/s$. Distance (D) is given in Table 8.

$$Efficiency \, (\mu) = \frac{1}{(1 + 2 \times Acceleration)};$$

$$Acceleration = \frac{Time \, of \, propagation \, (T_p)}{Time \, to \, transfer \, frame \, (T_f)};$$

Table 6 Notations of Proton-Key algorithm

Variables	Description
SCORE	Determines the ranking of the file
DAYS	Number of days the file is on the cloud
CLICK	Number of times the file is accessed
N	File which is processing for score calculation
DISTANCE	Distance from Noida to server

Table 7 Files classified by their scores

ID	File	Download	Days	Score
1	Data classification based on confidentiality in virtual cloud environment	452	20	22.6
2	Classification of big point cloud data using cloud computing	78	5	15.6
3	Classification of cloud data using Bayesian	130	9	14.44444
4	Performance enhancement of classifiers using integration of clustering	632	45	14.04444
5	Cloud classification for weather information	150	13	11.53846
6	Classification of gene expression data on public clouds	89	8	11.125
7	k-nearest neighbour classification	521	54	9.648148
8	Document clustering algorithm	69	8	8.625
9	Plagiarism detection based on SCAM algorithm	745	98	7.602041
10	Big data classification problems and challenges	52	7	7.428571
11	3D security	40	6	6.666667
12	Web document clustering approaches using K-Means algorithm	456	75	6.08
13	Implementation of text clustering using genetic	230	45	5.111111

Table 8 Servers listed in the order with the nearest distance from current location (Noida, India)

ID	Server	Latitude	Longitude	Distance from Noida (km)
1	China	35.3789	104.3335	3096.395
2	Russia	66.1844	105.4268	5221.908
3	Korea	35.2743	127.6527	5644.406
4	Germany	52.1786	10.2694	7906.550
5	Algeria	28.1731	2.5908	8311.058
6	Australia	−25.837	130.2848	8439.540
7	Brazil	−9.019	−53.7216	15159.13
8	U.S	41.5811	−100.3329	19808.08
9	Mexico	24.5375	−102.364	19985.81
10	Canada	59.5369	−112.2685	21361.85

$$T_p = \frac{D}{V}; T_f = \frac{L}{R};$$

Calculating efficiency of the server with respect to the client distance (Table 9):

Table 9 Server classified on the bases of their efficiency

ID	Server	Distance (km)	Efficiency (µ)%
1	China	3096.394624	94.1683524
2	Russia	5221.908315	90.54377425
3	Korea	5644.405888	89.85629229
4	Germany	7906.550311	86.34601739
5	Algeria	8311.057974	85.74702936
6	Australia	8439.540194	85.55851027
7	Brazil	15159.12841	76.73521918
8	U.S	19808.08337	71.62494311
9	Mexico	19985.8071	71.44305691
10	Canada	21361.84601	70.06545205

6 Conclusion and Future Scope

In this paper we have proposed Checker's and Pronto-Key algorithm. The Checker's algorithm only allows $<X\%$ (Admin decided percent) of similar content on cloud which reduced data redundancy and improved the performance of and Pronto-Key algorithm processed the data in such a way that it provides a better management of files by classifying the dataset on the basis of their calculated score and it enhancing the server accessing list by detecting the nearest server from the client location.

1. By analysing the checker's algorithm, we conclude that it reduces the duplicacy of files on the cloud and free up to $1,660$ KBs of space (comparing Table 1 files and Table 5 files).
2. By analysing the Pronto-Key algorithm, we conclude that

   i. The files are classified and ordered according to their scores (popularity).
   ii. Servers to access the files are also classified on the basis of their efficiency.

In future, the enhancement of checker's algorithm will processed not only text files, many other type of files will also processed like audio files, compressed files, image files and backups files. Checker's algorithm will able to reduce data redundancy by comparing other than text files also. Pronto-key algorithm will also enhancing the calculation of files score by adding few more factors like type of clients (student, business man, government employee, women), type of file, location and size. Pronto-key algorithm will use these factors also with days and access count to enhance score of the file and classify the files.

References

1. T.S. Dillon, C. Wu and E. Chang, "Cloud Computing: Issues and Challenges," Advanced Information Networking and Applications (AINA), 2010 24th IEEE International Conference on. IEEE, pp. 27–33, 2010.
2. S. Tirodkar, Y. Baldawala, S. Ulane and A. Jori, "Improved 3-Dimensional Security in Cloud Computing," International Journal of Computer Trends and Technology (IJCTT), Vol. 9, pp. 242–247, 2014.
3. J. Miao, Z. Fan, G. Chen, H. Mao, L. Wang, "A Private Cloud document management system with document clustering algorithm," National Conference on Information Technology and Computer Science (CITCS 2012). Atlantis Press.
4. S. Suthaharan, "Big Data Classification: Problems and Challenges in Network Intrusion Prediction with Machine Learning," Department of Computer Science, University of North Carolina at Greensboro, Greensboro, NC 27402, USA, 2012.
5. Dhanya P.M, Jathavedan M and Sreekumar A, "Implementation of Text clustering using Genetic Algorithm," International Journal of Computer Science and Information Technologies (IJCSIT), Vol. 5, pp. 6138–6142, 2014.
6. M. A. Zardari, L.T. Jung and M.N. Zakaria, "Data Classification Based on Confidentiality in Virtual Cloud Environment", Research Journal of Applied Sciences, Engineering and Technology, Vol. 8, pp. 1498–1509, 2014.
7. D. Anzelmi, D. Carlone, F. Rizzello, R. Thomsen and D.M. Akbar Hussain, "Plagiarism detection based on SCAM algorithm," in Proceedings of the International MultiConference on Engineers and Computer Scientists, Vol. 1, pp. 272–277, 2011.
8. Praveen K Gupta, Nitin Rakesh, "Different job scheduling methodologies for web application and web server in a cloud computing environment", 2010 3rd International Conference on Emerging Trends in Engineering and Technology (ICETET), pp. 569–572, 2010.
9. Nitin Rakesh, Vipin Tyagi, "Failure recovery in XOR'ed networks", 2012 IEEE International Conference on Signal Processing, Computing and Control (ISPCC), pp. 1–6, 2012.
10. Kinjal Shah, Gagan Dua, Dharmendar Sharma, Priyanka Mishra, Nitin Rakesh, "Transmission of Successful Route Error Message (RERR) in Routing Aware Multiple Description Video Coding over Mobile Ad-Hoc Network", International Journal of Multimedia & Its Applications (IJMA), Vol. 3, No. 3, 51–59, August 2011.
11. Z. Su, B. R. Ahn, K. Y. Eom, M. K. Kang, J. P. Kim, and M. K. Kim, "Plagiarism detection using the Levenshtein distance and Smith-Waterman algorithm," in Proc. 3rd Int. Conf. Innovative Computing Information and Control, Dalian, Liaoning, 2008.

Load Balancing Tools and Techniques in Cloud Computing: A Systematic Review

Mohammad Oqail Ahmad and Rafiqul Zaman Khan

Abstract The cloud computing is an up-and-coming computing standard, which offers large-scale distributed computing environment to share information and other resources as per users demand. Load balancing is a central challenge of the cloud computing, which requires distributing work load fairly across all the nodes for the better use of available resources. This research paper is a systematic review with a comparative study on existing techniques and tools for load balancing. Load balancing metric parameters (reaction time, fault tolerance, scalability, throughput, etc.) were used to evaluate the performance of existing techniques for comparison. Further, we have discussed virtualization of the cloud and its types for optimal resource utilization.

Keywords Load balancing · Cloud computing · Cloud virtualization · Techniques for load balancing

1 Introduction

Cloud computing has become the most popular way in the recent time to access online computing resources and user fulfillment in a low cost manner [1]. Because it allows user to access pool of resources and provides an easy way to storage of data according to its requirement pay per use concept without worry about the hardware needed [2, 3]. The cloud supports the virtualization concept which shows virtualized data centres. So when contrasted with old fashioned "own and use"

M.O. Ahmad (✉) · R.Z. Khan
Department of Computer Science, A.M.U, Aligarh, India
e-mail: oqail.jmu@gmail.com

R.Z. Khan
e-mail: rzk32@yahoo.co.in

© Springer Nature Singapore Pte Ltd. 2018
S.K. Bhatia et al. (eds.), *Advances in Computer and Computational Sciences*,
Advances in Intelligent Systems and Computing 554,
https://doi.org/10.1007/978-981-10-3773-3_18

concept in case all of us usage cloud computing, investing as well as maintenance charge of the setup is alleviated [4].

Virtualization is a significant as well as central cloud computing technology. This technology helps the abstraction of the primary components of processing like networking, storage, and hardware. It would facilitate the cloud data centers to efficiently boost resource consumption and decrease energy costs [5]. Load balancing is a technique in cloud computing to disburse workload regularly to every one nodes in the offered workspace so it convinced no single nodes in the system is overwhelmed or unused for each instant of time. A proficient load balancing algorithm would clarify that every single node in the system might have pretty much identical measure of work. The liability of the load balancing algorithm is truly to contract with the assignments which are placed in advance to the cloud region for the fresh services. Hence, that the whole open response time is improved and what's more it gives capable to resource utilization. Balancing the workload being one of the remarkable stresses in cloud computing, since we cannot make sense of the amount of requests that are released consistently within a cloud system. The unpredictability is credited to the always differing tendency of the cloud. The principal thought of load balancing in the cloud is in disseminating and designating the load progressively all through the nodes with a particular end objective to fulfill the client necessities and to give ideal resource utilize just by arranging the whole realistic load to differing nodes [6, 7].

2 Cloud Virtualization

Cloud virtualization refers to "something that is not real" however, it acts as real [2]. This is a software application on which diverse programs are able to perform like an actual system. It is a component of cloud computing, for the reason that diverse services of cloud can be utilized by user [8]. It is really useful to boost processing productivity as well as make use of optimal hardware resources [2].

Two main types of virtualization are noticed with regard to clouds as discussed below

2.1 Full Virtualization

In the context of full virtualization, the whole set up of a single system is carried out on another system. The major benefit of the entire application is its power transforming the platform for any user working at the server to virtual server. It has been successful for numerous reason such as allocation of PC between various clients, isolating clients from one another and from the program control, and imitating hardware on another machine [7, 9].

For example, KVM, VMware

2.2 Para Virtualization

In para virtualization, numerous operating systems are permitted to drive on a solitary system by utilizing framework resources such as processor as well as memory [9]. At this place, entire services are not completely out there, but the services are offered partly. It has some significant characteristics such as disaster recovery, capacity management, and migration.

For example, Xen, Denali

3 Existing Techniques for Load Balancing

Some of the following existing techniques for load balancing in the cloud are:

3.1 LBVS

H. Liu et al. [10, 11] projected a Load Balancing technique on Virtual Storage (LBVS) which offers a huge amount of data storing capacity on cloud. Storage virtualization is accomplished utilizing a design that is three-layered as well as load balancing is obtained utilizing two load balancing modules. These will allow during enhancing the productivity of simultaneous access by applying duplicate copy of balancing in advance for limiting the reaction time and improve the potential of tragedy rescue. It also helps in boosting the utilization ratio of stock resource, versatility, and the strength of the infrastructure.

3.2 Honeybee Foraging Behavior

M. Randles et al. [3, 8] have projected a distributed load balancing technique based on honeybee which is a nature motivated algorithm to obtain personal organization. It does overall load balancing using neighboring server activities. The efficiency of the system is improved by increased of infrastructure range, however throughput does not enhance with a boost inside infrastructure range. This is suitable in favor of circumstances in which distinctive types of service are needed.

3.3 CLBVM

A. Bhadani et al. [10, 12] have introduced a Central Load Balancing Strategy for Virtual Machines (CLBVM) which manages the job of the cloud lightly.

This approach boosts the entire system efficiency, however, does not look into the systems which are flaw tolerance.

3.4 SBLB for Internet Distributed Services

A. M. Nakai et al. [13, 14] projected a unique Server-based Load Balancing tactic of network servers to be spread across the globe. It supports in minimizing the facility reaction times with the use of a set of rule which restricts the redistribution of demands to the nearby distant servers short of overloading them. It helps web servers to persist overloads. Migration time and response time is low, that signifies the entire time expected in moving the tasks from node to node is low and the time period between sending a request and its reaction is also low, which enhance the entire efficiency.

3.5 Join-Idle Queue

Y. Lua et al. [10, 15] investigated a Join-Idle Queue technique for automatically extensible Internet resources. It offers a significant amount of load balancing through dispersed dispatchers by, initial load balancing blank CPUs all over dispatch that are intended for the presence of unused CPUs on every correspondent and after that, providing tasks toward CPUs to scale down normal line size on every CPU. Just avoiding the work load as of the crucial way of calling for handing out, it efficiently minimizes the framework load, gets refusal message in the clouds on task returns as well as does not enhance reaction time.

3.6 Decentralized Content Aware LB

H. Mehta et al. [10, 16] projected a unique Content Aware LB strategy titled as workload and Client Aware Policy (WCAP). It refers a parameter named as USP to recognize the Unique and Special Property of the demands and processing nodes. USP assists the scheduler to select the most ideal node for computing the demands. It included in a distributed approach with overhead below. Through utilizing the content detail to restrict the look for, it enhances entire searching system efficiency. It additionally supports during minimizing the unused computing times of the nodes, thus enhancing their usage.

3.7 Index Name Server

T. Yu Wu et al. [17] have proposed dynamic technique for load balancing known as Index Name Server (INS) that eliminates the information replication as well as repetition in cloud framework. It operates with a mix of repetition and access point maximization. To compute the best choice point, a few INS attributes are specified: hash code of chunks of information to be found, location of the server getting victimized, progress condition, and ideal Internet speed. In this way, the activity in the following minute can be anticipated all the more precisely and the execution of intensely loaded cloud storage framework can be incredibly upgraded.

3.8 Stochastic Hill Climbing

B. Mondal et al. [18] have introduced a technique for load balancing named Stochastic Hill Climbing taking into account delicate figuring for determining the streamlining issue. It supports to easy loop shifting in the way of growing value that is mounting. This makes slight modification into unique allotments regarding a quantity of parameter intended. It includes two parameter one is a contender creator set up feasible descendant and the other is an estimate which positions every applicable result.

3.9 HBB-LB

D. Babu et al. [19] have projected a technique for Honey Bee Behavior inspired Load Balancing (HBB-LB) that supports to gain smooth load balancing all around VM to increase throughput. This look at the precedence of job is ready in line for carrying out in a VM. Later on when load on virtual machine measured, judges framework is overwhelmed and below loaded. VMs are categorized according to this load. The task is scheduled on VMs according to the new load on the VM. Tasks which are taken off much earlier, which are beneficial to discover the proper below loaded virtual machine for the present job. Forager bee is applied like Inspect bee in the upcoming actions.

3.10 Cloud Server Optimization

Y. Sahu et al. [20] have projected a Dynamic Compare and Balanced based technique that advances the cloud framework utilizing load adjusting as well as server solidification. It supports server to utilize the cloud services inside DCs that is watched and the cloud servers are advanced utilizing transfer process as a part of request to adjust the load of the group and minimize the essential energy. This technique uses two standards of cloud improvement. Initial one uses to advance the cloud framework at host machine level and next one uses to improve the cloud framework utilizing active limit values taking into account activity of client request.

3.11 Response Time Based LB

A. Sharma et al. [21] have proposed response time based Load Balancing Algorithm that needs a defensive technique of load balancing focusing on reaction time of every demand. Depending upon the reaction time, the projected technique judges the allotment of upcoming newly arrived demand. The technique is not really effective in quality, but additionally minimizes the interaction and excess computing on every server. The algorithm removes meaningless interactions of the load balancer with the VMs by not inquiring related to their present resource accessibility. It brings into consideration the reaction time, that is conveniently obtainable with the load balancer like every demand as well as reaction goes by the load balancer, thus removes demand to obtaining more records from numerous resources, therefore misusing the network speed.

3.12 Ant Colony Optimization

E. Gupta et al. [22] proposed Ant Colony Optimization(ACO) technique for load balancing concept that observes overwhelmed and below loaded servers and thus operates load balancing operations among observed servers of data center. ACO is motivated from the ant colonies foraging behavior which is to collectively look for fresh food source and cumulatively utilize the presented food sources to move the food return to the nest. These ants make a pheromone path ahead relocating from node to node. Through using the following pheromone paths, the ant consequently returned to the food. The magnitude of the pheromone may differ on several aspects including the superiority of resources of food, remoteness of the food, etc. This technique ensures accessibility, accomplishes productive resource use, improves a number of demands took care by cloud, and decreases time needed to fulfill several demands.

3.13 PLBS

M. Shahid et al. [23] have proposed a centralized load balancing policy for a set of tasks in which every application is symbolized by DAG having the goal of reducing the load unevenness on the nodes as enhancing the reaction time for the tasks. It looks for more suitable computing node for the modules as for the implementation time and allocates the modules relating to chosen node over to its load as far as limit record. In this way, relying upon the dynamic framework situation, the most ideal portion of allotment for the task of application is achieved at a given amount of time.

3.14   A2LB

A. Singh et al. [24] have projected an Autonomous Agent Based Load Balancing (A2LB) technique that offers dynamic nature to support cloud platform. This technique provides optimistic load computation of virtual machine in a data center and anytime work of a virtual machine arrives close to edge value, load provider triggers look for an applicant VM using other DCs. Keeping the data of applicant VM early minimizes response time. Whenever a virtual machine becomes over-whelmed, the vendors have to disperse the services in such a way that the accessible resources are going to be widely used in an appropriate way and load at every virtual machines will continue to be moderate.

4   Metrics of Cloud Load Balancing

There are quite a number of metrics that can be enhanced to obtain a more desirable cloud load balancing [6, 25].

4.1   Throughput

This is the whole tasks which have completed execution within provided amount of time limit. This time is required to have better throughput for high efficiency of the system.

4.2   Overhead

It defines the calculation of operating cost, which is made while execution of the load balancing techniques. It should minimize for better execution of the technique.

4.3 Fault Tolerance

It is the capacity to manage effectively as well as continuously in the condition of breakdown at any one node in the framework. A fault-tolerance parameter is useful in order to change to different node which resolves faulty nodes.

4.4 Transfer Time

This is the time to relocate or move the resources of one specific node to another node for execution. This time will need to be minimized for better efficiency of the system.

4.5 Reaction Time

The measure of time in use by a specific load balancing framework to react in a distributed environment. This time length will need to be minimal to help for better execution.

4.6 Resource Utilization

This specifies the measure to ensure the resources of the framework to be used [6]. This parameter should be maximized for resource utilization.

4.7 Scalability

It specifies the capability of the framework to realize the technique for load balancing using a negligible amount of CPUs [6]. This parameter can be progressed for more attractive system execution.

4.8 Performance

It stands to the entire efficiency of the system in the wake up of delivering load balancing [25]. In case all the above metric are fulfilled, it will help better execution of the system.

5 Comparison of Existing Techniques for Load Balancing

Using this comparison, efficiency of existing techniques of load balancing can be enhanced by implementing some unique thoughts as Table 1, gives an idea about a technique what is present and what is not there, if present metrics, that is utilizing properly or not [26] (Table 2).

Figure 1 depicts number of parameters considered in existing techniques of load balancing for better performance of the system but none of the technique considered almost all the parameters.

6 Tools for Load Balancing

6.1 CloudAnalyst

The tool is valuable to developers as it is used to evaluate large scale cloud application requirement in respect to user workloads and computing server's geographical distribution [10]. The later was developed to assist developers in having insight on ways to adopt for application distribution within cloud infrastructures including value added services while the former was modeled for the objective of studying application behaviors within several deployment configurations [6].

6.2 GroudSim

GroudSim has proposed as an autonomous event based scientific simulator tool on Cloud and Grid infrastructures. This tool offers difficult simulation outcomes from normal task executions on rented computing services to computation cost, and load of cloud environment [1, 27]. GroudSim provides a few simple statistics and analysis ideas subsequent to runtime to permit the client to simply author more mind boggling examination [28].

6.3 GreenCloud

GreenCloud is developed as an advanced packet level simulator tool for power conscious ability of cloud computing data centers with a concentrate on cloud communications. This tool transmits a complete high quality modeling of the power used by the server farm IT instruments like network switches, communication links, and computing servers [29]. It basically applied to build up original solutions in controlling, distribution of resources, arrangement of workload, and maximization of interactions protocols and system frameworks [1] (Table 3).

Table 1 Existing techniques for load balancing

S. no.	Techniques	Author/Year	Source	Environment	Observations
1	LBVS [11]	H. Liu et al./ 2010	IEEE conference	Cloud storage	Archive dummy load balancing feature that can manage changes while the access load balancing using the Fair-Share Replication strategy
2	Honeybee foraging behavior [3]	M. Randles et al./2010	IEEE conference	Large scale cloud systems	Attains worldwide load balancing by using nearby server performance
3	CLBVM [12]	A. Bhadani et al./2010	COMPUTE conference	Cloud computing	Primarily maps the load to virtual machine and from that virtual machines to host machine
4	Server-based load balancing [14]	A. M. Nakai et al./2011	IEEE LADC conference	Distributed web servers	Applies a set of rule to restrict transfer costs to prevent distant servers overloading. Utilizes a heuristic to permit rapid load variations
5	Join-idle-queue [15]	Y. Lua et al./2011	IJPE conference	Cloud data centers	Primarily loads unused CPUs to carriers to obtain the presence of the unused CPUs at every dispatcher
6	Decentralized content aware [16]	H. Mehta et al./2011	ICWET conference	Distributed computing	Utilizes a (USP) parameter of demands as well as processing nodes to guide scheduler to elect more suitable node for accepting the demands
7	Index name server [17]	T. Yu Wu et al./2012	IEEE conference	Cloud storage	To sort out the data duplication and redundancy in the system

(continued)

Table 1 (continued)

S. no.	Techniques	Author/Year	Source	Environment	Observations
8	Stochastic Hill Climbing [18]	B. Mondal et al./2012	C3IT conference	Load distribution in cloud computing	A local maximization technique Stochastic Hill climbing is applied for allotment of newly arriving tasks to the virtual servers
9	HBB-LB [19]	D. Babu et al./2013	Elsevier journal	Cloud computing	Defines load balancing all over virtual machine to increase throughput
10	Cloud server optimization [20]	Y. Sahu et al./2013	IEEE conference	Green computing	Decreases the set amount of host machines to be powered on for minimizing the expense of cloud resources
11	Response time based [21]	A. Sharma et al./2014	IEEE conference	Cloud computing	Cutting out require of interactions among the load balancer as well as back end Simulated Machines which includes the facts of open services obtainable with the server
12	Ant Colony Optimization [22]	E. Gupta et al./2014	IEEE conference	Cloud data center	Performance as well as throughput of the system are improved due to availability of resources and optimal resource usages
13	PLBS [23]	M. Shahid et al./2014	IEEE conference	Compute intensive heterogeneous environment	Precedence and capacity index based allocation improve utilization of resources with minimum response time
14	A2LB [24]	A. Singh et al./2015	ICACTA conference	Cloud computing	Keeps detail of client of VM in advance, to minimize response time

Table 2 Comparison of existing techniques for cloud LB based on metrics

Metrics/Techniques	Throughput	Overhead	Fault tolerance	Transfer time	Reaction time	Resource utilization	Scalability	Performance
LBVS	Less	No	Yes	Less	More	Less	Yes	High
Honeybee Foraging	Less	No	No	Less	Less	High	No	Less
CLBVM	More	No	No	More	Less	High	No	More
SB LB	Less	No	No	Low	Less	Low	Yes	High
Join-Idle-Queue	Less	Yes	No	More	Less	Less	No	More
Decentralized content aware	More	No	No	More	Less	High	Yes	High
INS	Less	No	Yes	Less	High	Less	No	High
Stochastic Hill Climbing	More	No	No	More	Less	High	No	High
HBB-LB	More	No	No	More	Less	Low	No	Less
Cloud server optimization	Less	No	No	More	More	High	No	Less
RTBLB	Less	No	No	More	Less	Less	No	Less
ACO	Less	No	No	More	High	High	No	Less
PLBS	Low	No	No	Less	Less	High	Yes	More
A2LB	Low	No	No	High	Less	High	Yes	More

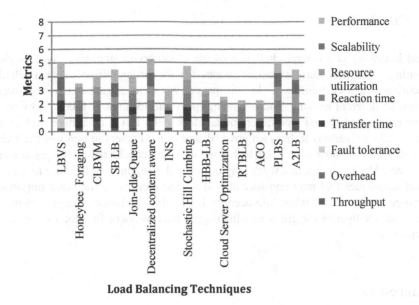

Load Balancing Techniques

Fig. 1 The number of metrics considered in different load balancing techniques

Table 3 Comparison of CloudSim, GroudSim and GreenCloud

S. No.	Tools/Parameters	CloudAnalyst	GroudSim	GreenCloud
1	Platform	CloudSim	ASKALON	NS2
2	Availability	OpenSource	OpenSource	OpenSource
3	Graphical support	Limited	None	Limited
4	Language	Java	Java	C++/OTel
5	Support of TCP/IP	None	Full	Full
6	Communication model	Limited	Limited	Full
7	Simulation time	Seconds	Seconds	10's Minutes
8	Physical models	None	None	Memory, storage and networking
9	Energy models	None	None	Precise
10	Power saving modes	None	None	DVFS, DNS and both
11	Application mode	Computation and data transfer	Execution of jobs on VMs and data transfer	Computation and data
12	Workload	Full	Full	Full
13	S/W or H/W [27]	S/W	S/W	S/W
14	Simulation speed	Fast	Low	Low

7 Conclusion

Load balancing is a central challenge of the cloud computing, that involves distributing work load fairly across all nodes for the better utilization of available resources and user fulfillment. In this paper, we have carried out a systematic review of different existing load balancing tools and techniques followed by their comparative study based on different parameters. After this comparative study, it is concluded that various techniques cover different parameters to analyze the techniques. However, none of existing technique has been considered all the parameters as revealed by work. So, this work may promote the beginner of the field to design novel techniques that may consider almost all the metrics. Some other important parameters also can be taken into account like carbon emission, energy consumption, and intelligence for green and intelligent load balancer for cloud computing environment.

References

1. Sinha, U., Shekhar, M.: Comparison of Various Cloud Simulation tools available in Cloud Computing, International Journal of Advanced Research in Computer and Communication Engineering, Vol. 4, Issue 3 (2015).
2. Ahmad, M.O., Khan, R.Z.: The Cloud Computing: A Systematic Review, International Journal of Innovative Research in Computer and Communication Engineering (IJIRCCE), Vol. 3, Issue 5 (2015).
3. Randles, M., Lamb, D., Taleb-Bendiab, A.: A Comparative Study into Distributed Load Balancing Algorithms for Cloud Computing, 24th International Conference on Advanced Information Networking and Applications Workshops, pp. 551–556 (2010).
4. Desai, T., Prajapati, J.: A Survey Of Various Load Balancing Techniques And Challenges In Cloud Computing, International Journal Of Scientific & Technology Research Volume 2, Issue 11 (2013).
5. Kulkarni, A.K., Annappa, B.: Load Balancing Strategy for Optimal Peak Hour Performance in Cloud Datacenters, Signal Processing, Informatics, Communication and Energy Systems (SPICES), pp. 1–5, IEEE (2015).
6. Nwobodo, I.: Cloud Computing: A Detailed Relationship to Grid and Cluster Computing, International Journal of Future Computer and Communication, Vol. 4, No. 2 (2015).
7. Khan R.Z., Ahmad M.O. (2016) Load Balancing Challenges in Cloud Computing: A Survey. In: Lobiyal D., Mohapatra D., Nagar A., Sahoo M. (eds) Proceedings of the International Conference on Signal, Networks, Computing, and Systems. Lecture Notes in Electrical Engineering, vol 396. Springer, New Delhi.
8. Aditya, A., Chatterjee, U., Gupta, S.: A Comparative Study of Different Static and Dynamic Load Balancing Algorithm in Cloud Computing with Special, Vol. 5, No. 3, pp. 1898–1907 (2015).
9. Padhy, R.P., 2011. Load balancing in cloud computing systems (Doctoral dissertation, National Institute of Technology, Rourkela).
10. Kansal, N.J., Chana, I.: Existing Load Balancing Techniques In Cloud Computing: A Systematic Review, Journal of Information Systems and Communication, ISSN: 0976-8742, E-ISSN: 0976-8750, Vol. 3, Issue 1, pp. 87–91 (2012).

11. Liu, H., Shijun, Liu, Meng, X., Yang, C., Zhang, Y.: LBVS: A Load Balancing Strategy or Virtual Storage, IEEE International Conference on Service Science (ICSS), pp. 257–262 (2010).

12. Bhadani, A., Chaudhary, S.: Performance Evaluation of Web Servers using CentralLoad Balancing Policy over Virtual Machines on Cloud, COMPUTE (2010).

13. Begum, S., Prashanth, C.S.R.: Review of Load Balancing in Cloud Computing, IJCSI International Journal of Computer Science Issues, Vol. 10, Issue 1, No. 2 (2013).

14. Nakai, A.M., Madeira, E., Buzato, L.E.: Load balancing for Internet Distributed Using Limited Redirection Rates, LADC, pp. 156–165 (2011).

15. Lua, Y., Xie, Q., Kliot, G., Geller, A., Larus, J.R., Greenberg, A.: "Join-Idle-Queue: A novel load balancing algorithm for dynamically scalable web services", An International Journal on Performance Evaluation (2011).

16. Mehta, H., Kanungo, P., Chandwani, M.: Decentralized Content Aware Load Balancing Algorithm for Distributed Computing Environments, International Conference and Workshop on Emerging Trends in Technology (ICWET), pp. 370–375 (2011).

17. Tin-Yu Wu, Wei-Tsong Lee, Yu-San Lin, Yih-Sin Lin, Hung-Lin Chan and Jhih-Siang Huang: Dynamic load balancing mechanism based on cloud storage, Computing, Communications and Applications Conference (ComComAp), pp. 102–106, IEEE (2012).

18. Mondal, B., Dasgupta, K., Dutta, P.: Load Balancing in Cloud Computing using Stochastic Hill Climbing-A Soft Computing Approach, CEIT, Vol. 4, pp. 783–789 (2012).

19. Dhinesh, L.D., Babu, Krishna, P.V.: Honey bee behavior inspired load balancing of tasks in cloud computing environments, Appl. Soft Computing, Vol. 13 (2013).

20. Sahu, Y., Pateriya, R.K., Gupta, R.K.: Cloud Server Optimization with Load Balancing and Green Computing Techniques Using Dynamic Compare and Balance Algorithm, 5th International Conference on Computational Intelligence and Communication Networks, pp. 527–531, IEEE (2013).

21. Sharma, A., Peddoju, S.K.: Response Time Based Load Balancing in Cloud Computing, International Conference on Control, Instrumentation, Communication and Computational Technologies (ICCICCT), IEEE (2014).

22. Gupta, E., Deshpande, V.: A Technique Based on Ant Colony Optimization for Load Balancing in Cloud Data Center, International Conference on Information Technology, IEEE (2014).

23. Shahid, M., Raza, Z.: A Precedence Based Load Balancing Strategy for Batch of DAGs for Computational Grid, International Conference on Contemporary Computing and Informatics (IC3I), pp. 1289–1295, IEEE (2014).

24. Singh, A., Juneja, D., Malhotra, M.: Autonomous Agent Based Load Balancing Algorithm in Cloud Computing, International Conference on Advanced Computing Technologies and Applications (ICACTA) (2015).

25. Gupta, R.: Review on Existing Load Balancing Techniques of Cloud Computing, International Journal of Advanced Research in Computer Science and Software Engineering, ISSN: 2277, Vol. 4, Issue 2 (2014).

26. Rastogi, D., Khan, F.U.: Techniques of load balancing in cloud computing: A survey, International Journal of Engineering Science and Computing (IJESC), pp. 335–338 (2014).

27. Ostermann, R.P.T.F., Plankensteiner, S.K.: Groudsim: an event based simulation framework for computational grids and clouds, in CoreGRID/ERCIM Workshop on Grids and Clouds. Springer Computer Science Editorial, Ischia (2010).

28. Khairunnisa and Banu., M.N.: Cloud Computing Simulation Tools - A Study, Intern. J. Fuzzy Mathematical Archive, Vol. 7, No. 1, 13–25 (2015).

29. Kliazovich, D., Bouvry, P., Khan, S.U.: GreenCloud: a packet-level simulator of energy-aware cloud computing data centers, Journal of Supercomputing, Vol. 62, no. 3, pp. 1263–1283 (2012). Available: http://greencloud.gforge.uni.lu/.

A Hybrid Optimization Approach for Load Balancing in Cloud Computing

Apoorva Tripathi, Saurabh Shukla and Deepak Arora

Abstract Cloud computing is characterized as a technology which is recent and has an important effect on the field of IT in the nearby future. Load balancing is one of the fundamental issues in cloud computing. There could be various types of load such as CPU load, delay or network load, and capacity of the memory. Load balancing is the procedure of dispersing the load of work among numerous nodes in a system which is distributed for better response time in the job and also for better utilization of resources. Load balancing guarantees that all the processors in the system perform equalized amount of work at any particular instant of time. In this paper, a hybrid algorithm of ant colony optimization and bee colony algorithm is proposed. The proposed mechanism takes the characteristic of a complex network into consideration. The strategy is combined with the setting of other parameters of the upgraded bee colony algorithm to form a new ACO method. The new hybrid approach has been simulated using the Cloud Analyst tool.

Keywords Cloud computing · Load balancing · Cloud analyst · ACO · ABC

1 Introduction

With the progression in the period, the technology is getting increasingly more extensive. In these days, individuals get the administrations as per their demands subsequently, they do not need to pay unnecessarily for the things which they do not

A. Tripathi (✉) · S. Shukla · D. Arora
Department of Computer Science & Engineering, Amity University,
Lucknow 226028, India
e-mail: apoorvatripathi1992@gmail.com

S. Shukla
e-mail: saurabh3611@gmail.com

D. Arora
e-mail: darora@lko.amity.edu

© Springer Nature Singapore Pte Ltd. 2018
S.K. Bhatia et al. (eds.), *Advances in Computer and Computational Sciences*,
Advances in Intelligent Systems and Computing 554,
https://doi.org/10.1007/978-981-10-3773-3_19

need. Cloud computing is one of the technologies which is emerging which offers the administrative rights appropriately to the clients. Cloud computing is scalable and efficient yet keeping up the stability for processing various jobs in the environment of cloud computing is an exceptionally complex issue with load balancing accepting much consideration for researchers. Be that as it may, cloud computing is confronting a dispute of load balancing which is challenge of a real time. The principle explanation behind this challenge is the expansion in the interest of users for services of the cloud. So it is impossible practically to keep up the one or all the free services for satisfying the interest. Providing every server with one demand for fulfilling will come about into traffic on the server and at last the system crash [1].

Recent years have seen the advancement of cloud computing, and a standout among the hugest problems in cloud computing is load balancing. In cloud computing environment, data centers are clusters intensively with network devices and computers with high performance. Given that the prerequisites of clients are different and dynamically varying, and the services and resources are heterogeneous ordinarily, it is basic that clusters workload is unbalanced particularly. In particular, few physical machines are over-loaded and there is an effect on the efficiency, whereas others are generally idle and consequently there is wastage of resources. Hence, how to guarantee the load balance is the main issue in the environment of cloud [2].

The rest of the paper is organized as follows: Sect. 2 describes related work. The description of our problem and the objectives are in Sect. 3. Section 4 describes ant colony optimization and Sect. 5 describes artificial bee colony algorithm. Section 6 discusses the proposed scheme, Sect. 7 is the experimental results and Sect. 8 is the conclusion.

2 Related Work

There has been lots of work which is already done in the section of load balancing in cloud computing. Some important work in this area is given as follows:

In paper [3], authors gave a survey on techniques of soft computing dependent upon load balancing in cloud computing. Cloud computing has turned out into a new method of computing all over the Internet. Cloud computing has numerous problems over the Internet like fault tolerance, load management, and also few issues of security are there. Among them, load balancing is one of the fundamental issues to be resolved so as to enhance the cloud efficiency. The load balancing portrays that system's dynamic workload is dispersed among different nodes in such a way that no node is underutilized or overloaded. The load might be network load, memory capacity, or CPU load. The new concept was earlier discussed that was utilizing techniques of soft computing for performing load balancing. Numerous algorithms are additionally discussed in this paper.

Reference [4] provided a study on effective load balancing for dynamic resource allocation in cloud computing. The resources are dynamic, so a load of resources differs with a change in cloud configuration and hence tasks' load balancing in the environment of cloud can impact the performance of cloud. In the approach already available, the main issue is distributed dispatching of a task to a resource. In this paper, the fundamental objective of load balancing is to give a scheme which is low cost and distributed that adjusts the load over every processor. For enhancing the throughput of resources of cloud globally, efficient and effective algorithms of load balancing are significant. In this paper, a combination of algorithm known as ACBLA with an algorithm of a queue is presented which is applied to computation jobs which are scheduled effectively.

Paper [5] presented partitioning and load balancing in cloud computing. The implementations of cloud computing provide apparently infinite pooled resource of computing over the network. In cloud computing, resources are virtually given for applications related to developing, storage and computation. Cloud computing is made on the base of virtualization, grid computing, and distributed computing. Cloud computing is characterized as paradigm of distributed computing at a large scale. There are various algorithms of load balancing like ant colony optimization, equally spread current execution, and round robin algorithm. This paper presented numerous schemes of load balancing for the better enhancement of the cloud's performance.

In Ref. [6], authors proposed an implementation of load balancing which was dependent upon partitioning in cloud computing. Load balancing in the environment of cloud computing has a significant effect on the performance. This paper presents a better model of load balance for the public cloud dependent upon the concept of cloud partitioning with the mechanism of a switch for picking various methodologies for diverse circumstances. The algorithm applies the theory of game to the strategy of load balancing for improving the efficiency in the environment of a public cloud. A cloud computing which does not utilize load balancing has various drawbacks. Presently in these days, the utilization of Internet and its related resources has broadly increased.

Paper [7] discussed ant colony optimization. The primary point of interest in this paper is load balancing in cloud computing. This load could be network load or delay, the load of CPU, or capacity of the memory. Load balancing is the procedure of circulating the load among numerous nodes of a system which is distributed for enhancing response time of job and resource utilization while additionally keeping away from a situation in which few nodes are doing very less work or may be idle. Numerous techniques have been appeared such as genetic algorithms, has the method, Particle Swarm Optimization, and also many more algorithms of scheduling. In this paper, a method relied upon ant colony optimization is proposed for resolving the issue of load balancing in an environment of cloud.

3   Motivation

The enormous measure of calculations a cloud can satisfy in a particular measure of time cannot be carried out by the best supercomputer. However, the performance of cloud could be still enhanced by ensuring that all the accessible resources in the Cloud are optimally utilized by utilizing a good algorithm of load balancing. Statistical results of various research papers in the literature review indicate that intensive local search improves the quality of solutions found in such dynamic load balancing algorithms. The proposed methodology is to ensure that every one of the processors in the system or each node in the network does around the equivalent measure of work at any particular instant of time. Furthermore, enhances numerous aspects of the related algorithm of ant colony optimization which is implemented to comprehend load balancing in a system which is distributed. In addition, the proposed mechanism takes the attribute of Complex Network into concern. We first propose an adaptive strategy for load balancing as per the quality of the solution discovered by artificial ants. Second, the procedure is joined with the setting of different parameters of the upgraded bee colony algorithm like to form a new ACO method.

The research is based on following objectives:

1. To study the Ant colony optimization (ACO) and Artificial Bee Colony (ABC) optimization algorithms.
2. Implement a hybrid algorithm using ACO and ABC algorithms.
3. To compare and analyze the results of the original state of the art with the proposed algorithm on the basis of following metrics: Overall Response time, Data Center Service Time, and Transfer Cost.

4   Ant Colony Optimization

The remembrance of ants is exceptionally constrained. In the gathering of ants, the behavior of an individual shows up as an element which is massively different. The ants perform common strategy for accomplishing an assortment of difficult assignments with good reliability and steadiness, which has turned into the field of ACO (Ant colony optimization). They have particularly limited memory and show activities related to the entity and also have an informal section. In spite of the fact that it is fundamentally vital self-organization, they utilize phenomenon which is like overtraining in the strategies of learning. The complicated social practices of ants have been tremendously concentrated on by science, and scientists are presently finding that these patterns of behavior can give models for finding solutions to difficult problems of combinatorial optimization. The essential steps connected with an algorithm of ACO are as follows [8]:

1. Initialization of pheromone.
2. While no satisfaction of criteria, then repeat.
3. At first situate position of all the ants on the state of an entry.
4. Next state selection.
5. While not come to the state which is final then do again from step 4, if it is achieved then step 6.
6. Stages of pheromone (deposit, daemon and evaporate).
7. Check whether satisfaction of criteria or not, if fulfilled then end, but if not then again perform from step 2.
8. End.

5 Artificial Bee Colony

The technique of ABC is intensified on the aspects of the social behavior of bees which means how do they exchange the over of data among all honey bees. By watching the entire bees hive, we can separate few components that more often survive in all hives. The vital idea of the hive as per exchanging information and knowledge is the region of dancing. Entire contacts of all honey bees correspond to the quality and location of food resources in the dancing area. The related Boogie is known as waggle dance. Since a bee which is onlooker has information of good sources of food and exists on the floor of dance, they potentially may come across at numerous boogies and after that, she selects to make utilize source which is beneficial [9]. The steps engaged in the algorithm of artificial bee colony are as given below [9]:

1. Initialization of population which is distributed randomly over solution space.
2. Consider the value of fitness of every individual.
3. Compute the probability of movement of an outlooker bee.
4. Try for the value of fitness for iteration repeated.
5. If there is no improvement in the value of fitness then do.

   a. Reject all sources of food.
   b. Convert employee bees to Scout bees.

6. Choose the best value of fitness and also its location.
7. Check for a criterion of stopping, if satisfied then stop and exit, otherwise go to step 1.

6 Proposed Scheme

The basic motivation of the proposed approach is to improve the ant colony optimization load balancing algorithm in terms of overall response time, data center processing time, total virtual machine cost, and total data transfer cost.

We have hybridized modified ant colony optimization and modified bee colony optimization algorithm.

Modified ACO will have different forward and backward movement for the selection of underloaded and overloaded nodes in the cloud.

The ants in proposed algorithm will continuously originate from the Head node. These ants traverse the width and length of the network in the way that they know about the location of underloaded or overloaded nodes in the network. These ants along with their traversal will be updating a pheromone table, which will keep a record of the resources utilization by each node. The main task of ants in the algorithm is to redistribute work among the nodes. The ants traverse the cloud network, selecting nodes for their next step based on the pheromones.

6.1 Modified ACO Algorithm

Create data centers, virtual machines, user bases using GUI package of cloud analyst tool. Assign the resources and initialize ants and pheromones. Assign threshold values to virtual machine (VM) nodes like a stopping condition and randomly select head node through which traversing of ants through VM will initialize. If the ant/cloudlet is valid then check the virtual machine load with the threshold value and search for underloaded and overloaded node and update the FP (Foraging pheromones) and TP (Trailing Pheromones) list. If an overloaded or underloaded node is found then perform the load shifting of nodes according to those conditions.

6.2 Modified Bee Colony

The enhanced bee colony algorithm which will be then hybridized in ant colony algorithm is defined as

- For calculating the fitness function, we will take parameters named: virtual machine memory, bandwidth, CPU processing power, and MIPS searching.
- By taking these parameters for calculating the fitness value, we will get the better available virtual machine for the task to be performed on a cloud.

$$\text{fit}_{ij} = \frac{\sum_{i=1}^{n} \text{cloudlet\_length}_{ij}}{\text{Vm}_j\_\text{mips}},$$

where Vm_j_mips is defined by millions of instructions per second for each processor of Vm_j, n is the total no of scout foragers, fi_{ij} defines the fitness function of population of bees (i) for Vm_j or say capacity of Vm_j with ith bee number, cloudlet_length is the task length that has been submitted to Vm_j.

The virtual machine (Vm$_j$) capacity is being calculated using the following parameters:

$$Capacity\_Vm_j = Vm_j\_cpu \; *Vm_j\_size + Vm_j\_bandwidth,$$

where Vm$_j$_cpu is the number of processors in a virtual machine Vm$_j$, Vm$_j$_size is the virtual machine memory size, Vm$_j$_bandwidth is the network bandwidth of Virtual Machine Vm$_j$.

6.3 Hybrid Algorithm

Initialize the population with a set of data centers and on host servers. Virtual machines (VM) will run in host servers. The resources will be allocated to Virtual Machines (VM). The threshold value will be set for each Virtual Machine (VM). Modified ACO and ABC algorithm will run to produce the output. On the completion of a single iteration, the best common nodes with the minimum load, maximum resources will be selected as optimized solutions to balance the load. The algorithm is as given below

- Create Datacenters, virtual machines, userbases, i.e., no of ants and bees [Population initialization].
- Set Minimum and a Maximum threshold of the resources [Threshold Set].
- Run the modified ACO and check the pheromone persistence, Run the modified BCO algorithm and select the initial head node, i.e., initial available VM randomly and check the load on that resource.
- If Load < threshold and bees fitness satisfy the threshold level.
- If yes then set underload = true, schedule the task by allocating vms and search for the overloaded nodes. If overload node is found then schedule the task if not found then do allocate the vm in same continuous loop.
- If no then set overload = true, and search for the underloaded nodes. If an underloaded found is true the schedule the task if not found then allocate the vm in the same continuous loop.
- Repeat the procedure until all the request is being fulfilled.

7 Results and Discussions

The simulation and performance analysis has been performed using the cloud analyst toolkit [10]. To analyze the performance of ACO and proposed Hybrid Algorithm, we need to set the configuration of the various components of Cloud

Analyst tool. We have set the parameters for user base configuration, application deployment configuration, data center configuration. We have defined six user bases located in six different regions of the world. We have taken three data centers, i.e., DC1, DC2, and DC3 having 25 numbers of VMs each. Here VMs having the unequal number of processors. The duration of the simulation is 24 h.

Following are the statistical metrics derived as the output of the simulation in the cloud analyst version of the simulator:

- Overall response time of the system.
- Total Data Center Processing time.
- Overall Processing Cost (Sum of Total virtual Machine Cost and Total Data Transfer Cost).

The results of the simulation are as follows:

7.1 Response Time

The response time for each user base and overall response time calculated by Cloud Analyst for ACO and proposed Hybrid load balancing algorithm as can be seen from Tables 1 and 2.

7.2 Data Center Processing Time

Data center request servicing time for each data center and data center processing time for ACO and proposed Hybrid load balancing algorithm are calculated by Cloud Analyst has been shown in the Tables 3 and 4.

Table 1 Response time by region

User base	ACO (ms)	Hybrid (ms)
UB1	68.93	68.37
UB2	75.56	75.68
UB3	78.00	77.94
UB4	350.52	342.19
UB5	373.00	374.04
UB6	283.07	283.46

Table 2 Overall response time

	ACO (ms)	Hybrid (ms)
Overall response time	259.06	256.94

Table 3 Data center request servicing time

Data center	ACO (ms)	Hybrid (ms)
DC1	55.73	55.63
DC2	24.01	24.10
DC3	52.88	49.90

Table 4 Data center processing time

	ACO (ms)	Hybrid (ms)
Data center processing time	49.33	47.09

7.3 Cost

The cost is calculated by Cloud Analyst for ACO is given below

- Total Virtual Machine Cost ($): 177.54
- Total Data Transfer Cost ($): 81.06
- Grand Total: ($) 259.36

The cost is calculated by Cloud Analyst for Hybrid algorithm is given below

- Total Virtual Machine Cost ($): 178.11
- Total Data Transfer Cost ($): 80.47
- Grand Total: ($) 259.04

The various results of ACO and proposed hybrid algorithm are shown in the tables and comparison is done which shows that hybrid algorithm gave better results.

8 Conclusion

The main aim of this research is to improve the ant colony optimization load balancing algorithm in terms of overall response time, data center processing time, total virtual machine cost, and total data transfer cost. The bee colony algorithm is hybridized with ant colony optimization algorithm and the results obtained are better than the individual ACO algorithm.

References

1. Abhinav Hans, Sheetal Kalra.: Comparative Study of Different Cloud Computing Load Balancing Techniques. 2014 International Conference on Medical Imaging, m-Health and Emerging Communication Systems (MedCom). IEEE. 2014.

2. Yang Xianfeng, Li Hong To: Load Balancing of Virtual Machines in Cloud Computing Environment Using Improved Ant Colony Algorithm. International Journal of Grid Distribution Computing. Vol. 8, Nol. 6, pp. 19–30. 2015.
3. Mayur S Pilavare, Amish Desai: A Survey of Soft Computing Techniques based Load Balancing in Cloud Computing. International Journal of Computer Applications. Vol. 110, No. 14, January 2015.
4. K Prasanna Kumar, S. Arun Kumar, Dr Jagadeeshan: Effective Load Balancing for Dynamic Resource Allocation in Cloud Computing. International Journal of Innovative Research in Computer and Communication Engineering. Vol. 2, Issue 3, March 2013.
5. Esha Sarkar, Ch. Sekhar: Partitioning and Load Balancing in Cloud Computing. IJCST. Vol. 5, Issue 4, October-December 2014.
6. Adiseshu Gupta S., Srinivasa Rao K.V.: Implementation of Load Balancing Based on Partitioning in Cloud Computing. International Journal of Advanced Research in Electrical, Electronics and Instrumentation Engineering. Vol. 3, Issue 10, October 2014.
7. Ratan Mishra, Anant Jaiswal: Ant Colony Optimization: A Solution of Load balancing in Cloud. International Journal of Web & Semantic Technology. Vol. 3, No. 2, April 2012.
8. Madhurima Rana, Saurabh Bilgaiyan, Utsav Kar: A Study on Load Balancing in Cloud Computing Environment Using Evolutionary and Swarm Based Algorithms. International Conference on Control, Instrumentation. Communication and Computational Technologies (ICCICCT). IEEE. 2014.
9. Kun Li, Gaochao Xu, Guangyu Zhao, Yushuang Dong, Dan Wang: Cloud Task Scheduling based on Load Balancing Ant Colony Optimization. 2011 Sixth Annual China Grid Conference. IEEE. 2011.
10. Cloud Analyst: A Cloud-Sim-based tool for Modeling and Analysis of Large Scale Cloud Computing Environments. MEDC Project Report Bhathiya Wickremasinghe, 2009.

A Comparative Analysis of Cloud Forensic Techniques in IaaS

Palash Santra, Asmita Roy and Koushik Majumder

Abstract With the growth of cloud users, various criminal activities, attacks, suspicious, or malicious activities in cloud environment are growing day by day. It is the responsibility of cloud service providers (CSP) to secure the user data residing on cloud. Cloud forensic is the field of study that helps in investigating the criminal activities taking place in the cloud environment. Proper identification of the threat and presenting it in the court of law is desired. Traditional digital forensics method is not applicable in the multi-tenant cloud environment which gave rise to cloud forensics. There are certain proposed cloud forensics methods which help in identifying the malicious users. These approaches have certain advantages and limitations. This paper focuses on the various cloud forensics methods that are proposed for Infrastructure as a Service (IaaS) model. IaaS model is more vulnerable to attacks because virtual machine (VM) works in that layer. Generally most of the crimes take place using VM. A detailed comparative analysis is done in this paper highlighting each of the existing cloud forensics methods, their advantages, and drawbacks. Further, the future research areas are pointed out which will help in developing a better cloud forensic model by overcoming the drawbacks of the various existing methods.

Keywords Cloud computing · Cloud forensic · Investigation · Evidence · IaaS · Virtual machine · Introspection

1 Introduction

In the modern world of computation, cloud computing is one of the most trending technology. This technology has brought a revolution in the area of parallel computing. Cloud computing has gained a lot of interest in both medium and large

P. Santra · A. Roy · K. Majumder (✉)
Department of Computer Science & Engineering, West Bengal University
of Technology, BF 142, Sector 1, Kolkata 700064, India
e-mail: koushik@ieee.org

© Springer Nature Singapore Pte Ltd. 2018   207
S.K. Bhatia et al. (eds.), *Advances in Computer and Computational Sciences*,
Advances in Intelligent Systems and Computing 554,
https://doi.org/10.1007/978-981-10-3773-3_20

companies. This type of computation enables the companies to deliver services to its clients through Internet facility. This enables the cloud service provider (CSP) in gaining lot of business in the market. CSP provides services like pay per use model, measured services, on demand service, rapid elasticity, and many more. While providing such services to its users, it must shoulder the responsibility for providing security to its end users so that their personal data is not prone to various attacks. There are some intrusion detection system and security framework for providing security in cloud computing environment. Some processes are there to identify malicious or suspicious activity in the cloud environment [1]. These processes are referred to as cloud forensics that protects the cloud service providers (CSP) from future attacks, provides justice to victims, and even helps in troubleshooting various services. To provide such services, a forensic investigator has to analyze various suspicious evidences in cloud environment. Generally, most complex computation is done in IaaS model of cloud. Here, users are allowed to create their own virtual machine (VM) for computation. Needed resources are supplied by cloud service provider to create virtual machine. These resources are physically located in a large server, and virtualized instances are provided to the end users for their computation. It has been observed that illegal activities in the virtual machine (VM) are growing gradually hence security needs to be improved by CSP in this model. This paper focuses on some forensic investigation techniques on VM Introspection. VM Introspection is an investigation approach where illegal activity related evidences are collected by monitoring VM. Evidence acquisition, collection, analysis, and presentation techniques are discussed here to obtain a complete model of cloud forensic [2]. Still, there are some limitations over these earlier proposed models. These limitations have been identified which helps in exploring future research areas.

2   Cloud Forensics Investigation Challenges

Different challenges of forensic investigation in cloud environment are identified by authors. By overcoming these challenges, an effective cloud forensic system can be designed. The main challenges [3, 4, 5] are as follows:

- Forensic Data Collection: Collection of digital evidence is the primary task in forensic investigation. For high level abstraction and security purposes, cloud service provider hides the data location. This makes it difficult for the cloud investigators to collect data and analyze.
- Semantic Integrity: The semantics of a data set describe a whole snapshot of the system. Data sets are often broken or destroyed due to various reasons. It is very important to design a mechanism to gain full semantics of a data set.
- Stability of Evidence: A data set may be distributed in nature. So the data set can be stored partially in different servers. In this situation stability of evidence should be maintained. If a partial data set is acquired, then the other portions may be altered. This leads to instability of evidences.

- Elastic, Static, and Live Forensic: Due to the distributed data, it is difficult to discover data and its original source. Time synchronization of resources is also important to find source of evidences. Deleted real-time data is a threat for cloud environment. It is a challenge to recover deleted data and reconstruct the incident.
- Evidence Segregation: Different instances of resources running on a single system are isolated via virtualization. Customers use only the virtualized resources instead of physical resources. So accessing log of a particular resource may violate confidentiality of multiple tenants. It is a challenge for cloud service provider to segregate the different virtualized evidences without breaching the confidentiality.
- External Dependency Chain: Chain of dependency is one of the factors of forensic investigation in cloud. Cloud service provider generally uses third party host or provider to implement their services. In the time of investigation, dependency may create a barrier; investigating another host is a challenge in forensic examination.
- Service Level Agreement: Lack of international regulation in cloud computing and sometime service level agreement (SLA) itself is a barrier on cloud forensic. Investigator should not violate SLA during investigation in cloud environment.

3 Literature Survey of IaaS-Based Cloud Forensics Techniques

3.1 Enabling Cloud Forensic [6]

Cloud forensic architecture and different techniques for forensic analysis are discussed in [6]. In this paper, traditional four key steps of digital forensic are described with the flavor of cloud aspect. A general suggestion is also given to overcome the challenges in each stage of cloud forensic. Authors suggest that evidences from virtual machine (VM) and virtual machine monitor (VMM) be inspected for identification of malicious activity.

In Fig. 1, data source selection and acquisition module collects and preserves important evidences for investigation. Original evidences are cloned to avoid tampering of original sources. Integrity and authenticity are ensured in preservation module using different encryption techniques such as MD5, Hash algorithms. Analysis module reduces time consumption and workload of forensic investigation by optimizing the search space. File carving, keyword search, metadata extraction, and virtual disk reconstruction are some popular analysis techniques which are suggested in this aspect.

Finally, presentation module enables the investigators to present the evidence or result from evidence analysis in the court of law. It should be in a format by which a judicial non-technical person can understand.

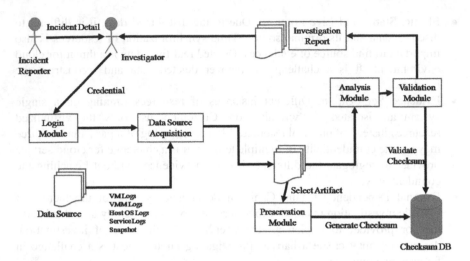

Fig. 1 A diagram of the cloud forensic process [6]

3.2 Cloud Forensic Using VM Snapshot [7]

Here, intrusion detection system is incorporated with each VM to monitor malicious activity. This technique is inspired from VM introspection where VMs are analyzed from the hypervisor point of view. Continuously, VM Monitor (VMM) and snapshots are stored in a sequential manner [8]. It helps in regeneration of event for identifying associated malicious host [9]. Also using digital provenance, the owner of the digital data is identified [10]. Isolation of cloud instance is suggested in this technique. This helps in prevention of data contamination and corruption. Hence integrity and authenticity of digital evidence [11] is preserved. Figure 2 describes the entire process flow for this model.

3.3 Introspection of Virtual Machine [12]

Authors identified that virtual machines may be one of the sources of criminal activity. This approach helps to investigate criminal evidence or suspicious activity in different virtual machines. Here, introspection refers to a thorough examination or analysis of different components of cloud. Authors have suggested their approach using these three main techniques:

- Swap space analysis: Swap spaces of each virtual machine are analyzed here. It monitors swap space continuously to get malicious information. When an attack occurs, swap space amount increases abnormally. This abnormality can be used

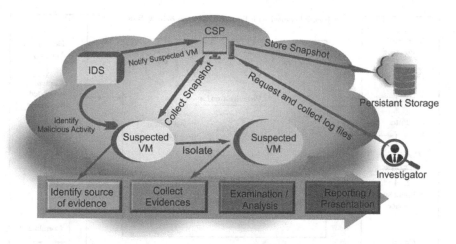

Fig. 2 Diagram for cloud forensics using VM snapshots as digital evidence [7]

as triggering technique. Thus continuous monitoring helps in identifying malicious event.

- Conjunctive delivery model: When a VM is closed or terminated, the resources acquired by that VM are reallocated to another VM. During reallocation, VMM can inspect the previously used resources. In this derivation technique [12], investigator may get useful evidences.
- Terminated process based approach: Here outband technique [12] is used. Monitoring all processes in a particular VM is done continuously. When a forcefully or abnormally terminated process is detected, it indicates a suspicious event may have occurred. Using this terminated process indication, suspicious events can be identified.

Using the above techniques, a hybrid approach is drawn by the authors. In Fig. 3, the main working principle of the hybrid approach is shown.

3.4 Forensically Enabled IaaS Cloud [13]

In this paper, authors have designed a model for forensically enabled IaaS cloud architecture. Agent-based approach for forensic investigation is introduced here. Agent coordinators are generally incorporated with each VM and it communicates to external Cloud Forensic Acquisition and Analysis system (Cloud FaaS). In cloud FaaS, the available agents are—nonvolatile memory agent, volatile memory agent, network traffic agent, activity log agent. In Fig. 4, the model for agent-based approach is described. In this approach, cloud FaaS is the centralized processing point from where the forensic data can be collected. Agent coordinator

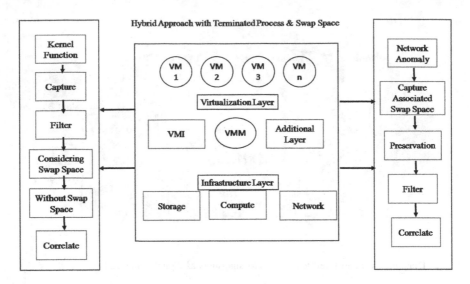

Fig. 3 A model for virtual machine introspection [12]

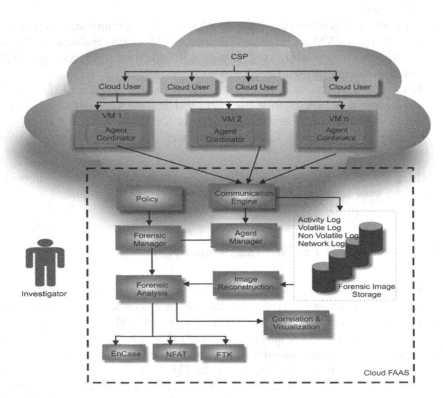

Fig. 4 Model for data acquisition within IaaS [13]

communicates with the central FaaS to provide forensic data. Agent coordinator and agent manager should include Hash algorithm to provide data integrity. Image reconstruction module is also used to collect related old image from image repository.

4 Comparative Study

See Table 1.

Table 1 Benefits and limitations of earlier proposed models

Approach	Benefits	Limitation
Enabling cloud forensic [6]	1. Search space reduction and data filtering is suggested. This optimizes workload and time consumed for forensic investigation 2. Data or evidence extraction using VMI technique is suggested 3. Data integrity is maintained by preventing tampering of evidences	1. There is no suggestion given for live forensics 2. No particular algorithm is suggested for search space reduction and filtering technique 3. Data privacy is not covered in this approach 4. Suggestion for cloud instance isolation is not given
Cloud forensic Using VM snapshot [7]	1. Incorporation of intrusion detection system in VMM and VM fires a trigger to take a snapshot 2. Monitoring resources for longer time provides better result 3. Transfer of suspected VM to another network node (cloud instance isolation) prevents data contamination 4. Regeneration of event is suggested	1. Uncontrolled continuous monitoring on a particular resource may affect the cloud server performance 2. No search space reduction and filtering technique is suggested here. Hence search space analysis takes long time 3. Data integrity and privacy are not suggested
Introspection of virtual machine [12]	1. Data collection is done from internal architecture of cloud and VM. This provides reliable data 2. Suggestion of swap space analysis is given. It may be a good source of evidence. Because swap space is extended part of virtual memory 3. Here forensic investigation is followed by intrusion detection making the analysis easier 4. Suggestion for search space reduction and filtering is given	1. Continuous swap space monitoring may reduce cloud service performance 2. Isolation technique for swap space is not defined properly 3. Forcefully or abnormally killed process is a metric of investigation here. But every time forcefully killed or abnormally suspended process may not be malicious 4. There is no suggestion given for data integrity and privacy

(continued)

Table 1 (continued)

Approach	Benefits	Limitation
Forensically enabled IaaS cloud [13]	1. Evidences are stored in another server, and retrieved from it. This provides better security to evidences 2. Forensic tasks are modularized. Hence reducing the complexities of forensic process 3. Event and image reconstruction methods are suggested 4. CSP and consumer access the centralized server for evidences without disturbing main server thus enhancing system performance 5. For security purpose, data integrity is incorporated in the model 6. Incremental backup is suggested for all digital objects in cloud	1. Data analysis techniques are not suggested here 2. In the image acquisition technique, amount of search space may be huge for analysis there by hampering the forensic analysis 3. Incremental backup for all objects in the system affects cloud performance. Some controlling parameters of incremental backup helps to overcome such combinatorial explosion 4. Suggestion for data integrity is given but how to achieve it is not clearly defined

5 Conclusion and Future Scope

Different cloud forensics approach and models are discussed clearly pointing out their advantages and limitations. Based on the limitations some future research areas are identified such as:

- Identification of evidence source: To investigate in cloud environment, we must collect suspicious evidences from the cloud. As investigation is going in IaaS architecture (Virtualized environment), logical to physical resource mapping is difficult. First, we have to identify the resources or part of resource where the malicious event has occurred. According to the type of evidence and type of attack, an algorithm can be designed to identify proper source of evidence.
- Collection or extraction of evidence from source: When source of evidences is identified, another challenge is extraction of proper information from identified source. Generally, VMI techniques are used to extract data from identified sources. Advancement on general VMI technique can be done using memory, I/O and system call introspection. Genetic algorithm and other optimization techniques can be used to enhance the performance of VMI technique.
- Evidence analysis: Analysis should be done in optimized way so that it will take minimum time and will not hamper cloud service performance. Search space reduction, filtering, searching, and other techniques can be used here. An effective evidence analysis technique can be developed with data clustering and classification algorithms. It can reduce search space to optimize the time of forensic investigation.

- Evidence isolation and preservation with data integrity, privacy and authenticity: Evidences are very sensitive even before and after it is produced in court of law. So a well-secured data preservation technique can be developed using advanced encryption techniques such as quantum and Boolean cryptography.
- Intrusion detection as a trigger: A trigger for forensic data collection can be designed using advanced intrusion detection technique. Fuzzy association rules and artificial neural network can be used in the advancement of the intrusion detection system.

By analyzing the limitations in the earlier forensic techniques, future scopes have been identified. With enhancements in these areas, a strong cloud forensic framework can be designed.

References

1. Reilly, D., Wren, C., & Berry, T.: Cloud computing: Forensic challenges for law enforcement. In Internet Technology and Secured Transactions (ICITST), 1–7 (2010).
2. Zawoad, S., & Hasan, R.: Cloud forensics: a meta-study of challenges, approaches, and open problems. arXiv preprint arXiv:1302.6312 (2013).
3. Saxena, A., Shrivastava, G., & Sharma, K.: Forensic investigation in cloud computing environment. In The International Journal of forensic computer science, 2, 64–74 (2012).
4. Ruan, K., Carthy, J., Kechadi, T., & Crosbie, M. Cloud forensics. In Advances in digital forensics VIII Springer Berlin Heidelberg 35–46 (2011).
5. Zargari, S., & Benford, D.: Cloud forensics: Concepts, issues, and challenges. In Third International Conference on Emerging Intelligent Data and Web Technologies 236–243 (2012).
6. Meera, G., Kumar Raju Alluri, B. K. S. P., Powar, D., & Geethakumari, G.: A strategy for enabling forensic investigation in cloud IaaS. In Electrical, Computer and Communication Technologies (ICECCT), IEEE International Conference 1–5 (2015).
7. Rani, D. R., & Geethakumari, G.: An efficient approach to forensic investigation in cloud using VM snapshots. In Pervasive Computing (ICPC), International Conference 1–5 (2015).
8. Dykstra, J., & Sherman, A. T.: Acquiring forensic evidence from infrastructure-as-a-service cloud computing: Exploring and evaluating tools, trust, and techniques. Digital Investigation, 9, 90–98 (2012).
9. Belorkar, A., & Geethakumari, G.: Regeneration of events using system snapshots for cloud forensic analysis. In India Conference (INDICON), Annual IEEE 1–4 (2011).
10. Muniswamy-Reddy, K. K., & Seltzer, M.: Provenance as first class cloud data. ACM SIGOPS Operating Systems Review, 43(4), 11–16 (2010).
11. Delport, W., Köhn, M., & Olivier, M. S.: Isolating a cloud instance for a digital forensic investigation. In ISSA (2011).
12. Kumar Alluri, B. K. S. P., & Geethakumari, G.: A digital forensic model for introspection of virtual machines in cloud computing. In Signal Processing, Informatics, Communication and Energy Systems (SPICES), 2015 IEEE International Conference 1–5 (2015).
13. Alqahtany, S., Clarke, N., Furnell, S., & Reich, C.: A forensically-enabled IaaS cloud computing architecture (2014).

Cloud Detection: A Systematic Review and Evaluation

Harinder Kaur and Neelofar Sohi

Abstract Cloud detection plays an important role in numerous remote sensing applications and in meteorological research. It is the process of systematic consideration of individual pixels, in order to classify them either as cloud or sky element. Many state-of-the-art methods are available for cloud detection but there is lack of universal method which can be used for all types of clouds. The biggest challenge in detecting clouds is non-uniform illumination, broken, and thin (cirrus) clouds. This paper presents a systematic literature review of cloud detection presenting the basic concepts, challenges, research areas, and available methods with their research gaps. Performance evaluation of state-of-the-art cloud detection methods is performed both subjectively and objectively using parameters like Correlation Coefficient, Mean Square Error, Precision, Recall, and F-measure. Experimental results demonstrate that Two-Step Segmentation and Hybrid Thresholding Algorithm (HYTA) algorithms attain highest performance for cloud detection.

Keywords Cloud detection · Otsu · Fixed Thresholding · Mean and Hybrid Method · Hybrid Thresholding (HYTA) · Local Thresholding and Two Step Segmentation

1 Introduction and Motivation

Methodically assessing the weather images comprises cloud classification, cloud detection, cloud brokenness, cloud cover, and so on. But detection of cloud is the prerequisite for numerous remote sensing applications. Cloud detection is the

H. Kaur (✉) · N. Sohi
Department of Computer Engineering, Punjabi University,
Patiala 147002, Punjab, India
e-mail: narain1293@gmail.com

N. Sohi
e-mail: sohi_ce@yahoo.co.in

© Springer Nature Singapore Pte Ltd. 2018
S.K. Bhatia et al. (eds.), *Advances in Computer and Computational Sciences*,
Advances in Intelligent Systems and Computing 554,
https://doi.org/10.1007/978-981-10-3773-3_21

process of systematic consideration of individual pixels, in order to classify them either as cloud or sky element. Cloud Properties such as obscure contour, illumination variations, brightness, color, texture reduces the feasibility to determine clouds manually; pixels may be misclassified with manual methods. So, Automatic techniques to detect clouds plays illimitable role in cloud research.

Thresholding techniques are often proven as a benchmark to distinguish cloud and sky. Shang Liu, Zhong Zhang, Baihua Xiao (2015) et al. come up with the Superpixel Segmentation and Automatic Graph Cut Algorithm, which are milestone for cloud field research. Also, continuous increased proportion of research in this field emanates the need for systematic literature review as well as critical evaluation of techniques. This paper reports the same and helps to identify basic concepts as well as key areas of research on segmentation of weather images for cloud detection.

1.1 Motivation for Work

- Despite the vast amount of review work exists for segmentation methods but after assaying the work, lack of systematic literature review and performance evaluation of existing techniques for segmentation of weather images for cloud detection is realized.
- It will explore the research gaps and statistical knowledge for future researches.

This paper is organized as follows: Sect. 2 describes the background, basic methods of segmentation. Section 3 reports the key research areas, research gaps, and review of existing techniques for unfamiliar readers. In Sect. 4 we monitor the evaluation and comparison studies. Section 5 discusses the findings and Sect. 6 concludes the evaluation results and future work for further researches.

2 Background

To interpret an image automatically, segmentation divides the image into regions on the basis of common characteristics such as color, texture, etc., and extract the homogeneous region of interest for future analysis [1]. As cloud is a region of interest in weather images, segmentation is consummated to detect the cloud.

2.1 Various Segmentation Methods

Thresholding Method. Pixel position (i, j) will be binarized with optimal threshold value, which is computed either manually or automatically [1]. Selection of optimum threshold value based on Histogram method is as given below (Fig. 1):

Edge-Based Segmentation. Edges are the points in an image where intensity changes abruptly; these edges are then connected to form object's outline [2]. The result will be transformed to segmented image by filling the object boundary.

Region Growing Method. It is the process to build larger regions from sub-regions. Select the starting seed and annex the neighborhood pixels on the basis of homogeneity, i.e., defined similarity properties [3].

Clustering. It is unsupervised learning technique which defines similarity criteria to classify pixels into clusters [2]. Different clustering methods are K-means Clustering, Hard Clustering, and Fuzzy Clustering [4] and so on.

2.2 Types of Cloud

Clouds can be classified based on their shape, altitude, intensity level and so on. Generally, there are ten types of clouds according to International Cloud Classification System (ICCS) [5], but they can be merged into three fundamental classes as shown:

Cumulus Clouds. They are low level, puffy clouds with sharp contour and distinguished intensity levels of sky and cloud pixels [5] (Fig. 2).

Stratus Clouds. Clouds having sheet like layering structure or shape are stratus clouds. They occur at low level altitude.

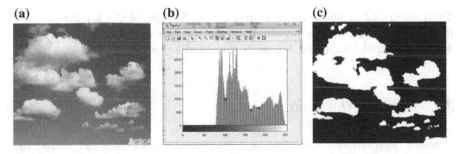

Fig. 1 **a** Original image of cloud. **b** Histogram showing the candidate threshold values with *red dots*. **c** Segmented binary image with optimal threshold value among candidate values

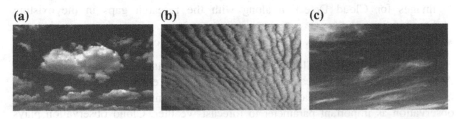

Fig. 2 Examples of **a** Cumulus cloud, **b** Stratus cloud, **c** Cirrus cloud (Thin cloud)

Cirrus Clouds. They are also known as thin clouds. Cirrus clouds are having little variation in intensity of sky and cloud elements; thus, they are difficult to recognize.

2.3 Why Segmentation

Analyzing Weather images is a continuous, data intensive, multidimensional, and turbid process [6]. Challenges and issues encountered in this errand are listed here

- Complex and large dataset in weather images makes cloud detection chaotic and difficult [7].
- Highly tedious mathematical computations are involved [6].
- Cloud is the dominant part of weather image and primary obstacle in analysis due to uncertainty in shape or structure.

Segmentation restricts the surroundings to reliably extract only that region, which is to be assessed further; hence it reduces the large data and computation complexity.

3 Review Method

The pathway taken into account to develop the Review Model for the literature modeling is: Conduct the organized review of existing techniques for cloud detection, pinpoint the research gaps of study and key areas of research.

3.1 Research Questions

Research questions are the building blocks for the scientists to plan and conduct any research; therefore it is fundamental to frame such questions. Key questions encountered during our study are listed below

- What are existing techniques, tools, algorithms for segmentation of weather images for Cloud Detection along with the research gaps in the existing literature?
- Which are the key areas of research in the field of cloud detection?

Key Research Areas. Clouds have immense importance in hydrological cycle. Along with the microphysical properties like pressure, reflexivity, wind speed, etc., meteorological weather radar Doppler stations also considers ground based cloud observation as important parameter to forecast weather. Cloud observation plays

Table 1 Cloud detection techniques

Study	Technique	Principle	Research gaps
Liu S, et al. (2015)	Automatic Graph Cut (AGC)	AGC detects cloud in two stages: 1. Automatically acquiring the Hard Constraints by using Otsu thresholding and by maximizing the variance between cloud and sky. 2. These hard constraints will act as input to the graph model. The cost function E (L) for each segmentation is defined as: $E(L) = \lambda \cdot R(L) + B(L)$ Where $\lambda \geq 0$, specifies the relative importance between the region, cost R (L) and the boundary cost B (L). The min-cut/max flow algorithm in is applied for solving the above equation [12]	Aerosol optical depth (AOD) is not considered while detecting cloud elements, which plays vital role in visibility. Therefore, this method is not well suited under low visibility conditions
Lee I and Mahmood MT (2015)	Two Step Segmentation	It provides the combination of fixed and adaptive thresholding to detect thin clouds as well. In this method global initial threshold T for thick clouds, is computed by averaging the intensity values of the whole image. Local threshold value for Thin clouds is given as follows: **Step1:** Local contrast C_{local} for the P_{center} within local window $U(x, y)$.is computed as: $C_{local} = (U_{max} - U_{min})$ **Step2:** User provided T'' If $C_{local} \geq T''$ then $T_{local} = (U_{max} + U_{min})/2$ Else $T_{local} = T$ **Step3:** T_{local} acts as a local threshold value for pixel P_{center}. If T_{local} is greater than pixel position (i, j) then P_{center}.is considered to be cloudless otherwise cloudy [13]	Do not work well for low visibility conditions

(continued)

Table 1 (continued)

Study	Technique	Principle	Research gaps
Liu S, et al. (2015)	Superpixel Segmentation	Superpixel Segmentation (SPS) Algorithm partitions the image on the basis of texture similarity, brightness similarity and contour continuity; which forms irregular image blocks, each block is known as superpixel. Normal Cut Algorithm is used to determine local threshold value for each superpixel. By interpolating the all local threshold values of superpixels local threshold matrix will obtain. Finally, Cloud is determined by comparing the featured image with threshold matrix pixel by pixel [14]	
Dev S, et al. (2014)	Systematic Study of Color Spaces and Components	This method addresses the evaluation of color models for segmentation of sky/cloud images. This mainly consider the PCA technique, Clustering and color components for analysis such as RGB, HSV, YIO, CIE L*a*b* [8]	This method considers only color components to determine homogeneity. Properties such as texture and contour etc. can be considered for better accuracy
Ramesh B, et al. (2013)	Based on Mean and Hybrid Methods (Hybrid)	Proposed method find the column wise and row wise mean for input image. This method works well for both high and low brightness clouds with preserving the quality of an image. Firstly, Each pixel is compared with the column wise mean value. Eliminate such pixels of image whose value is less than column mean. Also, remaining pixels are compared with row wise mean to detect clouds [15]	Preserves quality of image but it is not able to detect thin clouds. Also, This method shows inaccuracy in cloud detection
Li Q, et al. (2011)	Hybrid Thresholding Algorithm (HYTA)	Unimodal images consist of only either sky or cloud. Bimodal images comprises both sky and cloud	It is not well suited for real world applications as well as does not shows

(continued)

Table 1 (continued)

Study	Technique	Principle	Research gaps
		elements. First Step in HYTA is to detect weather image is unimodal or bimodal. To accomplish this objective, compute standard deviation; if it is greater than $T_s = 0.03$ then image is considered as bimodal otherwise unimodal. Fixed Thresholding and MCE cross thresholding techniques are applied for unimodal and bimodal images respectively [11]	significant results to detect Cirriform or Cirrus clouds
Huo J and Luo D (2009)	Cloud Determination Under Low Visibility	Sky color and visibility features are directly proportional to the Aerosol optical depth (AOD) parameter. It will strongly effect the detection of cloud pixels; ignorance may lead to misclassification. So, This method considers the Aerosol Optical Depth Parameter. First, Fast Fourier Transform (FFT) is applied to determine the homogeneity of image. Histogram of the B/R ratio is used to set the threshold value and standard deviation and Edge searching method are used to determine final threshold [16]	Although, this works under low visibility, but it is not effective approach for thin cloud detection
Kreuter A, et al. (2009)	Fixed Thresholding	Fixed Thresholding uses the single threshold value for entire image to detect cloud. Long et al. sets R/B threshold to 0.6. Pixel value greater than this value will considered as cloud otherwise sky. Further this criterion was changed to R-B = 30 [5]	This can only work for thick clouds

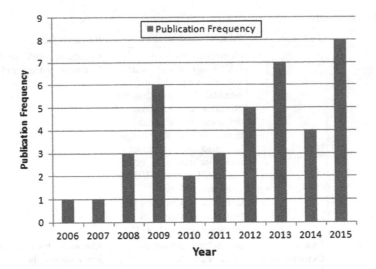

Fig. 3 Time based count of publications

important role in solar irradiation measurement, climate modeling, and analysis of signal attenuation in satellite [8]. Suitable planning of farm in agriculture is directly affected by the forecasting of weather. Weather warnings are important forecasts for applications such as rainfall prediction, flood, storm, hurricane detection, and various other remote sensing applications. Table 1 reports the recent work on cloud detection techniques.

3.2 Systematic Count of Publications

Our studies examine the critical time based count of cloud detection approaches. It also covers the cloud removal techniques applied on different aspects of satellite imagery such as MODIS satellite imagery, Multitemporal, LandSat Satellite imagery, and so on. Organized Frequency of publications is provided in order to help the future researchers with statistical results (Fig. 3).

4 Evaluation of Cloud Detection Techniques

This section monitors the evaluation trends, perform corresponding qualitative and quantitative evaluation.

4.1 Evaluation Trends

Quantitative Evaluation: It is the systematic computation and empirical investigation of statistical metrics for cloud detection techniques [9]. Selected metrics for critical evaluation of techniques are correlation coefficient, mean square error; entropy and application specific metrics are precision, recall, and F-score.

Qualitative Evaluation: It is the realization of cloud detection techniques' outcomes based on human observations. We figure out the performance for three types of basic clouds; Cirrus, Cumulus, and Stratus Clouds in Sect. 5.

4.2 Evaluation Metrics

Traditional framework of evaluation consist of correlation coefficient, mean square error but in the era of image pattern recognition and machine learning, parameters such as Precision, Recall, and F-score are fundamental for cloud specific applications. In accordance with the above facts, these five criterions give the satisfactory provision towards scrutinize of the results and comparison studies.

Correlation Coefficient: In 1895, Karl Pearson comes up with the product-moment correlation coefficient. It epitomizes degree of correlation and linear dependence exists between measured quantities [10]; hence depicts how strong the relationship is. It returns value lies between −1 and 1, value above 0.5 is considered as a candidate of better performance.

Mean Square Error (MSE) measures the cumulative means of squares of error between the estimator, i.e., original image and what is estimated. It signifies the squared error loss in computations.

Possible Outcomes of Confusion Matrix are True Positive (TP), False Positive (FP), True Negative (TN), False Negative (FN); TN and TP are correctly segmented cloud and sky pixels, FP pixels are incorrectly detected as cloud and FN are falsely

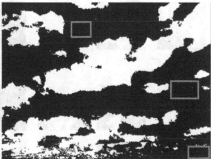

Fig. 4 Example of Recall, *Red squares* are depicting misclassified false negative pixels (Matlab 7.10.0)

detected as sky element [11]. Precision, Recall, and F-measure are derived from confusion matrix.

Precision measures the exactitude of quality, i.e., fragments of fetched pixel elements that are admissible. It is Type 2 Error. Mathematically, $Pr = (TP)/(TP + FP)$.

Recall measures the completeness of quantity which means fragment of admissible pixel elements that are fetched. It is also known as sensitivity or Type 1 Error. Figure 4 shows the pixels that are falsely classified as sky but in actual weather image they are cloud. Such measurement is performed in Recall. Mathematically, $Rc = (TP)/(TP + FN)$.

F-measure is measurement of accuracy which is computed in terms of precision and recall. It provides trade-off between precision and recall. Mathematically,

$$F - measure = \frac{2 \cdot recall \cdot precision}{recall + precision}$$

4.3 Experimental Results

To perform evaluation and comparison studies of above-mentioned methods, experiments are set up in MATLAB 7.10.0 (R2010a) on i3 Processor. Segmentation

Table 2 Experimental results for cloud detection techniques

Cumulus Clouds					
Otsu	Fixed	Hybrid	HYTA	Local	Two Step Segmentation

Stratus Clouds					

Cirrus Clouds					

Table 3 Quantitative results for various methods

		Otsu	Fixed	Hybrid	HYTA	Local	Two step
Correlation coefficient	Cumlus	0.8172	0.555	−0.75	0.91246	0.8320	0.882
	Stratus	0.8518	0.167	−0.53	0.667	0.5962	0.777
	Cirrus	0.835	0.4221	−0.56	0.3993	0.59	0.62
Mean square error	Cumlus	2.037e+004	1.632e+004	2.23e+004	1.685e+004	1.005e+004	0.2504
	Stratus	2.077e+004	2.17e+004	2.22e+004	2.180e+004	0.1835	0.1290
	Cirrus	2.163e+004	8.33e+004	1.84e+004	8.99e+003	0.236	0.2268
Precision	Cumlus	0.7779	0.9386	0.082	0.8332	0.7212	0.799735
	Stratus	0.8943	0.98	0.315	0.6945	0.6779	0.8361
	Cirrus	0.70	0.7112	0.02	0.60	0.43	0.54
Recall	Cumlus	0.7874	0.5868	0.1886	0.7598	0.7831	0.830
	Stratus	0.7114	0.563	0.1560	0.7617	0.7584	0.722
	Cirrus	0.6269	0.51	0.12	0.52	0.78	0.78
F-measure	Cumlus	0.7776	0.7227	0.150	0.79316	0.7311	0.821
	Stratus	0.7883	0.6546	0.078	0.71034	0.6823	0.767
	Cirrus	0.6198	0.5532	0.30	0.511	0.53	0.64

results of few images from the tested image set are presented for both subjective and objective evaluation. The tested image set consists of images with cumulus, stratus, and cirrus clouds to evaluate suitability and universality of algorithms.

Subjective Evaluation: Human observations are primary channel for evaluation in meteorological research Graphical Results of experiments are presented in Table 2.

Objective Evaluation: Objective assessment provides the justification and eliminates inaccuracy of human observations based results Objective evaluation of algorithms is done using performance parameters presented in Sect. 4.1. Numerical results are presented in Table 3.

5 Discussion

Non-Uniform illumination, broken and thin clouds are primary challenges in cloud detection. Algorithm may perform well for one type of cloud but may not suitable for other cloud type. So, studied algorithms are tested on cumulus, stratus, and cirrus clouds (thin clouds) to evaluate their universality. Experimental results in Table 2 demonstrate that all algorithms work good for cumulus clouds. Otsu and Two-Step Segmentation deliver best results for stratus clouds followed by HYTA algorithm. Although Local thresholding algorithms also behave better for stratus clouds but it consumes larger time for segmentation. Fixed Thresholding and

Hybrid algorithms are not adaptable for stratus clouds. For Cirrus Clouds, Two-Step Segmentation performs the best among all. Otsu and HYTA also achieves good results for some images. Recognition of local, fixed thresholding, and hybrid algorithms scaled down for cirrus clouds.

Furthermore, various conclusions can be drawn from quantitative results represented in Table 3. First, Correlation Coefficient values of Two-Step Segmentation, HYTA, and local thresholding algorithms that they are good candidate to detect cumulus clouds with least mean square error. Second, Precision and recall must be balanced to give linear precision–recall curve. Results illustrate that Otsu and Two-Step Segmentation algorithms gives best segmentation results for stratus clouds. F-measure is the way to compensate the trade-off between precision and recall. Two-Step Segmentation and HYTA techniques gives uniform and stabilized precision and recall values. All the methods give poor results for calculated parameters for cirrus clouds (thin clouds).

6 Conclusion

In this study, weather image segmentation techniques for detecting clouds are reviewed presenting their achievements and limitations. Experimental results for Performance evaluation of the prominent techniques are presented. Experimental Results and discussion signifies that Two-Step Segmentation and HYTA algorithms attains highest performance for cloud detection. One of the challenges for future researchers is to achieve more accurate results for thin or cirrus clouds.

References

1. R.C Gonzalez and R.E. Woods, Digital Image Processing, 3rd ed., India: Dorling Kindersley, 2008, pp. 689–743
2. Kandwal, R., Kumar, A. and Bhargava, S.: Review: Existing image Segmentation Techniques. International Journal of Advanced Research in Computer Science and Software Engineering, vol. 4, pp. 153–156 (2014)
3. Kumar, J., Dr. Kumar, G.V.S.R and Kumar, R.V.: Review on Image Segmentation Techniques. International Journal of Scientific Research Engineering & Technology, vol. 3, pp. 992–997 (2014)
4. Sohi, N., Kaur, L. and Gupta, S.: Performance Improvement of Fuzzy c-mean Algorithm for Tumor Extraction in MR brain images. International Journal of Computer Applications (IJCA), vol. 59 (2012)
5. Heinle, A., Macke, A. and Srivastav, A.: Automatic cloud classification of whole sky images. Atmospheric Measurement Techniques, vol. 3, pp. 557–567 (2010)
6. Dr. Roomi, M.M., Bhargavi, R. and Banu, T.M.H.R.: Automatic Identification of Cloud Cover Regions Using Surf. International Journal of Computer Science, Engineering and Information Technology, vol. 2, pp. 159–175 (2012)

7. Kaur, A. and Dr. Randhawa, Y.: Image Segmentation Using Modified K-means Algorithm and JSEG Method. International Journal of Engineering and Computer Science, vol. 3, pp. 6760–6766 (2014)
8. Dev, S., Lee, Y.H. and Winkler, S.: Systematic Study of Color Spaces and Components for the Segmentation of Sky/Cloud Images. In: IEEE ICIP, pp. 5102–5106 (2014)
9. Quantitative Research, https://en.wikipedia.org/wiki/Quantitative_sresearch. [Accessed: 15-Jun-2016]
10. Kaur, A., Kaur, L. and Gupta, S.: Image Recognition using Coefficient of Correlation and Structural Similarity Index in Uncontrolled Environment. International Journal of Computer Applications, vol. 59, pp. 32–39 (2012)
11. Li, Q., Lu, W and Yang, J.: A Hybrid Thresholding Algorithm for Cloud Detection on Ground-Based Color Images. Journal of atmospheric and oceanic technology, vol. 28, pp. 1286–1296 (2011)
12. Liu, S., Zhang, Z., Xiao, B and Cao, X.: Ground Based Cloud Detection Using Automatic Graph Cut. In: IEEE Geoscience and remote sensing letters, vol. 12, pp. 1342–1346 (2015)
13. Lee, I.H. and Mahmood, M.T.: Robust Registration of Cloudy Satellite Images using Two Step Segmentation. In: IEEE Geoscience And Remote Sensing Letters, vol. 12, pp. 1121–1125 (2015)
14. Liu, S., Zhang, L., Zhang, Z., Wang, C., and Xiao, B.: Automatic Cloud Detection for All-Sky Images Using Superpixel Segmentation. In: IEEE Geoscience And Remote Sensing Letters, vol. 12, pp. 354–358 (2015)
15. Ramesh, B. and Dr. Kumar, J.S.: Cloud Detection and Removal Algorithm Based on Mean and Hybrid Methods. International Journal of Computing Algorithm, vol. 2, pp. 121–126, (2013)
16. Huo, J. and Lu, D.: Cloud Determination of All-Sky Images under Low-Visibility Conditions. Journal of atmospheric and oceanic technology, vol. 26, pp. 2172–2181 (2009)

Sentiment Classification for Chinese Micro-blog Based on the Extension of Network Terms Feature

Fei Ye

Abstract Sentiment analysis is widely used in product reviews, movie reviews, and micro-blog reviews. Micro-blog review is different from general commodity or movie reviews, which often contains the user's randomness and lots of network terms. So the micro-blog reviews emotional analysis is not a small challenge. Network terms generally express strong emotions or the user's point of view, the traditional bag words model and machine learning method do not use the network terms features. In the face of ever-changing micro-blog reviews manifestations, forecast accuracy may be affected. Therefore, in this paper our study focuses on the micro-blog emotional analysis through the extended network terms features and integration with other features. We are taking experiments to compare prediction performance under the different feature fusions, to find out which feature fusion can get the best results. Our results show that by the extended network term feature integration with other features ways to improve the accuracy of predictions, especially some of the most popular micro-blog reviews.

Keywords Text sentiment classification · Support vector machine · Machine learning

1 Introduction

With the development of Internet, network online user reviews have become an important way people express opinions and disseminate information. The reviews are shared, real-time, and interactive. The dissemination of diversification reflects the profound degree of public preference for a certain type of theme or a certain commodity. Reviews of one kind of topic have become an important reference on the choosing of purchase goods. Therefore, emotion analyses of reviews gradually become a hot research direction. Micro-blog is a relationship based on user

F. Ye (✉)
Southwest Jiaotong University, Chengdu, China
e-mail: 122404504@qq.com

© Springer Nature Singapore Pte Ltd. 2018 231
S.K. Bhatia et al. (eds.), *Advances in Computer and Computational Sciences*,
Advances in Intelligent Systems and Computing 554,
https://doi.org/10.1007/978-981-10-3773-3_22

information sharing, dissemination, and access platforms. Analyzing the sentiment tendency of micro-blog is very useful to company. You can get the user's attention on a topic by studying these reviews, as well as the degree of preference for a movie or a commodity. Companies can use them to study consumer satisfaction with the products, so as an important reason for product improvement [1]. The sentiment classification of the micro-blog reviews can also predict the future fluctuations in stock prices [2]. Governments can use them to survey public opinion.

In this article, the main focus of our study is the importance of networking terms in micro-blog reviews to judge the emotional tendencies, and analysis on the relationship between network terms and other emotional words.

The contributions of this article are as follows:

- Propose the optimization method of word segmentation and use word2vec tool to extend the original lexicon, and to construct expansion lexicon.
- By collecting popular online language server as the lexicon of network terms, and to extract network term features according to the network term appearing in the sentence.
- Generate the vector of every word of sentences by word2vec tools, and these word vectors by weight calculations and combinations to represent the sentence feature.

2 Related Work

Sentiment classification is divided into two methods, the first lexicon-based approach, and the other is based on machine learning methods. For the lexicon method, it usually takes a huge sentiment how to create library lexicon, and according to each micro-blog appear sentiment vocabulary to calculate sentiment. For example, Pak [3] constructed a sentiment classifier through Twitter data, and applied it to sentiment classification of documents, experiment results show that this method improves the prediction accuracy than previous methods. Augustyniak [4] proposed a method to automatically create sentiment lexicon and were compared with the traditional lexicon method, experiments show that the new method significantly improves the accuracy. Jun [5] creates a Chinese lexicon library through improved SO-PMI algorithm. This lexicon is used to carry out the reviews of hotel for the sentiment classification. Kon-stanty [6] proposed a lexicon based on fine-grained to Polish text sentiment classification. Hui [7] proposed a method for solving the determination of the sentiment word tendencies and calculate their sentiment weight. The method is mainly based on the conditional selection from a word synonyms and antonyms to serve as seed. Then directed to measure the weight of each word in the sentiment semantics, it is mainly used in the method of semantic similarity values to determine the similarity between the words.

Machine learning approach to text sentiment classification has become a hot research direction. General machine learning method divided into two categories, one of which is a supervised learning method, the other is unsupervised learning methods. Currently there are supervised learning method based on naive Bayes, maximum entropy, and support vector machine (SVM). Zheng [8] proposed a new method to select features and calculate the weights. Then use of these three classifiers were trained and tested. Thus the results for the three classifiers were evaluated. Zhang [9] provides a survey and analysis of the effect on sentiment classification, respectively, in machine learning and lexicon methods. The results also show that support vector machines (SVM) and Naive Bayes method has higher accuracy in the supervised learning method, Meanwhile lexicon-based method also has some competition because it does not advance human well marked each text category. And its prediction effects are less susceptible by a different training set. In addition to supervised and unsupervised learning methods, there are semi-supervised learning methods for sentiment classification, for example Shusen [10] proposed a new semi-supervised learning method known as AND. It is mainly used to solve supervised learning with the condition of insufficient training set. AND architecture is based on a small amount marked reviews that under the semi-supervised learning and a lot of unmarked reviews that under the unsupervised learning. The method is particularly suitable for only a few marked reviews and a large number of unmarked reviews at the same time. It can be said semi-supervised learning method has gradually become an increasingly hot research field. Zhai [11] used different types of features to enhance sentiment classification performance in Chinese texts, and several groups of experiments indicated that effective features are usually at varying lengths rather than fixed lengths and play an important role in the improvement of sentiment classification performance. Zhang [12] employed word2vec tool to cluster the similar features for obtaining the semantic features. And then, it used the SVMperf for training and testing. The experiment was constructed on the Chinese reviews on clothing products and the results indicated that the method had achieved superior performance in Chinese sentiment classification.

3 Method Overview

In the process of micro-blog development, network terms are always accompanied by the development of new words, and these words often express author's strong emotions. The micro-blog reviews are unlike product reviews or movie reviews and they have often more direct expressions and tend to use more network terms. So our main job is to study how to use the network terms and combine other features on the micro-blog sentiment analysis. Our approach is divided into three steps. The first step is to build a network term lexicon and extract emotional features from micro-blog reviews by some rules. The second step is using the proposed optimized algorithm of word segmentation for text pretreatment, and use word2vec tool to expand sentiment lexicon. The third step is using word2vec tool to generate word vector of each word. The sentence feature is represented by calculated and combined word vectors. The last step is to generate a uniform feature profiles through different combined of features set, and using the support vector machines for training and testing.

3.1 Construction of Network Terms Lexicon

We collect the popularity of network terms from Sina micro-blog in recent years. We ultimately retained the 100 common network terms that were added to the lexicon by heat and usage frequency.

For network terms, it is evolving as a new thing. It has often been difficult to accurately determine its polarity through the word in context and word segmentation. Here we rely on an artificial way to label each network terms given a polarity (Table 1).

3.2 Word Segmentation Operation and Optimization

We use Southwest Jiao Tong University Chinese word segmentation system for word segmentation operation. The system not only can segment words for Chinese texts but also tag the POS of words. After the word segmentation operations, the meaninglessness words and punctuation are removed.

Through the word segmentation operation, we find some phrases are able to express the author's feelings. For example "非常美丽" ("very beautiful"), "很好" ("very good"). However, if some long phrases are separated into several words, the original polarity will change. For example, "不麻烦" ("No trouble"),"不太好" ("not very good") are split into "不" ("not") and "麻烦" ("trouble") and "不" ("not") and "太好" ("very good"). Therefore, we need to set some rules to optimize word segmentation. We define some labels as POS of words which are lists in Table 2. Adverbials usually consist of the one of individual character or the two of individual character. It is used to modify adjectives usually located in front of the adjective. For example, "不太好" ("not very good") consists of the two of individual character and "很好" ("very good") consists of the one of individual character. Conjunctions include "and," " as well as," "or" and so on, it may consist of

Table 1 Some examples of network language

Network terms	Examples
也是醉了	哇, 这大神的技术, 我也是醉了
然并卵	你数学足够好? 然并卵, 只要你碰到葛军
你妹的	你妹的, 这东西真不好用
有钱, 就是任性	复旦大学食堂居然出了辣条炒饭, 有钱, 就是任性!
给力	这就是天竺吗? 不给力啊老湿

the one of individual character or the two of individual character. For example, "并且" ("in additon") consists of two individual characters and "和" ("and") consists of one individual character.

The words are separated which will become a new phrase by some combination of rules. These new phrase can express correctly the author's feelings. These rules are shown in Table 3.

The first rule is very clear that adverb used to modify sentiment degree of adjectives. And if it is sentiment directional adverbial, then adjective's polarity will change. In the third rule, the intermediate directional word can change the polarity of the back of the word, but only the first word of the phrase to modify the sentence. Rules 3 and 4, conjunctions combine two phrases together, and if the conjunctions "and," "as well as," etc., both sides of the word is the same polarity. If the conjunction"or" both sides of the word is the opposite polarity.

Table 2 Word POS acronym definition

POS shorthand	Meaning expression
D	It represents a single adverbial
DO	A single adverbial having emotional direction
DD	Adverbial by the two words
A	Adjectival
J	Conjunctions
V	Verb

Table 3 Word combination rules

Combination of words	Example
(D, DD, DO) + (A, V)	很好, 非常好 (Very good)
DO + D + A	不太好 (Not very good)
D + DO + A	很不错 (Pretty good)
A + J + A	好看又好吃, 美丽或丑陋 (Good-looking and good to eat)
(A, V) + J + D + (A, V)	难看又不好吃, 不懂还装懂 (Ugly and not good to eat)

Table 4 Features of the fusion

Various combinations	Explain
dic + sen	lexicon feature and sentence feature
dic + net	lexicon feature and network terms
dic + net + sen	lexicon feature, network terms feature, and sentence feature

3.3 Extended Lexicon and Creation of Sentence Feature

Step 1: Use the Chinese word segmentation system of Southwest Jiao Tong University for word segmentation, and get part of speech of each word. We select some word as the word seed in accordance with Table 3 rules generally adjectives, verbs, and conjunctions. Select the word around the seed and its continuous-word phrases according to the rules in Table 3, and then calculate its polarity. Polarity calculation methods we use PMI method.

In order to verify the reasonableness of the new words, we put the word with the original micro-blog content that no word segmentation to match. If the match is correct that it is present in the micro-blog content, but also that it is reasonable. We can add new words into the original lexicon according to these rules.

Step 2: We use word2Vec tool, word2vec is open source tool form the Google based on the depth of learning. It will be taking word converted into a vector, and can calculate the distance between the vectors to represent the similarity of two word semantics. In this article, we can use word2Vec distance command to get the nearest word list corresponding to each word. Thus get more words as an extended lexicon. The lexicon as a candidate feature set. Specific operation of the tool is as follows:

We use the micro-blog content that have been divided serve as input text of word2Vecc and then trained. Its output text contains the vector of each word. So that we can get the vector of each word and the distance between words. Word2vec Training commands in Linux are:

./word2Vec –train myTrain.txt –output vectors.bin –chow 0 –size 200 –window 5 –negative 0 –hs 1 –sample le-3 –threads 12 –binary 1.

Here we need to be concerned mainly parameters, it includes size, window. The parameter size represent dimension vector size of each word. The parameter window represents training window size of each word. For example, "window 5" training window range in the first five words and five words after the word.

Step3: We constructed sentences feature by the vector of each word in the sentence. It is represented by the following formula sentence feature vector.

$$s_i = \sum_{i=n}^{n} \left(\sum_{j=n_i}^{n_i} W_j \right) /n \tag{1}$$

s_i represents the ith sentence feature, n represents the number of vectors word of the sentence, n_i represents the dimension of each word vector, w_i represents jth item of every word vector. Sentence feature is sum of all the word vectors and then averaged.

3.4 Fusion Sentiment Features

Feature fusion is to extract features from different angles, and in some way or with the calculating methods to be combined form to a new feature file. Before the operation we extract feature selection for sentiment classification advantageous feature items. It is collected and screened from three aspects. These three features are the lexicon feature, the network term feature and the feature set that is consisting of feature vector of each word of in-depth learning by word2vec. Specific feature fusion as follows:

Fusion features may use more attributes as feature items, which can improve the ability to distinguish categories. For example, in the dic + net features, we can get sentiment words and semantic similarity between words server as features, but can not achieve good results in a short review. If only in the dic + net features, it can not achieve good results when facing long review or sentences which has complex structures. Therefore, we consider the integration of three feature set to improve the ability to distinguish categories.

Feature fusion can make good use of semantic features contained in the text under deep learning, also used in the cyberspeak features, which can complement each other which lack effect of a single characteristic on machine learning.

4 Experiment

Through two experiments to validate our method can improve the accuracy of sentiment classification. The first data set is the first natural language processing and Chinese Computing Conference (2012) provided. It contains a total of 20 topics and 17,518 and 31,675 micro-blog sentences. The dataset of this experiment have been manually tagged polar micro-blog reviews. Therefore, we screened 3500 micro-blog reviews with sentiment by random selection as the experimental data set. Wherein the number of positive reviews is 1037, the number of negative reviews is 2463. We use SVM as the classification method, which we use libsvm tool, it is an easy to use and efficient SVM package that Taiwan University professor lin Chih-jen development and design. We use a linear function as the kernel function. Sentence divided is using the Chinese word segmentation system of Southwest Jiao Tong University combined with our optimized word segmentation algorithm. We conducted the extended lexicon by word2vec tools. The primitive emotion lexicon is NTUSD, it collected by National Taiwan University, contains a total of 11,086 words, of which 2810 belong to the positive trends of words, 8276 words belonging to the negative attributes. We gained 1432 new words through distance command of the word2vec tool and added to the original lexicon, then we remove duplicate words, the number of the original lexicon reaches 12,518.

In this paper, the criteria of methods performance evaluation contain the recall, precision and F value.

Table 5 The results of experiment 1

Method		dic + sen	dic + net	dic + sen + net
Our method	Accuracy	75.20%	76.20%	82.50%
	Recall	70.16%	71.10%	78.40%
	F	77.13%	73.56%	80.39%
Lexicon	Accuracy	69.90%	–	–
	Recall	62.67%	–	–
	F	66.09%	–	–

4.1 Experiment 1

In experiment 1, we were selected the 3500 micro-blog reviews as experimental data, micro-blog reviews are 1037 positive reviews, negative reviews is 2463. We randomly selected 2000 contains both positive and negative tendencies micro-blog reviews as a training set, the rest of 1500 as a testing set. In the first experiment, we use the lexicon method as a comparison method, we set the three tests by selected the integration depending on different features of the fusion, respectively, lexicon + sentence features (abbreviated as dic + sen), lexicon + network terms features (abbreviated as dic + net) lexicon + sentence + network terms features (abbreviated as dic + net + sen). These three tests are mainly used to determine which features fusion combination can making the sentiment classification accuracy is maximized. The test results are shown in Table 5.

4.1.1 Analysis of the Results of Experiment 1

The results in Table 4 show that our method of network terms + lexicon + sentence features have highest accuracy rate, indicating that the combination of three feature sets is the best solution. In contrast the accuracy of feature fusion between lexicon and network terms is lower than the accuracy of feature fusion between lexicon and sentence. This is not to say that the importance of low network terms features, but the current data set does not use the latest micro-blog reviews, each review contains less network terms and is not representative of the current Chinese micro-blog prevalent style. Therefore, we have chosen the latest content and expression of a more general micro-blog review data set as the experimental data in experiment 2.

4.2 Experiment 2

We chose the popular Chinese micro-blog reviews as an experimental data set. We randomly selected reviews with sentiment from Sina micro-blog. These reviews

Table 6 The results of experiment 2

Different fusion		200	300	400	500
dic	Precise	66.82%	–	–	–
	Recall	59.20%	–	–	–
	F	62.78%	–	–	–
dic + net	Precise	79.24%	–	–	–
	Recall	74.45%	–	–	–
	F	76.78%	–	–	–
dic + sen	Precise	78.53%	77.32%	75.87%	74.25%
	Recall	73.45%	71.60%	68.23%	68.50%
	F	75.90%	74.35%	71.85%	71.26%
dic + sen + net	Precise	83.71%	82.92%	82.95%	81.56%
	Recall	78.40%	79.65%	76.78%	75.80%
	F	80.97%	81.25%	79.75%	78.57%

total of 5000, in which the positive trends are 2500, 2500 negative tendencies there. Sina micro-blog reviews generally more representative of the prevailing Chinese micro-blog style.

We also set the different feature fusion and sentences feature according to different features dimensions of 200, 300, 400, and 500 were tested 10 times. Observations to see if the fusion of three features set can achieve higher accuracy. The results show in Table 6.

4.2.1 Analysis of the Results of Experiment 2

From the experimental results, our method achieved the best result is 83.71. In this case the sentence dimension 200 is the best result. Lexicon + network terms accuracy is 79.24%. Because there is no features dimension in the sentence. The best accuracy of the combination of lexicon and sentence is 78.53%. The semantic distance between words is also very important for sentiment classification. We observed most of the Chinese micro-blog reviews, many reviews have contained network terms. By results of this experiment found that, the network terms as feature items can increase the effect of sentiment analysis to some extent. The traditional bag of words model were difficult to determine some very short and emotional lexicon word or phrase that does not exist. For example, a micro-blog review, "Oh," which is usually expressed in author's negative side, the bag of words model depends on the size of the lexicon and facing some of the new online language or combination of words which classification results will be poor. When a single bag of words model and network terms together, then that can significantly improved classification results.

But if just a simple integration of network terms and lexicon together as a set of features, then ignores the contributions of other words of the micro-blog review on the article's sentiment. A large number of micro-blog reviews not only contain content network terms, there is also the network terms with the emotional tendencies similar words. For example, a micro-blog review, "你妹的, 这件衣服太难看了," ("Your sister's, this dress is ugly") network terms "你妹的" ("Your sister's") strongly bring out the author feelings at this time. It is similar with the "太难看" ("too ugly") also belongs to negative words, the probability of them together more large than positive word and the network terms mix together. This also shows that the network terms has a more general fixed polarity, especially for applications like the micro-blog content which are usually only 2 sentences constitute a direct expression of micro-blog review.

5 Conclusion

Based on the characteristics of micro-blog reviews, our approach focuses on the expansion of network terms features. In the face of increasingly diverse style of micro-blog comments, coupled with the flexibility of the Chinese language. Through this network terms features combination with other feature of the kind sentiment word can easily be converted into a feature items of machine learning. In this paper we use SVM as a classifier, and achieved certain results. The experimental results indicate the following conclusions:

- This paper proposed extended network terms feature-based approach combined with other features can effectively improve the accuracy of Sentiment classification.
- Network terms feature is particularly suitable for application in the popular Chinese micro-blog reviews, especially some short reviews. These reviews usually contain a large amount of network terms.
- A network terms and with the same polarity sentiment words simultaneously appearing in the Chinese micro-blog reviews have large probability than with the opposite polarity sentiment words.

References

1. Kangale A, Kumar S K, Naeem M A, et al.: Mining consumer reviews to generate ratings of different product attributes while producing feature-based review-summary. International Journal of Systems Science, 1–15, 2015.
2. Nguyen T H, Shirai K, Velcin J: Sentiment analysis on social media for stock movement prediction. Expert Systems with Applications, 42(24): 9603–9611, 2015.
3. Pak A, Paroubek P: Twitter as a Corpus for Sentiment Analysis and Opinion Mining. In: The International Conference on Language Resources and Evaluation. Malta: Springer Berlin Heidelberg, 2010: 1320–1326.

4. Łukasz Augustyniak, Piotr Szymański, Tomasz Kajdanowicz, et al. Fast and Accurate - Improving Lexicon-Based Sentiment Classification with an Ensemble Methods. In: Asian Conference on Intelligent Information and Database Systems, Da Nang, Vietnam: Springer Berlin Heidelberg, 2016, 14–16.
5. Jun Zhang, Zhijian Wang, Shuren Zhu: Research on building a Chinese sentiment lexicon based on SO-PMI. Applied Mechanics and Materials, 263:1688–1693, 2013.
6. Haniewicz K, Rutkowski W, Adamczyk M, et al: Towards the Lexicon-Based Sentiment Analysis of Polish Texts: Polarity Lexicon. In: Computational Collective Intelligence Technologies and Applications—5th International Conference. Craiova, Romania. Springer Berlin Heidelberg, 2013: 286–295.
7. Hui Cao, SenZhang: Research on building Chinese semantic lexicon based on the concept definition of HowNet: In: Asian Language Processing. Kuching, Malaysia: IEEE, 2014: 29–33.
8. Niu Z, Yin Z, Kong X: Sentiment classification for microblog by machine learn-ing In: Computational and Information Sciences. Chongqing, China: IEEE, 2012: 286–289.
9. Gan Wen yan, Jiang. Bo: Machine learning and lexicon based methods for sentiment classification. In: A survey. Web Information System and Application Conference (WISA). Tianjin, China: IEEE, 2014:262–265.
10. Zhou S, Chen Q, Wang X.: Active deep learning method for semi-supervised sentiment classification. Neurocomputing, 120: 536–546, 2013.
11. Zhai Z, Xu H, Kang B, et al. Exploiting effective features for chinese sentiment classification [J]. Expert Systems with Applications, 2011, 38(8): 9139–9146.
12. Zhang D, Xu H, Su Z, et al. Chinese comments sentiment classification based on word2vec and SVM perf [J]. Expert Systems with Applications, 2015, 42(4): 1857–1863.

Implementation of Stress Measurement System Based on Technology of Internet of Things

Qingshuai Wang, Hui Cao, Ailin Li and Tao Xu

Abstract This project is small production which is made based on STC89C52 single chip, small production, and great wisdom. Although production is small, it embodies basic idea and basic function of Internet of Things. For this reason, we must use: sensor and wireless module. Application of sensor is to make all things "open their mouths and speak". Object tells its story through sensor. In terms of key sensing layer, we apply the newest and state-of-the-art monitoring method known so far. In terms of wireless module, we apply popular Wi-Fi technology. Application of wireless Wi-Fi technology makes it possible that data can be connected to Internet in real time. In network layer, joining of Wi-Fi guarantees confidentiality, reliability, and stability.

Keywords Internet of things · Stress measurement · Stress sensor of four-pole magnetic survey · Stress · Solar energy · Wi-Fi · Convertor

1 Introduction

Application of sensor technology and standardization of product is of great importance, so we can firstly gain the advantage and realize strategic breakthrough in these aspects, thus getting speaking right and initiative of development of Internet of Things. Key point of our project is a sensing, i.e., sensing layer. This production is starting from sensing layer, layer-layer progressive to realize Internet of Things. Compared with existing sensors, this system has more advantages than other sensors in terms of sensor accuracy, degree of precision. Compared with other stress measurement equipment, this system realizes Internet of Things [1].

Q. Wang (✉) · H. Cao · A. Li · T. Xu
Key Laboratory of Chinese National Languages Information Technology,
Ministry of Education in Northwest University for Nationalities,
Lanzhou 730030, China
e-mail: 7498859@qq.com

© Springer Nature Singapore Pte Ltd. 2018
S.K. Bhatia et al. (eds.), *Advances in Computer and Computational Sciences*,
Advances in Intelligent Systems and Computing 554,
https://doi.org/10.1007/978-981-10-3773-3_23

From "comprehensive sensing" in Internet of Things, we get inspiration to let steel "speak" and let steel itself to tell its "story" to us. Let bridge "speak" through sensing layer of Internet of Things. Use this technology to make industrial machinery, electrical appliance and instrument, and meter more intelligent [2]. Realize online testing of stress through circuit consisting of sensing module, signal amplification module, analog–digital conversion module, control module, displayer, and signal transmission module, etc. Collect data and upload data in real time. It is network system transmitting information from object to object, and from object to people.

2 "Comprehensive Sensing" in Internet of Things

Describe in terms of stress measurement of steel researched by us, i.e., let steel "speak". Implementation of research on stress measurement of steel is later than that of research on damage inspection of steel, because stress of steel is rather abstract, and research on testing technology is insufficient [3].

Piezomagnetic sensor is excellent application case in magnetoelasticity phenomenon of electromagnetism. In terms of ferromagnetic material affected by static-state mechanical stress or dynamic-state mechanical stress, its hysteresis loop changes with changing of stress on test piece. Change rule is as shown in Fig. 1. Magnetic conductivity u of ferromagnetic material will change with changing of hysteresis loop. According to comparison of changing proportional relation of hysteresis loop and magnetic flux by checking (table), we can calculate actual stress of ferromagnetic material tested through numeration [4].

Sensor consists of two-layer coils. Sensor will not affect any mechanical and physical property of test piece except for magnetization of test piece because of long time effect. The biggest advantage of this stress test method of ferromagnetic material is that it will not change original structure situation of test piece. Even if test piece is sealed in cement outer cover or cement duct, we can still perform test to stress situation of ferromagnetic material. Mathematical expression of relationship of stress received by ferromagnetic material and its magnetic conductivity σ is:

$$\frac{\Delta\mu}{\mu} = \frac{2\lambda_n}{B_m^2}\sigma\mu \tag{1}$$

In formula: B_m is saturation flux density.

Seen from formula (1), $\frac{\Delta\mu}{\mu} = f(\sigma, B_m)$, to make $\frac{\Delta\mu}{\mu} = f(\sigma)$ unique, B_m must be constant.

Where formula (1) is theoretical model of piezomagnetic-type sensor, seen from formula, material constituting framework of piezomagnetic-type inductive sensor must have:

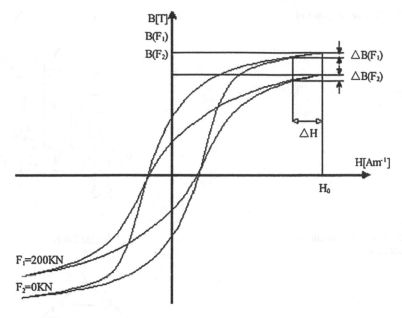

Fig. 1 Change figure of hysteresis loop

(1) Acting force being able to load large-tonnage and high magnetic conductivity;
(2) To enlarge variation range of magnetic conductivity $\Delta\mu/\mu$ of test piece, it needs to be ensured that saturation flux density B_m is smaller.

When structure support material of sensor is chosen, necessary conditions shall be considered, for example this material must have strong loading capacity and its magnetism is good and saturated magnetic intensity of this material is small, etc. This system chooses to use silicon steel sheet torn down from pulse transformer, for its low loss, high permeability, and stronger anti-jamming capability lowers working condition error in the maximum degree [5]. In terms of sensor structure, choose I-type structure which has good symmetry in structure. The design makes sensor introduced with standard compensatory piece. Through comparison of difference value of test data of test piece, it can be calculated that actual stress received by test piece is larger [6]. Figure 2 is structure diagram of sensor applied in this system. As Fig. 2 shows, sensor has four magnetic poles and each magnetic pole echoes to one magnetic test coil. Parameters of four magnetic test coils are consistent. Connecting element of structure is twined to magnetic test coil to balance induced voltage generated by four magnetic poles [7]. In terms of connecting way of four magnetic test coils, way of tandem connection of positive and negative directions of each group coil is adopted. Perform reverse direction connection to two coils after they are connected in series. Connection form of reverse direction and differential motion is formed then.

Fig. 2 Structure diagram of sensor

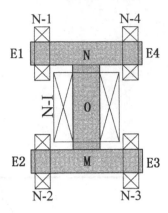

Fig. 3 Structure diagram of HLK-RM04

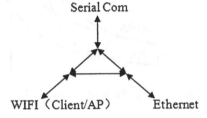

3 "Interconnection" of the Internet of Things

The realization of networked module; this chapter involves many of the techniques, so I only make a brief overview.

As we all know, the interconnection is the blood vessels and nerves of the Internet of Things, and its main function is to transmit information and information [8]. In order to achieve this goal, this production uses the module of Serial Com WIFI.

HLK-RM04, which is adopted there, could meet the requirements of this production with its powerful function; its function structure diagram is as follows: (Fig. 3).

WIFI supports for 802.11b/g/n. Small size: 40 × 29 (which is even smaller than one third of a credit card); working temperature: industrial grade 25 ~ 75 °C. Support for the current common WIFI encryption methods and algorithms available in the market: WEP/WAP-PSK/WAP2-PSK/WAPI; optional Server/TCP Client/UDP TCP working mode; support network protocol: TCP/UDP/ARP/ICMP/HTTP/DNS/DHCP; adjustable serial port speed: 1200–230400. As for the TTL and RS232 conversion circuit only, there are a lot of types, and I will not enumerate them all for saving time and energy. Those who are interested can check by themselves. We directly take RS-232 to TTL (3.3 V), which is an industrial grade passive RS-232 to TTL level converter [9].

4 "Smart Operation" of the Network of Things

Smart operation solves the problem of computing, processing, and decision-making and it is the brain and pivot of the Internet of Things, which mainly includes network management center, information center and intelligent processing center, etc., and the main functions are the deep analysis and the effective processing of information and data [10]. In the real Internet of Things, this step includes a lot of technologies, such as our familiar big data and cloud computing, etc.

The feedback control system is completed in this layer. When the steel works under normal stress condition, the device does not send out alarm signal [11]. When the steel stress is beyond the safe range of stress value, the device sends out alarm signal, and the application layer of the system informs people to take the necessary measures.

5 System Test

The sensor of this system has been precisely tested by the torsion test machine, of which the test piece is the steel plate in size of φ 70 × 500. According to the principle of the piezoelectric effect, when the other conditions are unchanged, the stress received by the same material is directly related to its own magnetic properties of the material [12]. So, in order to obtain the magnetic force torque value, you need to construct the mathematical model between the magnetic torque and output voltage. No. 45 steel is used in the test as the specimen material, so the regression equation of magnetic torque and output voltage is built according to the graph; the experimental data of different torque loading for many times are shown in Table 1, and the experimental data of the same torque loading for many times are shown in Table 2.

Differential sensor (Note: the test sensitivity or sensor refers to the ratio of the magnetic output voltage increment of Aoshi sensor and the torque measurements).

The standard load of this system is designed to be 15,000 N, sufficient for the detection of mechanical materials. The system performance test uses two jacks to produce the tensile stress to the test pieces, and then test the test piece through the sensor; at the same time, a liquid pressure sensor is placed at the top of each jack [13]. We can get the linear degree 2.1% FS of the sensor by the test data calculation by comparing the outputs of the two kinds of sensors (Table 3). The experimental results are relatively close to the standard tensile force value; the greater the test sensor measurement value is, the closer it is to the real stress value [14]. The sensor used in this system can be used to measure the loading condition of steel structure with large load. In the table, A represents the tensile force $\sigma 0$ (kN), and B represents the actual measured value $\sigma 1$ (kN).

Table 1 Experimental data of different torque loading for many times

Load torque (N•m)	Magnetic voltage (mV)	Magnetic torque (N•m)	Relative error (%)	Test sensitivity MV/N · M
50	25	46	8.0	0.50
100	64	95	5.0	0.64
150	85	144	4.0	0.57
200	128	192	4.0	0.64
250	164	241	3.6	0.67
300	210	289	3.3	0.70

Table 2 Cross core sensor

Load torque (N•m)	Magnetic voltage (mV)	Magnetic torque (N•m)	Relative error (%)	Test sensitivity MV/N · M
50	31	46	8.0	0.62
100	75	96	4.0	0.75
150	98	147	2.0	0.65
200	145	196	2.0	0.73
250	185	244	2.4	0.74
300	225	293	2.3	0.75

Table 3 Stress measured by sensor and test data table

A	0.02	2.00	4.00	6.00	8.00	10.0	12.0	14.0	15.0
B	0.03	1.92	4.05	6.21	8.09	9.90	12.0	13.9	14.9

6 Conclusions

The essence of the matter is to intelligentize things, which is the conception of Internet of Things. Sensor is an important and key link to realize the interconnection of all things in the internet, and it is also one of the bottlenecks restricting the development of the Internet of Things.

Although we have carried out careful and rigorous analysis of the steel stress testing, new problems still come out all the time, such as steel wear, accelerating wear of curve steel, theory and practice of seamless line and the use and repair of old steel [15]. These detection theories are based on certain physical principles, having their own advantages and disadvantages, so they can not consider all the aspects and be widely used.

Cover range of various stress and damage detection methods should be continuously increased, so that it can be applied to various cases of stress and damage detection, in particular, it can be used to speed up the upgrading of testing technology in the industry and construction field to protect citizens' personal and property safety and to benefit the development of the national economy in a better way.

References

1. Chen Liuqin. Internet of Things: Developing Trends at Home and Abroad and the Key Issues to Be Solved Urgently, Decision advisory communication (2010).
2. Huang Yulan. Introduction to Internet of Things. People's Posts and Telecommunications Press (2011).
3. Huang Dongyan. Research on Nondestructive Testing Method based on Magnetic Technology. Jilin: Jilin University of Science and Technology, 2012.
4. Zeng Jiewei, Su Lanhai, Xu Liping, et al. Journal of Mechanical Engineering. Research on Internal Stress Detection Technology of Steel Plate with Counter-magnetostriction Effect, 2014, 50 (8): 18–22.
5. Guo Peifei, Jia Zhenyuan, Yang Xing, et al. Piezomagnetic Effect and Its Application in Sensor. Piezoelectrics & Acoustooptics, 2001, 23 (1):27–29.
6. Liu Lili. Summary of Research on Internet of Things, Conference paper, Beijing Institute of communications, wireless and mobile communications (2010).
7. Xiao Weisi. Research on the application of Internet of Things in the Field of Railway Safety Monitoring, Beijing Jiaotong University.
8. Liu Kaixu, Duan Yubo. Experiment Research on Nondestructive Testing for Unidirectional Stress of Ferromagnetic Material. Mechanics in Engineering, 2015, 37 (2): 227–229.
9. Xiong Ergang. Theoretical and Experimental Research on the Full Magnetic Flux Stress Testing Technology of Steel Structure Based on the Magnetic Effect. Xi'an University of Architecture and Technology (2007).
10. Deng Rui. Research on Stress and Magnetic Testing Technology for Three-dimensional Management System of Sea. Qingdao: Ocean University of China, 2011.
11. Wang Ruibao. A Method Based on Magnetic Variation of Material to Measure Stress of Ferromagnetic Material. [Master's thesis] Changchun: Jilin University, 2005.
12. Shi Zhibo. Design of Warehouse Management System Based on RFID Technology, Soochow University, (2013).
13. Chen Juan. Design of Magnetic Bridge Type Stress Sensor. Sensor Technology, 1997,16 (2):30–35.
14. Li Yanfeng. Experiment Research on Magnetoelectric Effect of Steel Structure Material Stress. Shenyang: Northeastern University, 2006.
15. Yi Dabin. Rail Stress and Nondestructive Testing, Northern Jiaotong University, China Railway phase 6, (1991).

Social Media Big Data Analysis for Global Sourcing Realization

Shi-Feng Huang, Chuan-Jun Su and Maria Belen Vargas Saballos

Abstract The supplier selection is a multi-criteria decision making process determined by a comparison of suppliers using a shared set of characteristics. The adequate selection of suppliers makes a strategic difference to an organization's ability to reduce cost and improve the quality of its product. Social media has become a relevant source of information within the supply chain, this vast portfolio of data can assist in the supplier selection process by providing relevant and up-to-date information regarding the different risks suppliers might present. This research proposes Twitter Enabled Supplier Status Assessment (TESSA) that assists in the supplier selection process by assessing the risks and uncertainty each supplier holds. We aim to open a new window on the research field regarding supplier selection, as a result, companies can be more prone to exploit the extensive data social networks has to offer and utilize it in the decision making process.

Keywords Global supplier selection · Supplier risk · Big data · Social media

1 Introduction

The supplier selection is a multi-criteria decision-making process determined by a comparison of suppliers using a shared set of characteristics. One of the main objectives of the supplier selections is to establish a long and stable relationship of mutual value and support between suppliers and buyers. Over the years companies have become more reliant on their suppliers, as a result more attention has been

S.-F. Huang (✉) · C.-J. Su · M.B.V. Saballos
Department of Industrial Engineering and Management, Yuan Ze University,
Taoyuan, Taiwan (ROC)
e-mail: s1028901@mail.yzu.edu.tw

C.-J. Su
e-mail: iecjsu@saturn.yzu.edu.tw

M.B.V. Saballos
e-mail: s1038908@mail.yzu.edu.tw

© Springer Nature Singapore Pte Ltd. 2018
S.K. Bhatia et al. (eds.), *Advances in Computer and Computational Sciences*,
Advances in Intelligent Systems and Computing 554,
https://doi.org/10.1007/978-981-10-3773-3_24

251

paid to the decision-making. Any deficiency in the process could be enough to deteriorate the whole supply chain's financial and operational position [1].

Globalization has made a big change in the way companies source and manufacture products, particularly in the retail and consumer sector. Globalization has transformed how manufacturers operate, offering an opportunity to reach new customers in new markets and new sourcing options. Global sourcing is no longer an option but a threshold requirement for some organizations. This practice has been embraced due to the cost savings it generates. Organizations can choose suppliers from anywhere in the world, developing countries are becoming more competitive given their low labor and operating costs. As organizations turn to global sourcing the selection of global suppliers is the key to their success, organizations need to establish a long-term relationship with unfamiliar and unproven suppliers. This makes the selection process a very complicated and risk prone process.

Global sourcing implies long distance supply chain with extended lead times, which have major implications for demand responsiveness and security of supply. It involves new challenges and complexities, the risk and uncertainty increases. Liu et al. explain that the modern trend is to reduce the number of suppliers and establish a long and stable relationship of mutual trust, benefit, and support [2]. This aggravates the supplier selection process because organizations not only should measure suppliers by quality, price, service, but also by the risk factors.

In regards of supply chain management, social media offers a wide set of opportunities and challenges. Information gathered from social media can provide insight and knowledge about participants that is usually not available. The information in social media is updated rapidly and spreads virally at an exceptional speed; this provides us with first-hand information. We now have the opportunity to analyze this vast portfolio of information to assist the supplier selection process.

This research main objective is to provide a tool that can assist the procurement team when selecting a supplier, Twitter Enabled Supplier Status Assessment (TESSA) can provide information on the risk and uncertainty for each supplier helping the decision team to reduce their potential supplier list; this tool combines both, a risk and uncertainty classification and social media data. In this research a risk classification is defined in different categories such as socio-political, economic, financial, and environmental. This is defined from the existing literature reviews regarding the external supply chain risks and supply chain risk management.

2 Research Methodology

In this research we proposed TESSA, to work as a tool and assist in the supplier selection process. TESSA's architecture is formed by four modules: Ontology Building, Extraction, Preprocessing, and Classifier as shown in Fig. 1.

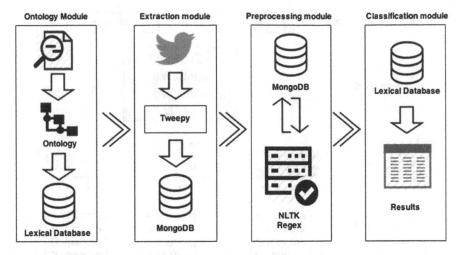

Fig. 1 TESSA architecture

The first task of TESSA involves building a Risk and Uncertainty ontology (RUO) which encapsulates risk and uncertainty factors regarding global supplier. A lexical database can then be derived from the ontology, this lexical database encompasses all the aspects and attributes from the ontology and serves as the classifier for the tweets. The Data Extraction function is responsible for retrieving related tweets using Twitter's streaming API through a python script crawler. The extraction is done using the supplier name as keyword. Afterwards a series of process are subsequently performed to normalize (cleanse) the retrieved tweets. Once the tweets are preprocessed they are classified using the lexical database.

2.1 Risk and Uncertainty Ontology (RUO)

In this phase RUO is developed based on the Formal Concept Analysis (FCA), which is principled way of deriving a concept hierarchy or formal ontology from a collection of objects and their properties [3]. FCA explicitly formalizes extension and intension of a concept and their mutual relationship. A strong community of academic, government, and corporate users have utilized tool Protégé when generating an ontology; this tool is used to develop the RUO.

In this research the data set of risk factors that we use is based on various literatures including supplier risk, global risk, and other ontologies. FCA distinguishes three levels: object, concept, and representation levels as depicted in Fig. 2.

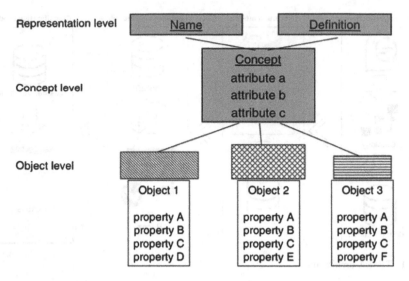

Fig. 2 FCA architecture

2.2 Data Extraction

This research relies primarily on data from Twitter, in order to retrieve Tweets we must access Twitter API. A tweet script crawler is built with the programming language Python in order to extract the tweets. The tweet script crawler is designed to retrieve the Tweets based on selected keywords, which are the names of the targeted supplier in our case. With the tweet script crawler, the tweets can be retrieved and subsequently stored in a NoSQL database called MongoDB. In order to allow the communication between MongoDB and Python, we implement the Python driver called Pymongo into the tweet script crawler.

The retrieved tweets have to be cleansed by removing the undesired (noisy) data for further analysis. Two Python modules, the Natural Language Toolkit (NLTK) and the Regular Expression module are used for accomplishing the preprocessing.

Once the tweets are preprocessed they go through the lexical database, in the lexical database reside all the aspect and attributes from the ontology. Here all the tweets go through the database and are matched within the aspects and attributes. If there is a match a number 1 is reflected and the tweet is saved in a collection under the name of the aspect, if there is no match a number 0 is shown, the tweets that don't belong to any aspect are classified as trash.

Kohavi and Provost explained that the exactitude of a classifier can be evaluated by computing the number of correctly recognized class examples, the number of correctly recognized examples that do not belong to that class, and examples that either were incorrectly assigned to the class or that were not recognized as a class examples [4]. According to Kohavi and Provost, these four concepts are the entries of the confusion matrix as shown in Table 1.

Table 1 Confusion matrix

		Predicted	
		Positive	Negative
Actual	Positive	tp	fn
	Negative	fp	tn

Table 2 Classification evaluation metrics

Metric	Formula	Description
Accuracy	$\frac{tp+tn}{tp+fp+fn+tn}$	Proportion of the total number of predictions that were correct
Precision	$\frac{tp}{tp+fp}$	Proportion of the predicted positive cases that were correct
Recall	$\frac{tp}{tp+fn}$	Proportion of positive cases that were correctly identified
F-score	$\frac{2 \times precision \times recall}{precision + recall}$	Relations between data's positive labels and those given by a classifier

According to Sokolova and Lapalme, the most common evaluation metrics in text classification are accuracy, precision, recall, and f-score [5]. Given the confusion matrix we can evaluate TESSA's accuracy, precision, recall, and f-score, we will do it by implementing the equations presented in Table 2.

The next step is to extract information regarding the suppliers Yue Yuen and Li and Fung, the search of the tweets is done using the name of the supplier as keywords of the search, and the search is done by one supplier at the time.

3 Results

In this part, we show a scenario where a sample company called "ABC" is interested in two footwear manufacturing companies, Yue Yuen and Li and Fung.

We obtained 100 tweets from May 14 to May 17, in 2014, for Yue Yuen the tweets are stored in the database. Out of the 100 tweets for Yue Yuen only 1 category was detected which is labor strike, 68 tweets were identified in this category and 32 tweets were classified as trash. In the results, we find that TESSA scored high results in all the metrics and the tool has 0 false positives, showing to be an efficient classifier for this aspect with 100% score in precision. However, it missed 26 tweets from the strike category and label them as trash.

We obtained 200 tweets regarding for Li and Fung. Out of the 200 tweets for Li and Fung, only 1 category was detected which is labor strike, 114 tweets were identified within this category and 86 were identified as trash. In the results for Li and Fung, TESSA scored high in all the metrics and showed no false positives. The f-score is very close to 90% providing evidence of the good performance the classifier has.

4 Conclusions

TESSA exposed the risk that Yue Yuen and Li and Fung present, which is Labor Strike, this is due to the riots happening in Vietnam in the month of May. Since both companies have production facilities in Vietnam they present risk for the Company ABC interested in making business with them. Now the ABC Company can use this knowledge in order to make their final decision.

The development of TESSA can be critical for any company in need of an international supplier, the results obtained show high levels of accuracy and performance. This tool allows companies to have a better insight when selecting an international supplier. TESSA can give the supplier selection team more knowledge to make the appropriate decision regarding their needs.

References

1. Araz, C., Ozkarahan, I.: Supplier evaluation and management system for strategic sourcing based on a new multicriteria sorting procedure. International Journal of Production Economics. 106, 585–606 (2007).
2. Liu, L., Zhou, Y., Zhu, H.: A conceptual framework for vendor selection based on sup-ply chain risk management from a literature review. Journal of System and Management Sciences. 1, 1–8 (2011).
3. Wille, R.: Restructuring lattice theory: An approach based on hierarchies of concepts. NATO Advanced Study Institute Series. 445–470 (1982).
4. Sokolova, M., Lapalme, G.: A systematic analysis of performance measures for classifi-cation tasks. Information Processing & Management. 45, 427–437 (2009).

Based on Hidden Markov Model to Identify the Driver Lane-Changing Behavior of Automobile OBD Internet of Vehicles Research and Design

Yu Tu, Fengdeng Zhang and Zhijian Wang

Abstract In this paper, we mainly study and design a kind of OBD vehicle network which can help the driver to change the behavior of the driver, so as to reduce the traffic accident caused by incorrect operation of the driver's lane-changing behavior. First, get the hidden Markov model to identify driver's lane-changing behavior was studied by MATLAB simulation. Second, the model is applied to the software and hardware design of OBD vehicle networking. Finally through the use of the OBD port will be from the CAN bus acquisition to the driver lane-changing behavior data and transfer to the Hidden Markov Model to deal with and determine the driver lane-changing the threshold of steering wheel angular velocity and steering wheel steering angle.

Keywords OBD · Hidden markov model (HMM) · Lane-changing behavior · Vehicle networking

1 Introduction

On-Board Diagnostics (OBD) technology is the car's on-board diagnostic system, the diagnosis system is used to real-time monitor the running status of engine and exhaust gas treatment system working condition [1]. Usually, the cause of road traffic accident and the driver's driving strategy is concerned, the danger often occurs in the driver did not make the corresponding action response to the current traffic conditions. So it is very important to get the driver's behavior information as early as possible in the design of advanced driver assistance system.

Y. Tu (✉) · F. Zhang · Z. Wang
Department of Control Science and Engineering, Key Laboratory
of Modern Optical System, University of Shanghai for Science
and Technology, Shanghai, China
e-mail: tuyujiayou@163.com

F. Zhang
e-mail: FDZhang@usst.edu.cn

© Springer Nature Singapore Pte Ltd. 2018
S.K. Bhatia et al. (eds.), *Advances in Computer and Computational Sciences*,
Advances in Intelligent Systems and Computing 554,
https://doi.org/10.1007/978-981-10-3773-3_25

At present of car networking theory and technology research, Diewald, Möller, Roalter proposed the one based on the Android driver assistance systems DriveAssist. This system can be to obtain the information between the vehicle and the vehicle, in a graphical way to display, and can trigger the alert messages to remind the driver to prevent accidents [2]. University of Paderborn, Germany use CAN-Bluetooth intelligent main control node to establish a wireless communication system, the smart phone's internal sensors and automotive sensor information, through the GPS to achieve navigation, for the fusion of sensor data to the Android smartphones, improve the navigation performance of the automobile [3].

In this article, through CAN bus information will be collected by the driver's raw data, such as car steering wheel Angle, steering wheel rotation speed, yawing velocity, vehicle speed, etc., using HMM for driver's lane-changing behavior are identified. Finally the identification results are used to make the vehicle active safety OBD vehicle networking function development.

2 Hidden Markov Model to Identify the Driver Lane-Changing Behavior

2.1 The Pilot Lane-Changing Behavior HMM Structure Definition

In this paper, using the HMM model for driver's lane-changing behavior are identified, the lane-changing behavior is divided into three kinds of situations, namely Lane Keep (LK), Lane Change Normal (LCN), Lane Change Emergency (LCE). Because the CAN bus data of the LCE behavior is not obtained in the actual data collection, the LCE behavior is not considered in this paper. We use the HMM model structure from left to right, as shown in Fig. 1, which means that the three hidden states are only allowed to pass to their own or pass to the next state.

According to the practical experience of driving, each lane-changing behavior can be divided into: ready state (S1), action state (S2), and recovery state (S3). Corresponding to each state, we can get the corresponding CAN message from OBD diagnostic port, for the pilot lane-changing behavior, this paper uses the collected steering wheel angular velocity and steering wheel steering angle, vehicle yawing angular velocity data as the observed sequence of lane-changing behavior recognition model. Figure 2 is to use the CANoe software capture driver left lane-changing CAN message signal.

Fig. 1 Lane-changing behavior of HMM structure

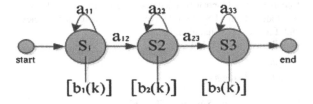

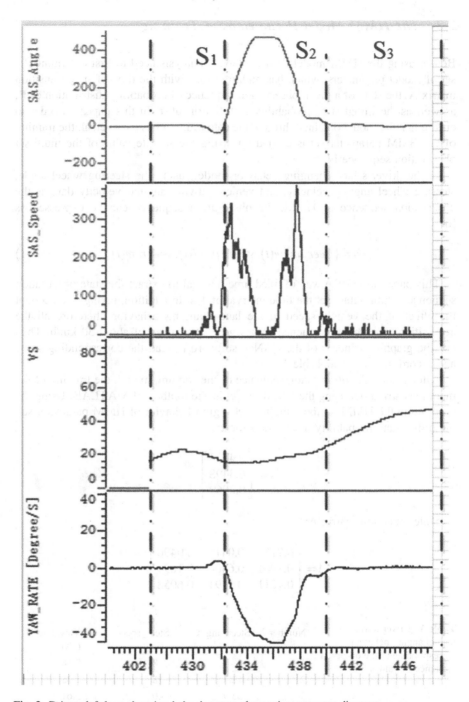

Fig. 2 Drivers left lane-changing behavior state observation sequence diagrams

2.2 The HMM Model Parameter Set Training

Before using the HMM model driver behavior analysis need to first determine the set of model parameters, which has nothing to do with the time of state transition matrix A, the state of a given observation sequence of probability distribution of B, as well as the initial state probability distribution of π. In this paper, in order to ensure that the model obtained through the data training has a universal, the training of the HMM parameter set is carried out using the sample value of the multi-set observation sequence [4, 5].

In the driver's lane-changing behavior model, using the steering wheel angle, steering wheel angular velocity, and vehicle yawing angular velocity data as the observation sequence of HMM, the observation sequence can be expressed as [6, 7]:

$$O_t = \{steer\_angle(t), steer\_speed(t), yaw\_rate(t)\} \tag{1}$$

This paper adopts the way of fixed time interval to extract the state observation sequence output value, set the time interval of 1 s. In addition, taking into account the effect of the vehicle speed to the lane-changing behavior, therefore all the acquisition of observed sequence values were at a speed of about 30 km/h. Then use the graphics window of the CANoe software to read the corresponding value and record, as shown in Table 1.

After getting the observation sequence of the training model, the training of the model is carried out using the HMM model in the toolbox of MATLAB. Using the program in the HMM toolbox, the lane-changing behavior of HMM parameter set, the initial state probability distribution for π:

$$x = \begin{bmatrix} 0.4433 \\ 0.3498 \\ 0.2069 \end{bmatrix}$$

State transition matrix A:

$$A = \begin{bmatrix} 0.9154 & 0.0415 & 0.0490 \\ 0.1901 & 0.6406 & 0.1699 \\ 0.1211 & 0.1791 & 0.6058 \end{bmatrix}$$

Table 1 Driver normal lane-changing behavior corresponding to the observed sequence of values	Number	Steer_angel (°)	Steer_speed (°/s)	Yaw_rate (°/s)
	1	−1	8	0.18
	2	−12.4	28	1.08
	3	−33.9	40	3.61
	4	193.2	208	−14.4
	5	397.6	160	−30.69

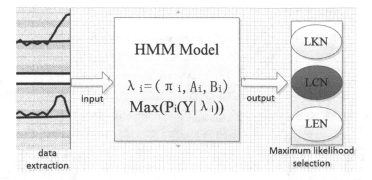

Fig. 3 Driver lane-changing behavior recognition

Under a given state observation value of the probability distribution of B

$$B = \begin{bmatrix} 0.4907 & 0.0272 & 0.4886 \\ 0.1050 & 0.6171 & 0.2779 \\ 0.2738 & 0.6409 & 0.1213 \end{bmatrix}$$

As shown in Fig. 3, using the HMM MATLAB toolbox, a new set of driver behavior data is calculated as the input value of the model, we can obtain the corresponding output value of the maximum likelihood estimation of the model.

3 Car OBD Car Network Design and Implementation

In order to ensure the real-time performance of the system task, this design uses the OSEK OS as system software architecture. The development of OBD vehicle networking function software mainly includes the OSEK/VDX real-time operating system software development, V850 microcontroller hardware driver layer software development, CAN message signal analysis software development and function implementation of the upper layer software development, etc. The main part of the whole software integration can be expressed in Fig. 4.

4 The System Evaluation

To ensure the real-time performance of OBD vehicle networking system, need of OSEK/VDX real-time operating system to evaluate the running effect of the test, there are three executive tasks in the software design, 1 ms task is mainly to

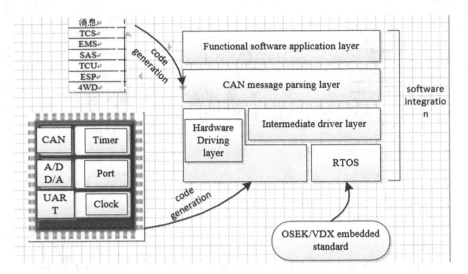

Fig. 4 OBD car networking software integration as a whole

Table 2 OSEK OS task definition of time and the actual execution time	Task definition of time (ms)	Average real time (ms)
	1	1.047
	5	5.076
	10	10.076
	20	20.055

complete the initialization work related to the CAN controller, the 10 ms task is mainly to complete the CAN message sending and receiving work, the 20 ms task mainly completes the CAN message and the UART serial message transmission work. Test results as shown in Table 2.

From the table above knowable, OSEK OS task definition of time and the actual execution time error in the MCU is small, which shows the validity of the software design method of OSEK/VDX operating system.

In the OBD car networking software testing and the overall function of the test process, using CANoe software online Logging function, all CAN messages of a vehicle in a stationary or moving state are recorded. When the need for vehicle information, only need to be in the CANoe software for the corresponding configuration, the data collected will be played back, through the CANcaseXL interface to access the CAN bus can get the vehicle information. Finally, the results of the Trace software CANoe window comparison, this OBD car networking system is operating normally.

5 Conclusion

In this paper, we study the use from the OBD port acquisition of can data bus. The driver for auxiliary channel behavior corresponding to the HMM model is established to determine lane-changing behavior of the steering angle and the steering wheel rotation speed threshold, its application to embedded system. As the alarm signal trigger conditions. Through the actual test and data analysis, the system can work normally.

Acknowledgements This work was supported by Natural Science Foundation of Shanghai under Grant No. 15ZR1429300

References

1. Diewald S, Möller A, Roalter Lss, et al. DriveAssist-A V2X-Based Driver Assistance System for Android[C]//Mensch & Computer Workshopband. 2012: 373–380.
2. Diewald S, Möller A, Roalter L, et al. DriveAssist-A V2X-Based Driver Assistance System for Android[C]//Mensch & Computer Workshopband. 2012: 373–380.
3. Walter O, Schmalenstroeer J, Engler A, et al. Smartphone-based sensor fusion for improved vehicular navigation[C]//Positioning Navigation and Communication (WPNC), 2013 10th Workshop on. IEEE, 2013: 1–6.
4. Choi S J, Kim J H, Kwak D G, et al. Analysis and classification of driver behavior using in-vehicle can-bus information[C]. Biennial Workshop on DSP for In-Vehicle and Mobile Systems. 2007: 17–19.
5. Habenicht S, Winner H, Bone S, et al. A maneuver-based lane change assistance system[C]. Intelligent Vehicles Symposium (IV), 2011 IEEE. IEEE, 2011: 375–380.
6. Kuge N, Yamamura T, Shimoyama O, et al. A driver behavior recognition method based on a driver model framework[R]. SAE Technical Paper, 2000.
7. Meng X, Lee K K, Xu Y. Human driving behavior recognition based on hidden markov models [C].Robotics and Biomimetics, 2006. ROBIO'06. IEEE International Conference on. IEEE, 2006: 274–279.

The Research on Key Technique of Raw Coal Management Information System

Xiaoyan Zhang and Lei Zhang

Abstract The author analyzed the business process of a large coal enterprise on raw coal quality. A raw coal quality MIS (Management Information System) was designed and implemented using four-tier structure which built by Freemarker + Struts2 + Spring + Mybatis. The system uses network as data transmission medium, and manages the audit flow of different mining's data and reports according to different permissions. It provides the visualization display of coal quality data, too. The paper analyzed the characteristics of raw coal quality data and established a reasonable database structure. Then it makes an exposition of function design scheme of the system and a method about permission control with configurable function. It also introduces the implementation strategy of reports and graphics. The system has been applied in the enterprise and work well. The work efficiency of coal quality management and audit are improved effectively.

Keywords Coal quality management · Four-tier structure · Report and graphic generation · Permission control

1 Introduction

Coal industry has been one of the important economy pillars in China, and coal quality management level affects the competition of coal industry directly. Due to the influence of coal mining conditions and coal filed geological conditions; it has certain differences in coal quality, which produced in different time and different mines [1, 2]. The yield and quality of clean coal are affected directly by the quality of raw coal. The accuracy and effectiveness of coal quality information of raw coal influenced the quality of the products and the economic benefits of coal enterprises.

X. Zhang (✉) · L. Zhang
Xi'an University of Science and Technology, Xi'an 710054, China
e-mail: 1161880978@qq.com

L. Zhang
e-mail: 542013793@qq.com

© Springer Nature Singapore Pte Ltd. 2018
S.K. Bhatia et al. (eds.), *Advances in Computer and Computational Sciences*,
Advances in Intelligent Systems and Computing 554,
https://doi.org/10.1007/978-981-10-3773-3_26

Large coal enterprises have many mining and departments scattered unevenly. In the actual business processes, coal quality data need to be transferred for statistics and audited inspection between multiple departments. In addition, the automatic management level of each unit is different, which is easy to cause inefficient statistical analysis and hysteretic information transfer, even the information island [3, 4].

Therefore, the main task of the coal enterprise is how to build the new management mode in the process of coal management [5]. For the inefficiency of management and information island, the author used information technology to achieve a raw coal quality MIS (Management Information System) of a large coal enterprises [6–8], which is based on analysis and research of coal quality business process of a coal enterprises.

2 System Design

The raw coal quality MIS designed in this paper is based on B/S structure [9–11]. The data and report produced by each mine are transferred over the internet and reviewed according to different permissions. Also, data is showed as graphics. It is an effective solution to many problems, such as scattered mining, complex administration, high requirements for information transfer, data security control, and so on. It has also greatly enhanced the information management level of coal enterprises.

2.1 System Structure Design

Considering the scattered mining and complex departments of coal enterprises, it used MVC (Model View Controller) design pattern combined B/S structure to design four-tier structure for the implementation of the system [12, 13]. The whole system is divided into presentation layer, control layer, business logic layer, and data persistence layer. FreeMarker as page template engine implements the user interface. Struts2 controls the distribution requests of business to achieve the jump. Mybatis does persistence operations on the data, and the whole structure is integration by Spring [14]. The structure diagram is shown in Fig. 1.

2.2 System Database Design

It uses the current mainstream Oracle database of IBM as the system database. Oracle database is considered because the system requires stability, data manipulation of versatility, safety, storage efficiency of data, and the portability of the system.

Fig. 1 System structure diagram

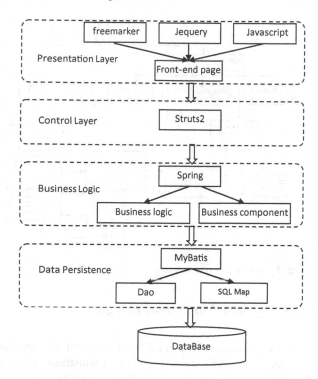

After the analysis of the business process of the raw coal quality management, the data can be classified into the following categories:

Input data: basic data of raw coal quality, device information, personnel data of coal quality, etc.

Data processing: intermediate results, output statements, graphics.

Output data: collecting statistical sampling about coal quality to make quality inspection report; carrying out the classification, statistics, and summary on the coal quality basic data of Quality Supervision Bureau and making raw coal report of mine, raw coal daily of group, raw coal ten days' report of group, raw coal assay sheet, raw coal monthly plan, raw coal year plan, monthly decomposing plan, coal quality line chart, and others.

According to the characteristics of the system, the database of coal quality management information system also contains: data dictionary, basic data table, temporary table of intermediate process, indexing table, etc. They can make data entry easily and enhance the flexibility of the system by setting data dictionary. It can storage the data after processing by the stored procedure through setting the temporary tables of intermediate process, which is convenient for other modules to read, and it can improve the response speed of the system. Because of the frequent access, and large amount of data, indexing some tables with a key can improve the data reading speed greatly for all basic data table of coal.

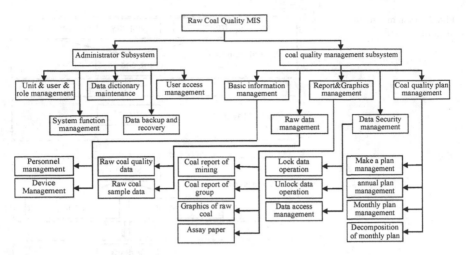

Fig. 2 System function diagram

2.3 System Functional Design

Raw coal quality MIS consists of the coal quality management subsystem and administrator subsystem [15]. The administrator subsystem mainly realizes background management function. There are units, users, and roles management; data dictionary maintenance; user permission management; system function management; data backup; and recovery. Coal quality management subsystem is the main part of the system to achieve specific business requirements of raw coal information management. It includes basic information management; raw data management; report and graphics management; data security management; coal quality plan management. System function block diagram given in Fig. 2.

3 Key Technologies

3.1 The Implementation of User Interface Based on Freemarker and Jquery

The web pages are processed using Freemarker, Jquery, Ajax, and some JS (JavaScript) plugins. Freemarker template interpolation technique can make the page change dynamically. Jquery technology can give some tags attributes dynamically to realize some functions. And the Ajax technology is mainly to achieve partial refresh of the page to improve the user experience. SlickGrid plugin allows the foreground becomes more flexible and compact.

```
<select style="width:90"name="facName" id="facName">
 <option value="" selected>selected</option>
  <#if output.gmresults?exists>
   <#list output.gmresults as gg>
    <option value=${gg['unitName']}>${gg['unitName']}
    </option></#list> </#if>
</select>
```

Freemarker template interpolation technique that code used can make the value of the drop down list changed when the "unitName" is changed.

3.2 The Implementation of Report and Graphics Based on Jxl and JfreeChart

Excel is a common format used for saving statistical data. Jxl (Java Excel) is a widely used and excellent open source tools to read and write Excel files.

These codes get the output path of the Excel report, then, create the Excel file with the "filePath".

```
sc = ServletActionContext.getServletContext();
filePath = sc.getRealPath("/temp/"+fileName);
WritableWorkbook wwb;
OutputStream os = new FileOutputStream(filePath);
wwb - Workbook.createWorkbook(os);
```

After the file is successfully established, then, the properties of the cell can be set using the object provided by Jxl. Such as the font, colors, and so on. Data can also be added by those Jxl object easily.

JFreeChart is an open graphics library on JAVA platform. It is completely written using JAVA language. JFreeChart can generate a pie chart, histogram, scatter diagram, sequence diagram, Gantt charts, and other diagrams.

Similar to the JXL, JFreeChart technology offers a variety of objects to achieve the drawing operation. For example JFreeChart object can set graphics title, the category of fonts, and so on. CategoryPlot object is used to operate the drawing area. LineAndShapeRenderer object can create a line graph object.

Statement analysis and graphics are required to read data from many tables for processing. Considering processing speed, it specifically designed the temporary table in previous database design. When the data is calculated by the stored procedure, it will be inserted into a temporary table, so that can be read directly from the temporary table while drawing or tabling. It can greatly reduce the waiting time for user and improve the speed of data processing.

Table 1 Function list

Function id	Function name	Parent id
10010	Basic information management	10000
10011	Person management	10010
10020	Report and Graphic	10000

3.3 The Implementation of User Permission Control with Configurable Function

Units may change at any time for large-scale coal enterprises, so there is a higher request to the flexibility of the system. That can increase the units and give certain function and access to the units neatly.

For functional design, all the functions of the management of raw coal are stored in database as a tree. Each unit can get the suitable function tree for their own, only in accordance to their own actual needs to cut down the function branch that is not related in the function tree.

This design greatly reduce data storage space, at the same time, there is an association in a similar way like foreign key between filtered information and units. So, the function of different units can be customized as required. Function list is stored as Table 1.

For the function tree querying, the code "start with... connect by prior..." provided by Oracle can be used. The basic syntax is:

select... from tablename start with condition 1
connect by condition 2

Condition 1 is limited statements of root node. It is Function ID in this system. Condition 2 is connection condition. It is "connect by prior Function ID = Parent ID" in this system. So that the whole function tree can be got.

4 Conclusion

For large-scale coal enterprises, mining area is scattered and it is easy to form the information island. Data sharing is more difficult and data security can not be guaranteed well. In order to improve coal quality information management level of enterprises, it is necessary to develop a coal quality MIS which based on B/S structure. In this paper, the raw coal quality MIS based on four-layer development framework and designed by the B/S model has been put into operation. The operation results show that the system can solve the data sharing of scattered mining and the circulation audit of report well. The graphic representation of coal quality data also greatly improved the readability of coal information to business leaders. Complete function module covers planning, mining processing, and report and statistics. It makes the management of business logic more effective and convenient, the centralized processing of data more quickly and rigorously. It promoted the enterprise information management and enhanced the core competitiveness of the enterprise greatly.

Acknowledgements This work was supported by Education Department of Shaanxi Provincial Government of China. Project No. 14JF016

References

1. Li Meiying, Wang Hang and Yin Shiyu. Resource Characteristics of China's Coal and Its Utilization. Petroleum & Petrochemical Today, 11:24–28 (2015).
2. ChangZhen Chen, Shangkun Jia and Hongmin Yang. Discussion on Improving the quality of raw coal. Coal Science & Technilogy Magazine. 4:150–151, (2015).
3. Lixia Wang. Application of Computer Technology in the Coal Quality Management. Digital Technology and Applications. 1:100 (2014).
4. Dejun Peng. Application of Workflow Technology in Coal Quality Management. Chinese hi-tech enterprises. 30:48–49 (2015).
5. Jiang Li. Construction of the coal enterprises of coal quality management information system. Chemical Industry Management. 54 (2015).
6. Xiaoliang Zhang. Information System design of coal quality testing management. Coal quality Technology. 2:6–12 (2015).
7. Zhanying Zhang, Ziping Zhao. A Logistics Information Platform Based on J2EE Lightweight Framework. International Journal of Digital Content Technology and its Applications, Vol.7 (9), pp. 395–401 (2013).
8. Bing-mei Zhao, Li-feng Wei. A Kind of Personalized Employment Information Retrieval System Based on J2EE. International Journal of Digital Content Technology and its Applications, Vol. 7 (6), pp. 174–181 (2013).
9. Xiaozhong Guo, Jian Pan, Ganguo Wu, Fanxiu Xue, Yuanqing Lu. The Research and Design of A Land Reclamation Supervisory Information System in A Mining Area. Irrig. and Drain., Vol. 64 (1) (2015).
10. Huanzhi Zhang. Web, Java technology research in the coal enterprise network management system. Coal technology, 32(7):267–268 (2013).
11. Xiaoyan Zhang, Panliang Cai. The framework research of large-scale coal enterprises marketing information. Industrial automation, 40(4):93–95 (2014).
12. Hui Zhou, Haijun Ren and Liang Ma, etc. MVC Design Pattern and Its Application in Information System Development. Software guide, 11(10):120–122 (2012).
13. Li Jing-zhao, GUO Wen-cheng, HU Ling-li, XIE Chui-yi, LIANG Guo-jun. Research and Practice of System Reconstruction Method Based on MVC Framework. Computer Knowledge and Technology. Vol. 11, No. 22 (2015).
14. Song Tao, Wang Hongxin, Xu qingzeng. An Overview of SPRING 3.0 Lightweight Framework Technology. Telecom World. 23:306–307 (2015).
15. Weidong Zhao, Xiaoqing Bi and Xinming Lu. Design and implementation of role-based fine-grained access control model. Computer engineering and design, 34(2):474–479 (2013).

Structural Modeling of Implementation Enablers of Cloud Computing

Nitin Chawla and Deepak Kumar

Abstract Cloud Computing is not the buzzword but it is a shift of IT departments to the outsourcing vendors without impacting the business efficiency. Some organizations are moving toward cloud computing but many are having a resistance in adopting cloud computing due to limitation of knowledge and awareness in classifying the elements, which affect decision for the acceptance of cloud computing. Therefore, this research paper has focused on accumulating the elements, which can act as enablers, by reviewing existing literature and study from professional and academic viewpoint. All the identified enablers have been structurally modeled to develop the relationship matrix and establish the driving power and dependence power of every element by employing Total Interpretive Structural Modeling (TISM) and Cross Impact Matrix Multiplication Applied to Classification (MICMAC) analysis.

Keywords Cloud computing · Total interpretive structural modeling (TISM) · Cross impact matrix multiplication applied to classification (MICMAC) analysis

1 Introduction

In this business world, organizations have become competitive and global. The aim of most of the organizations is long-standing endurance and it depends upon organization's ability to enable its business processes and needs by implementing applications with the help of IT at moderate upfront investment and minimum maintenance cost. Now is the time to initiate the deliberation about IT infrastructure and IT application which need to be more dynamic to address the significant infrastructure challenges. Organizations must start cognizant thinking to build or

N. Chawla (✉) · D. Kumar
AIIT, Amity University, Noida, Uttar Pradesh, India
e-mail: Nitin1203@gmail.com

D. Kumar
e-mail: deepakgupta_du@rediffmail.com

© Springer Nature Singapore Pte Ltd. 2018
S.K. Bhatia et al. (eds.), *Advances in Computer and Computational Sciences*,
Advances in Intelligent Systems and Computing 554,
https://doi.org/10.1007/978-981-10-3773-3_27

improve IT infrastructure to support flexible and on demand business needs. Flexibility and Agility are the key to tackle the business requirements, which are continuously changing or increasing. Cloud computing has given the option to use IT resources either infrastructure or application or both on private cloud or public cloud. From business perspective, self service offerings on SaaS model with pay-as-you-go pricing options are available. Services can be provisioned in very less time with the flexibility to scale up or down along with the prices. From IT perspective, lesser involvement of IT resources in infrastructure management and better service delivery with the improved SLAs are driving factors. Virtualization of IT infrastructure has become the backbone of cloud computing by fulfilling the need of shared usage of large pool of resources.

Organizations face difficulties in dealing with complex issues because of availability of ample impacted attributes or criterion and their relationships among each other. The existence of inter relationship between attributes complicates the articulation in an understandable manner. Thus, development of ISM methodology [1], which facilitates in recognizing a structure within a system, took place.

Usage of IT applications and infrastructure is going to increase so that business can run smoothly and that with the right way to manage its data and processes. Advances in Information and Technology has lead the need of cloud computing. Cloud computing is computing paradigm which offers infrastructure or platform or software as a service. Cloud computing backbone is the architecture of need-based resource distribution while leveraging multiple technologies such as Virtual Machines. Cloud computing also offers services such has levels of service management, lower downtime, and higher uptime window. Cloud computing provides services based on the needs of the organizations for example increased CPUs for faster processing, dedicating RAM to support high transaction volume [2].

Adoption of new technology or concept is always a challenge that is why there are still some reservations to adopt cloud computing. Various strategies and factors have been identified which helps the organizations in adoption of cloud computing. Factors not only cover the technical aspects but also cover other functional and cost related aspects to identify the right cloud computing environment. Some of the factors are Capital Expenditure required to use an application or infrastructure, lead time to enable the application or its related infrastructure, security related aspects or factors, availability of the application or infrastructure [3].

This paper is divided into multiple sections as follows: Sect. 2 highlights the review of some of the existing literature to identify the elements that might affect the adoption of cloud computing. Section 3 shows the key enablers that are used for structural modeling by establishing the contextual relationship among the enablers. After which, final reachability matrix has been defined which are used for portioning and then arrive to a MICMAC analysis. Last section presents the conclusion and suggestions for future research.

2   Literature Review

Cloud computing [4, 5] provides the platform to deploy infrastructure or applications which can be accessed using internet or intranet. The model of Cloud computing is different from the conventional model in which an organization needs to setup the IT infrastructure by procuring servers, hardware, switches, applications licenses, etc. Cloud computing offers everything in the form of services And it depends on the needs of IT department of the organization that IT infrastructure is required on cloud or application is required on cloud for example more storage power can be fulfilled by cloud as well as online email as a service is also offered by cloud. Generally, cloud-delivered services have following common attributes

- Clients will need to embrace standards to take advantage of Cloud Services
- Pay as you go model and low upfront investments give benefits in pricing.
- Higher flexibility to scale up or scale down as per business demands.
- Virtualized servers, network, and storage are pooled in a centralized location without impacting the customer's application or usage on sharing basis.

Multiple security problems [6] are discussed in cloud computing along with the approach to fight security problems. A study showed that data privacy in the cloud-based systems is the main concern and through comparative analysis security problem cannot be tackled by single security method. Many conventional and recent technologies and strategies are required to protect the cloud-based solutions. There are multiple views in terms of security is concerned such as cloud architecture view, delivery model view, characteristics view, and the stakeholder view. Some solutions are investigated through which the dynamic cloud model can be secured [7, 8].

A study, done by Young-Chan Lee, has presented a framework to choose the cloud computing deployment model suitable for an organization. This decision model is developed using Analytic Hierarchy Process (AHP) and benefit-cost-opportunity-risk (BCOR) analysis. It has four hierarchies: benefits, costs, opportunity, and risks. All the hierarchies have some attributes related to its category [9].

Another study showed key attributes with the advantages and threats of cloud computing influencing the adoption decision. This study has defined the type of attributes which affect the adoption of cloud computing. This study not only shows the attributes but also shows the motivation/benefits and concerns/risks related to those attributes. This literature explored the inclination of an enterprise to accept cloud computing services because key drivers for adoption were managed services by third party to adhere better SLAs and usage of advanced technology. Decision of Cloud was based on three theoretical perceptions: Transaction cost theory, resource dependence theory, and diffusion of innovation theory [10].

The Total Interpretive Structural Modeling (TISM) process is used for finding the relationship among the identified variables or elements that act as enablers and are the source of a problem or an issue. It is a well defined and comprehensive systemic model useful to structure variables that are directly or indirectly linked. A structure was defined to know the relationship of the elements. Lastly the relationship of the elements was represented in the form of diagraph model. All the elements were given the category of driving power or dependence power which helps to identify the complexity of the relationship [1].

Total Interpretive Structural Modeling (TISM) begins with the identification of various attributes, applicable to the issue, and then broadens with the team problem solving technique. After applicable secondary relation is decided, a pairwise comparison of variables is performed to develop structural self-interaction matrix (SSIM). As a subsequent stride, reachability matrix (RM) is arrived based on SSIM to obtain a matrix. Then, using partitioning and diagraph TISM model is derived [11].

3 Methodology

In this study, identification of key enablers, as mentioned in Table 1, has been performed by going through multiple existing studies and brainstorming session conducted with cloud computing experts. Then, multiple sessions were conducted to recognize the key elements which will act as enablers identified by literature review. Following enablers were decided for this study. This paper uses TISM and MICMAC analysis to do modeling in order to know the interrelationship between the enablers.

Table 1 Identification of key enablers

S.no	Key enablers	S.no	Key enablers
C-1	Upfront capital investment for software	C-10	Use of advanced technology
C-2	Upfront capital investment for infrastructure	C-11	SLA improvements
C-3	Deployment model	C-12	Agility to upscale anytime
C-4	Data security and privacy	C-13	Internet speed/latency
C-5	Data privacy	C-14	Outages
C-6	Data migration	C-15	Disaster recovery
C-7	Deployment/implementation	C-16	Automatic updates
C-8	Involvement of employees	C-17	Data ownership
C-9	Process improvement, reduction in cycle time		

3.1 Interpretive Structural Modeling Steps

The different steps involved in TISM modeling are as follows

1. Literature study and survey contributes the classification of elements relevant to the issue for this study and elements are mentioned in Table 1.
2. Institute a relative relationship between the enablers.
3. Based on the discussion with industry experts, structural self-interaction matrix (SSIM) is extended for elements that shows pair-wise relations between elements.
4. Initial reachability matrix is shaped based on structural self-interaction matrix by substituting 1 and 0 s in place of Y and N in the initial reachability matrix. Transitivity concept has been induced to obtain the reachability matrix.
5. Reachability set and antecedent sets are derived for each element and to discover the level of each element partitions are done.
6. Based on the driving power and dependence power, elements are placed on the Conical matrix clusters (Linkages, Dependent, Independent, and Autonomous).
7. The preliminary diagraph is obtained. After removing the transitivity, a final diagraph is developed.
8. Diagraph is converted into the TISM model by replacing nodes of the factors with statements.

3.2 Structural Self-Interaction Matrix (SSIM)

Elements are identified using literature study and integrated approach is required to build a structural relationship among the elements using various processes like brain storming with industry and academic experts. Nature of contextual relationships among the elements has been achieved.

We are using TISM methodology which guides to create the contextual relationship matrix between the elements. If one element influences another element then "Y" is marked otherwise "N" is marked as depicted in Table 2.

3.3 Development of Reachability Matrix

Reachability matrix from the SSIM is to be developed. The SSIM has been converted into a binary matrix, by substituting Y and N to 1 and 0. Subsequent step is to redefine few of the cells which are influenced by inference. It can be achieved by inducing the transitivity and final reachability matrix can be defined in Table 3.

Table 2 Structural Self-interaction matrix (SSIM)

Code	Elements	C-17	C-16	C-15	C-14	C-13	C-12	C-11	C-10	C-9	C-8	C-7	C-6	C-5	C-4	C-3	C-2	C-1
C-1	Upfront capital investment for software	N	N	N	N	N	N	Y	Y	Y	N	N	N	N	N	N	Y	Y
C-2	Upfront capital investment for infrastructure	N	Y	N	N	N	Y	Y	N	Y	Y	N	N	N	Y	N	Y	Y
C-3	Deployment model	Y	Y	Y	Y	Y	Y	Y	N	N	Y	Y	Y	Y	Y	Y	Y	Y
C-4	Data security	N	N	N	N	N	N	Y	Y	Y	N	Y	N	N	Y	Y	Y	N
C-5	Data privacy	N	N	N	N	N	N	N	Y	Y	N	Y	N	Y	N	Y	N	N
C-6	Data migration	N	N	N	N	Y	N	Y	Y	Y	N	N	Y	N	N	Y	N	N
C-7	Data ownership	N	N	N	N	N	N	Y	Y	Y	N	Y	N	Y	Y	Y	N	N
C-8	Deployment/implementation	N	Y	N	N	N	N	Y	Y	Y	Y	N	N	N	Y	Y	Y	N
C-9	Service model	Y	Y	Y	Y	Y	Y	Y	N	Y	Y	Y	Y	Y	Y	Y	Y	Y
C-10	Process improvement, reduction in cycle time	N	N	N	N	N	N	Y	Y	N	Y	Y	N	Y	Y	N	N	Y
C-11	Use of advanced technology	Y	N	N	N	N	N	Y	N	Y	Y	Y	Y	Y	Y	Y	Y	Y
C-12	SLA Improvements	Y	Y	Y	N	N	Y	Y	N	Y	N	N	N	N	N	Y	Y	N
C-13	Agility to upscale anytime	Y	N	N	N	Y	Y	N	N	Y	N	N	Y	N	N	Y	Y	N
C-14	Internet speed/latency	Y	Y	Y	Y	Y	Y	N	N	Y	N	N	Y	N	N	Y	N	N
C-15	Outages	N	N	Y	Y	N	N	N	N	Y	N	N	Y	N	N	Y	N	N
C-16	Disaster recovery	N	Y	Y	Y	N	Y	Y	N	Y	N	N	N	N	N	Y	Y	Y
C-17	Automatic updates	Y	N	Y	Y	N	N	Y	N	Y	N	N	N	N	N	Y	N	N

Table 3 Final reachability matrix

Code	Elements	C-1	C-2	C-3	C-4	C-5	C-6	C-7	C-8	C-9	C-10	C-11	C-12	C-13	C-14	C-15	C-16	C-17	Driver Power
C-1	Upfront capital investment for software	1	1	0	0	0	0	0	0	1	1	1	0	0	0	0	0	0	5
C-2	Upfront capital investment for infrastructure	1	1	0	1	0	0	0	1	1	0	1	1	0	0	0	1	0	8
C-3	Deployment model	1	1	1	1	1	1	1	1	0	0	1	1	1	1	1	1	1	15
C-4	Data security	0	1	1	1	0	0	0	0	1	1	1	0	0	0	0	0	0	6
C-5	Data privacy	0	0	1	0	1	0	1	0	1	1	0	0	0	0	0	0	0	5
C-6	Data migration	0	0	1	0	0	1	0	0	1	1	1	0	1	0	0	0	0	6
C-7	Data ownership	0	0	1	0	1	0	1	0	1	1	1	0	0	0	0	0	0	6
C-8	Deployment/implementation	0	1	1	1	0	0	0	1	1	1	1	0	0	0	0	1	0	8
C-9	Service model	1	1	1	1	1	1	1	1	1	0	1	1	1	1	1	1	1	16
C-10	Process improvement, reduction in cycle time	1	0	0	1	1	1*	1	1	0	1	1	0	0	0	0	0	0	8
C-11	Use of advanced technology	1	1	1	1	1	1	1	1	1	0	1	0	0	0	0	0	1	11
C-12	SLA improvements	0	1	1	0	0	0	0	0	1	0	1	1	1*	0	1	1	1	9
C-13	Agility to upscale anytime	0	1	1	0	0	1	0	0	1	0	1*	1	1	0	0	0	1	8
C-14	Internet speed/latency	0	0	1	0	0	1	0	0	1	0	1*	1	1	1	1	1	1	10
C-15	Outages	0	0	1	0	0	1	0	0	1	0	0	0	0	1	1	0	0	5
C-16	Disaster recovery	1	1	1	0	0	0	0	0	1	0	1	1	0	0	1	1	0	8
C-17	Automatic updates	0	0	1	0	0	0	0	0	1	0	1	0	1*	1	1	0	1	7

3.4 Level Partitions

After getting the final reachability matrix, reachability set, antecedent set, and intersection set has been found and shown in Table 4.

3.5 MICMAC Analysis and TISM Diagraph

Cloud Computing adoption enablers are divided into four groups:

- Autonomous enablers—These elements have weak driving and weak dependencies power so these are disconnected from the system or have fewer linkages available in the system.
- Dependent enablers—These elements have strong dependence but weak driving power.
- Linkage enablers—This group have strong driving and strong dependency power. These elements have the direct impact on other elements and are impacted from other elements.
- Independent enablers—These elements have strong driving power but weak dependence power. These elements impact other elements but are unaffected by other elements actions.

These enablers are segregated on depending upon their driving power and dependencies as shown in Table 4 and categorized in Fig. 1.

Final Reachability matrix, mentioned in Table 4, has been used to create the structural model. This graph is known as directed graph, or diagraph. This diagraph is converted into TISM-based model for the adoption of cloud computing and shown in Fig. 2.

4 Key Conclusions and Future Directions

The key enablers are essential in adoption of cloud computing for the organizations. There are some important enablers visible in this study and these enablers are put into an interpretive structure modeling model to explore the relationship among them. Key enablers need to evaluate the successful adoption of cloud computing. The TISM methodology was used to establish the driving power and dependencies of the enablers identified.

Table 4 Iteration 1–5/Level 1–5

Level partition—level1

Code	Elements	Reachability set	Antecedent set	Intersection set	Level
C-1	Upfront capital investment for software	1, 2, 9, 10, 11	1, 2, 3, 9, 10, 11, 16	1, 2, 9, 10, 11	1
C-2	Upfront capital investment for infrastructure	1, 2, 4, 8, 9, 11, 12, 16	1, 2, 3, 4, 8, 9, 11, 12, 13, 16	1, 2, 8, 9, 11, 12, 16	
C-3	Deployment model	1, 2, 3, 4, 5, 6, 7, 8, 11, 12, 13, 14, 15, 16, 17	2, 3, 4, 5, 6, 7, 8, 9, 11, 12, 13, 14, 15, 16, 17	2, 3, 4, 5, 6, 7, 8, 11, 12, 13, 14, 15, 16, 17	
C-4	Data security	1, 2, 3, 9, 10, 11	2, 3, 4, 8, 9, 10, 11	2, 3, 9, 10, 11	
C-5	Data privacy	3, 5, 7, 9, 10	3, 5, 7, 9, 10, 11	3, 5, 7, 9, 10	1
C-6	Data migration	3, 6, 9, 10, 11, 13	3, 6, 9, 10, 11, 13, 14, 15	3, 6, 9, 10, 11, 13	1
C-7	Data ownership	3, 5, 7, 9, 10, 11	3, 5, 7, 9, 10, 11	3, 5, 7, 9, 10, 11	1
C-8	Deployment/implementation	2, 3, 4, 8, 9, 10, 11, 16	2, 3, 8, 9, 10, 11	2, 3, 8, 9, 10, 11	
C-9	Service model	1, 2, 3, 4, 5, 6, 7, 8, 9, 11, 12, 13, 14, 15, 16, 17	1, 2, 4, 5, 6, 7, 8, 9, 11, 12, 13, 14, 15, 16, 17	1, 2, 4, 5, 6, 7, 8, 9, 11, 12, 13, 14, 15, 16, 17	
C-10	Process improvement, reduction in cycle time	1, 4, 5, 6, 7, 8, 10, 11	1, 4, 5, 6, 7, 8, 10	1, 4, 5, 6, 8, 10	
C-11	Use of advanced technology	1, 2, 3, 4, 5, 6, 7, 8, 9, 11, 17	1, 2, 3, 4, 6, 7, 8, 9, 10, 11, 12, 13, 14, 17	1, 2, 3, 4, 6, 7, 8, 9, 11, 17	
C-12	SLA improvements	2, 3, 9, 11, 12, 13, 15, 16, 17	2, 3, 9, 12, 13, 14, 16	2, 3, 9, 12, 16	
C-13	Agility to upscale anytime	2, 3, 6, 9, 11, 12, 13, 17	3, 6, 9, 12, 13, 14, 17	3, 6, 9, 12, 13, 17	
C-14	Internet speed/latency	3, 6, 9, 12, 13, 14, 15, 16, 17	3, 9, 14, 15, 16, 17	3, 9, 14, 15, 16, 17	
C-15	Outages	3, 6, 9, 14, 15	3, 9, 12, 14,15, 16, 17	3, 9, 14, 15	
C-16	Disaster recovery	1, 2, 3, 9, 12, 14, 15, 16	2, 3, 8, 9, 11, 14, 16	2, 3, 9, 14, 16	
C-17	Automatic updates	3, 9, 11, 13, 14, 15, 17	3, 9, 12, 13, 17	3, 9, 11, 13, 17	

(continued)

Table 4 (continued)

Code	Elements	Reachability set	Antecedent set	Intersection set	Level
Level partition—level1					Level
Level partition—level2					
Code	Elements	Reachability set	Antecedent set	Intersection set	Level
C-2	Upfront capital investment for infrastructure	2, 4, 8, 9, 11, 12, 16	2, 3, 4, 8, 9, 11, 12, 13, 16	2, 8, 9, 11, 12, 16	
C-3	Deployment model	2, 3, 4, 8, 11, 12, 13, 14, 15, 16, 17	2, 3, 4, 8, 9, 11, 12, 13, 14, 15, 16, 17	2, 3, 4, 8, 11, 12, 13, 14, 15, 16, 17	2
C-4	Data security	2, 3, 9, 10, 11	2, 3, 4, 8, 9, 10, 11	2, 3, 9, 10, 11	2
C-8	Deployment/implementation	2, 3, 4, 8, 9, 10, 11, 16	2, 3, 8, 9, 10, 11	2, 3, 8, 9, 10, 11	
C-9	Service model	2, 3, 4, 8, 9, 11, 12, 13, 14, 15, 16, 17	2, 4, 8, 9, 11, 12, 13, 14, 15, 16, 17	2, 4, 8, 9, 11, 12, 13, 14, 15, 16, 17	
C-10	Process improvement reduction in cycle time	4, 8, 10, 11	4, 8, 10	4, 8, 10	
C-11	Use of advanced technology	2, 3, 4, 8, 9, 11, 17	2, 3, 4, 8, 9, 10, 11, 12, 13, 14, 17	2, 3, 4, 8, 9, 11, 17	2
C-12	SLA improvements	2, 3, 9, 11, 12, 13, 15, 16, 17	2, 3, 9, 12, 13, 14, 16	2, 3, 9, 12, 16	
C-13	Agility to upscale anytime	2, 3, 9, 11, 12, 13, 17	3, 9, 12, 13, 14, 17	3, 9, 12, 13, 17	
C-14	Internet speed/latency	3, 9, 12, 13, 14, 15, 16, 17	3, 9, 14, 15, 16, 17	3, 9, 14, 15, 16, 17	
C-15	Outages	3, 9, 14, 15	3, 9, 12, 14, 15, 16, 17	3, 9, 14, 15	2
C-16	Disaster recovery	2, 3, 9, 12, 14, 15, 16	2, 3, 8, 9, 11, 14, 16	2, 3, 9, 14, 16	
C-17	Automatic updates	3, 9, 11, 13, 14, 15, 17	3, 9, 12, 13, 17	3, 9, 11, 13, 17	
Level partition—level3					
Code	Elements	Reachability set	Antecedent set	Intersection set	Level
C-2	Upfront capital investment for infrastructure	2, 8, 9, 12, 16	2, 8, 9, 12, 13, 16	2, 8, 9, 12, 16	3

(continued)

Table 4 (continued)

Level partition—level1

Code	Elements	Reachability set	Antecedent set	Intersection set	Level
C-8	Deployment/implementation	2, 8, 9, 10, 16	2, 8, 9, 10, 11	2, 8, 9, 10, 11	
C-9	Service model	2, 8, 9, 12, 13, 14, 16, 17	2, 8, 9, 12, 13, 14, 16, 17	2, 8, 9, 12, 13, 14, 16, 17	3
C-10	Process improvement, reduction in cycle time	8, 10, 11	8, 10	8, 10	
C-12	SLA improvements	2, 9, 12, 13, 16, 17	2, 9, 12, 13, 14. 16	2, 9, 12, 16	
C-13	Agility to upscale anytime	2, 9, 12, 13, 17	9, 12, 13, 14, 17	9, 12, 13, 17	
C-14	Internet speed/latency	9, 12, 13, 14, 16. 17	9, 14, 16, 17	9, 14, 16, 17	
C-16	Disaster recovery	2, 9, 12, 14, 16	2, 8, 9, 14, 16	2, 9, 14, 16	
C-17	Automatic updates	9, 13, 14, 17	9, 12, 13, 17	9, 13, 17	

Level partition—level4

Code	Elements	Reachability set	Antecedent set	Intersection set	Level
C-8	Deployment/implementation	8, 10, 16	8, 10, 11	8, 10, 11	
C-10	Process improvement, reduction in cycle time	8, 10, 11	8, 10	8, 10	
C-12	Sla improvements	12, 13, 16, 17	12, 13, 14, 16	12, 16	
C-13	Agility to upscale anytime	12, 13, 17	12, 13, 14, 17	12, 13, 17	4
C-14	Internet speed/latency	12, 13, 14, 16, 17	14, 16, 17	14, 16, 17	
C-16	Disaster recover	12, 14, 16	8, 14, 16	14, 16	
C-17	Automatic updates	13, 14, 17	12, 13, 17	13, 17	

Level partition—level5

Code	Elements	Reachability set	Antecedent set	Intersection set	Level
C-8	Deployment/implementation	8, 10, 16	8, 10, 11	8, 10, 11	

(continued)

Table 4 (continued)

Level partition—level1

Code	Elements	Reachability set	Antecedent set	Intersection set	Level
C-10	Process improvement, reduction in cyclet ime	8, 10, 11	8, 10	8, 10	
C-12	Sla improvements	12, 16, 17	12, 14, 16	12, 16	
C-14	Internet speed/latency	12, 14, 16, 17	14, 16, 17	14, 16, 17	
C-16	Disaster recovery	12, 14, 16	8, 14, 16	14, 16	
C-17	Automatic updates	14, 17	12, 17	17	

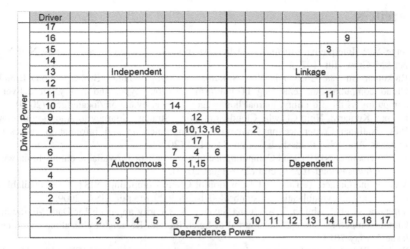

Fig. 1 Cluster of cloud computing enablers

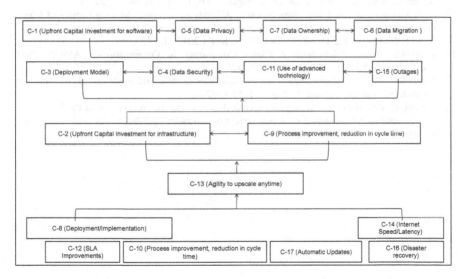

Fig. 2 Interpretive structural model of cloud computing adoption enablers

TISM proves that all the enablers play an important role in the successful adoption of cloud computing. In this research, some enablers have been used to develop the TISM model, but more enablers can be included to develop the relationship among them using the TISM methodology.

The limitation of this study is that the model developed here is not statistically verified so future study can be conducted to implement this model.

References

1. Sage A. P., Interpretive structural modeling: Methodology for large scale systems, New York, NY: McGraw-Hill (1977).
2. Rajkumar Buyyaa, Chee Shin Yeo, Srikumar Venugopal, James Broberg, Ivona Brandic, Cloud computing and emerging IT platforms: Vision, hype, and reality for delivering computing as the 5th utility, journal homepage: www.elsevier.com/locate/fgcs, (2009).
3. Deepak Kumar & Nitin Chawla, Cloud Computing Strategy: Journey to Adoption of cloud, Proc. National Conference on Computing, Communication and Information Processing, Itanagar, Arunachal Pradesh, India, pp. 112–122 (2015).
4. Final Version of NIST Cloud Computing Definition Published. Available online: http://www.nist.gov/itl/csd/cloud-102511.cfm (25 August 2013).
5. Mell, P and Grance, T. The NIST Definition of Cloud Computing, NIST, USA. available at: http://csrc.nist.gov/publications/nistpubs/800-145/SP800-145.pdf, USA, 2009.
6. Virendra Singh Kushwah & Aradhana Saxena, A Security approach for Data Migration in Cloud Computing, International Journal of Scientific and Research Publications, Volume 3, Issue 5, May 2013.
7. W. Liu. 'Research on cloud computing security problem and strategy.' IEEE, pp 1216–1219, (2012).
8. Behl K. & Behl A., An analysis of cloud computing security issues, IEEE Transactions on Information and Communication Technologies (WICT), 2(2) 109–114 (2012).
9. Lee, Young-Chan and Hanh, Tang Nguyen, A Study on Decision Making Factors of Cloud Computing Adoption Using BCOR Approach, Journal of the Korea society of IT services Vol.11, Issue 1, 2012, 155–171, doi:10.9716/KITS.2012.11.1.155.
10. Nuseibeh, Hasan, "Adoption of Cloud Computing in Organizations", AMCIS 2011 Proceedings - All Submissions. Paper 372 (2011).
11. Warfield, J. W., Developing interconnected matrices in structural modeling. IEEE transcript on systems, Volume 4, Issue 1, Pages 51–58 (1974).
12. Sabahi, F. Virtualization-level security in cloud computing. In Proceedings of the 2011 IEEE 3rd International Conference on Communication Software and Networks (ICCSN), Xi'an, China, 27– 29; pp. 250–254. (May 2011).

Labelling and Encoding Hierarchical Addressing for Scalable Internet Routing

Feng Wang, Xiaozhe Shao, Lixin Gao, Hiroaki Harai
and Kenji Fujikawa

Abstract Hierarchical addressing and locator/ID separation solutions have been proposed to address the scalability issue of the Internet. However, how to combine the two addressing schemes has not been received much attention. In this paper, we present an address encoding method to integrate hierarchical addressing and locator/ID separation. Our analysis and evaluation results show that the proposed encoding method could guarantee the scalability property, and alleviate the inefficiency of address space.

Keywords Variable length encoding · Hierarchical addressing · Internet routing

1 Introduction

The Internet is facing the accelerating growth of routing table size. Since the Internet has become integrated into our daily lives, the scalability of the Internet has become extremely important. There are two trends in improving the scalability of inter-domain routing: locator/ID separation solutions and hierarchical addressing scheme [1–5]. Combining these two trends—the address format follows locator/ID separation technology, and the location allocation is based on a hierarchical routing topology—can significantly reduce routing table size.

F. Wang (✉)
School of Engineering and Computational Sciences, Liberty University,
Lynchburg, VA 24515, USA
e-mail: fwang@liberty.edu

X. Shao · L. Gao
Department of Electrical and Computer Engineering, University of Massachusetts,
Amherst, MA 01003, USA

H. Harai · K. Fujikawa
National Institute of Information and Communications Technology,
Tokyo 184-8795, Japan

© Springer Nature Singapore Pte Ltd. 2018
S.K. Bhatia et al. (eds.), *Advances in Computer and Computational Sciences*,
Advances in Intelligent Systems and Computing 554,
https://doi.org/10.1007/978-981-10-3773-3_28

However, integrating hierarchical addressing and locator/ID separation is challenging [3–5]. A common way to encode hierarchical addressing is to use fixed-length addresses. However, fixed-length addresses do not scale well because it is very difficult to pre-assign address length to accommodate the future growth of the Internet. In addition, it is difficult to automatically allocate locators with minimal human intervention.

In this paper, we employ a prefix-based labelling scheme to label hierarchical addresses, and use variable-length encoding to implement the proposed addressing scheme. We investigate the performance of the encoding method by understanding the degree to which the Internet can benefit from hierarchical addressing. Based on the real routing data, our study shows that it is difficult to use fixed-length encoding to implement hierarchical addressing for today's Internet. Furthermore, fixed-length encoding uses address space inefficiently. Our analysis and evaluation results show that variable-length encoding could resolve both the scalability problem and the inefficiency of address spaces.

The rest of paper is organized as follows. We present a variable-length encoding scheme in Sect. 2. We evaluate the performance of the proposed encoding scheme in Sect. 3. We conclude the paper in Sect. 4 with a summary.

2 Variable-Length Encoding Hierarchical Addressing

We employ a locator/ID address format by separating the location and the identity of a host. An address of each host consists of two parts: a *host ID* and a *locator*. Since the location of a node may be dynamic and change with node movement, a locator is used to describe the network attachment point(s) to which a host connects. Each host is assigned a globally unique number, which would not change even if the host moved to another location. Each host may have multiple locators representing different routes toward it. Each AS obtains a locator from a provider in the upper layer. Locators are labelled by a prefix-based labelling scheme. Each locator consists of a *provider label* followed by a *self label*. A provider label identifies the provider, while a self label identifies a customer. We use delimiter '.' to join a set of provider labels and self labels. Because Tier-1 ISPs do not have any provider, their provider labels are empty. A centralized authority, such as IANA, can assign them a unique self label. So, their self labels are their locators. For other non-Tier-1 ASes, they first determine a self label for each customer. Then, each provider concatenates its own provider label and the self-label to generate a customer's locator. With the prefix-based labelling, customers that have the same providers should share a common prefix. To avoid the allocation loop, a provider is prohibited from advertising a locator back to a customer from which the locator was assigned. In addition, a provider can detect a locator containing a loop by examining whether the locator and any one of its locators share the same prefix.

The proposed addressing scheme is used on a loose underlying hierarchical topology, where the hierarchy is not strictly enforced. Even though providers are organized as a tiered hierarchy, it only provides a loose characterization of the AS hierarchy. There are interconnections between providers at the same tier, and a non-Tier-2 AS may obtain locators directly from a Tier-1 provider. For example, in Fig. 1, AS28917 (non-Tier-2) obtains a locator from Tier-1 AS3356.

The idea of using a variable-length representation is to assign each self label a variable-length number. The size of each self-label is essentially determined by a specific variable-length encoding scheme. Using a variable-length encoding method, we can use an infinite number of integers to represent self labels. That means, it does not have a restriction on the size of self labels.

Universal codes [6] are particularly suitable for our purpose because each universal code assumes an implied probability distribution. The implied distribution is an estimation of power law or Zipf distribution. Previous work has shown that the number of customers in today's Internet can be described as a power law distribution [7]. We believe that the distribution of self-labels would follow a power law as well. There are several possible universal codes for self labels: Elias' γ, Elias' δ, Elias' ω codes [8], and Baer code [9]. Table 1 shows some part of Baer codewords. Note that we do not claim that this code is the optimal code for self labels. For example, based on each AS's rank, which is determined by the number of descendants, we show the encoded locators in Fig. 1. From this figure, we see that the maximum length of AS34608's locators is 26 bits. On the contrary, fixed

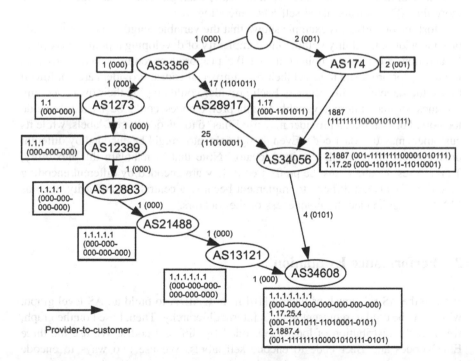

Fig. 1 An example of Baer encoded locators. The number in parentheses represents an encoded locator or self label. Dashes are used to make encoded locators more readable

Table 1 An example of Baer codeword tables

Value	Baer
1	000
2	001
3	0100
4	0101
5	0110
6	0111
7	10000
8	10001
9	10010
10	10011
11	101000
12	101001
13	101010
14	101011
15	101100

self-label encoding uses locator space inefficiently because the size of self-labels must be predetermined. As a result, a 32-bit number may not be sufficient to represent all locators in the Internet. In the same example, AS34608's locators have more than 32 bits using fixed self label encoding.

Most importantly, this example shows that the variable-length encoding methods provide a lot of flexibility without the complexity of developing a prefix mask plan. In addition, we need to point out that the proposed encoding scheme provides flexibility for providers to select their own universal codes. Providers are not forced to use the same encoding scheme. Each provider could employ their own encoding schemes so that different variable-length prefix codes could be used to encode locators. For example, a provider may use Elias' δ to encode its self-labels, while its customers may use Baer code. Even though a locator might be encoded by different universal codes, it is still a unique locator. Note that to maintain this flexibility, Tier-1 ASes' locators must be prefix-free if they are encoded by different encoding schemes. This is not difficult to implement because a centralized authority, such as IANA, can guarantee the uniqueness of the locators.

3   Performance Evaluation

We use the AS relationships dataset [10] from CAIDA to build an AS level graph, which can be used to represent today's Internet hierarchy. Then, based on the graph, we infer the distribution of locators encoded by different codes. We use the three Elias' codes and Baer code to encode self-labels. We use two ways to encode locators: *random assignment* and *ranking-based assignment*. Random assignment

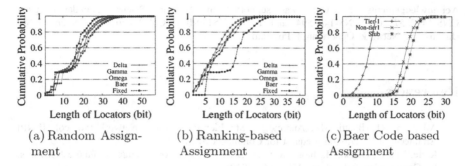

(a) Random Assignment   (b) Ranking-based Assignment   (c) Baer Code based Assignment

Fig. 2 Length distribution of locators encoded in different codes

means that we randomly assign a self label to a customer, and encode the number using one of the codes. Ranking-based assignment means that we assign a self-label according to the rank of a customer. The ranking is determined by the number of descendants of each customer. Then, we sort the ranking of the customers in non-increasing order. According to its order, a self label codeword is assigned to each customer. Meanwhile, we compare the length of encoded locators with that of fixed-length locators. We use the total number of customers for each provider to determine the number of bits to represent self-labels.

We present the results in Fig. 2. From Fig. 2a, we find that the locator length will be almost the same if we randomly select a codeword. The maximum length of fixed-length locators is 56 bits. Over 80% of the locators encoded in Baer code have no more than 24 bits. As shown in Fig. 2b, if we employ ranking-based assignment, the locators encoded by Elias' δ and Baer codes are always shorter than the others, including fixed-length method. In Fig. 2c, we redraw the length distribution of locators encoded by Baer code. It shows the length at different hierarchical layers. We can see locators at different layers might have different length distribution, and the majority of the locators are 10–25 bits long. In addition, the average length of variable-length locators is about 19 bits, while the average length of the fixed-length locators is about 28 bits. Hence, the variable-length locators are much shorter than fixed-length locators.

4 Conclusion

In this paper, we propose a variable-length addressing encoding scheme to mitigate the scalability issue of the current Internet. To support this new addressing architecture, we present an address encoding method to implement the addressing scheme. Our analysis and evaluation results show that the proposed encoding method could guarantee the scalability property of hierarchical addressing, and alleviate the inefficiency of address space due to fixed-length addresses.

Acknowledgements This work was supported by National Science Foundation grant CNS-1402857 and CNS-1402594 and under NSF-NICT Collaborative Research JUNO (Japan-U.S. Network Opportunity) Program.

References

1. Atkinson, RJ., Bhatti, SN., Andrews, U.St.: Identifier-Locator Network Protocol (ILNP) Architectural Description. Request for Comments 6740. (2012).
2. Kafle, Ved P., Otsuki, H., Inoue, M.: An ID/locator Split Architecture for Future Networks. In: Comm. Mag. vol. 48, pp. 138–144. (2010).
3. Tsuchiya, P.: Efficient and Flexible Hierarchical Address Assignment. In: INET92. pp. 441–450. Kobe Japan (1992).
4. Zhuang, Y., Calvert, K.L.: Measuring the Effectiveness of Hierarchical Address Assignment. In: GLOBECOM. pp. 1–6. IEEE Press, Miami (2010).
5. Lu, X., Wang, W., Gong, X., Que, X., Wang, B.: Empirical analysis of different hierarchical addressing deployments. In: 19th Asia-Pacific Conference on Communications (APCC). pp. 208–213. IEEE Press, Denpasar (2013).
6. Ziv, J., Lempel, A.: A Universal Algorithm for Sequential Data Compression. In: IEEE Transactions on information theory. vol. 23, pp. 337–343. IEEE Press (1977).
7. Shakkottai, S., Fomenkov, M., Koga, R., Krioukov, D., Claffy, kc: Evolution of the Internet AS-Level Ecosystem. In: Complex Sciences. vol. 5, pp. 1605–1616. Springer, Berlin Heidelberg (2009).
8. Elias, P.: Universal Codeword Sets and Representations of the Integers. In: IEEE Transactions on Information Theory. vol. 21, pp. 194–203. IEEE Press (1975).
9. Baer, M.B.: Prefix Codes for Power Laws. In: 2008 IEEE International Symposium on Information Theory (ISIT 2008). pp. 2464–2468. IEEE Press, Toronto (2008).
10. The IPv4 Routed/24 AS Links Dataset, http://www.caida.org/data/active/ipv4_routed_ topology_aslinks_dataset.xml.

A Cuckoo Search Algorithm-Based Task Scheduling in Cloud Computing

Mohit Agarwal and Gur Mauj Saran Srivastava

Abstract Recently, Cloud computing emerges out as a latest technology which enables an organization to use the computing resources like hardware, applications, and software, etc., to perform the computation over the internet. Cloud computing gain so much attention because of advance technology, availability, and cost reduction. Task scheduling in cloud computing emerges out as new area of research which attracts the attention of lots researchers. An effective task scheduling is always required for optimum or efficient utilization of the computing resources to avoid the situation of over or under-utilization of such resources. Through this paper, we are going to proposed the cuckoo search-based task scheduling approach which helps in distributing the tasks efficiently among the available virtual machines (VM's) and also keeps the overall response time (QoS) minimum. This algorithm assigns the tasks among the virtual machines on the basis of their processing power, i.e., million instructions per seconds (MIPS) and length of the tasks. A comparison of cuckoo search algorithm is done with the first—in first—out (FIFO) and greedy-based scheduling algorithm which is performed using the CloudSim simulator, the results clearly shows that cuckoo search outperforms the other algorithms.

Keywords Virtual machines · Cloud computing · QoS · Task scheduling · Cuckoo search · Genetic algorithm · Makespan

M. Agarwal (✉) · G.M.S. Srivastava
Department of Physics & Computer Science, Dayalbagh Educational Institute,
Agra, India
e-mail: rs.mohitag@gmail.com

G.M.S. Srivastava
e-mail: gurmaujsaran@gmail.com

© Springer Nature Singapore Pte Ltd. 2018
S.K. Bhatia et al. (eds.), *Advances in Computer and Computational Sciences*,
Advances in Intelligent Systems and Computing 554,
https://doi.org/10.1007/978-981-10-3773-3_29

1 Introduction

Exponential growth in the accessibility of internet and its based technologies, Cloud Computing emerges out as a new form of computing paradigm which becomes very much popular in both industry and academia in short duration of time [1, 2]. As cloud is a collection of the large number of computing resources like processing unit, storage units, and networking components, etc., to service the user. Cloud computing has the features of grid computing, distributed computing, and utility computing [2] but it differs from them as it uses the concept of virtualization for the resource management [3]. Some important features which attract the users and businesses to migrate to the cloud-based technology are: scalability, economic in use, rapid elasticity, on demand accessibility of the resources, and its ubiquitous property [4, 5]. In cloud computing, users do not have to bother about the location of services, i.e., where the resources are located, but the only need is the internet access to fulfill their demands as the resources are available across the globe over the internet. The user needs to use the resources and pay as and when used [6]. Scheduling refers to the allocation of the jobs or tasks to the processing units like virtual machine's (VMs) for their processing. An efficient task scheduling is always needed in cloud computing environment for the proper utilization of the supporting resources. Task scheduling not only helps maintain the makespan of the virtual machines but also helps in maintaining the conditions of service level agreement (SLA) of better service to the consumers by giving the good Quality of services (QoS).

In the proposed CS-based approach, we have utilized the robust search property of Cuckoo search in order to find out the minimum makespan of the tasks in distributed environment.

The remaining portion of this paper is divided into the following sections: next section presents the literature survey with brief discussion of task scheduling in cloud computing. In Sect. 3, cuckoo search algorithm is introduced while Sect. 4, brings our proposed approach for cloud scheduling. Section 5 helps in presenting the experimental results and evaluation work. Section 6 presents our conclusion.

2 Literature Survey

In this section, various task scheduling methods in cloud computing are reviewed and analyzed. Task scheduling problem lie in the category of NP—Hard problem and many scheduling algorithms including the heuristics, which have been applied to solve them. David et al. [7] proposed distributed negotiation-based resource scheduling algorithm and the proposed resource scheduling mechanism is used in heterogeneous and distributed computing environment for the improvement in features like resource utilization capacity, execution cost, and time required for the completion of the tasks. Pandey et al. [8] presents the Particle Swarm-based scheduling mechanism. But this algorithm has some disadvantage like cost of transferring and deadline for the QoS were not taken into consideration. Song et al. [9] come with a Load balancing-based task scheduling mechanism for balancing the entire system load and trying to reduce the makespan for a given set of tasks. They have implemented two different load balancing algorithms which are based on the Ant-Colony optimization (ACO) technique, with aim of minimizing the required completion time on the basis of pheromone. Li et al. in [10] presents the job oriented resource scheduling mechanism for cloud paradigm. This model involves jobs assignment to the computing resources on the basis of the job rank. We also analyze scheduling algorithm such as Round Robin resource scheduling, Pre-emptive Priority scheduling method, and Shortest Remaining Time First. Yang Xu et al. [11] proposed the intelligent load distribution technique from the performance improvement point of view of the cloud system. Mohit et al. [12] proposed genetic algorithm-based task scheduling mechanism for the efficient utilization of the resources implemented on CloudSim. Al-maamari et al. [13] proposed the task scheduling in cloud computing-based PSO, which aims to reduce the makespan and helps in maximum utilization of the resources. Panda et al. [14] presents efficient task scheduling mechanism for multi-cloud environment which aims to minimize the makespan and better utilization of the cloud.

3 Proposed Methodology

Cuckoo search (CS) lie in the category of nature inspired meta-heuristic search algorithms and was developed and proposed by Yang and Dev [15–17]. The proposed algorithm proves very fruitful in function optimization and engineering design problems. Recent research proves that Cuckoo Search-based optimization techniques are very effective and efficient in comparison to the genetic algorithm and swarm techniques like PSO. CS algorithm is based on aggressive reproduction strategy of the birds, especially cuckoo birds used to lay their eggs in the nests of the other host birds. Each egg represents the solution while a cuckoo bird egg represents a new solution. If the host bird discovers that the eggs is not of its own then they either destroy the egg or abandon the nest. This results in the evolution of cuckoo eggs.

Cuckoo Search Methodology

Yang and Dev describes the Cuckoo search with the help of the following three important idealized rules:

- One cuckoo bird at a time lays one egg only, which is used to represent the solution, and the same egg is dumped into the randomly chosen nest.

- The nest with high quality of eggs is termed as best nest and will be carried over to the next generation.

- The number of hosts nests are fixed and $p_a \varepsilon$ [0, 1] be the probability of the finding the cuckoo's egg by the host bird.

Begin
Objective function f(x)
Randomly generate initial population with n host nest
Evaluate the fitness of each initial solution
        while (t<maxGeneration)
                t = t + 1
                Randomly pick a cuckoo (say a)and generate new solution
                using levy flights
                Evaluate fitness say F_a
                Randomly choose a nest (say b) among n nest
                If($F_a > F_b$)
                then
                        Replace b by the new solution
                end if
                Abandoned or discard a fraction (say P_m) of nest
                Generate new solutions from cuckoos randomly
                Calculate fitness
                Retain the better solutions
                Rank the solutions and determine the current best one.
        End of while

Fig. 1 Pseudocode for cuckoo search algorithm

Based on the above three rules, the host bird can get rid of the cuckoos eggs either by throwing them away or by abandoning the existing nest and building the completely new nest (Fig. 1).

For generating a new solution $X(\tau + 1)$, a lévy flight [18] is performed.

$$X(\tau+1) = X(\tau) + \alpha * \tau^{-\lambda} \tag{1}$$

Where $\alpha > 0$ represents the step size and is related to the size of the problem under consideration and $1 < \lambda \leq 3$. Here, in above pseudocode, n represents the population size.

4 Problem Formulation and Our Solution

4.1 Task Scheduling

For any cloud computing system to be successful, an efficient task scheduling is always much needed. Task scheduling is a mechanism of distributing the tasks or jobs among the computing resources for the computation. Task scheduler also needs to take of the situation if no VM is available to execute the particular task then such task should need to migrate to another VM for its execution.

Let there be m VM's, i.e., $(V_0, V_1, ..., V_m)$ and with the help of these VM's n jobs $(J_0, J_1, ..., J_n)$ need to be executed. So there will be n^m possible ways of allocation on such n jobs among m machines. For example, for 8 tasks and 5 VM's one of the solution (nest) using vector representation is {2, 1, 1, 0, 3, 4, 2, 4}. So we are going to deploy Cuckoo search-based mechanism for task scheduling so that the overall execution time of the system will remain as less as possible.

4.2 Mathematical Model

The objective of task scheduling is to reduce the Makespan; Makespan may be defined as overall time taken by a task to complete its execution.

$$VT_i = \sum_{j=0}^{m} ET_{ji} + C_{ji} \qquad (2)$$

where

VT_i = finish time for ith VM
i = ith Virtual Machine
j = jth Task
E_{ji} = execution time of jth task when processed on ith VM
C_{ji} = (TaskIsize + TaskOsize)/Bandwidth

5 Performance Evaluation

We have compared the Cuckoo search-based task scheduling mechanism with the FIFO (First In First out) scheduling and the greedy-based scheduling mechanism and found that CS scheduling really outperforms the remaining two. In this work, we have used the CloudSim simulator [19] developed and designed by the University of Melbourne, Australia. Table 1 shows the average execution time taken by the CS, Greedy, and FIFO based task scheduling mechanism for 5 VM's while 8, 10, 15, 20, and 25 task for the execution (Fig. 2).

Table 1 Average execution time

Task number	CS	FIFO	Greedy
8	65	103.6	80
10	88.7	121	97.5
15	101	143.4	107.8
20	119.7	250	141
25	143.6	320.1	159

Fig. 2 Average execution time

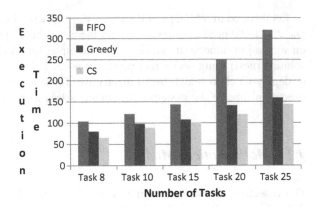

6 Conclusion

In this work, we present the Cuckoo search-based task scheduling mechanism in cloud computing environment. The objective is to map the task to the VM's in such a way that the execution time of the task in distributed environment should be as less as possible. Comparison results have shown that proposed cuckoo search algorithm clearly outperforms the other algorithms, which are taken into consideration in term of makespan. As a future work, there is a scope to optimize other QoS with this mechanism and to develop the hybrid version of above algorithm, which might give better results and improve the cloud system performance significantly.

References

1. Toosi, A.N., Calheiros, R.N. and Buyya, R. 2014. Interconnected cloud computing environments: Challenges, taxonomy, and survey. *ACM Computing Surveys* 47, 1, 1–47.
2. Sadiku, M., Musa, S., Momoh, O. 2014 Cloud computing: opportunities and challenges, IEEE Potentials 33 (1) 34–36.
3. Mezmaz, M., Melab, N., Kessaci, Y., Lee, Y.C., Talbi, E.G., Zomaya, A.Y., Tuyttens, D. 2011. A parallel biobjective hybrid metaheuristic for energy aware scheduling for cloud computing systems, Elsevier, Journal of Parallel and Distributed Computing, 71(11), 2, pp. 14971508.
4. Duan, Q., Yan, Y. and Vasilakos, A.Y. 2012. A survey on service-oriented network virtualization toward convergence of networking and cloud computing, IEEE Trans. Netw. Service Manage. 9 (4) 373–392.
5. Abbas, A., Bilal, K., Zhang, L. and Khan, S.U. 2014. A cloud based health insurance plan recommendation system: a user centered approach, Future Gener. Comput. Syst., http://dx.doi.org/10.1016/j.future.2014.08.010.
6. Dikaiakos, M.D., Katsaros, D., Mehra, P., Pallis, G., and Vakali, A. 2009. Cloud computing: distributed internet computing for IT and scientific research. IEEE Internet Computing, 13(5), pp. 1013.

7. An, B., Lesser, V., Irwin, D., Zink, M.: Automated negotiation with decommitment for dynamic resource allocation in cloud computing. In: Proceedings of the 9th International Conference on Autonomous Agents and Multiagent Systems: volume 1, vol. 1, pp. 981–988. International Foundation for Autonomous Agents and Multiagent Systems (2010).
8. Pandey, S., Wu, L., & Buyya, R. (2010). A particle swarm optimization-based heuristic for scheduling workflow applications in cloud computing environments. In Advanced Information Networking and Applications (AINA), 2010 24th IEEE International Conference (pp. 400–407). Perth, WA: IEEE. doi:10.1109/AINA.2010.31.
9. Song, X., L. Gao, and J. Wang. Job scheduling based on ant colony optimization in cloud computing. In Computer Science and Service System (CSSS), 2011 International Conference on. 2011. IEEE.
10. Li, J., Qian, W., Cong, W., Ning, C., Kui, R. and Wenjing L. 2010. Fuzzy Keyword Search over Encrypted Data in Cloud Computing, IEEE INFOCOM, pp. 15.
11. Yang Xu, Lei Wu, LiyingGuo, ZhengChen, Lai Yang, Zhongzhi Shi, "An Intelligent Load Balancing Algorithm Towards Efficient Cloud Computing", in Proc. of AI for Data Center Management and Cloud Computing: Papers, from the 2011 AAAI Workshop (WS-11–08), pp. 27–32, 2008.
12. Agarwal, M., & Srivastava, G.M.S. (2016). A genetic algorithm inspired task scheduling in cloud computing. *In the proceedings of 2nd IEEE Conference on Computing, Communication and Automation 2016.*
13. Al-maamari, A. and Omara, F.O. 2015. Task Scheduling Using PSO Algorithm in Cloud Computing Environments, International Journal of Grid Distribution Computing, Vol. 8, No. 5, pp. 245–256.
14. Panda, S. K. and Jana, P.K. 2015. Efficient Task Scheduling Algorithms for Heterogeneous Multi-Cloud Environment, Journal of supercomputing 71:1505–1533.
15. Yang, X.S. and S. Deb, 2009. Cuckoo search via Lévyfligh. Proceeding of World Congress on Nature and Biologically Inspired Computing (NaBIC 2009), December 2009, India, IEEE Publications, USA, pp: 210–214.
16. Yang XS, Deb S (2010) Engineering optimization by cuckoo search. Int J Math Modell Num Opt 1(4):330–343.
17. Yang XS, Deb S (2012) Multiobjective cuckoo search for design optimization. Comput Oper Res. Accepted October (2011). doi:10.1016/j.cor.2011.09.026.
18. Burnwala, S. and Deb, S. 2013. Scheduling Optimization of Flexible Manufacturing System Using Cuckoo Search Based Approach, Intl. J. Adv. Manuf. Technol., vol.64, pp. 951–959.
19. Calheiros, R.N., Ranjan, R., Beloglazov, A., De Rose, D.A.F. and Buyya, R. 20111. CloudSim: a toolkit for modeling and simulation of cloud computing environments and evaluation of resource provisioning algorithms, Software—Practice and Experience, vol. 41, no. 1, pp. 23– 50, 2011.

Performance Optimization in Cloud Computing Through Cloud Partitioning-Based Load Balancing

Sonam Srivastava and Sarvpal Singh

Abstract Cloud computing is one of the today's largest hearing fields and exciting technologies, because it is flexible and scalable, also it reduces the cost and complexity of applications. Consequently, arises the concept of large scale computing where there is geographical distribution of data centres and end users all across the globe. Thus, it becomes challenging issue for these data centres how to efficiently handle the huge number of incoming requests from different user bases. So, evaluation of performance of the cloud and its proper analysis is of utmost importance, so that the users can take right decisions. One of the very important issues in association with this domain is that of load balancing. The major object of algorithms concerned with balancing of load is how to efficiently designate the task to nodes, such that the overall response time of request is reduced and processing of request is brought about effectively. Thus, in this paper we have integrated the concept of Cloud partitioning along with the central load balancing algorithm so as to balance the load effectively. Due to the existence of partitions in the system, it is possible to apply good load balancing strategy and use optimal algorithms based on the state of the partition. We have carried out the implementation and simulation of our proposed work on CloudSim simulator. Thus, the task execution is done effectively and the results are better for the proposed modified central load balancing algorithm as compared to previous algorithm in large scale cloud computing environment.

Keywords Cloud computing · Load balancing · Data centre · Virtual machine

S. Srivastava
Department of Computer Science and Engineering, Madan Mohan Malaviya University
of Technology, Gorakhpur, Uttar Pradesh, India
e-mail: sonamsrivastava114@gmail.com

S. Singh (✉)
MMMUT, Gorakhpur, Uttar Pradesh, India
e-mail: spsingh@mmmut.ac.in

© Springer Nature Singapore Pte Ltd. 2018
S.K. Bhatia et al. (eds.), *Advances in Computer and Computational Sciences*,
Advances in Intelligent Systems and Computing 554,
https://doi.org/10.1007/978-981-10-3773-3_30

1 Introduction

Cloud computing basically refers to various services and applications that are delivered from distinguished data centres throughout the world. Such application services are delivered via Internet. Three types of services that are rendered by cloud computing are IaaS, PaaS and SaaS [1]. Cloud computing is a rising field and due to its emerging status there are a vast number of organisations and deluge of enterprises that are drifting towards using this environment [2]. Thus, comes into picture the concept of large scale computing where there is vast number of user bases and data centres geographically distributed all across the world. Thus, it becomes a great challenge for these data centres to effectively handle the huge number of incoming requests from different user bases. So, evaluation of performance of the cloud and its proper analysis is of utmost importance, so that the users can take appropriate decisions. The key technology in support of cloud computing is virtualization. Here the virtual machines are particularly abstract form machine running on physical machines. Since, the tasks that are co-located do not intervene with each other and ingress only their own data, it feels to the user as working in a completely isolated environment. There are several factors that can be effective on cloud performance. As far as cloud environment is concerned there are several techniques to take care of the huge services along with the various operations that are performed upon them. One of the very important issues in association with this domain is that of load balancing. There exist distinctive algorithms for balancing of load in distinctive environments. The major object of algorithms concerned with balancing of load is how to efficiently designate the task to nodes, such that the overall response time of request is reduced and processing of request is brought about effectively. Although virtualization has made remarkable attempts for balancing the load of the entire system there still occur a possibility of the underutilisation or overutilisation of resources [3]. Underutilisation of server results in the increase in the power consumption of entire system thereby raising the operational cost and is also not environment friendly. Overloaded servers bring about degradation of performance. Due to imbalance in load distribution excess heat would be produced by heavily loaded servers which would affect the cooling system cost. So, it is mandatory to properly balance the load on different servers by choosing appropriate strategy of assigning the incoming requests from end users to existing VMs. The performance of cloud scenario depends largely on execution time of tasks. The observed limitation of the existing central load balancing algorithm considered here in this thesis is that it does not state any remedy to the fact that when the status of all the VMs is busy what can be done instead of placing all the cloudlets in the queue one by one and then allocating the cloudlets when any one of them is available. So, this would take a lot of time for execution. Our major objective is to design and develop a modified central load balancing algorithm that ameliorates the identified problem and optimises the performance based on the parameter values of the overall response time.

2 Load Balancing

Load balancing in cloud is concerned primarily with job of distributing workloads across various computing resources. The very prominent significance of balancing load in cloud computing is the resource availability is ameliorated and also the document management systems are benefitted with reduced costs. Through load balancing in cloud the transfer of loads can be brought about at a global level. Load balancing strategically helps resources and networks by providing reduced response time and hike in throughput [4, 5]. It also divides the traffic between various servers so that no delay is incurred while sending and receiving the data. It fairly allocates the resources to a computer resulting in ultimate user satisfaction and appropriate resource utilisation which consequently minimises resource consumption and minimises bottleneck problem.

A. Types of Load Balancing

The two broad classifications of load balancing algorithms are static and dynamic load balancing algorithms.

Static load balancing algorithms: The allocation of tasks to processors is done during compile time that is prior to program execution. It is done based on processing capabilities of machines. The static scheduling algorithms are non-pre-emptive and their major objective is to reduce overall execution time. They have nothing to do with dynamic modifications at run time.

Dynamic load balancing algorithms: They deal with redistributing processes amongst the processors during the run time of execution. It dynamically re-designates the load amongst machines by collecting information, knowing the run-time conditions and gathered characteristics. The communication dropdown associated with the task is the main reason behind the run time overhead in balancing load.

The dynamic load balancing further has two classifications **centralised** and **distributed**. If the single node is responsible for handling the distribution within the entire system (centralised). On the other hand if work involved in making decisions is distributed amongst different processors (distributed). Although centralised approach is simple, it involves single point of failure and bottleneck problems.

Distributed algorithms are free from it. Distributed dynamic scheduling can again be either **cooperative** or **non-cooperative**. In the former, every processor is responsible for carrying out its own part of task scheduling whereas in the latter one the independent processors independently arrive at a decision of their resource usage, without affecting the rest of the system.

3 Literature Review

Sidhu et al. [6] have discussed the Round Robin strategy in this paper. Here the data centre controller handles all the requests and allocates them to the Virtual machines in rotating manner. The very first request is assigned to randomly picked

Virtual machine. Subsequently, the requests are allocated in the circular fashion. Even though the authors distribute the load of work amongst processors equally but still the processing time of different processes is not identical. Although this approach does the equal distribution of load, there appears the heavy loading on certain virtual machines whereas on the other hand remaining is idle.

Zhang et al. [7] have discussed the Weighed Round Robin strategy in this paper. It is the updated Round Robin version in which every virtual machine is assigned weight according to their processing capacity. In this manner the powerful virtual machine is assigned more requests. In such a situation the two requests are assigned by the controller to the one which is the powerful virtual machine in comparison to one request assigned to the weaker one. It provides better resource utilisation but at the same time it does pay heed to the time of processing each request.

Ahmad et al. [8] have proposed an ESCEW algorithm which scans the queue of jobs and virtual machine list. In this algorithm any available VM is allocated a request which can be handled by it. If the VM is overloaded then it is the responsibility of the balancer to distribute some tasks to the idle VM. This would ameliorate the response time and the overall time of processing tasks.

James et al. [9] gave a Throttled algorithm of balancing load that tries to evenly scatter the load amongst several virtual machines. When the request arrives, the status of virtual machines is consulted from recorded list, if it is idle then the request is accepted otherwise it returns −1 and request gets queued up.

Lua et al. [10] propose a Join-Idle-Queue distributed algorithm for large systems. Here, firstly idle processors are load balanced across dispatchers and then decrease average length of queue at each processor jobs are designated to processors. Thus, JIQ algorithm does not incur any kind of communication overhead between processors and dispatchers at job arrivals.

Wang et al. [11] have proposed a two phase algorithm of balancing load by amalgamating the two phase load balancing algorithms namely opportunistic load balancing algorithm and load balancing min-min algorithm to have better execution efficiency and load of the system is well maintained. The phases of the algorithm are that the job is allocated to the service manager by the Opportunistic Load Balancing scheduling manager in the first phase. Service manager uses the Load balancing min-min to select the appropriate service node for executing the subtask in the second phase. However, the algorithm is only applicable in case of static environment.

Ghafari et al. [12] propose a decentralised nature inspired honey-bee-based load balancing technique for self-organisation which works best under the availability of heterogeneous resources. Local server action is something through which it achieves the global balancing of load. The approach is similar and adopted from real life behaviour of honey bees for harvesting their food and foraging. In the algorithm the servers are set certain virtual servers where each of it has its own virtual server queues. As per the approach as specified and implemented in the paper works very well as far as the resources under consideration are heterogeneous. The problem related to this approach is that it does not improve throughput equally with the increase in the resources.

Zhang et al. [13] have proposed an ant colony and complex network theory-based mechanism of balancing load in open federation of cloud computing. It enhances numerous aspects of ant colony related algorithms which dealt with only balancing loads in distributed environment.

Ren et al. [14] have proposed a Fractal algorithm of balancing load. It is dynamic in nature and assures that the unnecessary migrations are not triggered by the peak instantaneous load. Timing of virtual machine migration is decided by the load forecasting method. The results in the paper demonstrate that the proposed achieves considerably better load balancing. The comparison of the algorithm has been done with existing honey-bee and ant-colony algorithms. The authors conduct the testing of the existing algorithm and experimentally conclude that the algorithms upgrades system performance and achieve load balancing.

Achar et al. [15] have proposed a weighted least connection algorithm that assigns the load to the nodes with the minimum that is the least number of connections. Although it efficiently balances the load, it still does not consider the storage capacity and processing speed.

Randles et al. [16] have stated that the complexity and scalability of the systems brings about the infeasibility in the assignment of jobs to specific servers in a centralised manner. Thus, the paper examines the three possible solutions proposed for balancing load. So, the authors introduce a self aggregation using the concept of active clustering. It works on the tenet of aggregating together similar nodes and then working on them by efficient load balancing. The limitation with this algorithm is that it works poorly in heterogeneous environment.

4 Proposed Approach for Load Balancing

We have adopted cloud partitioning strategy to balance the load. The load balancing then starts, immediately after the cloud partitioning has been done. Once the job arrives at the appropriate system, it is the responsibility of the main controller to decide which job will be moved to which cloud partition. The load balancer of that partition then decides as to how the job has to be assigned to the nodes. If the load status of a cloud partition is detected as normal, this partitioning can be achieved locally. If the load status of cloud partition is not normal, this job should be transferred to another partition. In our proposed approach each and every cloudlet will pass through different stages. And ultimately each cloudlet will be assigned to the suitable virtual machine.

- Cloudlet to broker.
- From broker to the most suitable data centre.
- From data centre to its controller.

- From controller to best partition.
- Finally cloudlet is assigned to the virtual machine.

Figure 1 shown is the diagrammatic representation of how our proposed approach works. In our proposed approach the tasks or the requests or jobs from the users firstly arrive at a broker, from there the requests are forwarded to appropriate data centre. The data centre has a data centre controller which decides on which partition the requests will be sent to. Every data centre is divided into partitions and each partition has a number of virtual machines. The activities of each partition are handled by its own partition balancer. Thus, the Data centre controller sends the request to particular partition on the basis of their status. If the status of first partition is normal the request is assigned to the available node under that partition otherwise it checks for the next partition and repeats the process for the next partition.

A. *Proposed Algorithm*

The flowchart of the proposed algorithm has been shown in Fig. 2. The step-wise description of the proposed algorithm is given below

1. User sent a request.
2. All user requests are collected at the point of the broker and broker sends these requests to the Data centre controller.
3. A table is maintained with the central load balancer that contains id of virtual machine (VMid), availability of status of states like (BUSY or AVAILABLE) and VMs priority.

These states are completely dependent on the cloud partition, like if one cloud partition is normal then other two partitions are idle and overloaded, we have set this condition inside the cloud. Initially, all Virtual Machines are in available state.

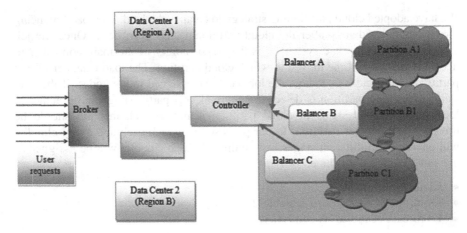

Fig. 1 Working of proposed algorithm

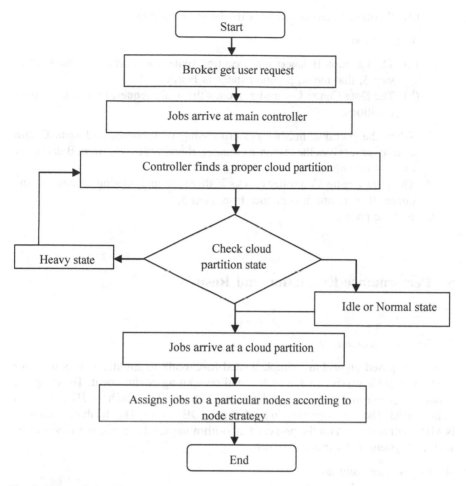

Fig. 2 Proposed algorithm

4. Data Centre Controller receives the new request from the broker.
5. Data Centre Controller finds a proper cloud partition for next allocations, every cloud partition has its own balancer.
6. Partition Balancer parses the table from top in order to find the highest priority partition virtual machine and the state of that cloud partitions is available.

   If found:

   (a) The Partition Balancer returns the VMid to the Data Centre Controller.
   (b) The Data Centre Controller forwards the request to the VM identified by that VMid.
   (c) Data Centre Controller notifies the Partition Balancer of newer allocation. Central Load Balancer refreshes the table accordingly.

(d) Partition Balancer refreshes the table accordingly.

If not found:

(a) The Partition Balancer not accepting request and process again back to step 5, that means partition states is heavy.
(b) The Data Centre Controller queues the again request to find the proper partitions.

7. When the partition finishes the processing of requests, and Data Centre Controller receives the response cloudlet, it notifies the Partition Balancer of the VM de-allocation.
8. The Data Centre Controller checks if there are any waiting requests in the queue. If there are, it continues from step 5.
9. End the process.

5 Performance Evaluation and Results

A. *System Requirements*

The proposed algorithm is implemented in CloudSim simulator. It is used for modelling and analysis of large scale cloud computing environment. To set up the simulation environment we need Java development kit which is JDK 1.6 and Eclipse IDE. The language used in the Eclipse IDE is Java. The database used here is Microsoft access. With the proposed algorithm the database connectivity is done with the system and window 7 is used.

B. *Performance analysis*

The certain metrics of performance have to be considered while balancing load in cloud environment. The scalability, utilisation of resources, response time, etc. are certain parameters on the basis of which performance can be computed and analysed. Here, in our work we have seen that the performance of our proposed approach is better than the existing algorithm. The considered existing algorithm is central load balancer [4]. We have modified the existing algorithm by introducing the concept of partitioning [5]. The algorithm in the considered base paper is compared with Round Robin and Throttled. Round Robin and Throttled algorithms give good performance with smaller size requests but when huge data is arriving for allocation the central load balancer would give better results. In addition to this if very large scale computing is to be handled, that can now be done with the help of modified central load balancing algorithm.

Table 1 shown above gives the parametric considerations that have been taken for evaluating the overall performance of modified central load balancer as compared to central load balancer. Figure 3 clearly portrays the overall response time

Table 1 Parametric considerations

S. No.	Parameters	Central load balancer	Modified central load balancer
1	Data centres	2	2
2	Service broker policy	Closest data centre	Closest data centre
3	User base	2	2
4	Request per user per hour	60	60
5	VM	5	5
6	Average RT (ms)	0.35	0.25
7	Minimum RT (ms)	0.01	0.01
8	Maximum RT (ms)	0.62	0.48

Fig. 3 Comparison of two algorithms on the basis of closest data centre policy

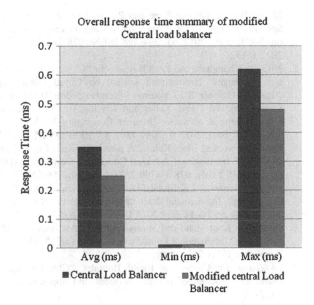

summary of modified central load balancer and thus from these results we conclude that the performance is better optimised with the help of our algorithm.

6 Conclusion

The performance of a cloud computing system depends on the task execution time. In our proposed approach we have integrated the concept of cloud partitioning with central load balancing algorithm to achieve efficient load balancing. In the proposed algorithm prior to the allocation of request to virtual machine the request has to move to suitable partition which is decided by the data centre controller.

Thus, the future work is to develop an enhanced load balancing algorithm considering the fact that existence of partitions in the system, it is possible to apply good load balancing strategy and use optimal algorithms based on the state of the partition. Thus, the task execution is done effectively.

Acknowledgements The completion of this paper would be incomplete unless we express our deep gratitude to the authors whose work we have taken for reference. Their papers and books have provided us with constant guidance and encouragement without which the successful completion of this paper could not have been made possible.

References

1. Shyam Patidar, Dheeraj Rane and Pritesh Jain "A survey paper on cloud computing" 2012 Second International Conference on Advanced Computing and Communication Technologies on 7–8 January 2012 pp. 394-398 ISBN: 978-1-4673-0471-9 doi:10.1109/ACCT.2012.15 2012 IEEE.
2. Niloofar Khanghahi and Reza Ravanmehr "Cloud computing performance evaluation: Issues and challenges" International Journal on Cloud Computing: Services and Architecture (IJCCSA) October 2013 Volume 3 Number 5 doi:10.5121/ijccsa.2013.3503.
3. Drnitry Drutskoy, Eric Keller and Jennifer Rexford "Scalable Network Virtualization in software-defined networks" Internet Computing IEEE on 27 November 2012 pp. 20–27 Volume 17 Issue 2 ISSN: 1089-7801 doi:10.1109/MIC.2012.144 IEEE.
4. Soni, Gaurav, and Mala Kalra. "A novel approach for load balancing in cloud data centre." 2014 IEEE International Advance Computing Conference (IACC), on 21–22 February 2014 pp. 807–812 ISBN: 978-1-4799-2571-1 doi:10.1109/IAdCC.2014.6779427 IEEE.
5. Antony Thomas and Krishnalal G, "A Novel Approach of Load Balancing Strategy in Cloud Computing" International Journal of Innovative Research in Science, Engineering and Technology on 16–18 July 2014 ISSN: 2319-8753.
6. Amandeep Kaur Sidhu and Supriya Kinger, "Analysis of load balancing techniques in cloud computing" International Journal of Computers and Technology March-April 2013 Volume 4 Number 2, ISSN 2277-3061.
7. Qi Zhang, Lu Cheng and Raouf Boutaba, "Cloud computing:State of art and research challenges" Journal of Internet Services and applications May 2010 Volume 1 Issue 1 pp. 7–18.
8. Tanvee Ahmad and Yogendra Singh, "Analytic study of load balancing techniques using tool cloud analyst" International Journal of Engineering Research and Applications (IJERA) March-April 2012 pp. 1027–1030 Volume 2 Issue 2 ISSN: 2248-9622.
9. Jasmin James and Dr. Bhupendra Verma, "Efficient VM load balancing Algorithm for a cloud computing environment" International Journal on Computer Science and Engineering September 2012 Volume 4 Issue 9 pp. 1658–1663.
10. Y. Lua, Q. Xiea, G. Kliotb, A. Gellerb, J.R. Larusb and A. Greenber, "Join-Idle-Queue: A novel load balancing algorithm for dynamically scalable web services" Performance Evaluation November 2011 Volume 68 Issue 11 pp. 1056–1071 doi:10.1016/j.peva.2011. 07.015.
11. Shu-Ching Wang, Kuo-Qin Yan, Wen-Pin Liao and Shun-Sheng Wang, "Towards a load balancing in a three level cloud computing network" 2010 3rd IEEE International Conference on Computer Science and Information Technology (ICCSIT) on 9–11 July 2010 pp. 108-113 ISBN: 978-1-4244-5537-9 doi:10.1109/ICCSIT.2010.5563889 IEEE.

12. Seyed Mohssen Ghafari, Mahdi Fazeli, Ahmad Patooghy and Leila Rickhtechi, "Bee-MMT: A load balancing method for power consumption management in cloud computing" 2013 Sixth International Conference on Contemporary Computing on 8–10 August 2013 pp. 76-80 ISBN: 978-1-4799-0190-6 doi:10.1109/IC3.2013.6612165 IEEE.

13. Z. Zhang and Xu Zhang, "A load balancing mechanism based on ant-colony and complex network theory in open cloud computing federation" 2010 second International Conference on Industrial Mechatronics and Automation (ICIMA) on 30–31 May 2010 pp. 240–243 ISBN: 978-1-4244-7653-4 doi:10.1109/ICINDMA.2010.5538385 IEEE.

14. Haozheng Ren, Yihua Lan and Chao Yin, "The load balancing algorithm in cloud computing environment" 2012 Second International Conference on Computer Science and Network Technology (ICCSNT) on 29–31 December 2012 pp. 925–928 ISBN: 978-1-4673-2963-7 doi:10.1109/ICCSNT.2012.65226078 IEEE.

15. Raghavendra Achar, P. Santhi Silagam, Nihal Soans, P. V. Vikyath, Sathvik Rao and Vijeth A. M., "Load balancing in cloud based and on live migration of virtual machines" 2013 Annual IEEE India Conference (INDICON) on 13–15 December 2013 pp. 1–3 ISBN: 978-1-4799-2274-1 doi:10.1109/INDICON.2013.6726147 IEEE.

16. Martin Randles, David Lamb, A. Taleb Bendiab, "A comparative study into distributed load balancing algorithms for cloud computing" 2010 IEEE 24th International Conference on Advanced Information Networking and Application Workshop (WAINA) on 20–23 April 2010 pp. 551–556 ISBN: 978-1-4244-6701-3 doi:10.1109/WAINA.2010.85 IEEE.

Part III
Intelligent Image Processing

An Optimistic Approach of Locking Strategy in Progress Fourth Generation Language

Neha Prabhakar and Abhishek Singhal

Abstract While working in the multi-user environment, the records should not be kept in the record buffer for the long duration of time; doing so will prevent the other users from accessing the record. Optimistic Locking is a strategy wherein transactions access the records without actually locking them and making them available for other users. Before committing, each transaction read a record, take note of a database records, and check that the record has not changed before you write the record back to the database. This paper is tailored towards implementing Optimistic Locking Strategy in the Progress Fourth Generation Language (4GL) and explaining various reasons of transaction failure in the Progress 4GL. Lastly our idea is compared with the existing locking strategy in the database management system on the basis of various issues that occurs in the locking mechanisms.

Keywords Transaction processing · Concurrency control · Records · Database deadlock · Record scope

1 Introduction

The Progress Fourth Generation Language (4GL) indicates the outcome of numerous Foundation technologies combined into a particular programming situation. These technologies include: A block-structured syntax with control structures (including blocks, procedures, and functions) that let you encapsulate flexible and powerful application objects, an integrated front-end manager featuring an event-driven architecture that you can use to create and manage flexible user interfaces for both

N. Prabhakar (✉) · A. Singhal
Department of Computer Science & Engineering, ASET, Amity University,
Uttar Pradesh, India
e-mail: nehaprabhakar91@gmail.com

A. Singhal
e-mail: Asinghal1@amity.edu

© Springer Nature Singapore Pte Ltd. 2018
S.K. Bhatia et al. (eds.), *Advances in Computer and Computational Sciences*,
Advances in Intelligent Systems and Computing 554,
https://doi.org/10.1007/978-981-10-3773-3_31

315

graphical and character Environments. An integrated back-end manager includes both a native Progress Relational Database Management System (RDBMS) and a Data Server facility. This can be used to access many Third-party RDBMSs and flat file systems.

DATABASE framework is indispensable for few applications, extending from space station operations to Automated Teller Machines. A database state addresses the estimations of the database addresses that address some genuine substance. The database state is changed by the execution of a client transaction. Particular transactions running in isolation are thought to be right. Right when different clients get to different database objects living on various regions in a distributed database structure, the issue of concurrency control rises [1].

This paper is segregated into seven sections. Section 1 consists of introduction to Progress 4GL and overview of the other basic concepts involved in this study. Section 2 consists of literature Survey which describes the method adopted for short listing the papers and describes the outcomes of various research works already carried out in this area. Section 3 presents the Overview of transaction management in Traditional DBMS. Section 4 consists of description of Database locking and Mechanism of locking in Progress 4GL. Section 5 elaborates the existing locking strategy and proposed strategy followed by Simulation Experiments in Sect. 6. Lastly, the conclusion of our research is portrayed in Sect. 7.

2 Literature Survey

A meticulous literature review has been performed to analyze the gaps in the existing technologies. Following databases were referred for gathering research papers for the review purpose like IEEE Xplore, ACM digital library and other online sources like Google scholar and open access journal.

Following keywords were used for searching relevant research papers (Table 1):

Leis et al. [2] presented hardware transactional memory (HTM). In this paper they have demonstrated that HTM grants for accomplishing practically without lock handling of database transactions by astutely checking the information format and the entrance frames.

We understand that the usage of phones would rise within the near forthcoming. A couple ensures for better transaction precision and beneficial information administration techniques will be required at the point when clients will request more personality boggling errands to be performed on their cellular telephones [3].

Table 1 Keywords Used in searching papers

Serial number	Keywords used
1	Optimistic locking strategy
2	Transactions and locking in database
3	Concurrency control in database management system

Execution of distributed database were balanced with stores of difficulties inspected in past investigates, few of such breaker conflicting framework advancement, high cost of Computers, and precariousness among customers. [4] presents the challenges experienced in appropriated database exchange and proffered likely approaches.

Through concentrating on the standard segment for securing with 2PL distributed database framework, Li et al. [5] proposed another design for the transactions arranging in Distributed Database System. The examinations demonstrate that the new structure has redesigned the synchronization level of transactions and reduced the expense of correspondences which appear to be another locking structure for transactions.

Jayanta and Somananda [6] proposed an instrument to upgrade the execution of transactions logically distributed database systems. Synchronizer is utilized to keep up the serialization request among the arriving and executing transactions. For the transactions held up by one or more bolts, a prioritization is utilized to deal with the clashing among them.

Liu et al. [7] depicted the arrangement of utilizing the database as a bit of solicitation to guarantee the rightness of data control, we fathomed issues. Transaction is a game-plan of astute operation unit, the data change starting with one state then onto the accompanying state, to guarantee consistency of information in the database, data control ought to be discrete parties of reason cells, and when everything completed, Data consistency can be kept up, and when a touch of this unit comes up short, the whole transaction ought to be completely considered as a mistake, every last resulting operation from the most punctual beginning stage in the event that all fall back to the beginning state.

Lu et al. [1] portrayed another sort of database Anomaly, which is called Cumulated Anomaly, is tended to in this paper. Another affirmation method is proposed based on Dubiety-Determining Model (DDM) expected for it. Both of Cumulated Anomaly and DDM are portrayed and delineated in motivations behind interest formally. They have laid out programming framework working to reinforce the DDM for checking database transactions.

3 Overview of Transaction Management

Transaction Management manages the issues of continually keeping the database in predictable state [8]. A transaction is a gathering of activities that make stable changes of framework states while safeguarding framework consistency [9]. Conversely a distributed transaction is a transaction that upgrades information on two or more arranged PC frameworks. Distributed transaction extends the advantages of transactions to applications that should upgrade conveyed information. The consistency and dependability quality parts of transaction are because of four properties called ACID (Atomicity, Consistency, Isolation, and Durability) [9].

The subsequent actions would begin a transaction. Any FOR, DO, REPEAT procedure block that consists of the TRANSACTION keyword. An iteration of FOR EACH, DO ON ERROR, or REPEAT block that either directly reads or updates the database with the EXCLUSIVE-LOCK. Statements like UPDATE, DELETE, CREATE that directly updates the database means they contains a set of statements that can modify the database. According to the need of the application, we can explicitly control the size of the transaction by using the keyword TRANSACTION. By using this keyword we can either make the transactions large or small, according to the requirements or scenario [10].

3.1 Transaction Failures

There are diverse sorts of transaction failures. Failures can be because of a fault in the transaction brought on by inaccurate information and the recognition of a present or possible deadlock [11]. Moreover, some concurrency control procedures do not allow a transaction to continue or even to hold up if the information that they endeavor to get to are at present being used by another transaction. The standard way to deal with the instances of transaction failures is to prematurely end the transaction, in this manner resetting both database to its state preceding the begin of this transaction.

4 Database Locking

With a particular objective to guarantee atomicity, consistency, isolation, and durability (ACID) types of transactions in Database Management System (DBMS), the locking framework is utilized. Reference [5] exhibits that lock are a product framework. It suggests that a customer has held a couple of benefits (data, information sheets, record, etc.) to insist the integrity and consistency of data when transaction peruses the identical asset. Mechanism [12] present that lock is an extraordinarily basic strategy for fulfilling the control of the synchronization. Its accurate control is picked by type of the lock.

Mechanism of locking in Progress 4GL: Consider a scenario wherein 2 users are performing a task of updating in the multiuser environment. Suppose user A is shipping department and user B is receiving department. User A subtracts the item from the inventory maintenance table in the database when the item is shipped and the User B adds the item inventory maintenance table in the database. If User A updates the record in the database and at the same time user B fetches the record then user B would get the old values of the inventory tables. **Replication of database records**: In the multi-user environment, the same record has to be accessed by the multiple users at the same time. In the above scenario, things did not work as it was expected. Since user A has not used the appropriate lock on the inventory table while updating the record. So this resulted in the inconsistent data.

5 Traditional Mechanism

When developing a 4GL application for deployment in the multi-user environment, the developer must design a strategy that avoids keeping the records locked in the record buffer for the long duration of time. If records are held by a user for the long duration of time, then this will prevent other users from accessing the same record and thus leads to various problems such as Lock contention, Long term blocking, Database deadlock, etc.

The following mentioned algorithm shows the scenario wherein the record is accessed using the exclusive-lock:

1. PROMPT-FOR <var_name>.
2. UPDATE <var_name>.
3. FIND <table_name> WHERE <any matching criteria> EXCLUSIVE-LOCK.
4. IF AVAILABLE <table_name> THEN DO: <Query to execute any task>
5. UPDATE <field1>.
6. END.

In the above specified algorithm, the record of the table <table_name> is fetched in the EXCLUSIVE-LOCK, so this will prevent other users from accessing the records till the time that record is released by the previous user. Such a strategy is not effective in the multi-user environment.

5.1 Optimistic Locking Strategy

While working in the multi-user environment, the records should not be kept in the record buffer for the long duration of the time, doing so will prevent the other users from accessing the record and thus impacting the performance of the system.

Optimistic locking is a methodology where you read a record, note down the description, and verify that the description hasn't changed before you transcribe the record back. When you write the record back you check the update to ensure the atomicity of the data (i.e., it has not been changed by the user in between before committing back the changes to the database) and update the version in one hit [13].

1. FIND <rec> NO-LOCK NO-ERROR NO-WAIT.
2. IF AVAILABLE <rec> THEN

Table 2 Comparison of existing locking strategy and proposed strategy

Issue	User 1	User 2	User N
Existing locking strategy			
Database deadlock		✓	✓
Record availability	✓		
Consistent data	✓		
Parallel processing			
Proposed strategy			
Database deadlock			
Record availability	✓	✓	✓
Consistent data	✓	✓	✓
Parallel processing	✓	✓	✓

```
    <query to perform a required operation>. PAUSE.
    /*if user wishes to modify the record*/
 3. IF upd THEN DO:
 4. FIND CURRENT <rec> EXCLUSIVE-LOCK NO-ERROR NO-WAIT.
 5. IF AVAILABLE <rec> THEN DO:
 6. IF CURRENT-CHANGED <rec> THEN DO:
    <query to perform a required operation>.
 7. UPDATE <rec>.
 8. END.
 9. ELSE DO:
    <query to perform a required operation>.
10. END.
11. ELSE IF NOT AVAILABLE <rec> THEN DO:
12. IF LOCKED <rec> THEN DO:
13. DEFINE TEMP-TABLE <ttLockTable>
    <query to list the locked records>.
14. END.
15. ELSE DO: /*depends on the scenario*/
```

The following table represents the various issues that generally occur; these issues are gathered from the various research papers that are analyzed. The table shows whether mentioned issue arises for User 1, User 2 and so on. The tick mark shows the listed factor is satisfied by that corresponding user (Table 2).

6 Simulation Experiments

Before simulating, we implemented the algorithm with the new locking strategy. An excel file is used for storing meta-data of the data object. Then we implemented the generally used locking strategy, another excel file is maintained for storing the meta-data of the old locking strategy. The locking systems are tested in the

Fig. 1 Waiting time when user is continuing

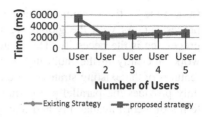

Fig. 2 Waiting time when user is waiting for release

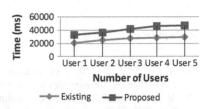

multi-user environments. In the existing algorithm, when user 1 access the record for the first time, at that point of time currently no record is present in the record buffer so PROGRESS allocates the records to the User 1 in comparatively lesser amount of time. When user 1 is modifying that record and at the same time if user 2, user 3 and so on requests the same record for the reporting purposes, then PRO-GRESS gives a message that current record is locked by the user. In such a scenario user 2, user 3 and so on has to wait until the user 1 does not release the record and the waiting time for the users currently in the queue keeps on increasing. In the proposed strategy, if user 2, user 3 wants to access the record for the reporting purposes then they can certainly carry out the processing and fetch the record from the record buffer. If they wants to modify the record and that record is locked by the previous user then according to the proposed algorithm, PROGRESS issues the message that this record is locked by the user and if the user continues then gives the name of the user along with the name of the Locked table. Results are represented in Fig. 1 when user opts for fetching the record and getting the name of the user and locked.

When in the second algorithm, if user opts to wait till the previous user who has locked the record releases that record, then it is noticed that second algorithm behaves almost like the first algorithm and users has to wait until previous users release the records (Fig. 2).

7 Conclusions

After a careful and detailed study of traditional mechanism for locking a record in progress 4GL, this paper proposes a new architecture of locking in the multi-user environment that provides an additional feature of parallel processing in almost the same time as the existing strategy takes without offering the feature of parallel processing. As with the global organizations and multi-tier architecture, parallel processing is the necessity. This feature of parallel processing enables the user to work efficiently in the multi-user environment without compromising any of the features of the existing strategy. The experiments shows that the novel strategy not only provides the parallel processing but also controls concurrency and excels in terms of record availability and data consistencies.

References

1. G. Lu, J. Y. Lü, "A New Database Transactions Monitoring Approach", International Conference on Mechatronics and Automation, Harbin, China, 2007.
2. A. Kemper, and T. Neumann. "Scaling HTM-Supported Database Transactions to Many Cores", IEEE Transactions on Knowledge and Data Engineering, Vol. 28, No. 2, February 2016.
3. O. Amer, A. Salem, A. al-Qeerm, "Classification of Transaction Models in Mobile Database System", Web Applications and Networking (WSWAN), 2015 2nd World Symposium on, 978-1-4799-8171-7, March 2015.
4. S. Al Idowu, S. Maitanmi." Transactions- Distributed Database Systems: Issues and Challenges", International Journal of Advances in Computer Science and Communication Engineering (IJACSCE) Vol 2 Issue I, March, 2014.
5. Z. Li, Z. Zhu and S. Zhang, "A Locking Mechanism for Distributed Database Systems", Journal of Networks, Vol. 9, No. 8, August 2014.
6. Y. Jayanta, Y. Somananda, "Conflicting Management of Transactions in Real Time Database System", First International Conference on Informatics and Computational Intelligence, 2011, pp. 327-331.
7. S. Liu Min, C. Xinyu, C. RuiZhang. "Research Based on Database Transactions", 3rd International Conference on Communication Software an Networks (ICCSN), IEEE, 2011, pp 550–553.
8. A. Waqas, A. Mahessar, N. Mahmood, Z. Bhatti, M. Karbasi and A. Shah, "Transaction Management Techniques and Practices in Current Cloud Computing Environment: A Survey", Shah International Journal of Database Management Systems (IJDMS) Vol. 7, No. 1, February 2015.
9. P. Ozsu, "Principles of Distributed Database Systems", ISBN 978-1-4419-8834-8 Springer, 2011.
10. Progress Programming Handbook, Progress Software Corporation.
11. J. Gray and A. Reuter, "Transaction Processing: Concepts and Techniques", Morgan Kaufmann, Publishers Inc. San Francisco, 1993.
12. A. Thomasian, "Two-Phase Locking Performance and its Thrashing Behavior", ACM Transactions Database Syst, 1993, 18(4): 579–625.
13. http://knowledgebase.progress.com "Progress Knowledge Base".

Combating Clickjacking Using Content Security Policy and Aspect Oriented Programming

Rakhi Sinha, Dolly Uppal, Rakesh Rathi and Kushal Kanwar

Abstract Clickjacking is a highly deceiving technique to make users click on some UI element either surreptitiously or by enticing them with offers. Clickjacking is disguised and it cannot be easily detected because it makes use of some statutory features in a web application like iframes. This paper aims at Clickjacking prevention by incorporating Content Security Policy directives at development stage of a web application to white list loaded resources and applying these headers to each web page by making use of Aspect Oriented Programming features in J2EE which offers some utilitarian features to enhance the security functionalities in a web application. By combining Content Security Policy with Aspect Oriented Programming we have tried to demonstrate a proactive approach for averting Clickjacking with minimum programming effort. First, we have discussed how an aspect can be included in a web application without modifying source codes. Then, the experimental results demonstrate minimum to maximum levels of protection that can be achieved using Content Security Policy.

Keyword Clickjacking · Web application security · Browser security · Content security policy · Aspect oriented programming J2EE

R. Sinha (✉) · D. Uppal · R. Rathi · K. Kanwar
Department of Computer Engineering & IT, Government Engineering College,
Ajmer, India
e-mail: rakhi168.sinha@gmail.com

D. Uppal
e-mail: dollyuppal290@gmail.com

R. Rathi
e-mail: rakeshrathi4@gmail.com

K. Kanwar
e-mail: rathorek30@gmail.com

© Springer Nature Singapore Pte Ltd. 2018   323
S.K. Bhatia et al. (eds.), *Advances in Computer and Computational Sciences*,
Advances in Intelligent Systems and Computing 554,
https://doi.org/10.1007/978-981-10-3773-3_32

1 Introduction

Clickjacking etymologically means "hijacking a click" that is, diverting a user click intended on some innocuous element such as a like button to some malicious element for initiating unwanted action without the user's knowledge or consent. The attacker who wishes to exploit this vulnerability very artfully loads a malicious page over an innocent page by framing this harmless page in an iframe. According to [1], a vulnerable UI element is put out of context in front of the user so she is forced to react out of context. An overview of the Clickjacking technique in [2] explains that the coordinates and opacity of the malicious UI element are adjusted in such a manner that it is transparent and is exactly positioned on top of the benign UI element so that the user knows she is clicking on the trusted element say, a button but unknowingly she is also contributing a click on the hidden harmful button.

Clickjacking requires the skilful use of "User Interface Redressing" with the aid of common HTML elements like iframes and other JavaScript and CSS features leading to potential damages like posting spam content on websites like Facebook and Twitter (Tweet bombs) to hijacking victim's camera and microphones through Adobe Flash Player vulnerability.

The upcoming sections of this paper are categorized as follows. Section 2 summarizes the related work and existing defenses for Clickjacking, Sect. 3 covers motivations, principles, introduces Content Security Policy, explains why Aspect Oriented Programming (AOP) should be used and comparative analysis, Sect. 4 discusses Proposed Work, Sect. 5 covers implementation of the proposed work, results, and implementation issues and finally, Sect. 6 concludes this research work and suggests its future scope.

2 Related Work and Existing Clickjacking Defenses

Clickjacking was first illustrated in a white paper by Paul Stone [3] with the help of a naïve CJ tool. After that, a thorough study on Amazon Mechanical Turk was conducted and possible attacks violating context integrity were demonstrated and its countermeasure InContext to eliminate the root cause of Clickjacking was designed and implemented. A popular technique for defending Clickjacking known as Frame Busting was clearly explained in [4].

Other significant works include Clickjuggler [5] and Clicksafe [6]. Clickjuggler is a very handy tool which can be used by inexperienced developers to check whether their website is vulnerable to Clickjacking. Clicksafe is a browser-based tool to detect and mitigate Clickjacking during web browsing by rating websites.

Apart from the above-mentioned contributions, in [2], most of the existing defenses are listed. Accordingly, Clickjacking defenses can be broadly classified as follows.

2.1 Frame Busting

Frame Busting is a JavaScript-based client side technique to stop a web page from being embedded in the iframe of an attacker's page. The simple JavaScript code that can be included with each page is:

$$\text{If}(top \,! = window) \text{ then } top.location = window.location;$$

If a malicious page will try to frame this page, the above code will take the user to the source website to which the target webpage originally belongs.

But, this basic script can be easily bypassed by the techniques listed in [4] and [5]. Also, frame busting is applied to an entire webpage at a time and it is not applicable if individual elements must be protected on a page from Clickjacking overlays.

2.2 X-Frame-Options HTTP Response Header

This method of including X-frame-options in an HTTP response [7] is a relatively newer technique than frame busting but presently, it is the most widely used and trusted option supported by a large number of modern browsers listed in [8]. Using this header, the browser is instructed in three ways: DENY, SAMEORIGIN and ALLOW-FROM<url>. But similar to frame busting, X-frame-options can be applied to an entire web page only and not to the individual elements on that page.

2.3 Browser Plugins and Add-Ons

NoScript [9] with ClearClick technology is an extensively used add-on for detecting Clickjacking attempts by screenshot capturing and comparing with parent page and warning users of Mozilla based web browsers. Other add-ons include ScriptSafe [10] plugin for Chrome users.

3 Motivation and Principles

3.1 Content Security Policy

Content Security Policy is a declarative policy [11] which contains its roots in same origin policy. Whenever some content is requested by a web page and is loaded by the browser, the latter is incapable of distinguishing between the content which

Table 1 URI directives enlisted in content security policy specifications

CSP directives	Intended use
default-src	Defines default behaviors of directives with suffix "src"
frame-src	Defines sources from where frames can be embedded. But this directive is deprecated
report-uri	Sends reports of violated policies to the given URL
frame-ancestors (CSP 2.0)	This directive is used to prevent Clickjacking as it restricts and specifies the sources which can frame the web page. It cannot be defined in the http-equiv of <meta> tag like other directives

comes from a trusted source and the one which comes from a suspicious source. So, it can even execute an injected script assuming it as webpage's own. Content Security Policy offers a rich set of directives [12] some of which are listed in Table 1 so that we can tune the policy according to particular needs of different web pages of our web application and for different resources like iframes, scripts, multimedia content, objects, etc., by clearly stating what type of content should be loaded and white listing all the sources from where the content can be loaded [13]. Therefore, we can frame a customized policy like:

Content Security Policy: default-src "self"; script-src https://www.examplecsp. com which means scripts from examplecsp.com can be executed but other resources like font, image, media, etc., can be loaded from same origin only as default-src is set to "self". So, in order to use third-party widgets like Facebook's like button, Twitter's tweet button and Google's +1 button, which require scripts and iframes, Content Security Policy for the website should be designed carefully [14].

3.2 Aspect Oriented Programming

The widespread use of Content Security Policy implies that frame ancestors directive is adequate to block Clickjacking attempts. But the effectiveness can be further improved by creating aspects (Aspect Oriented Programming) [15]. Aspect-Oriented Programming is used with the aim to:

- Reduce code scattering and code tangling that is security code is not spread and mixed with other code over the entire web application instead; the security code is segregated from the rest of the web application code and is placed in independent and isolated modules called aspects [16]. So, Aspect-Oriented Programming allows separation of security logic from application logic.
- Allow cross cutting of concerns that is, web application code need not to be modified or disturbed for evolving security code, aspect code is just interwoven with source code.

3.3 X-Frame-Options HTTP Header Versus Content Security Policy Header

In Sect. 2, we have discussed that X-Frame-Options is a reliable approach to prevent Clickjacking but the implementation of X-Frame-Options is limited to the entire webpage. To impose different security policies on individual elements of a webpage, we should look for element-customized approaches so that some elements should be permitted to be embedded in cross origins and some security-critical elements should be handled with more vigilance. Also, in some situations in which there are possibilities of attacks emanating from admissible origins, there is a need of more refined control over the resources that a browser can load and strict white listing of the domains from where to load. This paves way for Content Security Policy headers and its wide variety of directives.

Also, if browser engines come across more than one headers of X-Frame type which can happen if the header is applied in the source code as well as in the Apache or IIS server, then *HTTP RFC 2616* Sect. 4 states that multiple header fields should be joined together to form a single header field by merging the field values and is ultimately discarded as invalid by several browsers like Safari 5.1.7. [17]

4 Proposed Work

Our main objective is to design and implement an aspect for Clickjacking Prevention in such a way that it will bind the Content Security Policy HTTP header with GET and POST method calls so that the developer is relieved from adding the header manually to each page, saving a lot of development time. This is what we have tried to achieve in the following implementation section by demonstrating the use of various combinations of Content Security Policy directives which allows a developer to restrict the browsers to load content only from trusted sources.

5 Implementation and Results

5.1 Designing Content Security Policy for Blocking Clickjacking Attempts

As described in Table 1, frame ancestors directive is used to prevent framing of our webpage by cross origin web pages.

Response Check Algorithm for frame-ancestors directive

Input: A Request with a URL list and a target browsing context, a Response with a URL list and a Policy with directive set.

Output: URL "BLOCKED" OR "ALLOWED".
Procedure:

1. If the target browsing context is a nested browsing context,

   1.1 Then for each ancestor, repeat steps 1.1.1 and 1.1.2

      1.1.1 Analyze the origin's URL by carrying out UNICODE serialization and pass this value to the URL parser.

      1.1.2 Now match this origin's URL and specified directive's value to the source list. If no match found, return "BLOCKED".

2. Return "ALLOWED".

Thus, in case of multiple layers of framing, frame-ancestors directive matches the URL's of all parent origins and if at any layer, there is a mismatch (according to specified directive values: "self", "none" or url) then framing is blocked. [18].

Here, the content security policy is framed for different values of frame-ancestors directive and default-src directive (Sect. 5.3) such as:

Content Security Policy: default-src "self"; frame-ancestors "none"

These nine combinations of CSP headers (Table 2) are added to the aspect one by one and tested on different browsers.

5.2 Adding the Content Security Policy Header to a J2EE Aspect and Defining Point Cuts to Bind the Header to GET and Post Calls

Test Environment (software requirements): Eclipse Indigo (version 3.7) and Apache Tomcat 6 with AspectJ Plugin AJDT 2.2.2 for OWASP WebGoat (J2EE).

WebGoat by OWASP is a deliberately vulnerable web application developed for security testing [19]. So, an aspect was created and added to WebGoat. The flowchart in Fig. 1 explains the steps in creating AspectJ file. First, aspect is created by defining join points which define points at which AOP code will be included in source code. Then pointcuts are defined to specify at which point advice will be executed. Advice is the action taken. So, when joint point is reaches, doGet() or doPost method is called and response header is inserted at required place.

These four steps are repeated for doPost() method. Pointcut is defined to tie the header to these method calls in such a way that whenever one of these methods is called, header will be automatically inserted. CSP Header is added to response object using addHeader() method [20] in the following way:

   response.AddHeader("Content Security Policy", **"default-src https:; frame-ancestors 'self'"**);

The highlighted second parameter in this method can be varied with different combinations of policy directives to ensure maximum protection (Table 2).

Table 2 Results—different content security policies for Clickjacking prevention

Directives and values	frame-ancestors "none"	frame-ancestors "self"	frame-ancestors url
default-src "none"	No framing allowed but embedding from no source **Clickjacking Prevention**	Framing only by same origin resources but resources can be embedded from nowhere **Clickjacking Prevention**	Framing only by listed "url (s)" but resources can be embedded from nowhere **Clickjacking Prevention if "url" is a trusted source**
default-src "self"	No framing allowed but embedding from same origin **Clickjacking Prevention**	Framing only by same origin and same origin resources can be embedded **Clickjacking Prevention**	Framing only by listed "url (s)" but resources can be embedded from same origin **Clickjacking Prevention if "url" is a trusted source**
default-src url	No framing allowed but embedding from listed "url(s)"only **Clickjacking Prevention**	Framing only by same origin but resources can be embedded from listed "url (s)" **Clickjacking Prevention**	Framing only by listed "url (s)" but resources can be embedded from listed "url (s)" **Clickjacking Prevention if "url" is a trusted source**

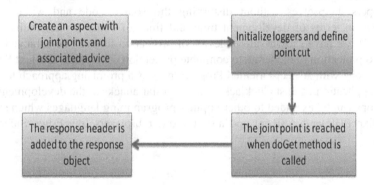

Fig. 1 Binding CSP header with doGet() and doPost() method calls

5.3 Testing Different Policy Combinations on Different Browsers

Table 2 summarizes the results for successful directive combinations that can lead to Clickjacking Prevention by disallowing the framing of our web page or its restrictive framing.

The results are for Google Chrome version 50.0 and Mozilla Firefox 46.0. For Firefox versions <45, frame-ancestors directive is non-recognizable. Similarly, tests for Internet Explorer 8 were unsuccessful as Content Security Policy is not supported. Any restricted framing attempts would generate the warning "blocked by Content Security Policy" failing the Clickjacking attempt.

5.4   Implementation Issues

While creating test environment, any software with J2EE support can be used but compatible AOP plugin should be chosen carefully according to the software versions. Other implementation issue can be that though all popular and modern browsers support CSP which is a feature of HTML5 but in case of outdated browser versions with no CSP support (like Internet Explorer less than 8) some browsers like Firefox automatically default the CSP header to X-frame-Options header.

6   Conclusion and Future Scope

We have demonstrated an approach to enhance the security of a web application by protecting it against Clickjacking attacks. Also, this required code can be applied by developers themselves without disturbing the source code and without much overhead in terms of development time and lines of code. In the results section (Table 2), we have shown various levels of protection that can be achieved that is from no protection to moderate to complete protection. Therefore, Content Security Policy along with Aspect-Oriented Programming is a promising approach to secure a web application against Clickjacking and similar attacks at the development time. The scope can be extended to other popular programming languages which support AOP like PHP and ASP.net and also for other browsers like Edge, Safari, and Opera.

References

1. Huang, L.S., Moshchuk, A., Wang, H.J., Schechter. S., Jackson C.: Clickjacking: Attacks and Defenses. In: USENIX Sec. Symposium. USENIX Association (2012).
2. Sinha, R., Uppal, D., Singh, D, Rathi, R.: Clickjacking: Existing Defenses and Some Novel Approaches. In: IEEE International Conference on Signal Propagation and Computer Technology, India (2014).
3. Paul Stone, "Next Generation Clickjacking: New attacks against framed web pages", Black Hat Europe Talk 2010.
4. Rydstedt, G., Bursztein, E., Boneh, D., Jackson, C.: Busting Frame Busting: A study of clickjacking vulnerabilities at popular sites. In: IEEE Oakland Web 2.0 Security and Privacy (W2SP 2010).
5. Takamatsu, Y., Kono, K.: Clickjuggler: Checking for Incomplete Defenses against Clickjacking. In: IEEE Twelfth Annual Conference on Privacy, Security and Trust (PST) (2014).

6. Shamsi, J., Hameed, S., Rahman, W., Zuberi, F., Altaf, K., Amjad, A.: Clicksafe: Providing Security Against Clickjacking Attacks. In: IEEE 15th International Symposium on High-Assurance Systems Engineering (2014).
7. OWASP, https:l/www.owasp.org/index.php/Clickjacking_Defense_Cheat_Sheet.
8. The X-Frame-Options-HTTP Response Header, https://www.developer.mozilla.org/en-US/docs/Web/HTTP/X-Frame-Options.
9. Balduzzi, M., Egele, M., Kirda, E., Balzarotti D.: A solution for the automated detection of clickjacking attacks. In: Proceedings of the 5th ACM Symposium on Information, Computer and Communications Security, ASIACCS '10 (2010).
10. ScriptSafe for chrome web browser, https://chrome.google.com/webstoreldetaillscriptsafe.
11. X-Frame-Options is dead, long live Content Security Policy, https://appsec-labs.com/portal/anti-clickjacking/.
12. Stamm, S., Sterne, B., Markham, G.: Reining in the Web with Content Security Policy. In: ACM WWW, North Carolina, USA (2010).
13. An introduction to Content Security Policy, http://www.html5rocks.com/en/tutorials/security/content-security-policy/.
14. OWASP, https://www.owasp.org/index.php/Content_Security_Policy_Cheat_Sheet.
15. Słowikowski, P., Zieliński, K.,: Comparison Study of Aspect-oriented and Container Managed Security For: AGH University of Science and Technology.
16. Mace, E., Kodra, L., Vrenozaj, F., Shehu, B.: Protection of Web Applications Using Aspect Oriented Programming and Performance Evaluation. In: BCI'12, Serbia.
17. Clickjacking: A common implementation mistake can put your websites in danger, https://blog.qualys.com/securitylabs/2015/10/20/clickjacking-a-common-implementation-mistake-that-can-put-your-websites-in-danger.
18. Content Security Policy Level 3, Editor's Draft, 27 April 2016: https://w3c.github.io/webappsec-csp/.
19. OWASP WebGoat, https://www.owasp.org/index.php/Category:OWASP_WebGoat_Project.
20. Interface HttpServletResponse (Servlet API Documentation), tomcat.apache.org.

A Conceptual Framework for Analysing the Source Code Dependencies

Nisha Ratti and Parminder Kaur

Abstract Component-based approach has been used for programming widely. Some problems creep up as softwares are developed using commercial-off-the-shelf components (COTS). Important issues related to component-based development has been addressed in this paper. As COTS components are used for developing software, there is a problem of coupling and cohesion. Another problem to be handled is to track the changes made in the source code during the development. Different available tools for the same purposes are studied and analysed. It is found that there is a need to design a tool which can manage all the already discussed problems in an efficient manner. The proposed tool will try to give the solution in order to manage the configuration of the system efficiently.

Keywords Component-Based systems · Version control system · Component · Configuration management · Code dependencies

1 Introduction

Component technology has given a new dimension to software development. Component ware allow the development of software not from scratch but by selecting consistent, reusable and robust software components and assembling them using appropriate software middleware. With component technology, the major emphasis has shifted from developing applications to assembling applications. Larsson [1] has discussed five major tasks to be done while working with components: Find, Select, Adapt, Deploy and Replace. Each task needs extensive expertise to develop the desired component as well as component-based system.

N. Ratti (✉)
Department of CSE, RIEIT, Railmajra, India
e-mail: nisharatti@gmail.com

P. Kaur
Department of CSE, GNDU, Amritsar, India
e-mail: parminder.dcse@gndu.ac.in

© Springer Nature Singapore Pte Ltd. 2018
S.K. Bhatia et al. (eds.), *Advances in Computer and Computational Sciences*,
Advances in Intelligent Systems and Computing 554,
https://doi.org/10.1007/978-981-10-3773-3_33

In Component-Based Systems (CBSs), management of components is done during their entire life span [2]. Only those components are used to configure a system which are tested rigorously according to various parameters. As the evolution of the systems progresses with new versions of components, there is no mechanism that the system can make out about the installation of new versions. There should be some mechanism to check that a comparison is performed between the new version to be installed and the older version which is already present in the system. This approach may predict whether it is advisable to work with old components or not. But it does not guarantee that the functionality will not be disturbed when new components are installed. Some questions have been proposed regarding component ware and dependencies. Component configuration management helps to counter these queries, such as: [3].

- No. of components which are being added or deleted after a reconfiguration.
- When one component is to be reused by another application, new requirement of the components is to be ascertained.
- Changes in Dependencies due to the new and improved configuration of the system [3].
- What will be the effect of change on the system when one component is changed?
- At least how many test cases are to be executed when one component is changed? [4].
- What will be the minimum number of components must be inspected, when a failure of the system occurs?
- What is the effect on a system if a new system of component is installed?
- What is the difference between two configurations?

All these queries [3] are related to Component Development Life Cycle (CBLC). So, a better understanding of CBLC is required to give right insight into the matter when upgrading systems [5–7]. The different stages of component dependency life cycle can be defined as in Fig. 1:

Component-based development begins with identifying the candidate component to be used in the development. From a number of components identified only a few are selected to be used. The selection criteria includes lesser conflicts while composing the systems [8]. When pre-fabricated components are used, a number of issues are to be dealt with. Such important issues are configuration management, version management, change management, etc. [9–11]. These issues help in reducing the risk of interoperability and testing processes [12].

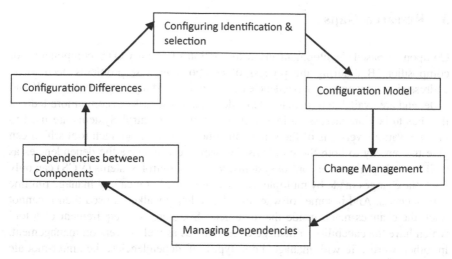

Fig. 1 Component dependency life cycle [8]

2 Related Work

The first Software Configuration Management (SCM) workshop [13] has given an initiation to version models. It has been followed by Tichy et al. [14], Feiler et al. [15], Estublier [16], Conradi et al. [17]. In the past, a number of version control tools have been proposed in the literature [18, 19]. Conradi et al. [20] has made a survey on version models. Another survey has been made regarding version oriented configurations [17]. In [17], after giving the review of the existing approaches, an approach has been proposed to construct consistent configurations for large and complex software products. Tichy [14] has given the basic terms like software object, source and derived object, version graph, etc. Feiler [15] has classified the model into four categories, i.e. Checkout/Check-in model, Composition Model, Long Transaction model and Change set Model. Some of the already existing version control tools are SCCS [21], RCS [22], DSEE [23], Adele [24], Clear Case [25], Aide-de-Camp [26] and VOODOO [27].

In the process of configuration management, after version management, managing dependencies is the crucial task. Larsson [28] has proposed a tool for managing the dependencies. This tool can parse the exe file and fetch the.dll files and generate the report for the dependencies [12]. Earlier Microsoft has also developed the similar tool named Dependency Walker (DW) [29]. DW is such a tool which scans 32-bit or 64-bit windows module and builds a hierarchical tree diagram of all modules. Another initiative has been taken by the team which developed the similar tool named "NDepend" [30]. This tool helps in generating important information regarding dependencies.

3 Research Gaps

Component based development gives the option of using COTS components for composition. But during the process of composition, few problems need to be addressed like managing dependencies and versions. Many software like DW and NDepend are available to manage dependencies. These tasks become more tedious if it has to be done across various versions. Version Control Systems are used to manage various versions of the same component. There is no such tool which can give the remedy to both the problems, i.e. which can manage the dependencies as well as work as a version control systems. Version Control Systems (VCS) can only track the changes made by multiple user over the time but it cannot manage runtime dependencies. And if some software like dependency walker is used then it cannot track the changes made inside the document. So, there is a requirement of a tool which have the capability to manage dependencies as well as version management. In other words, it will manage both types of dependencies, i.e. inter-module dependencies and intra-module dependencies.

4 Need of the Proposed Framework

In configuration management, while configuring the components, managing the dependencies is the most tedious tasks. An automated tool may be required to study these dependencies. But management of dependencies can not be done in isolation. Version management goes hand by hand with managing dependencies. So this whole process is divided into various phases. In the first phase, dependencies for various versions of the same files are generated. The tool responsible for this purpose has been designated as SCDA (Source Code Dependency Analyser). It will basically parse the source code of the file and generate the report regarding which files are dependent on which other files. In the second phase, a comparative analysis of the dependency reports of different versions is done. This comparison is done on the basis of some selection process. The entities to be analysed may be class, assembly and namespace. In the third phase, a number of metrics are used to measure the complexity, dependency and coupling/cohesion aspects of the software components [2].

5 Scope of Proposed Framework

The proposed tool can be used for following tasks:-

- When off-the-shelf components are used, there is always a probability of contradiction because of the dependencies between the components. In order to solve this problem of contradiction, the SCDA may be helpful. This tool will give the list of dependencies, which may be helpful in selecting the right component.

- Another advantage of SCDA may be that developer encounters some problems while installing new versions of the same software. Then also if the user has the source code of the particular software then also the dependencies may be analysed.
- Coupling and cohesion plays an important role while dealing with dependencies. So, it is important to calculate coupling and cohesion in the source code. The more the coupling is there, the more the dependencies will be there. In the similar fashion, the more the cohesion will be there, the less the dependencies will be there.
- Analysis of complexity in the code with the help of metrics.
- To compute Change Impact Analysis.
- Clean up of unused code.

6 Working of the Framework

The complete working of the functionality has been divided into three phases.

1. Generation of dependency report.
2. Comparison of dependency report of different versions.
3. Applying the suitable software metrics on the reports.

6.1 Phase1

The tasks to be done by this phase (Fig. 2) are accomplished by SCDA which has been proposed here only.

Parser and Resolver (P&R).

P&R component will take the input in the form of source code of file of some program and give the output in the form of XML file format. P&R component may be different for different languages. It analyses the inter-module dependencies. Inter-module dependencies are the dependencies between various modules/files,

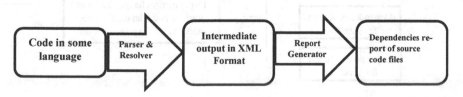

Fig. 2 Diagrammatic representation of source code dependency analyser: phase 1

i.e. which file is used by which file. Parser & Resolver (P&R) will be different for different languages, but the report generator can be used for any language as long as the output generated will conform to XML format required by this report generator. Hereby it achieves the component-based software development. As in Component-based software engineering, the user does not develop all the components from scratch. So, here report generator can be used again and again. P&R will generate the report in XML format. This report will be used by report generator component.

Report Generator.

This component will generate the report in html format. It will give the details about which file is using this particular file.

Dependency Report.

The final output of the phase 1 is the dependency report generated by SCDA. It includes the list of files where the code defined in a particular file is used.

6.2 Phase 2

In phase 2 the dependency report of different versions are compared. Output of phase1 will be input to phase 2. Output of phase 2 (Fig. 3) will show the changes among various versions. A tool has been used for comparing the reports of different versions. The user of the tool can make out the progress made over the time in different versions of the software. NDepend [30] has been used for this purpose. User of the tool can make out the progress made over the time in different versions of the software.

6.3 Phase 3

In the last phase, a number of metrics related to measurement of component dependencies, their coupling as well as cohesion will be used to determine the complexity of software component. With the help of various metrics, software

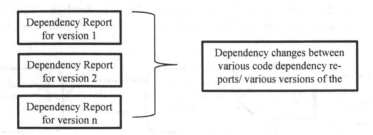

Fig. 3 Diagrammatic representation of phase 2

components can be analysed for their fitness for reusability with less coupling and high cohesion.

7 Applications of the Tool

- To act as a version control system, which has its significance when record-keeping and performance analysis for legal purposes is done.
- To keep the track of development of open source software.
- To keep the track of parallel developments. If you are a member of a software system team then it is in all probability simple to envision that while not version management chaos would result. However, version control is not helpful simply for groups. Many of the edges apply to solo developers further.
- To ensure that systems integration solutions can be rebuilt "from scratch" if required in case of a major disruption.
- To work with Student Management systems in an efficient manner.

8 Conclusion and Future Work

Component-Based Software Development basically promotes reusability of components which, in turn, results in lesser development costs. It is more effective when working with open source software components. Reusability of open source software components can be ensured with the help of metrics. Metrics give complete insight to the quantitative as well as qualitative aspects of the system.

In this paper, such framework has been proposed which will measure the dependencies and compare across various versions. Then metrics will be used for predicting the coupling, cohesion attributes, etc. The future work would include generating the dependencies report for all the languages instead of only one as in this paper. At this point of time, a small set of data has been used for analysis. The validation of the framework with various versions of the software is under process.

References

1. Larsson M., "Applying Configuration Management Techniques To component based Systems", IT Licentiate Thesis, 2000-007, Department of Information Technology, Uppsala University, 2000.
2. Kumar, V., Sharma, A., Kumar, R. and Grover, P.S. (2012), Quality aspects for component-based systems: A metrics based approach. Software Practice Experience, 42: 1531–1548. doi:10.1002/spe.1153.

3. http://www.it.uu.se.
4. http://www.imperial.ac.uk/computing.
5. Cook J. E. and Dage J. A., "Highly Reliable Upgrading of Components", In Proceedings of 21st International Conference on Software Engineering, ACM Press, 1999.
6. Hoek A. v. d., "Capturing Product Line Architectures", In Proceedings of 4th International Software Architecture Workshop, ACM Press, 2000.
7. R.Bialek, The architecture of a dynamically updatable component-based system. COMPSAC, 2002.
8. Kaur P. and Singh H., A Metric-Based Analysis of Component Dependencies, Journal of the CSI, 2010, ISSN-0555-7631, Volume 38, No. 4, pp 29–37.
9. Heinemann, G.T. and Council, W.T.; Component-Based Software Engineering: Putting the Pieces Together, Addison-Wesley Pub Co: ISBN: 0201704854, June 2001.
10. Szyperski C., Component Software - Beyond Object-Oriented Programming, Addison-Wesley, 1998.
11. A. W. Brown, Large Scale Component Based Development, Prentice Hall, 1st edition December, 2000.
12. M. Larsson, I. Crickovic, New Challenges for Configuration Management, System Configuration Management, SCM-9, Springer 1999, ISBN 3-540-66484-X.
13. Winkler, J. F. H. Ed. 1988. Proceedings of the International Workshop on Software Version and Configuration Control (Grassau, Germany, 1988). Teubner Verlag.
14. Tichy, W. F. 1988. Tools for software configuration management. In J. F. H. Winkle Ed., Proceedings of the International Workshop on Software Version and Configuration Control (Grassau, Germany, 1988), pp. 1–20. Teubner Verlag.
15. P. Feiler, Configuration Management Models in Commercial Environments, Technical Report CMU/SEI-91-TR-7, Software Engineering Institute, Pittsburgh, Pennsylvania, March 1991.
16. Estublier, J. and Casallas, R. 1995. Three dimensional versioning. In J. Estublier Ed., Software Configuration Management: Selected Papers SCM-4 and SCM-5, LNCS 1005 (Seattle, Washington, April 1995), pp. 118–135. Springer-Verlag.
17. Conradi, R. and Westfechtel, B. 1996. Configuring versioned software products. In I. Somerville Ed., Software Configuration Management: ICSE'96 SCM-6 Workshop, LNCS 1167 (Berlin, Germany, March 1996), pp. 88–109. Springer-Verlag.
18. Mei, H., Zhang, L., Yang, F., A component-based software configuration management model and its supporting system, Journal of Computer Science and Technology, July 2002, Volume 17, Issue 4, pp 432–441.
19. Kaur, Parminder and Singh, Hardeep. "Configuration Management Issues in Software Process Management", American Journal of Engineering & Applied Sciences, 2012.
20. R. Conradi and B. Westfechtel, Version Models for Software Configuration Management, ACM Computing Surveys, Vol. 30, No. 2, June 1998.
21. M. J. Rochkind, The Source Code Control System. IEEE Trans. Software Engineering 1, 4 (Dec.), 364–370. 1975.
22. W. Tichy, RCS – A System for Version Control, Software Practice and Experience 15, 7, 637–654. 1985.
23. D. B. Leblang and D. G. McLean, Configuration Management for Large-Scale Software Development Efforts, In GTE Workshop on Software Engineering Environments for Programming in the Large, pages 122–127. June 1985.
24. J. Estublier, A Configuration Manager: The Adele Data Base of Programs, In Proceedings of the Workshop on Software Engineering Environments for Programming-in-the-Large, pages 140–147. June 1985.
25. Leblang, D. 1994. The CM challenge: Configuration management that works. In W. F. Tichy Ed., Configuration Management, Volume 2 of Trends in Software (New York, 1994), pp. 1–38. John Wiley and Sons.
26. Cronk, R. D. Tributaries and deltas. BYTE 17, 1 (Jan.), 1992, 177–186.

27. C. Reichenberger, VOODOO—a tool for orthogonal version management. In Software Configuration Management: Selected Papers SCM-4 and SCM-5 (Seattle, WA, April), J. Estublier, Ed., LNCS 1005, Springer-Verlag, 61–79. 1995.
28. Larsson M., Crickovic I., Component Configuration Management, in proceedings ECOOP conference, workshop on component oriented Programming Nice, France, and June 2000.
29. www.dependencywalker.com.
30. www.ndepend.com.

DWT-SVD-Based Color Image Watermarking Using Dynamic-PSO

Nitin Saxena, K.K. Mishra and Ashish Tripathi

Abstract The main objectives of digital image watermarking schemes are to provide acceptable level of both imperceptibility and robustness for wide variety of host and watermark images. To support watermarking scheme in accomplishing these objectives, suitable watermark strength is identified in the form of scaling factor using Dynamic-PSO (DPSO) for colored images. A non-blind, DWT-SVD-based watermarking scheme is used to embed watermark in host image. Results achieved through DPSO are compared with PSO and other widely accepted variants of PSO. Experiment results demonstrate that performance of underline watermarking scheme when used with DPSO is better than other variants of PSO.

Keywords Particle swarm optimization (PSO) · Image watermarking · Optimization · Scaling factor · Imperceptibility · Robustness

1 Introduction

The last decade shows rapid increase in use of digital content such as images, audio, or video due to their easy availability through Internet. It is safer to apply techniques that provide ownership protection, content authentication, etc., for digital data before their distribution on Internet. Digital watermarking [1] is one promising and widely used solution to guard digital media including digital images from being illegally used.

In digital watermarking, authorized information (watermark image) is embedded imperceptibly into the original data (host image) to prove their authenticity and

N. Saxena (✉) · K.K. Mishra · A. Tripathi
CSED, MNNIT Allahabad, Allahabad, India
e-mail: nitinsaxena777@gmail.com

K.K. Mishra
e-mail: kkm@mnnit.ac.in

A. Tripathi
e-mail: ashish.mnnit44@gmail.com

© Springer Nature Singapore Pte Ltd. 2018
S.K. Bhatia et al. (eds.), *Advances in Computer and Computational Sciences*,
Advances in Intelligent Systems and Computing 554,
https://doi.org/10.1007/978-981-10-3773-3_34

343

ownership. The embedded information can be retrieved by applying contrary process. The embedding process of the watermark should not affect the quality of the image [2]. At the same time watermarked images are also prone to different modification attacks. These attacks destroyed or removed watermark from watermarked images, result to failure of recovering watermark at receiver end. Hence digital watermarking technique should be robust enough to overcome the effect of different modification attacks on watermarked image.

There are several criteria to classify watermarking schemes such as based on the availability of original content during the detection of watermark Blind or Non-blind, based on underlined domain spatial domain and frequency domain schemes. The frequency domain includes Discrete Cosine Transform (DCT), Discrete Fourier Transform (DFT), and Discrete Wavelet Transform (DWT) [3–5]. Further singular value decomposition (SVD)-based watermarking techniques are proposed to improve performance [6–8]. Watermarking schemes keep the balance between robustness and perceptual quality by controlling the percentage of the watermark embedded into host image, i.e., scaling factor. Scaling factor represents the strength of watermark used while embedding it in the host image.

A group of schemes embed principle components of the watermark in the host image [9]. The appropriate value of scaling factor is image dependent. Optimization algorithms such as genetic algorithm (GA) [10], particle swarm optimization (PSO) [11], etc., can be used to the find optimal value of scaling factor for watermarking schemes [12, 13]. These optimization algorithms are improved with the proposal of new variants [14–19]. This paper demonstrates the effectiveness of one variant of PSO, named Dynamic-PSO (DPSO) [20] for finding scaling factor in watermarking scheme. For this purpose, scheme proposed in [21] is used to perform image watermarking. The scheme defines non-blind watermarking for colored images which uses SVD in DWT domain. The performance of underline watermarking scheme, in terms of Peak Signal-to-Noise ratio, with DPSO algorithm is compared with other variants of PSO by calculating appropriate scaling factor. Further, scaling factor value calculated through different PSO variants are tested under various image-processing attacks to justify our claim.

In this paper, sections are arranged as: Sect. 2 presents PSO and its variants. In Sect. 3, DPSO is applied in watermarking scheme to find optimal value of scaling factor for given host and watermark images. Section 4 includes results and discussion. Finally, Sect. 5 draws the conclusion.

2　Particle Swarm Optimization and Its Variants

Particle swarm optimization (PSO) is metaheuristic population-based optimization algorithm developed by Kennedy and Eberhart [11, 14]. PSO maintains current position $X_i = (x_i^1, x_i^2 \ldots x_i^D)$, velocity $V_i = (v_i^1, v_i^2 \ldots v_i^D)$, and personal best $pbest_i = (pbest_i^1, pbest_i^2 \ldots pbest_i^D)$ vectors for each ith particle along with row vector for group best $gBest = (gbest^1, gbest^2 \ldots gbest^D)$ where D is the dimension of problem. PSO uses Eqs. (1) and (2) to update velocity and position of each particle in tth iteration (t = 1, 2, 3 ...).

$$v_i^d(t+1) = v_i^d(t) + c_1 * rand_1^d * \left(pbest_i^d(t) - x_i^d(t)\right) + c_2 * rand_2^d * \left(gbest^d(t) - x_i^d(t)\right)$$

$$(1)$$

$$x_i^d(t+1) = x_i^d + v_i^d(t+1) \tag{2}$$

where c_1 and c_2 are the acceleration coefficients, $rand_1^d$ and $rand_2^d$ are randomly generated numbers uniformly distributed in the range [0, 1]. Shi and Eberhart [14] introduce inertia weight ω (range 0–1) to control velocity component in velocity vector as given in Eq. (3).

$$v_i^d(t+1) = \omega * v_i^d + c_1 * rand_1^d * \left(pbest_i^d(t) - x_i^d(t)\right) + c_2 * rand_2^d * \left(gbest^d(t) - x_i^d(t)\right)$$

$$(3)$$

Shi and Eberhart further proposed PSO-TVIW, time dependent linearly varying inertia weight ($\omega = 0.9$ to 0.4) [15]. Similarly Ratnaweera et al. formulated PSO-TVAC scheme by varying acceleration coefficients linearly on time line. PSO-TVAC initially accentuates more on cognitive part and later shift focus on social component to update velocity vector. PSO-TVAC is further extended to form two variants Hierarchical PSO with TVAC (HPSO-TVAC) and Mutation PSO with TVAC (MPSO-TVAC) [17]. Particle Swarm Optimization with Aging Leader and Challengers (ALC-PSO) proposed by Chen et al. [18] defines leader to replace swarm best (gBest).

3 Application of Dynamic-PSO in DWT-SVD-Based Watermarking for Color Images

SVD-based watermarking schemes are sensitive to scaling factor while embedding watermark in host image. Different watermark image may require different value of scaling factor to achieve acceptable level of robustness and imperceptibility, even if being embedded in same host image. Identifying suitable value of scaling factor is a difficult problem. The complexity of problem increases drastically with the size of host and watermark image. Hence efficient optimization algorithm is required to obtain optimal value of scaling factor for different combination of host and watermark images. This section shows the suitability of Dynamic-PSO (DPSO) [20] in digital watermarking by finding suitable scaling factor in DWT-SVD-based watermarking scheme.

3.1 DWT-SVD-Based Watermarking Scheme

For embedding and extracting watermark, watermarking scheme for colored images in RGB space defined in [21] is used. Scheme uses low frequency sub band of DWT to embed watermark. Fractional (controlled by scaling factor) principal component of watermark image is added to singular value of host image identified through SVD. Modified singular values are used to form watermarked image. Reverse process extract the watermark from watermarked image.

3.2 Dynamic-PSO (DPSO)

Dynamic-PSO (DPSO) [20] focuses on stagnation of pBest and gBest occurred in PSO. Premature convergence takes place when PSO trapped in local optima, i.e., particle's pBest start stagnating. DPSO does not affect the basic concept of PSO, hence fast convergence characteristic of PSO remain intact. DPSO identifies stagnated particles of the population and provide dynamicity to stagnated particles explicitly such that stagnated particles explore new regions. This will provide required diversity and increases chance to recover from stagnation. Also swarm global best (gbest) is also tracked for local optima problem. If gbest does not improve for fixed number of iterations, new guidance is provided influencing from previous best position attained by group.

3.3 Fitness Function

Fitness function judges the survivability of particles in DPSO similar to PSO. The similarity between two images I and I* can be measured using normalized correlation coefficient (NCC) and is calculated as:

$$\text{NCC} = \frac{\sum_{i=1}^{M} \sum_{j=1}^{N} (I(i,j) - \bar{I})\left(I^*(i,j) - \bar{I}^*\right)}{\sqrt{\sum_{i=1}^{M} \sum_{j=1}^{N} (I(i,j) - \bar{I})^2 * \sum_{i=1}^{M} \sum_{j=1}^{N} (I^* - \bar{I}^*)^2}} \tag{4}$$

DPSO and other variants use following fitness function to identify suitable scaling factor for watermarking and includes imperceptibility and robustness characteristics [22].

$$f = \max\left(Corr\left(A, A^*\right) + Corr\left(W, W^*\right)\right) \qquad (5)$$

where A and A^* are host and watermarked image whereas W and W^* are watermark and extracted watermark images. $Corr(A, A^*)$ is the normalized correlation coefficient between host image and watermarked image while $Corr(W, W^*)$ denotes the normalized correlation coefficient between watermark image and extracted watermarked image.

3.4 Finding Scaling Factor Using Dynamic-PSO

Steps to find optimal value of scaling factor using Dynamic-PSO are:

1. Initialization

    (a) Initialize particles position (values of scaling factor in [−03, 0.3]) and velocity (values in [−0.1, 0.1]) randomly.
    (b) Initialize counters used to check stagnation of individual particle and group.

2. Compute fitness of each particle using Eq. (5). Identify pbest and gbest (scaling factor that produced best fitness value for particle and group).
3. Update velocity and position using Eqs. (3) and (2).
4. Accumulate fixed number of most recent gbest's as historical values.
5. Update pbest and gbest.
6. Identify particles which start stagnating and check gbest for group stagnation.
7. Reconstruct pbest for stagnated particles and gbest if group stagnates.
8. Check acceptability of reconstructed pbest of stagnated particles.

    (a) If improvement in pbest, accept reconstructed pbests for stagnated particles otherwise restored old respective pbest.

9. Similarly check acceptability of new gbest if reconstructed earlier.

    (a) If improvement in gbest, accept reconstructed gbest for group otherwise restored old gbest.

10. Stop iterations if termination condition satisfied otherwise go to step 3.

4 Results and Discussion

To measure the performance of watermarking scheme with different variants of PSO including DPSO metrics are calculated. Peak Signal-to-Noise Ratio (PSNR) metric evaluates the visual quality of the watermarked image as:

$$PSNR(in\ db) = 10\ \log_{10}\frac{255^2}{MSE} \qquad (6)$$

where, MSE is mean square error and calculated as:

$$MSE = \frac{1}{MN}\sum_{m=1}^{M}\sum_{n=1}^{N}[a_{mn} - a_{mn}^{w}]^2 \qquad (7)$$

Here M and N are height and width of host and watermarked images whereas, a_{mn} and a_{mn}^{w} are the pixel values of coordinate (m, n) in host and watermarked images, respectively.

Further its performance is tested against six image-processing attacks Gaussian noise (GN), salt-and-pepper noise (SPN), rotation (RO), median filter (MF), Gamma correction (GC), and Blurring the image (BF) to observe robustness of scheme.

Population size in each PSO variant including DPSO is fixed to 40. Maximum number of iterations permitted to each PSO variant is set to 50. The decision variable is initialized randomly between (−0.3, 0.3). The maximum velocity for particles V_{max} is limited to 0.1.

The image 'Lena' in Fig. 1a is taken as host image and image 'Mandrill' in Fig. 1b is used as watermark image. Both host and watermark images are of size 512 × 512.

In order to measure the viability of DPSO, watermarking is applied with PSO and its popular variants including DPSO. Watermarked image and extracted watermark obtained by applying watermarking scheme with DPSO are shown in Fig. 1c, d. The fitness value of f defined in Eq. (5) ranges from 0 to 2. Higher the value of f, higher is the corresponding similarity between two pair of images. The value 2 means that the two pair of images are exactly similar and 0 means totally

(a)   (b)   (c)   (d)

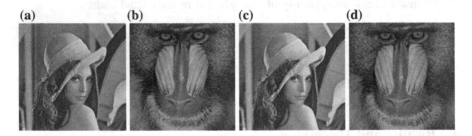

Fig. 1 **a** Host image **b** Watermark image **c** Watermarked image using DPSO **d** Extracted watermark using DPSO

Table 1 Comparison of obtained PSNR and fitness value using DPSO and other PSO variants

Method	PSNR	Fitness value (f)
Without PSO	23.523981	1.526723
PSO [14]	28.768498	1.768262
PSO-TVIW [15]	24.075918	1.557071
HPSO-TVAC [17]	25.302159	1.625536
MPSO-TVAC [17]	30.641393	1.840190
ALC-PSO [18]	33.416903	1.955121
DPSO [20]	**36.877771**	**1.977786**

Table 2 PSNR obtained using DPSO and other PSO variants under different attacks

Method	Attacks					
	GN 2% noise	SPN 5% noise	RO 10°	MF value = 3	GC value = 2	BF value = 20
Without PSO	13.2331	12.1298	10.2128	12.9127	11.2621	15.3219
PSO	17.6650	18.5198	14.1928	15.9183	13.9227	19.6921
PSO-TVIW	16.0511	15.1102	13.0782	15.1321	12.9012	18.1521
HPSO-TVAC	18.1873	17.8352	12.31904	18.3519	12.9103	21.3153
MPSO-TVAC	19.1378	19.3176	15.7163	20.7113	20.6129	21.7821
ALC-PSO	25.1612	24.6221	**22.6520**	23.2927	**21.9528**	23.6394
DPSO	**27.3731**	**27.1326**	22.6319	**25.3149**	21.9106	**26.3182**

Table 3 Fitness value obtained using DPSO and other PSO variants under different attacks

Method	Attacks					
	GN 2% noise	SPN 5% noise	RO 10°	MF value = 3	GC value = 2	BF value = 20
Without PSO	1.2729	1.2521	1.1231	1.1432	1.1351	1.3592
PSO	1.4928	1.5063	1.2591	1.3481	1.2490	1.5131
PSO-TVIW	1.4729	1.4672	1.2729	1.4691	1.2409	1.5012
HPSO-TVAC	1.5032	1.4862	1.1732	1.5132	1.1982	1.5592
MPSO-TVAC	1.5138	1.6251	1.4692	1.5418	1.5389	1.5603
ALC-PSO	1.7013	1.6522	**1.6020**	1.6123	**1.5559**	1.6492
DPSO	**1.7381**	**1.7129**	1.5962	**1.7089**	1.5401	**1.7025**

dissimilar. Table 1 includes the Peak Signal-to-Noise Ratio (PSNR) and value of fitness function through watermarking scheme with DPSO and other variants of PSO. PSNR and fitness values obtained by DPSO and other PSO variants-based watermarking scheme under different attacks are included in Tables 2 and 3, respectively. It is clear that the robustness of watermarking scheme is better with DPSO in comparison with other used PSO variants.

5 Conclusion

In this paper, method to find appropriate watermark strength (image dependent) in terms of scaling factor using Dynamic-PSO is proposed for DWT-SVD-based image watermarking scheme. Performance of DPSO is compared with other PSO variants using PSNR metric that show the similarity of host and watermarked image. Robustness is measured by applying various image-processing attacks on watermarked image. Experimental results demonstrate significant improvement in imperceptibility as well as robustness under various image-processing attacks.

References

1. G. Langelaar, I. Setyawan, and R. Lagendijk, "Watermarking digital image and video data", IEEE Signal Processing Magazine, vol. 17, no. 5, pp. 20–46, 2000.
2. M. Barni, F. Bartolini, A. De Rosa, and A. Piva, "Optimal decoding and detection of multiplicative watermarks", IEEE Trans. Signal Processing, vol. 51, no. 4, pp. 1118–1123, 2003.
3. M. Barni, F. Bartolini, A. De Rosa, and A. Piva, "Optimal decoding and detection of multiplicative watermarks", IEEE Trans. Signal Processing, vol. 51, no. 4, pp. 1118–1123, 2003.
4. A. Briassouli and M. G. Strintzis, "Locally optimum nonlinearities for DCT watermark detection", IEEE Trans. Image Processing, vol. 13, no. 2, pp. 1604–1617, 2004.
5. A. Nikolaidis and I. Pitas, "Asymptotically optimal detection for additive watermarking in the DCT and DWT domains", IEEE Trans. Image Processing, vol. 12, no. 5, pp. 563–571, 2003.
6. Discrete Wavelet Transform and Singular Value Decomposition", IEEE Transactions on Instrumentation and Measurement, vol. 59, no. 11, pp. 3060–3063, 2010.
7. Emir Ganic and Ahmet M. Eskicioglu, "Robust DWT-SVD Domain Image Watermarking: Embedding Data in All Frequencies," in Proc. Workshop Multimedia Security, Magdeburg, Germany, pp. 166–174, 2004.
8. G. Bhatnagar and B. Raman, "A new robust reference watermarking scheme based on DWT-SVD" Computer Standards Interfaces, vol. 31, no. 5, pp. 1002–1013, 2009.
9. Thai Duy Hien, Yen-Wei Chen, and Zensho Nakao, "Robust digital watermarking based on principal component analysis', IJCIA, Volume 04, Issue 02, 2004.
10. Goldberg, D. E., "Genetic Algorithms in Search, Optimization, and Machine Learning", Reading MA: Addison-Welsey, 1989.
11. J. Kennedy and R. C. Eberhart, "Particle swarm optimization," in Proc. IEEE Int. Conf. Neural Netw., Perth, Australia, vol. 4, pp. 1942–1948, 1995.
12. Ziqiang Wang, Xia Sun, and Wexian Zhang, "A Novel Watermarking Scheme Based on PSO Algorithm," LSMS 2007, LNCS 4688, PP. 307– 314, 2007 @ Springer-Verlag Berlin Heidelberg 2007.
13. Veysel Aslantas, A. Latif Dogan and Serkan Ozturk, "DWT-SVD based Image Watermarking Using Particle Swarm Optimizer," ICME 2008.
14. Y. Shi and R. C. Eberhart, "A modified particle swarm optimizer," in Proc. IEEE Congr. Evol. Comput., pp. 69–73, 1998.
15. Y. Shi and R. C. Eberhart, "Empirical study of particle swarm optimization," in Proc. IEEE Congr. Evol. Comput., pp. 1945–1950, 1999.
16. Y. Shi and R. C. Eberhart, "Fuzzy adaptive particle swarm optimization", in Proc. IEEE Congr. Evol. Comput., vol. 1, pp. 101–106, 2001.

17. R. Mendes, J. Kennedy, and J. Neves, "The fully informed particle swarm: Simpler, maybe better", IEEE Trans. Evol. Comput., vol. 8, no. 3, pp. 204–210, 2004.
18. Chen, W., Zhang, J, Lin, Y., Chen, N., Zhan, Z.-H., Chung, H.S.-H., Li, Y. and Shi, Y.-H., "Particle Swarm Optimization with an Aging Leader and Challengers", IEEE Trans. Evol. Comput., vol. 17, no. 2, pp. 241–258, 2013.
19. J. J. Liang, A. K. Qin, P. N. Suganthan, and S. Baskar, "Comprehensive learning particles swarm optimization for global optimization of multimodal functions", IEEE Trans. Evol. Comput., vol. 10, no. 3, pp. 281–295, 2006.
20. N. Saxena, A. Tripathi, K. K. Mishra and A. K. Misra, "Dynamic-PSO: An Improved Particle Swarm Optimizer", IEEE Congress of Evolutionary Computation (CEC 2015), 2015.
21. A. Srivastava, "Performance Comparison of Various Particle Swarm Particle Swarm Optimizers in DWT - SVD watermarking for RGB Images" ICCCT '15, 2015.
22. Ray-Shine un, Shi-Jinn Horng, Jai-Lin Lai, Tzong-Wang Kao, Rong Jian Chen, "An improved SVD based watermarking technique for copy right protection", Expert Systems with Applications 39, pp. 673–689, 2012.

Semi-supervised Spatiotemporal Classification and Trend Analysis of Satellite Images

Avinash Chandra Pandey and Ankur Kulhari

Abstract Classification of satellite images can be used for land information extraction, i.e., land cover maps, forest maps, industrial maps, residential maps, flooded maps, etc. The classification can be performed using any of the two methods, namely supervised classification method and unsupervised method. However, supervised classification methods require extensive training with existing training datasets. For satellite images, it is difficult to generate training dataset for all the land cover types. Therefore, this paper proposes a novel semi-supervised classification method to classify satellite images. The efficiency of proposed method is tested on satellite images of Delhi and Himalayan regions. Experimental results validate that the proposed method outperforms the existing methods.

Keywords Self-organizing feature map · Artificial neural network · Semi-supervised classification · Trend analysis · Vector quantized temporal associative memory · Accuracy assessment

1 Introduction

Satellite Images are the major sources of data and information, which are used in environmental studies, forest management, urban change detection, etc. [1]. So far, a lot of efforts have been made to extract information from remotely sensed images and various methods have also been developed in this field. Digital interpretation (quantitative analysis) [2] is one of the main approach for the identification and measurement of various targets (i.e., point line area feature) in satellite image. Classification is a common and powerful information extraction method, which is used in remote sensing. Classification of satellite images is a process in which

A.C. Pandey (✉) · A. Kulhari
Jaypee Institute of Information Technology, Noida, India
e-mail: avinash.pandey@jiit.ac.in

A. Kulhari
e-mail: ankur.kulhari@jiit.ac.in

© Springer Nature Singapore Pte Ltd. 2018
S.K. Bhatia et al. (eds.), *Advances in Computer and Computational Sciences*,
Advances in Intelligent Systems and Computing 554,
https://doi.org/10.1007/978-981-10-3773-3_35

353

different areas, viz., water, soil, vegetation, etc., are separated. The classified images are used to gain information about areas of interest and also used as base maps for other change detection analysis [3]. There are many classification methods, which have their own merits and demerits. Among classification methods, maximum likelihood approach [4] has been used more frequently. Srivastava et al. [5] provided a method to select most suitable classification method for land use/cover from maximum likelihood classification (MLC), support vector machine (SVM), and artificial neural network (ANN) and found that classification with ANN results far better than SVM and MLC. Artificial neural network (ANN) based methods provide better classification results for satellite images [5].

There are two classification methods based on ANN, namely supervised [6] and unsupervised [6]. In supervised classification [7], training datasets (representative sites) for each land cover type is used. Representative sample site of known cover types, also called as training areas, are used to compile a numerical interpretation key that describes the spectral attributes of each feature type of interest. In unsupervised classification [8], no supervisor is used for pixel categorization of satellite images. The automated system analyzes properties of each pixel and based on common characteristics of pixels, groups are formed without providing representative sites. However, in supervised methods [9] classification structure upon data is imposed by selecting appropriate training sites, which sometimes may not present in dataset [10] and it affects accuracy of overall system due to unclassified regions in satellite image(s). In unsupervised methods, natural groups are identified by spectral homogeneity, which sometimes may not correspond to a class of interest [10]. Hybrid classification methods [11, 12] perform better in most of the cases. However, its accuracy is decreased drastically when applied to image having heterogeneous regions. object-based classification [12] methods group a number of pixels with homogeneous properties into an object. Moser et al. proposed [13] Spectra-based classification method for the classification of satellite images which are simple and easy to be implement. However it neglects the spatial components therefore, its accuracy may decrease immensely. Blaschke [3, 14] incorporated spatial information in satellite images and used it for satellite image classification, i.e., generally called spatiocontextual image classification. Due to the limitation of above classification methods for satellite images, this paper introduces a novel semi-supervised method based on self organizing feature map (SOFM) for the classification of satellite images. The experimental results depict that the proposed method outperforms the existing methods. The rest of the paper is organized as follows: Sect. 2 describes the proposed method. Section 3 discusses experimental results and Sect. 4 concludes the paper.

Fig. 1 Flowchart of the
proposed semi-supervised
method

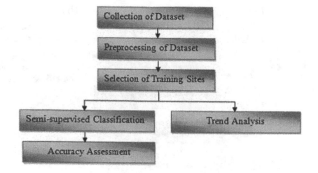

2 Proposed Work

In this article, classification of satellite images has been carried out using self-organized feature map followed by trend analysis of snow in Himalayan region using VQTAM algorithm. Figure 1 shows the flowchart of proposed method. For preprocessing, classification, and trend analysis of multispectral and multitemporal satellite images following steps are performed:

(a) Collection of Dataset
(b) Preprocessing of Dataset
(c) Selection of Training Dataset/Sites
(d) Classification using Semi-supervised Learning
(e) Accuracy Assessment using ERDAS IMAGINE
(f) Trend Analysis using Temporal SOFM

The proposed work is carried out in two phases:

(a) Classification of satellite images using semi-supervised classification method.
(b) Trend analysis of snow in Himalayan region.

Satellite images [15] have been taken from Delhi and Himalayan region. Delhi is located in North-West of Uttar Pradesh. It has a geographical extent of 028n40 latitude and 077e20 longitude. After acquisition of the dataset, raw satellite images are preprocessed using preprocessing steps in which satellite images are georeferenced [16] with the help of 25 GCPs (the number of GCPs used in process depends on the nature of the image). As a result, a Geo-Ti image is obtained which will be used as an input in the classification and trend analysis process.

Figure 2 shows the flowchart of semi-supervised classification and trend analysis. Semi-supervised method [17] overcomes the problems of supervised and unsupervised method by using the large amount of unlabeled data (data without training sites) along with some labelled data (data with training sites). However, semi-supervised method requires fewer amounts of human efforts; as most of the data are analyzed using unsupervised methods in most of the cases. Therefore, it outperforms supervised and unsupervised methods.

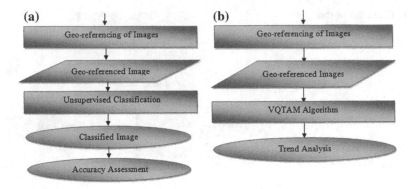

Fig. 2 **a** Flowchart of semi-supervised classification, **b** Flowchart of trend analysis process

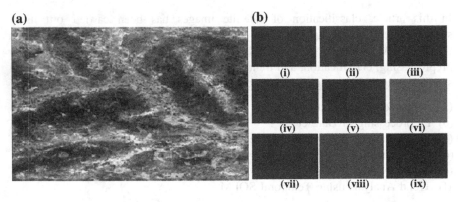

Fig. 3 **a** Reference map for training sites of Delhi region **b** Training sites for Delhi region **i** Crop area **ii** Residential area **iii** Fallow land **iv** Agriculture area **v** Industrial area **vi** Salt affected area **vii** Grazing land **viii** Open scrub **ix** Water body

Semi-supervised classification of the satellite images requires some training sites. The number of training sites depends on the major land cover in satellite image. In the proposed work the major land of Delhi area has been covered under the nine regions in the satellite image. These nine training sites such as Crop area, Residential area, Fallow land, Agriculture area, Industrial area, Salt affected area, Grazing land, Open scrub and Water body have been chosen as shown in Fig. 3. Training sites of Delhi region are taken from the reference map shown in the Fig. 3. The yellow points shown in Fig. 3a are the training sites that are used for semi-supervised classification methods. Each training site is represented with a square regions of size 5 5 3 which is taken from the reference map. Median value of each training site has been used to train SOFM network. Thus, we have nine median values, one for each training site and the average of median is used to train the SOFM network for semi-supervised classification which reduces the overall processing time.

Fig. 4 Training sites for
snow trends in Himalayan
region

September_2001 September_2006 September_2007

After training SOFM, input image is simulated using unsupervised (clustering) method and as a result of simulation, classified map having nine classes is obtained as an output. Different classes in the classified map are shown in different colors as red for crop area, pink for salt effected area etc. After semi-supervised classification, Trend analysis, is performed using the Vector Quantization Temporal Associative Memory (VQTAM) model [18]. Snow trends in the month of September 2008 in Himalayan Region has been predicted using VQTAM. To predict the snow trends in month of September, 2008 three satellite images of the Himalayan Region of month September 2001, September 2006, and August 2007 have been taken.

Images are cropped to select the same area, which is used as training sites shown in Fig. 4. Index values of training sites are used as a time series for the input. In this model the SOFM is trained using the sequentially training function that selects the random value from the input time series as a winner and then computes all the winning neurons that are closest to the winner and performs some calculations on the winning neurons to predict the value of snow trends as an output.

3 Experimental Results

3.1 Semi-supervised Classification of Satellite Images

A SOFM network of size 3 9 and weight vector used in the classification process have been shown in Fig. 5. Weight vector is generated after training SOFM network which is further used in classification process.

Figure 6 shows the satellite and classified image of Delhi area respectively. The satellite image is classified into nine different color classes as shown in Fig. 6c. The crop area is classified in red color whereas salt effected area is shown in pink color and water body is in black color. Purple, orange, pale green and green are for Residential area, Fallow land, Agriculture area, and Industrial area respectively. Yellow is for open scrub and Grazing land is shown in blue color.

(a)   (b)

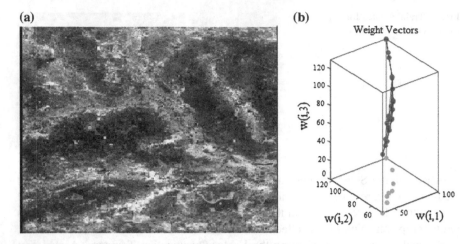

Fig. 5 **a** Network topology **b** Weight vector

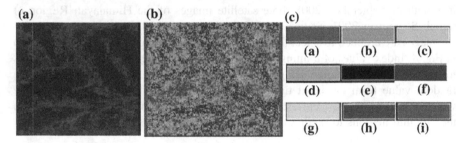

Fig. 6 **a** Satellite image **b** Classified satellite image **c** Reference color map for classified satellite image

3.2 Accuracy Assessment

To measure the accuracy of the semi-supervised classification method, an error matrix or confusion matrix is used. The accuracy assessment matrix E of size m x m represents that there are m number of classes and E_{xy} shows the number of patterns of class x predicted in class y. The diagonal values represent the correctly clustered patterns. The confusion matrix is being used to calculate accuracy by using Eq. (1).

$$\text{Accuracy} = \frac{\sum_{x=1}^{m} E_{xx}}{\sum_{x=1}^{m} \sum_{y=1}^{n} E_{xy}} \quad (1)$$

Table 1 Accuracy assessment of January 2003

	Undefined	Pink	Purple	Orange	Sky blue	Pale green	Yellow	Red	Green	Blue
Undefined	1	0	0	0	0	0	0	0	0	0
Pink	0	5	0	1	0	0	0	0	0	0
Purple	0	0	9	1	1	0	1	0	0	0
Orange	0	0	0	7	0	0	0	0	0	1
Sky blue	0	2	0	0	9	0	0	0	0	0
Pale green	0	0	2	8	3	12	2	0	0	0
Yellow	0	0	0	1	0	0	3	0	0	0
Red	0	0	0	0	0	0	0	0	0	0
Green	0	0	0	3	0	0	2	2	10	1
Blue	0	0	0	0	1	1	0	0	0	11

Table 2 Comparison of proposed method with the existing methods in terms of accuracy, computational time

Sr. no	Dataset	Method	Mean accuracy (%)	Mean computational Time
1	Satellite image dataset	Supervised method	60.87	316.01
2		Unsupervised method	61.81	342.85
3		Semi-supervised method	67.10	303.13

The accuracy assessment Table for the satellite image of Delhi region of January 2003 is shown in Table 1. From the accuracy assessment Table it is observed that the overall Accuracy of IRS-1D satellite image using semi-supervised classification is 67.10%. The reason of having low accuracy is due to lots of valleys and hills in study area, which results in more shadows in satellite imagery. These shadows result in misclassification. Mixed pixel classification information shows that around 18% area is with high certainty.

Table 2 shows the comparative results of the proposed method and existing considered methods in terms of accuracy, computational time. From the Table, it is visualized that the proposed method gives the best accuracy among all the considered methods. From the Table 2 it has been observed that supervised method has least accuracy among all the methods due to a lot of unclassified regions in satellite images.

3.3 Kappa Statistic

The Kappa statistic was derived to include measures of class accuracy within an overall measurement of classifier accuracy [19]. It provides a better measure of the accuracy of a classifier than the overall accuracy, since it considers interclass agreement. Kappa analysis yields a statistics that is a measure of agreement or accuracy [19]. The Kappa coefficient expresses the proportionate reduction in error generated by a classification process compared with the error of a completely random classification. The value of Kappa statistic for the proposed method is 0.6238, which implies that the classification process is avoiding approx 62% of the errors that generate a completely random classification.

The total number of reference points used in accuracy assessment is 100. The Kappa statistics [20] for January 2003 is 0.6238 as shown in Table 3.

3.4 Trend Analysis

Out of four bands available of IRS-1D satellite images only three bands are taken into consideration for satellite images of Himalayan region. To find out the snow trends in dataset of Himalayan region same area (as taken in phase-1) has been cropped from the original satellite images as training sites to analyze the snow trends.

The matrix values of all the training sites as shown in Fig. 7 are fed to the VQTAM algorithm [21] to find out the trends of next year based on the values of the input.

Figure 8 shows the snow trends in the satellite image of Himalayan region in the month of September 2008. The red color shown in Fig. 8 is the one step ahead values whereas the blue color shows the predicted values for September 2008. The overall normalized mean square error computed for the complete process is 0.5879%.

Table 3 Kappa statistics for January 2003

Class	Kappa
Undefined	10000
Pink	0.8208
Purple	0.7191
Orange	0.8418
Sky blue	0.7886
Pale green	0.3614
Yellow	0.7383
Red	0
Green	0.5062
Blue	0.8232

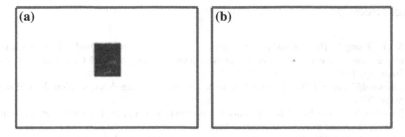

Fig. 7 Training sites for Himalayan Region **a** rock area, **b** snow area

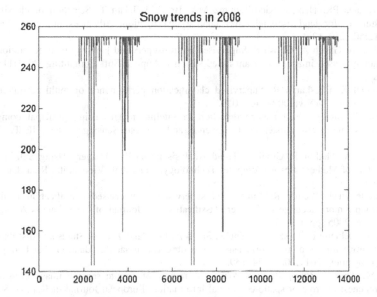

Fig. 8 Snow trend graph of Himalayan Region in September 2008

4 Conclusion

In this paper, semi-supervised classification method has been proposed for the classification of satellite images. The classification has been performed on the basis of the median feature of the training classes. The overall accuracy and the kappa statistics in the classification are 67.10% and 0.6238 respectively. Trend analysis has been carried out on dataset of Himalayan region's satellite images using Vector Quantization Temporal Associative Memory (VQTAM) model.

In future the image can be classified in more classes to get more accurate results. Other methods, i.e., harmonic mean and mode can also be used for training. Instead of VQTAM model, regression formula can also be used on the raw satellite image to find out the trends.

References

1. Xu L, Zhang S, He Z, Guo Y. The comparative study of three methods of remote sensing image change detection. In: Geoinformatics, 2009 17th International Conference on. IEEE; 2009. pp. 1–4.
2. Lillesand T, Kiefer RW, Chipman J. Remote sensing and image interpretation. John Wiley & Sons; 2014.
3. Blaschke T. Object based image analysis for remote sensing. ISPRS journal of photogrammetry and remote sensing. 2010;65:2–16.
4. Otukei JR, Blaschke T. Land cover change assessment using decision trees, support vector machines and maximum likelihood classification algorithms. International Journal of Applied Earth Observation and Geoinformation. 2010;12:S27–S31.
5. Srivastava PK, Han D, Rico-Ramirez MA, Bray M, Islam T. Selection of classification techniques for land use/land cover change investigation. Advances in Space Research. 2012;50:1250–1265.
6. Halder A, Ghosh A, Ghosh S. Supervised and unsupervised land use map generation from remotely sensed images using ant based systems. Applied Soft Computing. 2011;11:5770–5781.
7. Perumal K, Bhaskaran R. Supervised classification performance of multispectral images. arXiv preprint arXiv:10024046. 2010.
8. Celik T. Unsupervised change detection in satellite images using principal component analysis and-means clustering. Geoscience and Remote Sensing Letters, IEEE. 2009;6 (4):772–776.
9. Mangal A, Mathur P, Govil R. Trend Analysis in satellite Imagery Using SOFM. Apaji Institute of Mathematics & Computer Technology, Banasthali Vidhyapith, Rajasthan, India. 2005;.
10. Enderle DIM, WeihJr RC. integrated supervised unsupervised classification method to develop a more accurate land cover classification. In: Journal of the Arkansas Academy of Science; 2005. pp. 65–73.
11. Giri C, Ochieng E, Tieszen LL, Zhu Z, Singh A, Loveland T, et al. Status and distribution of man-grove forests of the world using earth observation satellite data. Global Ecology and Biogeography. 2011;20(1):154–159.
12. Li M, Zang S, Zhang B, Li S, Wu C. A review of remote sensing image classification techniques: the role of spatio-contextual information. European Journal of Remote Sensing. 2014;47:389–411.
13. Moser G, Serpico SB, Benediktsson JA. Land-cover mapping by Markov modelling of spatial-contextual information in very-high-resolution remote sensing images. Proceedings of the IEEE. 2013;101(3):631–651.
14. Ceccarelli T, Smiraglia D, Bajocco S, Rinaldo S, De Angelis A, Salvati L, et al. Land cover data from Landsat single-date imagery: an approach integrating pixel-based and object-based classifiers. European Journal of Remote Sensing. 2013;46:699–717.
15. Satellite image; 2005 (accessed December 15, 2015). www.nrsa.gov.in.
16. Van Laere O, Schockaert S, Dhoedt B. Combining multi-resolution evidence for georeferencing Flickr images. In: Scalable Uncertainty Management. Springer; 2010. pp. 347–360.
17. Zhu X. Semi-supervised learning literature survey. 2005.
18. Barreto GA. Time series prediction with the self-organizing map: A review. In: Perspectives of neural-symbolic integration. Springer; 2007. pp. 135–158.
19. Mather P, Tso B. Classification methods for remotely sensed data. CRC press; 2009.

20. Malpica A, Matisic JP, Van Niekirk D, Crum CP, Staerkel GA, Yamal JM, et al. Kappa statistics to measure interrater and intrarater agreement for 1790 cervical biopsy specimens among twelve patholo-gists: qualitative histopathologic analysis and methodologic issues. Gynecologic oncology. 2005;99:S38-S52.
21. Abinader F, de Queiroz AC, Honda DW. Self-organized hierarchical methods for time series forecasting. In: Tools with Artificial Intelligence (ICTAI), 2011 23rd IEEE International Conference on. IEEE; 2011. pp. 1057–1062.

21. Maglid...
22. ...

Improved Content-Based Image Classification Using a Random Forest Classifier

Vibhav Prakash Singh and Rajeev Srivastava

Abstract Content-based image classification is being one of the important phase in the process of automatic retrieval and annotation of images. In this paper, we are focused on effective image classification using low-level colour and texture features. It is well proved that the classifier performances are good for the unknown images, if quite similar images are present in the training set. On the other hand, classifier performances could not be guaranteed for images that are very much dissimilar from training set. This generalization problem of classifier can bias the image retrieval and annotation process. This paper objective is to investigate the discrimination abilities for such different class standard images. For improved image classification, we extract the effective low-level colour and texture features from the images. These features include; local binary pattern (LBP) based texture features, and colour percentile, colour moment, and colour histogram based colour features. To overcome the generalization problem, we have used random forest classifier, capable for handle over-fitting situation. Experimental analysis on benchmark database confirms the effectiveness of this work.

Keywords Image classification · Decision tree · Random forest · Feature extraction · Content-based image retrieval

1 Introduction

Image classification is an emerging research area, mainly used for annotation, categorization, indexing and retrieval of the images [1]. Currently, the tremendous growth in digital technology generated a huge amount of information, in which

V.P. Singh (✉) · R. Srivastava
Department of Computer Science and Engineering, Indian Institute of Technology (BHU),
Varanasi 221005, Uttar Pradesh, India
e-mail: vpsingh.rs.cse13@iitbhu.ac.in

R. Srivastava
e-mail: rajeev.cse@iitbhu.ac.in

© Springer Nature Singapore Pte Ltd. 2018   365
S.K. Bhatia et al. (eds.), *Advances in Computer and Computational Sciences*,
Advances in Intelligent Systems and Computing 554,
https://doi.org/10.1007/978-981-10-3773-3_36

images are easily available to the public, from web pages to photo collection. As the number of images increases, the need to develop fast and efficient algorithms to perform a search query on large image databases also increases. Content-based image retrieval (CBIR) techniques analyse the associated content with the image to perform search and retrieval. Features are extracted from an image based on low-level descriptions such as colour, texture or shape [2]. Based on these features, a CBIR system attempts to retrieve images that are visually relevant to the query image. Image classification is quite similar to CBIR system, typically consists of a database of images whose features have been extracted as a pre-processing step. Once the features have been extracted for all the images in the database, these features will subsequently be used to represent the image. Further, images are classify using these features with any machine learning classifier, and for CBIR the user provides a query image whose features are extracted and then matched with the feature database to retrieve similar images. Since visual data requires large amounts of computing power and memory for processing and storage, there is a need to efficient indexing and retrieval of visual information from large image database. Image classification categorizes the images into various classes, which can improve the retrieval performance and searching time of retrieval.

The one important focus of this work is to find suitable features for the images. Numerous features of the image are used to implement feature vector in image classification and retrieval, most of them are low-level features, based on colour, texture, shape and spatial orientation [3, 4]. In which, colour is an important feature that is most often used for effective representation and classification of colour images. Due to simplicity and fast indexing, colour histograms (CH), colour correlogram and colour moments are commonly used as a colour features [5–7]. Texture is also a useful features mainly used for the discrimination of fine and coarse images. Grey level co-occurrence matrix (GLCM), local binary pattern (LBP), Wavelet and Gabor coefficients, etc., are mainly used for the extraction of texture features [8–10]. Basically, the main problem behind an efficient CBIR is semantic gap between low-level representation and high-level understanding [2, 11], which can be reduced by using the supervised concept of image classification. The goal of CBIR is to provide the best retrieval results, and content-based image classification can be treated as a pre-processing step and applied prior computing similarity measure to speed up the searching and retrieval performances [12]. Furthermore, image classification can improve the response time by reducing the searching space.

As we know that for successful usage in image annotation and retrieval, the generalization capabilities of classifier are most essential [1, 13]. So, to overcome the over-fitting problem we have used random forest classifier. The random forest [14] is an ensemble approach based on divide and conquer approach. The basic concept behind ensemble methods is that a group (forest) of classification tree can come together to form a strong learner, which means final classification of an individual is determined by voting over all trees in the forest.

In this work, we classify the images into different classes using random forest classifier with colour and texture features. In the next section, we demonstrate the proposed framework, start with feature extraction, followed by image classification. Results analyses are presented in Sect. 3, and finally Sect. 4 presents the conclusion of the work.

2 Materials and Model

In this section, we explore the proposed work, based on fusion of colour histogram colour moment, colour percentile and LBP feature sets. For better classification, random forest classifier is introduced, which are good enough to reduce the effect of over fitting and classify images effectively. This work presented in this paper is divided into two Sects. (1 and 2), where Sect. 1 gives the details and extraction procedure of used features. Section 2 gives the details of used classifier and proposed work. The outcomes of this work are significantly better in terms of accuracy and other factors, which are discussed in Sect. 3.

2.1 Image Features

Quantifying discrimination ability of colour and texture features for classification problem is not being so easy. Among many possible features for classification purpose, we restrict our self to local binary pattern based texture features, and colour percentile, colour moment and colour histogram based colour features. Because all these features are fast in indexing and have rich content information for the classification.

2.1.1 Local Binary Pattern

Local binary pattern (LBP) is a computationally light-weighted texture features, derived from the local neighbourhood of each pixel in the image [8]. This operator treats each pixel as a centre pixel, and calculates the difference of centre pixel with neighbourhood pixel and multiplies it with a binary image window. Here, LBP features are extracted from luminance plane. LBP operator is mathematically defined as:

$$LBP_{P,R} = \sum_{i=0}^{P-1} 2^i D_1(I_i - I_c) \tag{1}$$

$$D_1(x) = \left\{ \begin{array}{ll} 0 & x \leq 0 \\ 1 & \text{else} \end{array} \right\}$$

The grey scale values of neighbourhood pixel and centre pixel are denoted as I_i and I_c. The connectivity from neighbouring pixels is represented as P and the neighbourhood radius is denoted as R. Finally, for an image of size M*N with L intensity value, histogram of LBP image is represented as texture features, denoted as Hist(L)|LBP and defined in Eq. (2).

$$\text{Hist(L)}|\text{LBP} = \sum_{x1}^{M} \sum_{x2}^{N} D_2(LBP(x_1, x_2), L)$$

$$L = [0, 2^{P-1}], \tag{2}$$

where $D_2(x_1, x_2) = \begin{cases} 1 & x_1 = x_2 \\ 0 & \text{otherwise} \end{cases}$

In this study, we have extracted LBP features for P = 8 and R = 1.

2.1.2 Colour Percentiles

This feature is very simple and computationally faster, used for visual retrieval and inspection [15]. During experimental analysis, colour percentile discrimination performance has been shown to be quite independent of colour space. So, for taking the advantages of computational simplicity and performances, this paper uses RGB colour space. The colour features used in this study are based on percentiles of one dimensional R, G and B colour channels. Colour percentile for an image of matrix X, with percentages P in the interval [0–100] depends upon nature of data. If X is a vector, then percentile Y is a vector or a scalar with the same length, then it means Y(i) contains the P(i) percentile. In this work, three percentile intervals [25 50, and 75] are taken for each RGB channel. Basically, percentile (centile) works on the concept of linear interpolation. Linear interpolation is calculated as y = f(x) for a given percentile x between x_1 and x_2 as follows:

$$y = y_1 + \frac{x - x_1}{x_2 - x_1}(y_2 - y_1) \tag{3}$$

Here, x_1 and x_2 are minimum and maximum value of input data, and y_1 and y_2 are range interval between[0–100]. This is one way to select the percentile features by selecting equal portions in a range between the minimum and maximum percentile values. We have calculated the value of x as a percentile value by rewriting the Eq. (3).

Table 1 Colour moments

Features	Equation	Descriptions
Mean	$\mu = \frac{\sum\limits_{i,j} X_{i,j}}{N}$	Reflects average intensity (brightness) of an image, having N pixels
Standard deviation	$\sigma = \sqrt{\frac{\sum\limits_{i,j}(X_{i,j} - \mu)^2}{N}}$	The variance tells the intensity variation around the mean
Skewness	$\frac{\sum\limits_{i,j}(X_{i,j} - \mu)^3}{N\sigma^3}$	Shows degree of asymmetry in the distribution

2.1.3 Colour Moment

Colour moment (CM) shows close description of colour distribution for an image [5]. If we define the ith colour channel (i = 1, 2, and 3) at jth image pixel as $X_{i,\,j}$. Then colour moments of an image are calculated as descriptions given in Table 1.

2.1.4 Colour Histogram

Colour histogram (CH) is a most simple and effective features used for colour image retrieval and classification. Representation of colour histogram is just like a bar graph, where each bar (bin) denotes the level of a particular colour [6].

For an image of size M × N, let the colour quantization levels are represented as $Q = Q_1, Q_2...Q_L$, and histogram $H = H_1, H_2... H_L$, where, H_i holds the count of pixels in colour Q_i.

CH also reflects the possibility of any pixel, in image I, that have colour Q_i.

$$Probability(P \in Q_i) = \frac{H_i}{M*N} \qquad (4)$$

For saving of time and space, this paper has taken HSV colour space and only 32 quantized bins.

2.2 Image Classification

To classify the images, first, chromaticity moments, colour percentile, colour histogram and LBP features are extracted from each image of database, and after fusion these features represent the content of images. The proposed working diagram for image classification is shown in Fig. 1.

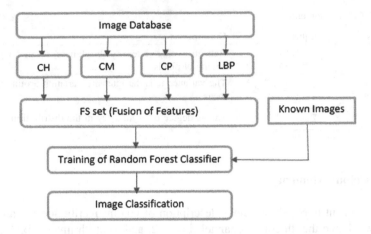

Fig. 1 Proposed working diagram

Let CM, CP, CH and LBP be feature sets of colour moment, colour percentile, colour histogram and local binary pattern, respectively, which are extracted independently and represented as:

$CM = [CM_1 \ CM_2 CM_3 \dots\dots\dots\dots\dots CM_9]$,

$CP = [CP_{10} \ CP_{11} \ CP_{12}\dots\dots\dots\dots\dots CP_{18}]$, and

$CH = [CH_{19} CH_{20}\dots\dots\dots\dots\dots\dots CH_{50}]$

$LBP = [LBP_{51} \ LBP_{52} \ LBP_{53}\dots\dots\dots\dots\dots\dots LBP_{86}]$

Final feature vector set (FS) are formed after fusion of all these feature sets

$FS = [CM \cup CP \cup CH \cup LBP]$

Final feature set for one image $= [F_1 \ F_2 \ F_3 \dots\dots\dots\dots\dots F_{86}]$

If there are M number of images then the size of feature matrix $= M \times 86$

So, we can write this feature vector into $M \times F$ representation, where M is the number of images and F is the number of features, and a feature vector can be expressed in a matrix such as:

$FS = [FS_{1,i} \ FS_{2,i}\dots\dots\dots FS_{J,i}\dots\dots\dots FS_{M,i}]$, where $i = 1$ to F.

Finally, feature vector set (FS) are stored into the database, used by the random forest classifier for image classification. The basic concept of random forest is based on the formation of weak decision trees in parallel, further we combine the trees to form a single, strong learner by averaging or taking the majority vote. For the classification of images, feature set (FS) are randomly divided into various subset for the training of decision tree individually, as shown in Fig. 2. Where, each node chooses small subset of features at random, for finding a feature which optimized the split. After training of each decision tree, final output is derived by ensemble approach using majority voting. Complete algorithm of random forest classification for training set S with F features is given below.

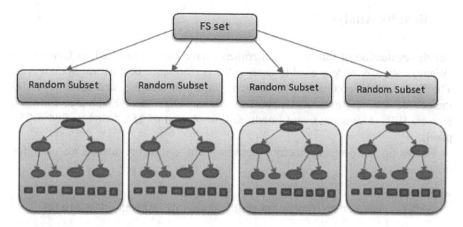

Fig. 2 Random forest classifier

Algorithm Random Forest
Input: A training set S with F features, and number of trees in forest B.
Steps:

    1.   function RF(S, F)
    2. $X \leftarrow \varphi$
    3. $for\ i \leftarrow$ 1 to B do
    4. $S^{(i)} \leftarrow$ A bootstrap sample from S
    5. $x_i \leftarrow$ RandomizedLearnTree ($S^{(i)}$, F)
    6. $X \leftarrow X \cup \{x_i\}$
    7.   end for loop
    8.   return X
    9.   end function

       Function RandomizedLearnTree ($S^{(i)}$, F)
    1.   At each node
    2. $f \leftarrow$ very small subset of F
    3.   Split on best feature in f
    4.   return learned tree
    5.   end function

Output:
    If the class prediction of bth random forest tree is $C_b(x)$ then final ensemble output of random forest classifier is:
$$C^B(x) = \text{majority } vote\{\ C_b(x)\ \}_1^B$$

3 Results Analysis

For the evaluation of this work, experiments have been performed on Corel-1000 (Wang) database, widely used to analyse the performance of retrieval system [16]. The Wang database contains, 10 different classes of images, each class have 100 images of Africa, Beaches, Monuments, Buses, Dinosaurs, Elephants, Flowers, Horses, Hills (Mountains), and Foods with resolution of 256 * 384 or 384 * 256 pixels.

Following performance metrics derived from confusion matrix are used for image analysis.

- Precision
  Reflects classifier discrimination ability among the different class images.

$$\text{Precision} = \frac{TP}{TP + FP} \tag{5}$$

- Recall
  Shows how well system recognized the different images.

$$\text{Recall} = \frac{TP}{TP + FN} \tag{6}$$

- Accuracy: Accuracy is defined as the proportion of instances that are correctly classified. It is calculated as:

$$\text{Accuracy} = \frac{TP + TN}{TP + FN + FP + TN} \tag{7}$$

- F-Measure is combined measure for precision and recall

$$F - \text{Measure} = \frac{2 \times \text{Precision} \times \text{Recall}}{\text{Precision} + \text{Recall}} \tag{8}$$

- MCC (Matthews Correlation Coefficient) is balanced measure mainly preferred for different sizes of classes.

$$MCC = \frac{(TP \times TN) - (FP \times FN)}{\sqrt{[TP + FP][TP + FN][TN + FP][TN + FN]}} \tag{9}$$

- AUC: Area under ROC curve of a classifier is the probability that the classifier ranks a randomly chosen negative image lesser than a randomly chosen positive.

Table 2 Confusion matrix

	Africa	Beach	Building	Bus	Dinosaurs	Elephant	Flowers	Horses	Mountain	Food
Africa	82	1	3	0	1	6	1	3	0	3
Beach	2	71	7	4	0	2	0	0	14	0
Building	4	3	83	0	0	4	1	0	4	1
Bus	0	0	0	100	0	0	0	0	0	0
Dinosaurs	0	1	0	0	98	1	0	0	0	0
Elephant	1	0	2	0	1	89	0	3	4	0
Flowers	0	0	0	0	0	0	98	1	0	1
Horses	0	0	0	0	0	0	1	98	0	1
Mountain	0	10	4	0	0	8	1	0	77	0
Food	2	1	1	0	0	0	1	0	3	92

Table 3 Performance analysis

Class	Precision	Recall	F-Measure	MCC	ROC area
Africa	0.901	0.820	0.859	0.845	0.976
Beach	0.816	0.710	0.759	0.737	0.969
Building	0.83	0.83	0.83	0.811	0.986
Bus	0.962	1.00	0.980	0.978	1.00
Dinosaurs	0.980	0.980	0.980	0.978	1.00
Elephant	0.809	0.890	0.848	0.831	0.989
Flowers	0.951	0.980	0.966	0.962	0.998
Horses	0.933	0.980	0.956	0.951	0.999
Mountain	0.755	0.770	0.762	0.736	0.981
Food	0.939	0.920	0.929	0.922	0.996
Average	**0.888**	**0.888**	**0.887**	**0.875**	**0.989**

Fig. 3 Random forest performance on different number of trees

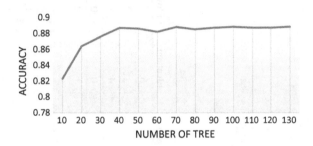

In order to evaluate the performance this work, we have used K-fold cross validation. In this paper, we have taken K = 10 for the evaluation. It means our dataset is divided into 10 equal sets, each have 100 distinct instances in which 9 sets (900 instances-90 images from each class) are used for training and one set (100 instances-10 images from each class) for testing. These processes are repeated 10 times and form a confusion matrix. Using this confusion matrix, we have evaluated various performance measures, which are given in Tables 2 and 3.

Figure 3 shows the performance of random forest for different number of trees, where it reflects maximum 88.8% accuracy for 70 and 100 number of trees. Random forest of 70 trees, each constructed while considering 7 random features takes 0.59 s for building model for classification, while 100 trees take 0.86 s. So, due to time constraint we have taken 70 number of trees for our result analysis. Confusion matrix and classification performances of this work are given in Tables 2 and 3.

From the confusion matrix of this classifier, it is clear that this classifier with fusion of colour and texture features reflects superior discriminative power for 8 class images of Wang database. This classifier is slightly failing to discriminate only between Beach and Mountain classes, 14% images of Beach classes are misclassified as Mountain and 10% images of Mountain are misclassified as Beach. The overall accuracy of this approach is 88.8%, which is 1000 Wang database

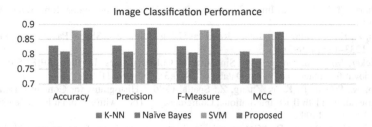

Fig. 4 Comparative analysis of performances

images; it correctly recognizes or detects 888 images for retrieval and indexing. Our proposed approach has significant encouraging results in terms of other performance factors such as; average precision = 88.8%, recall = 88.8%, area under ROC = 98.9%, MCC = 87.5%, and F-measure = 88.7%. Figure 4 shows the comparative analysis of performances, where overall accuracy, average precision, F-score and MCC of proposed approach is compared and found better than K-NN, Naïve Bayes and support vector machine (SVM) classifiers. So from all these experimental analyses, this is justified that using random forest classifier with concatenation of colour histogram, colour moment, colour percentile and LBP feature sets work very impressive for colour image classification.

4 Conclusions

In this paper, being inspired by the concept that similarity in the feature space should exhibit semantically similar images. This paper's objective was to investigate the discriminatory abilities of proposed feature set and look at efficient architecture (number of tree) of the classifier. For the classification of images, we have used random forest-based classifier with fusion of colour and texture features, and got significantly encouraging discriminative results. In future, we will used the classifier discriminative performance for the image retrieval, where this work is treated as a pre-processing step for the grouping of images, also this model helps in retrieval by detecting the class of query image.

References

1. Ng, W. W., Dorado, A., Yeung, D. S., Pedrycz, W., & Izquierdo, E.: Image classification with the use of radial basis function neural networks and the minimization of the localized generalization error. Pattern Recognition 40(1), 19–32 (2007).
2. Vailaya, A., Figueiredo, M. A., Jain, A. K., & Zhang, H. J.: Image classification for content-based indexing. IEEE Transaction Image Processing 10 (1), 117–130 (2001).

3. Acharya, T., & Ray, A.: Image Processing Principles and Applications, John Wiley & Sons, (2005).
4. Jain, A. K., & Vailaya, A.: Image Retrieval Using Colour and Shape. Pattern recognition *29* (8), 1233–1244 (1995).
5. Stricker, M. A., & Orengo, M.: Similarity of Colour Images. In: Proc. SPIE: Storage and Retrieval for Image and Video Databases 2420(3), 381–392 (1995).
6. Rao, A., Srihari, R. K., & Zhang, Z.: Spatial Colour Histograms for Content-Based Image Retrieval. In: 11-th IEEE International Conference on Tools with Artificial Intelligence pages, pp. 183–186 (1999).
7. Swain, M., & Ballard, D.: "Colour indexing" International Journal of Computer Vision 7(1), 11–32(1991).
8. Ojala, T., Pietikäinen, M., & Harwood, D.: A Comparative Study of Texture Measures with Classification Based on Feature Distributions. Pattern Recognition 29(1), 51–59 (1996).
9. Chang, T., & Kuo, C. J.: Texture analysis and classification with treestructured wavelet transform. IEEE Transactions on Image Processing 2(4), 429–441 (1993).
10. Turner, M.R.: Texture discrimination by Gabor functions. Biology, Cybernetics 55(4), 71–82 (1986).
11. Singh, V. P., & Srivastava, R.: Design & performance analysis of content based image retrieval system based on image classification usingvarious feature sets. In: Futuristic Trends on Computational Analysis and Knowledge Management (ABLAZE), pp. 664–670 (2015).
12. Yildizer, E., Balci, A. M., Hassan, M., & Alhajj, R.: Efficient content-based image retrieval using Multiple Support Vector Machines Ensemble. Expert Systems with Applications 39(3), 2385–2396 (2012).
13. Zhang, Y. J.: Image Classification and Retrieval with Mining Technologies. In: Handbook of Research on Text and Web Mining Technologies, pp. 96–110 (2008).
14. Breiman, L.: Random forests. Machine learning 45(1), 5–32 (2001).
15. Bianconi, F., Harvey, R., Southam, P., & Fernández, A.: Theoretical and experimental comparison of different approaches for colour texture classification. Journal of Electronic Imaging20(4), 1–20(2011).
16. Wang, J., Li, J., & Wiederhold, G.: Simplicity: Semantics–sensitive integrated matching for picture libraries. IEEE Transactions on Pattern Analysis and Machine Intelligence 23(9), 947–963 (2001).

An Advanced Approach of Face Recognition Using HSV and Eigen Vector

Santosh Kumar, Atul Chaudhary, Ravindra Singh, Manish Sharma and Anil Dubey

Abstract A constant framework for perceiving faces in a video stream gave by an observation camera was actualized, having continuous face identification. Along these lines, both face discovery and face acknowledgment strategies are synopsis displayed, without avoiding the imperative specialized perspectives. The proposed approach basically was to actualize and check the calculation HSV shading division for face identification and eigenfaces for Recognition, which tackles the acknowledgment issue for two dimensional representations of confronts, utilizing the main part investigation. The depictions, speaking to information pictures for the proposed framework, are anticipated into a face space (highlight space) which best characterizes the variety for the face pictures preparing set. The face space is characterized by the "eigenfaces" which are the eigenvectors of the arrangement of appearances.

Keywords Face recognition · PCA (Principal component analysis) · Skin color segmentation (HSV) · HUE

1 Introduction

Modernized face certification has industrialized a basic field of interest. Face recognition is the average procedure of distinguishing proof of a person by their facial picture. This strategy utilize the facial pictures of a man to validate him into

S. Kumar (✉)
Govt. Engineering College Ajmer, Ajmer, India
e-mail: sonu225914@gmail.com

A. Chaudhary · R. Singh · A. Dubey
Department of CS & IT, Govt. Engineering College Ajmer, Ajmer, India

M. Sharma
Department of Computer Engineering, Poornima Institute
of Engineering & Technology, Jaipur, India

© Springer Nature Singapore Pte Ltd. 2018
S.K. Bhatia et al. (eds.), *Advances in Computer and Computational Sciences*,
Advances in Intelligent Systems and Computing 554,
https://doi.org/10.1007/978-981-10-3773-3_37

377

an ensured framework, for criminal proof, for international ID confirmation, passageway control in structures, access control at automatic teller machines and in other applications for authentication purpose. This strategy utilizing principal component analysis for face acknowledgment and it was based on the hypothesis of face acknowledgement on a little arrangement of picture elements that best close arrangement of known pictures of face, without regards to that they relate to our non-judicious thoughts of facial parts and elements. The eigenface strategy and area base shading division coordinating (HSV) give a useful arrangement that is all around fitted for the issue of face acknowledgment. It is quick, generally simple, and functions admirably in a compelled domain. Head size and head introduction, the trade-offs between the quantity of eigen confronts fundamental for unambiguous order are matter of concern. A face acknowledgment framework would permit client to be distinguished by essentially strolling past an observation camera. Individuals frequently remember each other by one-of-a-kind facial attributes. Facial acknowledgment is turning into the best type of human observation. There are two dominating ways to deal with face acknowledgment: geometric (element based) and photometric (perspective based). The most broadly and regularly utilized calculation for face acknowledgment is shading division (HSV) and eigenvector (Pca). Acknowledgment rate of 85.113% is accomplished in our calculation.

2 Related Work

Before, numerous anomaly identification strategies have been proposed [1, 2, 3, 4, 5] Normally, these current methodologies can be separated into three classes: Knowledge-based template matching appearance-based and feature invariant based techniques. Factual methodologies [1, 3] expect that the information tails some standard or foreordained circulations, and this sort of methodology intends to discover the exceptions which digress frame such appropriations. Nonetheless, most dispersion models are accepted univariate, and consequently the absence of heartiness for multidimensional information is a concern. In addition, since these strategies are regularly executed in the first information space specifically, their arrangement models may experience the ill effects of the clamor present in the information. The presumption or the earlier learning of the information dispersion is not effectively decided for handy issues.

- Sung and Poggio [1] proposed and adequately realized Gaussian clusters to validate the assignment of facial part and non-facial part.
- Pattern Rowley et al. [2] utilized manufactured neural system for face discovery.
- Vijay Lakshmi proposed a segmentation algorithm for multiple face detection in color images with skin tone regions utilizing color spaces and edge detection techniques, in which varying shading space models, particularly, HSI and

YCbCr close by Canny and Prewitt edge ID procedures are utilized for better face detection [6, 7].

- Eyed Aldasouqi et al., exhibited a Smart Human Face Detection System, in which they clarified digital picture handling and quick identifying calculations taking into account HSV Color model without giving up the pace of face detection [8].
- Venu Shah et al., thought about and joined the traditional method of HSV Histogram Equalization with Versatile HSV division and Kekre Transform for Content-Based Image Retrieval [9].

3 Color Base Face Detection

Some human eyes have a tendency to see shading. RGB characterizes shading regarding a blend of essential hues, While, HSV portrays shading utilizing more recognizable examinations, for example, shading, energy, and shine. The shading camera, on computer, utilizes the RGB model to decide shading. Once the camera has perused these qualities, they are changed over to HSV standards. The HSV qualities are then utilized as a part of the code to decide the area of a particular article/shading for which computer is seeking. The pixels are independently tested to figure out whether they coordinate a foreordained shading limit (Figs. 1 and 2).

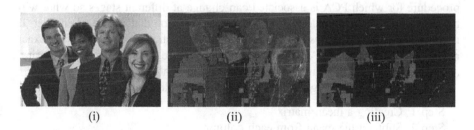

| (i) | (ii) | (iii) |

Fig. 1 i ii iii RGB, HSV and H component of image (2)

Original Image   gray image   Eigen image face

Fig. 2 Image conversion

$$H = cos^{-1}\left(\frac{\frac{1}{2}((R - G) + R - B)}{\sqrt{(R - G)^2 + (R - B(G - B))}}\right)$$

$$S = 1 - 3\frac{\min(R, G, B)}{R + G + B}$$

$$v = \frac{1}{3}(R + G + B)$$

Skin district in B shading place remains in the accompanying extent $0.79G$ $-67 < B < 0.78G + 42$. Non-skin region in B color place remains in the following range $0.836G - 14 < B < 0.836G + 44$. Non-skin region in H color space remains in the following range $19 < H < 240$. Skin region in Cb color space remains in the following range $102 < Cb < 128$.

4 Principal Component

PCA is a valuable present-day factual system that has utilized application as a part of fields, for example, face PCA is a useful front line true strategy that has been used in application such as a piece of fields, for instance, face affirmation and picture weight, and is an ordinary technique for finding plans in data of high estimation [10]. PCA is a commonly used strategy for article affirmation as its results, when used properly can be truly correct and flexible to disturbance. The procedure for which PCA is associated can change at different stages so what will be demonstrated is a sensible system for PCA application that can be taken after. It is up to individuals to investigate in finding the best strategy for conveying exact results from PCA and PCA Need less time to see the face, PCA can do desire, abundance departure, highlight extraction and data relationship.

Some steps of face recognition using PCA (Principle Component Analysis)

Step 1 Creating a mean matrix
Step 2 Subtract the mean from each column
Step 3 Calculate covariance matrix
Step 4 Get Eigen value and Eigen vector from covariance matrix
Step 5 Create Eigen image by multiplying mean subtracted data and Eigen value
Step 6 Creating weight matrix.

5 Proposed Methodology

In circumstances where shading representation assumes an essential part, the HSV shading model is regularly favored over the RGB model. The HSV model representation hues comparably to how the human eye has a tendency to see shading. RGB characterizes shading regarding a mix of essential hues, while, HSV represents shading utilizing more recognizable correlations, for example, shading, dynamic quality and brilliance. The color camera uses the RGB model to decide color. As soon as the camera has perused these qualities, they are changed over to HSV rates. The HSV qualities are then utilized as a part of the code to decide the area of a particular article/shading we are seeking for. Those eigen picture face is an extricated picture which is ought to be match with train database picture (Fig. 3).

The execution of the proposed face acknowledgment framework is tasted with various picture stances of same individual. In Fig. 4, we try to express our actual algorithm of how our system recognizes the face successfully. Initially a random image is captured by any static camera. After that image is obtained by color segmentation and than apply HSV and detect face in captured image. After that feature is extracted using PCA algorithm and obtain Eigen image. Same process is apply in our train database where feature is extracted and Eigen image is obtained by matching most similar image.

6 Experiment and Graphical Approach

The proposed methodology experimental process is operated with LG 13 mega pixel static camera and picture is not click more than five to six feet .captured image at a spatial resolution of 180 × 200 pixels frame rate is not more than 25 frame per second. A low rate camera is used for this experiment because our goal is also research design must be economical. When the code is executed in MATLAB 2013a than its take the train database and test database path respectively according to Fig. 5. The experimental result is shown in Fig. 5 where two different poses of image are matched successfully.

6.1 Efficiency Observed Over Proposed Algorithm

Six different pictures for each said condition were taken to test for ten different individuals. Light power is attempted to keep low (Fig. 6). Size variety of a test picture is not adjusted to much degree. We can watch that ordinary looks are

Fig. 3 Flow chart of
proposed algorithm

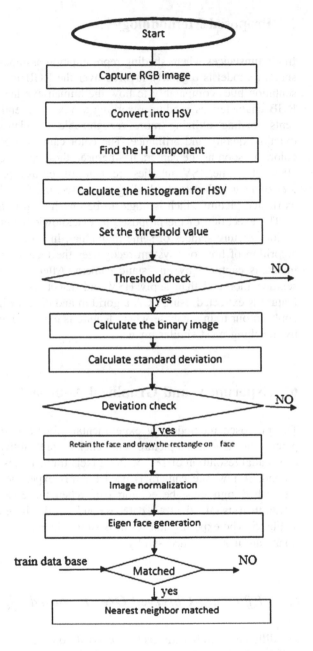

recognized as face effectively on the grounds that facial elements are not changed
much in that Case and in different situations where facial elements are changed
effectiveness is lessened in Recognition. The creator partitioned the state of pictures
submitted for acknowledgment into Six sorts, the countenances in the pictures were
of various sizes, positions and numerous subjects had even somewhat pivoted their

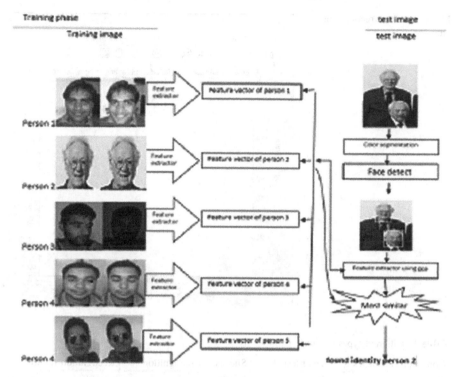

Fig. 4 Block diagram and approach of face recognition

Fig. 5 Matched image (3)

Fig. 6 Average recognition rate

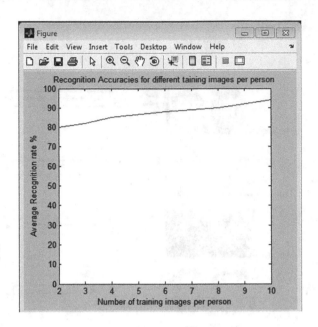

Table 1 Different types of image expressions

Condition of images	Total tested	Successful recognition	Recognition failures
Normal	30	27(90%)	3(10%)
Smile	30	26(85%)	4(15%)
Illumination	30	23(80%)	20(40%)
Size variation	30	27(89%)	3(21%)
Angry	30	25(84%)	5(16%)
Sad	30	24(81%)	6(19%)

Table 2 Recognition Analysis

Testing condition	Total tested	Successful recognition	Recognition failures
Controlled	30	25.5(85%)	7(23.66)

heads. Besides, lighting conditions were purposefully fluctuated by the analyst to strain the acknowledgment framework $(09 + 8.5 + 09 + 7.3 + 7.7 + 9.1)/06 = 8.5113$ (Tables 1 and 2).

Effectively the exactness level is approx $8.5113 \times 100 = 85.113\%$ found.

7 Conclusion

The vast majority of techniques for making PCs perceive appearances were constrained by the utilization of extemporized face shapes and highlight reports (matching straightforward separations), accepting that a face is close to the whole of its parts, the individual components. This truthful strategy Color division based programmed face location calculation. RGB-HSV to identify human appearances for understudies with negligible learning in picture preparing. The Skin area division was performed utilizing mix of RGB, HS Though there are some instances of false positives yet general distinguish rate of face location is acceptable and utilizing Principal Component Analysis for face acknowledgment was really activate by data hypothesis face acknowledgment on a little arrangement of picture elements that best close arrangement of known face pictures, without in regards to that they compare to our non-judicious ideas of facial parts and elements. The Eigen face technique gives a pragmatic arrangement that is all around built for the issue of face acknowledgment. It is quick, generally simple, and functions admirably in a compelled domain. Skull size, and Skull introduction, the adjustments between the quantities of Eigen confronts important for unmistakable grouping are matter of concern. The preparation pictures on which the calculation is tried are characteristic pictures at a spatial determination of 180 × 200 pixels, however test picture is taken in abundantly uncontrolled condition.

References

1. K K. Sung along with T Poggio, "Example-based learning for view-based human face detection," IEEE Transactions on Pattern Analysis and Machine Intelligence, vol. 20, pp. 39–51, Jan. 1998.
2. H Zhao along with P C Yuen, "Incremental linear Discriminant analysis for face recognition," IEEE Transactions on System Man and Cybernetics B, vol. 38, pp. 210–221, Feb. 2008.
3. H.C. Vijay Lakshmi a coupled with S. Patil Kulakarni, "Segmentation Algorithm for Multiple Face Detection in Color Images with Skin Tone Regions using Color Spaces and Edge Detection Techniques," International Journal of Computer Theory and Engineering, vol. 2, no. aug 2010.
4. Wankou Yang Laplacian "bidirectional PCA for face recognition, Neuro computing," Volume 74, elsevier 2010.
5. R.-L. Hsu, M. Abdel-Mottaleb, and A. K. Jain, "*Face detection in color images,*" IEEE Trans. PAMI, vol. 24, no. 5, pp. 696–707, 2002.
6. Dr. H.B. Kekre et. al./ (IJCSE) International Journal on Computer Science and Engineering Vol. 02, No. 04, 2010, 959–964.
7. J. Yang, D. Zhang, combined with A. F. Frangi. Two-dimensional PCA: A new approach to appearance-based face representation and recognition. IEEE Transactions on Pattern Analysis and Machine Intelligence, 26(1):131–137, 2004.

8. Chiunhsiun Lin, along with Ching-Hung Su, "Using HVS Color Space and Neural Network for Face Detection with Various Illuminations," Proc. of the 2007 WSEAS, Int. Conf. on Computer Engineering and Application, Gold Coast, Jan. 2007.
9. .Iyad Aldasouqi, a along with Mahmoud Hassan, "Smart Human Face Detection System, "International Journal of Computers, vol. 5, 2011.
10. Chiunhsiun L" Evaluation of face recognition technique using PCA, wavelets and SVM Pattern Recognition," Volume 44, Issues 10–11 elsevier 2011.

RMI Approach to Cluster Based Image Decomposition for Filtering Techniques

Sachin Bagga, Akshay Girdhar, Munesh Chandra Trivedi,
Yinan Bao and Jingwen Du

Abstract Logically programmed cluster provides the platform to compute certain complex problems which are not solvable on a single system. Parallel processing implemented on a distributed data, on a cluster with the help of multithreading is a base for the present work. It is very easy for an application based on parallelism to beat the results produced by the sequential program on the same platform. Taking this thing into consideration along with the cluster programming can be used to solve certain complex problems like processing of large size images to apply median filtering. In the present work, an image with large dimensions is break down into sub images with lesser dimensions and this breakdown is as per the number of nodes under consideration from a given cluster. For the purpose of communication between these nodes distributed object oriented programming remote method invocation (RMI) is used which creates a Single Instruction and Multiple Data (SIMD) model. At the master system there is actual breakdown of the image into smaller dimensions and at the slave systems there is virtual division of the sub image further into 3×3 or as per mask selected to apply median filter. Various

S. Bagga (✉) · A. Girdhar
Department of Information Technology, Guru Nanak Dev Engineering College,
Ludhiana, India
e-mail: sachin8510@gmail.com

A. Girdhar
e-mail: akshay1975@gmail.com

M.C. Trivedi
Department of Computer Science and Engineering, ABES Engineering College,
Ghaziabad, India
e-mail: Munish.Trivedi@gmail.com

Y. Bao
Department of Mechanical Engineering, Freshman, School of Engineering,
University of Wisconsin-Madison, Madison, USA
e-mail: ybao24@wisc.edu

J. Du
Dalian No. 24 High School, Dalian, China
e-mail: zoe.diamond@qq.com

© Springer Nature Singapore Pte Ltd. 2018
S.K. Bhatia et al. (eds.), *Advances in Computer and Computational Sciences*,
Advances in Intelligent Systems and Computing 554,
https://doi.org/10.1007/978-981-10-3773-3_38

performance metrics like Excessive parallel overhead (EPO), Overall Computation Time (OCT), Speed Up (SU) and Efficiency are calculated to find out the effectiveness of the proposed cluster.

Keywords Median filter · RMI · SIMD · Multithreading · Cluster · Nodes · Parallel processing

1 Introduction

Massively Parallel Processing (MPP) is a technique in which multiple processors work on the different parts of the program parallely, with each processor having its own controller i.e. operating system and separate memory [1]. MPP requires focus on partitioning of the different tasks in terms of dependent and independent for the concurrent execution of the different parts of the program. MPP also known as loosely coupled systems have distributed memory which provides multiprocessing where there is no sharing of physical memory and the input output channels [2]. Loosely coupled systems provides benefits over the tightly coupled systems by easy replacement of the individual node in case of occurrence of fault. RMI provides an intermediate network layer framework which can help in doing parallel processing. RMI allows calling of objects that are residing on some different virtual machine [1]. RMI along with the multithreading approach helps in executing parallel and distributed processing among all the nodes in the given cluster. A smallest unit for processing known as thread is a light weight sub-process means they can share the address space of the given process on which they are working. The context switching between the threads helps in implementing multithreading as a result multitasking can be achieved. With the help of RMI server and client communicate back and forth and thus provides distributed object application. Locating the remote objects, communication with the remote objects and loading of the class definition for the objects that are passed around are the main tasks in the RMI based distributed object application. The RMI registry provides a major role for obtaining a reference to the remote object. In the present work this facility of the RMI is used, to create a distributed application in which an image with large number of pixels is divided into number of sub images and then it is send to the different nodes in the cluster for denoising of the image [3]. Impulsive noise which comes into existence from number of sources like sensor temperature, switching, image acquisition, atmospheric disturbance while transmission etc. can result in degradation of the quality of image [3, 4]. With the help of MPP created using RMIto remove the degradation of the image with large number of pixels which occurs due to one of the impulse noise like salt and paper is the main objective of this paper. A popular nonlinear filter with high computational efficiency and denoising power known as Standard median filter (SMF) is used to achieve the objective. Although high performance can be achieved with parallel machines but they include tedious programming broadly the tasks related to parallel processing can be specified as [5]:

- Designing a Parallel Algorithm
- Partitioning of the data
- Controlling the communication between the systems
- Synchronization
- Scheduling
- Mapping of the various tasks
- Evaluating the various performance metrics

The given architect tries to efficiently implement all these activities to achieve high performance results. The mapping between the load generation and distribution among the nodes is done using the conversion from two dimension to three dimension arrays and vice versa. Interprocessor communication overhead (IPCO) in a parallel system which comes into existence when a single master system has to do message passing with the whole cluster [6]. To handle such constraints following arrangements were made:

- Multiple Network Interface Controller with the master system
- Dedicated separate paths for cluster nodes
- High speed switch
- The HTTP POST along with the Common Gateway Interface (CGI) for passing message through the firewall.

2 Literature Survey

Arora et al. [7] in their paper "Distributed cluster processing to evaluate interlaced run-length compression schemes" represents an idea to implement run-length compression schemes on an image using parallel computation. The author has used VB.6.0 TCP/IP socket programming using Mswinsock.ocx for creating a parallel system. The master node as per the scan line of the given image sub divide it and distributes it among the given machines of a cluster at specified port number. At the slave side given algorithm is applied on this sub image and is returned back to the master system. After the completion of the execution of the whole nodes in the cluster the master system creates the original image. Similarly, Kaur et al. [1] in their paper *"RMI Approach to Cluster Based Winograd's Variant of Strassen's Method"* implements Winograd's Variant of Strassen's Method for matrix multiplication with complexity of O ($n^{2.81}$) on the cluster programmed using the RMI technique. Author has proved that RMI based cluster system are much more efficient than Socket programming. Table 1 shows the facts regarding this which is based on the experimental results [1]. Same algorithm when implemented on the RMI takes lesser execution time than the Socket based programming. Without using the marshalling and unmarshalling facility of RMI, direct use of Socket programming is not much capable of sending large amount data in a single instance to whole cluster where as RMI based cluster is.

Table 1 Execution time on RMI and socket

No. of systems	Matrix order	RMI based execution time (s)	Direct socket based execution time (s)
2	256 × 256	0.635	2.447
4	256 × 256	0.262	1.369
2	1024 × 1024	7.194	183
4	1024 × 1024	4.457	74
8	1024 × 1024	3.192	40.47
2	1200 × 1200	11.65	232.2
4	1200 × 1200	7.054	119.1
6	1200 × 1200	5.914	81.65

Both these papers act as a base for the present work i.e. implementing image related operations on the cluster based systems programmed using the RMI technique. Other survey related to research include paper by Pasricha et al. [8] "*Analytical Parallel Approach to Evaluate Cluster based Strassen's Matrix Multiplication paper includes*" where author has implemented Strassen's Matrix Multiplication having complexity O $(n^{2.81})$ using socket based programming. Comparison of Strassen's divide & conquer algorithm and Traditional matrix multiplication algorithm is done which shows that parallel programming along with the algorithm with better complexity can give very efficient results.

3 Logical Programming Structure

The distributed master slave architecture is created using the Java based Remote Method Invocation (RMI) technique which makes use of Java remote method protocol (JRMP), whose working is over the TCP/IP. The communication between the master and slave systems is done through a specific TCP port number, by default it is 1099. There is a lease duration which is being granted to the virtual machines (VM) which represents the time period for which VMs can hold the remote reference to the objects [9]. By default, this value is about 10 min and can be changed accordingly. A bootstrap naming service known as rmiregistry is started by start rmiregistry command by all the slave systems and using rmic (rmi compiler) command stubs are created by all the slave systems which are being passed to the master system. After performing these tasks, the master system will distribute the workload among the whole cluster.

3.1 Median Filter

A nonlinear process known as Median filtering helps in reducing salt and pepper or other impulsive noise from the image. While reducing the random noise this technique preserves the edges of a given image. Certain random bits error which arises due to atmospheric disturbance, interference in channels, sensor temperature etc. can result in impulse noise in the image [9]. In the median filter approach given formula in Eq. (1) is applied on a given window size and the median of various intensity value of the pixels becomes the resultant intensity for the pixels under processing.

$$g(a,b) = median \left(\sum_{i=-1}^{1} \sum_{j=-1}^{1} f(a-i, b-j) \right) \tag{1}$$

A master system is having a complete image with large number of pixels from which noise has to be removed. The master system creates the subimages and sends to the whole cluster node which applies the median filtering upon it.

4 Workflow of Given System

RMI based cluster systems has the prerequisite that the master system must have the stub from all the slave systems passed to it. The workflow for proposed system can be explained as:

1. A master system also known as controller, controls the working of the whole cluster.
2. The master system has the complete image with given pixel values. It decomposes the image into subimages as per the number of nodes in the cluster and as per the dimensions of the mask to be applied.
3. The master system converts the whole image with m × n pixels into 2D array with m × n dimensions.
4. As per the nodes in the cluster this 2D array is further converted into the 3D array as per the requirements. The index values of any 3D array example x[a][b][c] represents a number of 2D arrays with b × c dimensions.
5. Each slave node receives a particular 2D array from a given 2D array.
6. The master system uses the multithreading approach to start multitasking with the whole cluster. Multithreading is to be explicitly add in the RMI based clusters otherwise only a distributed system is created without parallel processing.
7. All the slave nodes of cluster create the stub and pass them to the master system.
8. After this all the slave systems start the *rmiregistry* and they start listen for the incoming request by default at the port number 1099. There are various RMI properties available for checking the cluster status, some are listed below in Table 2.
9. All the slave nodes have implemented median filter algorithm and they denoise the image as per the pixels received from the master system as shown in Fig. 1.
10. The master node continuously checks the completion of the threads corresponding to each slave nodes of the cluster.

Table 2 Execution time on RMI and socket

Property	Description
java.rmi.activation.port	To set TCP port number for communication
java.rmi.dgc.leaseValue	Gives lease duration for holding remote reference to objects
java.rmi.server.codebase	Gives locations from which classes are published can be downloaded
java.rmi.server.hostname	Represents hostname string
java.rmi.server.logCalls	For logging incoming calls and exceptions to System err
java.rm.server. useCodebaseOnly	To publish automatic loading of classes
java.rmi.server.disableHttp	For disabling the HTTP request

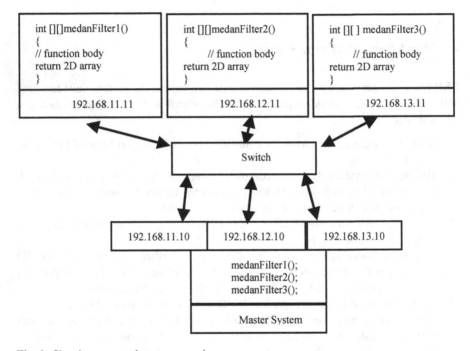

Fig. 1 Showing proposed system overview

11. After all the nodes in the cluster are finished with their execution and have sent their result back, the master system generates the resultant matrix.
12. Again the complete image is being generated from the partitioned matrix by the master system.
13. As there is dynamic creation of the array from a image with large number of pixels there is a requirement of explicitly reserving the heap space as the default heap space is not much sufficient and heap space error usually arises in such cases. For this purpose, the java program is executed in a special fashion as:

$$java - Xmx1400m\, file\, name$$

where 1400m is amount of memory reserved for the heap space dynamically in mb.

5 Benefits of Current Approach

Given cluster system provides number of benefits for image related operations using parallel processing. Some of these are as follows:

- Proposed work is capable of converting an image into a corresponding 2D array of the same dimensions [10].
- Proposed work is capable of decomposing an image with large number of pixels into sub images which can be distributed over the cluster for processing.
- Interprocessor communication overhead (IPCO) is decreased to a much greater extent using separate networks for nodes in a particular group of a given cluster. For this purpose, multiple network controllers are being used.
- The system is capable to dynamically include or exclude number of slave nodes in a given cluster.
- Multitasking using multithreading is also explicitly added which can result in creation of a parallel and distributed system at a same time.
- Most of the master systems activities are automatic with least human interaction.
- Master system can keep continuous watch on the working of various nodes in the cluster, by assigning a dedicated thread for each node.
- The proposed system provides number of facilities to track he communication in the give cluster.
- At the master system there is actual partitioning of the image and at the slave systems there is virtual partitioning of the image as per the mask selected for implementing median filter [11].
- RMI provides facility of penetrating even through the firewall.
- The proposed system is compatible to work on nodes with different versions of windows.

6 Experimental Setup and Workload Characterization

The experimental setup consists of following configuration of the whole cluster is shown in Table 3.

The workload used for the proposed system is shown in Table 4.

Table 3 Showing experimental setup of the cluster

Component name	No.	Configuration	OS support	Remarks
Master system in the cluster	01	• Architecture: x86_64 • CPU(s): 2 • Core Per Socket: 2 • Thread per Core: 1 • Socket: 1 • Vendor ID: genuine intel • CPU MHz: 3000 • L1d: 32 K • L1i: 32 K • L2: 256 K • L3: 3072 K • NUMA node0 CPU(s): 0, 1 •. RAM: 2 GB • Stubs for all the function body transferred from the slave systems	Windows 7 Home premium	JAVA (RMI) version: 1.7.0_31
LAN adapter	03	• IP address of 1st card 192.168.11.10 • IP address of 2nd card 192.168.12.10 • IP address of 3rd card 192.168.13.10	USB ethernet 10/100 Base-T	
Total slave systems	05	• Architecture: x86_64 • CPU(s): 2 • Thread per Core: 1 • Core Per Socket: 2 • Socket: 1 • Vendor ID: genuine intel • CPU MHz: 3000 • L1d: 32 K • L1i: 32 K • L2: 256 K • L3: 3072 K • NUMA node 0 CPU(s): 0, 1 • RAM: 2 GB • **IP address** 192.168.11.11–192.168.11.13 192.168.12.11–192.168.12.13 192.168.13.11–192.168.13.12	Windows 7 Home premium	JAVA (RMI) version: 1.7.0_31
Ethernet switch		Cisco 24 Port 10/100 Mbps		Provide non-blocking, wire-speed witching for 10,100 mbps network clients
Cat 5 UTP cable		Cat 5 UTP cable with transmission rate up to 100 MB/s		

Table 4 Showing workload characterization

No. of nodes of cluster used	Dimensions of the picture	Dimensions of sub image per node
2	4500 × 4500	2250 × 4500
2	5040 × 5040	2520 × 5040
3	2268 × 2268	756 × 2268
3	4509 × 4509	1503 × 4509
4	4500 × 4500	1125 × 4500
4	5040 × 5040	1260 × 5040
5	4500 × 4500	900 × 4500
5	5040 × 5040	1008 × 5040

7 Performance Measurements

The performance metrics like speedup, efficiency, excessive parallel overhead are evaluated in order to check the outcomes of the given cluster system:

(a) *Speedup:*

Gives the comparison of a multiprocessor system against a uniprocessor system [12]. It is calculated as the ration of Computation time on a single system and Computation time on a multiprocessor.

$$Speedup = \frac{\text{Time taken on a single system}}{\text{Time taken on a multiprocessor}} \tag{2}$$

(b) *Efficiency:*

It gives the measurement of the duration for which the cluster was busy in performing the given operation.

$$Efficiency = \frac{\text{Time taken on a single system}}{\text{Execution time on a multiprocessor} * \text{No. of Systems}} \tag{3}$$

(c) *Parallel processing overhead:*

It gives the measurement of extra time consumed in managing the cluster except of the actual computation time (Figs. 2, 3, 4, 5; Table 5).

$$POVR = P * \beta - \alpha \tag{4}$$

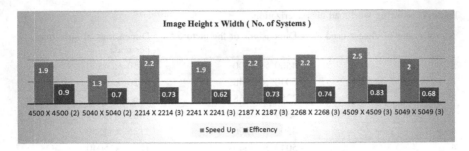

Fig. 2 Speed, efficiency values for different image size on 2, 3 systems

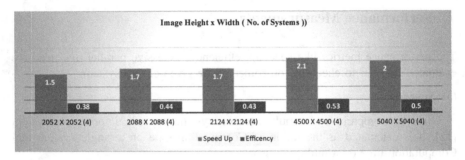

Fig. 3 Speed, efficiency values for different image size on 4 systems

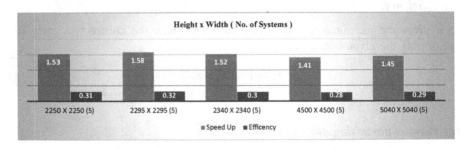

Fig. 4 Speed, efficiency values for different image size on 5 systems

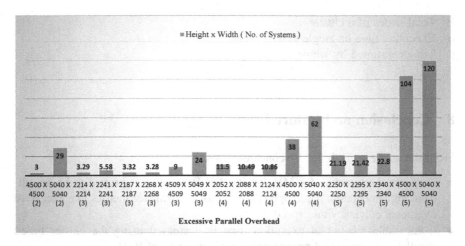

Fig. 5 Excessive parallel overhead values for different image size

Table 5 Showing values of performance metrics like speed up, efficiency, excessive parallel overhead for various systems in the cluster

No. of systems	Image size	Memory for heap space	Execution time on given no. of systems	Execution time on a single system	Speed up	Efficiency	Excessive parallel overhead
2	4500 × 4500	Xmx1400m	22	41	1.9	0.9	3.00
2	5040 × 5040	Xmx1400m	42	55	1.3	0.7	29.00
3	2214 × 2214	Xmx1400m	4.1	9.01	2.2	0.73	3.29
3	2241 × 2241	Xmx1400m	4.9	9.12	1.9	0.62	5.58
3	2187 × 2187	Xmx1400m	4.08	8.92	2.2	0.73	3.32
3	2268 × 2268	Xmx1400m	4.24	9.44	2.2	0.74	3.28
3	4509 × 4509	Xmx1400m	18	45	2.5	0.83	9.00
3	5049 × 5049	Xmx1400m	25	51	2.0	0.68	24.00
4	2052 × 2052	Xmx1400m	4.64	7.06	1.5	0.38	11.50
4	2088 × 2088	Xmx1400m	4.66	8.15	1.7	0.44	10.49
4	2124 × 2124	Xmx1400m	4.77	8.22	1.7	0.43	10.86
4	4500 × 4500	Xmx1400m	20	42	2.1	0.53	38.00
4	5040 × 5040	Xmx1400m	31	62	2.0	0.50	62.00
5	2250 × 2250	Xmx1400m	6.11	9.36	1.53	0.31	21.19
5	2295 × 2295	Xmx1400m	6.27	9.93	1.58	0.32	21.42
5	2340 × 2340	Xmx1400m	6.56	10	1.52	0.30	22.80
5	4500 × 4500	Xmx1400m	29	41	1.41	0.28	104.00
5	5040 × 5040	Xmx1400m	38	55	1.45	0.29	135.00

P   Total nodes of a cluster
α   Execution time on single system
β   Time consumed by whole cluster

8   Conclusion and Future Scope

From the above work it can be concluded that:

- Proposed Architecture for Massive Parallel Processing is proved to be an efficient implementation based upon the various performance metrics like excessive overhead, efficiency and speed.
- Comparable amount of execution time is decreased using the cluster based parallel and distributed programming with the help of RMI.
- Network traffic is perfectly managed using number of NIC cards as the communication is between the dedicated links for a particular group in a given cluster, which is not interfered by the other nodes in a given cluster.
- Multithreading helps implementing multicast service in a given cluster, without thread based programming processing can be done only in the distributed manner but not in the parallel manner.
- Excessive parallel overhead in some cases is very high which shows that extra number of nodes are deployed for computation. Thus more time is spent in managing the cluster communication than the time taken for actual computation done on the cluster nodes.
- The proposed system has bottleneck in some cases, when dynamic memory for the heap space cannot be allocated, as reservation of memory is dependent upon the configuration of the master system and the operating system used.

Future work includes use of some other programming techniques to solve this bottleneck, proposed technologies can be:

- Middleware Based Distributed Programming
- Web Based Distributed Programming
- Service Oriented Architecture (SOA)
- Enterprise Java Beans (EJB).

References

1. Kaur, Harmanpreet, Sachin Bagga, and Ankit Arora. "RMI approach to cluster based Winograd's variant of Strassen's method." In MOOCs, Innovation and Technology in Education (MITE), 2015 IEEE 3rd International Conference on, pp. 156–162. IEEE, (2015).

2. Arora, Swinky, Ankit Arora, and Gursharanjit Singh Cheema. "Scheduling simulations: An experimental approach to time-sharing multiprocessor scheduling schemes." *International Journal of Computer Applications* 63, no. 11 (2013).
3. Dhillon, Haryali, Gagandeep Jindal, and Akshay Girdhar. "A Novel Threshold Technique for Eliminating Speckle Noise in Ultrasound Images." In *International Conference on Modelling, Simulation and Control, IPCSIT*, vol. 10. (2011).
4. Kanwal, Navdeep, Akshay Girdhar, and Savita Gupta. "Region Based Adaptive Contrast Enhancement of Medical X-Ray Images." In *Bioinformatics and Biomedical Engineering, (iCBBE) 2011 5th International Conference on*, pp. 1–5. IEEE, (2011).
5. Ahmad, Ishfaq, Yu-Kwong Kwok, Min-You Wu, and Wei Shu. "Automatic parallelization and scheduling of programs on multiprocessors using CASCH." In *Parallel Processing, 1997, Proceedings of the 1997 International Conference on*, pp. 288–291. IEEE, (1997).
6. Singh, Arashdeep, Sunny Behal, and Ankit Arora. "Efficiency Measurement for Effective Stress Management in Heterogeneous 2-D Mesh Processor." *International Journal of Computer Applications* 81, no. 12 (2013).
7. Arora, Ankit, Sachin Bagga, and Rajbir Singh Cheema. "Distributed Cluster Processing to Evaluate Interlaced Run-Length Compression Schemes." *International Journal of Computer Applications* 46, no. 6 (2012).
8. Pasricha, Nidhi, Ankit Arora, and Rajbir Singh Cheema. "Analytical Parallel Approach to Evaluate Cluster Based Strassen's Matrix Multiplication." *International Journal of Computer Applications (IJCA)* 44.11 (2012).
9. Median Filtering. NPTEL. http://nptel.ac.in/courses/117104069/chapter_8/8_16.html (accessed May 10, (2016).
10. Bagga, Sachin, Deepak Garg, and Ankit Arora. "Moldable load scheduling using demand adjustable policies." In Advances in Computing, Communications and Informatics (ICACCI, 2014 International Conference on, pp. 143–150. IEEE, 2014).
11. Oracle. java.rmi Properties. http://docs.oracle.com/javase/7/docs/technotes/guides/rmi/javarmiproperties.html (accessed May 10, 2016).
12. Arora, Ankit, Amit Chhabra, and Harwinder Singh Sohal. "Cluster based Performance Evaluation of Run-length Image Compression."International Journal of Computer Applications33, no. 5 (2011).

Segregation of Composite Document Images into Textual and Non-Textual Content

Munesh Chandra Trivedi, Shivani Saluja, Tarun Shrimali
and Shivani Shrimali

Abstract Segregation of image into textual and graphical region is essential prior to feeding the document in optical character recognition. Text embedded in colored images, such as book covers, video frames, natural scene images and web images suffers variation in style, color, orientation, alignment, low-image contrast, degraded characters as well as multiplexed background. The proposed paper focuses on segregation of textual component in two junctures. Initially, textual blocks characterization is done followed by restoration of both textual and non-textual region into separate image sets. Textual component extraction is preceded by noise removal and color reduction which is followed by edge detection, detection of closely connected component which can be grouped together and identified as a separate blob. Basic morphological operation are used to restore the broken characters. Later on the textual and non-textual lobs are segregated into separate image sets, which can be further used for annotation, indexing, character recognition. Image in painting is also performed in order to restore the area in both the image sets from where the textual and non-textual blobs have been extracted in order to make the image sets appear like real-image sets. The proposed approach works irrespective of the font and language of the textual content in the image.

Keywords Segregation · Image · Edge detection · Textual content

M.C. Trivedi · S. Saluja (✉)
ABES–EC, Ghaziabad, India
e-mail: Shivani.saluja@abes.ac.in

T. Shrimali
Sunrise Group of Institutions, Udaipur, India

S. Shrimali
JRN Rajasthan Vidyapeeth University, Udaipur, India

© Springer Nature Singapore Pte Ltd. 2018   401
S.K. Bhatia et al. (eds.), *Advances in Computer and Computational Sciences*,
Advances in Intelligent Systems and Computing 554,
https://doi.org/10.1007/978-981-10-3773-3_39

1   Introduction

Text appearing in images [1, 2] and videos plays an important role in semantic video indexing, video [3–5] surveillance and security, multilingual video information access indexing, summarization, document analysis, localization and recognition of license plate, keyword-based image search, identification of parts in industrial automation, identification of objects, street sign recognition. A high-level complex image can be easily interpreted with the help of presence of text label below it or with the help of superficial text in the frame or video. Extracting text appears as a key clue for assimilating contents of video and for instance for automatically stratifying some videos.

Segregation of image into textual and non-textual [6, 7] content is a far more complex procedure. It is difficult to convert a complex dynamic video into static image frame. Multiple problems like text overlapping among two frames, blurred textual content, presence of multiple colors in text and background, overlapped image, and text region increases complexity of text extraction which in turn deteriorates the output rate.

Prior to analyzing the video the colored dynamic video is converted into frame set which are further processed by color reduction procedure document image structure and identifying its textual blobs/blocks, the image has to be converted into binarized form, i.e., colored or grayscale images are converted into a monochrome one (1 bit).

A document video imbibes text and graphical components. Sound content of the video is assumed to be neglected. Document videos are acquired by web, television recordings, etc.

2   Literature Survey

1. Bottom Up Methods

Fame segments generated from the image are divided into regions and group. In this approach we start with the base pixels which are grouped to form edges and groups of connected regions which are grouped together in a boundary box. The boxes generated were further used for classification into textual and non-textual content.

2. Top Down Methods

They first detect text regions in images using filters and then perform bottom-up techniques inside the text regions. These methods are generally based on heuristic filters or machine learning. Top down approaches are able to process more complex images than bottom-up approaches.

3. Thresholding Techniques

Thresholding approach plays an essential role in image binarization. There are two thresholding approaches.

One works globally whereas another works locally. Global approach works by selecting a single threshold value. All the image pixels are compared with the statically define threshold. Pixels which satisfy the criterion are set to black, otherwise white.

The threshold value is experimentally generated by scanning the image. Generation of threshold for a large-sized image is complex because it might involve scanning intensity of thousands of pixels.

Local thresholding imbibes dividing the image into small blocks. The threshold value is generated either statically by evaluating the average of intensities of pixel groups or dynamically using Otsu thresholding.

4. Noise Removal Techniques

Cleaning or smoothening filter is employed to remove noise. Noise can be defined as set of discontinuities, very small regions with negligible pixel count or several regions which comprise lip of single black or light pixels. Several filters are identified for eradication of the noisy pels. Multiple filters are available like Gaussian, butterworth etc. The approach employed Gaussian filter to smoothen out the noisy pels as well as smoothening of edges by considering the feature that only high intensity pels can form an edge. This improves efficiency because ambiguous edges are removed.

5. Edge Detection Technique

Edge detection can be traced as combination of statistical operations which identifies area of brightness in the image or the regions empowered with drastic fluctuation in intensity value. Edges are defined as the points at which brightness of the image brightness varies along the steep slope. Canny edge detector has been used in this approach.

3 Image Frame Sets from Videos

The input for the project is generated from prestored videos of news, clips from television, manuscript of existing paper-based documents from history, recordings of conferences, journals, scanned document, book covers, satellite television programs, advertisements, collection of hypertext pages (Figs. 1, 2, 3).

Fig. 1 Frame set generated
from book cover

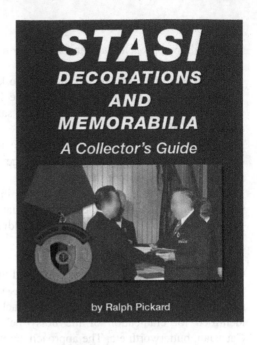

Fig. 2 Frame clip from a
video

కనిపించదు గాని మామయ్య గారింటికి వెళ్ళినప్పుడు విప్పితే పర్లక్ష్మి చూసిందంటే పరాభవమై పోతుంది. అందుకని ఒకటి కొందామని బాటా షాపుకి వెళ్ళాను.

"డూ యూ హావ్ మేజోర్స్?" అనడిగాను.

షాపువాడు తెల్లమొహం వేసి 'వాట్?' అన్నాడు. సారథి మళ్ళీ నవ్వడం మొదలెట్టాడు. అతనికి ప్రతిదీ నవ్వే.

షాపువాడు "వై ఆర్యూ లాఫింగ్, వాట్స్ది మేటర్" అనడిగాడు సారధిని.

"అవునోయ్ ఎందుకు నవ్వుతావు" అనడిగాను.

"నీ దౌర్భాగ్యపు ఇంగ్లీషు చూసి... మేజోర్స్ ఏమిటోయ్ నీ తలకాయ? సాక్స్ అనలేవూ" అన్నాడు అతను.

"అవునూ రెండూ ఒకటేగా" అన్నాను.

"ఓ... యూ వాంట్ సాక్స్" అంటూ అంటూ షాపువాడు మేజోళ్ళు చూపించాడు. వాటిల్లో నల్లవి చూపించి ధర అడిగాను. పద్దెనిమిది రూపాయలు చెప్పాడు.

"ఇటీజ్ వేరీ టూ మచ్" అన్నాను.

ధర తగ్గించనన్నాడు షాపువాడు. సరేనని తొమ్మిది రూపాయలిచ్చి "గివ్ మి వన్ సాక్" అన్నాను.

సారథి నోటికి కర్చీఫ్ అడ్డం పెట్టుకుని నవ్వుకుంటూ బైటకి వెళ్ళిపోయాడు. షాపువాడూ నవ్వుతూ "నో నో యూ కాంట్ హావ్ వన్. టేక్ ది పేర్" అన్నాడు.

ఒకటి వాలు మెట్రో అంటుంటే ఈ బలవంతం ఏమిటి?

"అయామ్ ఆల్రెడీ హావింగ్ యే బ్లాక్ పేర్. బట్ వన్ సాక్ ఈజ్ బిటన్ బై యే రాట్. సో ఐ

వాంట్ ఒన్లీ వన్ సాక్" అని చెప్పినా కన్విన్స్ కాలేదు షాపువాడు. "దిసీజ్ నథింగ్ బట్ ఎక్స్ప్లాయిటేషన్" అంటూ బైటికొచ్చేశాను.

ఈ పెద్ద షాపుల వాళ్ళకిదే రోగం - పేవ్ మెంట్ మీద కొనుక్కోవడం మంచిదనుకున్నాను. బైట సారథి ఇంకా నవ్వుతూనే ఉన్నాడు. పేవ్ మెంట్ మీద బేరం ఆడితే జత తొమ్మిది రూపాయలు చెప్పాడు. అంటే బాటా ధరలో సగం అన్నమాట. బాగానే ఉందనుకుని నాలుగున్నర ఇచ్చి సోక్ ఇవ్వమంటే వాడూ జత తీసుకోమని దబాయించడం మొదలెట్టాడు. పద్దు పొమ్మని వేరే వాళ్ళ దగ్గరికి వెళ్ళాను. వాళ్ళూ అదే మాట. పెద్దవలు... అందరూ కూడబలుక్కుని ఒకే కట్టుగా ఉంటారు. జనాన్ని దోచుకునేందుకు.

156

Fig. 3 Multilingual image

4 Challenges

- Resence of overlay texts present in the cluttered scene background.
- There is no consistent color distribution for texts in different videos. Consequently, the color-tone-based approach widely used in face or people detection application actually cannot be applied in text detection.
- The size of the text regions may be very small such that when the color segmentation-based approach is applied, the small text region may merge into the large non-text regions in its vicinity. Background and text may be ambiguous.
- Text color may change: text can have arbitrary and nonuniform color.
- Background and text are sometimes reversed. Text may move
- Unknown text size, position, orientation, and layout: captions lack the structure usually associated with documents.
- Unconstrained background: the background can have colors similar to the text color.
- The background may include streaks that appear very similar to character strokes.
- Color bleeding: Lossy video compression may cause colors to run together.
- Low contrast: low bit-rate video compression cannot cause loss of contrast (Fig. 4).

Fig. 4 Flowgraph view of segregation system

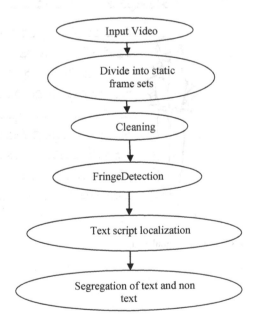

5 Proposed Approach

There are two phases in the entire approach. The first involves taking an input video cleaning it, detecting fringes, and finally text script localization.

Phase two involves segregation of entire image into textual and non-textual content.

Step 1: Input video

This step involves feeding the input video into project. Input video is combination of sequence of frames. Frames may be of different color and might contain overlapped images.

Step 2: Generation of frame sets

The video input provided in the previous step is divided into equisized image frames which are further processed turn by turn. The complexity of text extraction [8, 9] is very high in case the text is extracted directly from video. The static colored frames are converted to gray scale. Grayscale conversion is achieved using standard form for luminance

$$Y = 0.2126\,R + 0.7152\,G + 0.0722\,B,$$

where R, G, B represent the intensity value of red, green, and blue pixels in the image. For conversion, the entire image is divided into small block where each pixel is determined by a 24 bit value, each 8 bit individually representing R, G, and B intensities.

Pixel with RGB values of (20, 127, 255)
The red level R = 20.
The green level G = 127
The blue level B = 255.
Red' = Green' = Blue' = (R + G + B)/3 = (20 + 127 + 255)/3 = 134
so the pixel will get RGB values of:
(134, 134, 134)

Step 3: Cleaning:

Cleaning involves eradication of the noisy pixels from the image frame. Cleaning of image is used to smoothen out a set of data to create a quantified approximating function that tries to recognize essentials patterns in the data, while removing out noise.

In smoothing, the data points of a signal are modified so individual points (presumably because of noise) are reduced, and points that are lower than the adjacent points are increased leading to a smoother signal.

Smoothing may be used in two important ways that can aid in data analysis (1) by being able to extract more information from the data as long as the assumption of smoothing is reasonable and (2) by being able to provide analyses that are both flexible and robust. Grayscale conversion is followed by thresholding or binarization which converts the intensity value for each of the pixel to either 0 or 1, based upon the comparison with the threshold value. The threshold value is evaluated either on global basis or block-by-block.

Step 4: **Fringe Detection**

Fringe detection involves generation of lines or edges depending upon the intensity of the image. The idea is that text contrasts a lot with background.

Step 5: **Text Script Localization**

Text extraction involves detection, localization, tracking, binarization, extraction, enhancement, and recognition of the text from the given image. These text characters are difficult to be detected and recognized due to their deviation of size, font, style, orientation, alignment, contrast, complex colored, textured background. After the other to the entire frameset to remove the noisy content from the image. Next to noise removal morphological operations are applied to merge the broken areas (Fig. 5).

The main idea of this approach is that a letter can be considered as a homogeneous region (using our restrictions), and thus it could be very useful to divide the frame into homogeneous regions. To compute such a division, a split-and-merge algorithm seems to be very adequate.

Fig. 5 Text localization

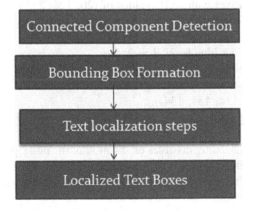

Its concept is: while there is a nonhomogeneous region, then split it into four regions. And if two adjacent regions are homogeneous, then they can be merged. Then, using some size characterizations of the text (not too big and not too small), the inadequate regions will be deleted. The same process will be executed for the different frames, and the results will be temporally integrated in order to keep only the elements which are present in all the frames.

Text in images can exhibit many variations with respect to the following properties:

1. **Geometry**: Size: Although the text size can vary a lot, assumptions can be made depending on the application domain.
   Alignment: The caption texts appear in clusters and usually lie horizontally, although sometimes they can appear as nonplanar texts as a result of special effects. This does not apply to scene text, which has various perspective distortions. Scene text can be aligned in any direction and can have geometric distortions.
   Inter-character distance: characters in a text line have a uniform distance between them.

2. **Color**: The characters tend to have the same or similar colors. This property makes it possible to use a connected component-based approach for text detection. Most of the research reported till date has concentrated on finding 'text strings of a single color (monochrome)'. However, video images and other complex color documents can contain 'text strings with more than two colors (polychrome)' for effective visualization, i.e., different colors within one word.

3. **Motion**: The same characters usually exist in consecutive frames in a video with or without movement. This property is used in text tracking and enhancement. Caption text usually moves in a uniform way: horizontally or vertically. Scene text can have arbitrary motion due to camera or object movement.

4. **Edge**: Most caption and scene text are designed to be easily read, thereby resulting in strong edges at the boundaries of text and background.

5. **Compression**: Many digital images are recorded, transferred, and processed in a compressed format. Thus, a faster TIE system can be achieved if one can extract text without decompression.

6 Results

See (Figs. 6 and 7).

Fig. 6 Segregated textual cover image

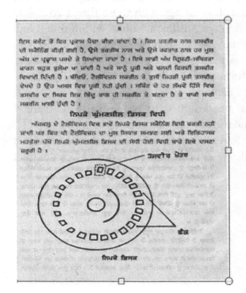

Fig. 7 Segregated Gurumukhi image

7 Applications

Vehicle license plate recognition, analysis of document Image, Identification of objects, recognition of street signs, recognition of name plate, Industrial automation, retrieval based on content, analysis of video and document images, page segmentation, search in the image using keywords, retrieval of documents, address block location, analysis of articles or blogs using graphs, maps, and charts.

References

1. B. Gatos, K. Ntirogiannis, and I. Pratikakis, "ICDAR 2009 document image binarization contest (DIBCO 2009)," in Proc. Int. Conf. Document Anal. Recognit. Jul. 2009, pp. 1375–1382.
2. I. Pratikakis, B. Gatos, and K. Ntirogiannis, "H-DIBCO 2010 handwritten document image binarization competition." in Proc. Int. Conf. Frontiers Handwrit. Recognit. Nov. 2010, pp. 727–732.
3. Ankur Srivastava, Dhananjay Kumar, *"Text extraction in Video"* International Journal of Computational Engineering Research, Vol 3, 2012.
4. Bolan Su, Shijian Lu, And Chew Lim Tan "Robust Document Image Binarization Technique For Degraded Document Images" 2013.
5. V.R. Vijaykumar, P.T. Vanathi, P. Kanagasabapathy "Fast and Efficient Algorithm to Remove Gaussian Noise in Digital Images" 2010.
6. Jayshree Ghorpade, Raviraj *"Extracting text from video Signal & Image Processing"* An International Journal (SIPIJ) Vol. 2, No. 2, June 2011.
7. Julinda Gllavata1, Ralph Ewerth1 and Bernd Freisleben1 "A Robust Algorithm for Text Detection in Images" http://www.mathematik.uni-marburg.de.
8. C.P. Sumathi, T. Santhanam *"a survey on various approaches of text extraction in images"* International Journal of Computer Science & Engineering Survey (IJCSES) Vol. 3, No. 4, August 2012.
9. Florence Kussener "Active contour: a parallel genetic algorithm approach", Proceedings of International conference on swarm intelligence (ICSI 2011), 2011.

Optimization of Automatic Test Case Generation with Cuckoo Search and Genetic Algorithm Approaches

Rijwan Khan, Mohd Amjad and Akhlesh Kumar Srivastava

Abstract Automatic test case generation is an optimization problem in software testing process. With the use of genetic algorithm we can generate the test cases automatically. Genetic algorithm alone does not give 100% accurate optimized test cases. Hence merging of genetic algorithm with Cuckoo search optimization technique produces better optimized test cases. The main aim of this paper is to customize the cost and time for the Testing process after the generation of test cases automatically. The two optimization techniques namely Cuckoo Search and genetic algorithm produce better result as compared to single one.

Keywords Genetic algorithm · Cuckoo search optimization (CSO) · Test cases · Software testing

1 Introduction

Software Engineering is one of the most important subjects of Computer Engineering. The Software development team always follows the process of software development life cycle. There are some important phases that have been defined in SDLC. The software testing is one of the most essential phases in SDLC. Software testing plays a very important role. The software testing phase is responsible for delivering the error free, reliable, and cost-effective Software to the customers.

R. Khan (✉) · M. Amjad
Department of Computer Engineering, Jamia Millia Islamia,
Jamia Nagar, New Delhi, India
e-mail: rijwankhan786@gmail.com

M. Amjad
e-mail: mamjad@jmi.ac.in

A.K. Srivastava
Department of Computer Science & Engineering, ABES Engineering College,
Ghaziabad, India
e-mail: joinakhilesh@yahoo.com

© Springer Nature Singapore Pte Ltd. 2018 413
S.K. Bhatia et al. (eds.), *Advances in Computer and Computational Sciences*,
Advances in Intelligent Systems and Computing 554,
https://doi.org/10.1007/978-981-10-3773-3_40

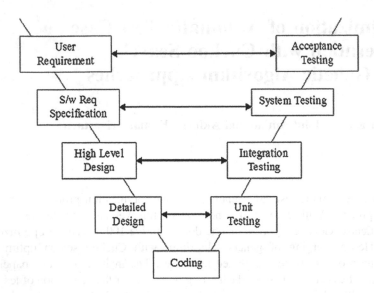

Fig. 1 V-model of software testing

There are many different kinds of the software testing. The basic types of Testing are unit testing, integration testing, system testing, and acceptance testing. In this paper, unit testing has been used among the different types of testing.

1.1 Software Testing Background

Software testing is a process of verification and validation of the Software. It is also responsible for delivering reliable Software to the customer. In each phase of Software development process, some kind of the testing has been applied. Software Testing is represented by V-Model. The V-Model of the software testing process has some kind of testing according the SDLC as shown in Fig. 1. In correspond to the user requirements, it is acceptance testing, on Software requirement specification it is system testing, on a high-level design, it is integration testing and on detailed design it is unit testing.

1.2 Unit Testing

Unit testing is one the important testing phase of the software testing process. Unit testing provides a way to check the functionality of the system with developer team and testing team. They take small parts (units) of Software independently to check its functionality. In the unit testing we check for boundary value analysis, decision table, equivalence partition, and path testing [1]. In this paper, we consider path testing and white box testing.

1.3 Path Testing

In path testing, we draw a data flow chart for the software/program. This data flow is working on all du (defined and use) (similar terms, c-use calculation (+, −, * etc), p-use—production use (>, <, <= etc)) path. The all du path are those in which variable is first defined and then used. The du path first defined and used as c-use or p-use later. Means use is calculation and in predicates. Here this entire process is explained:

An algorithm for finding the GCD of two numbers

1. GCD(x,y)
2. If(y>x)
3. {then interchange values of x and y}
4. T=x%y
5. while(T!=0)
6. { x=y
7. y=T
8. T=x%y}
9. return y

Fig. 2 Data flow diagram for GCD algorithm

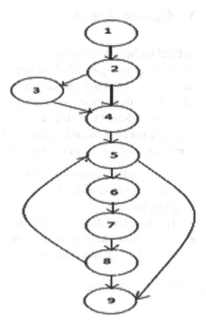

In Fig. 1, variables x and y are defined at node 1, means d-use at node 1. Variables x and y are p-use at node 2 and c use at node 3. Variables x and y are again defined at nodes 6 and 7. Variables x and y are c-used at node 8. Variable T is defined at node 4 and p-used at node 5. Again T is define at node 8 (Fig. 2).

2 Genetic Algorithm

Genetic algorithm is one of the nature inspired optimization technique. Genetic algorithm is developed by Holland who worked on the theory of survival of the fittest. Genetic and evaluation are two principles for genetic algorithm. In genetic algorithm, there are some operations as selection, reproduction, evaluation and replacement [2–4]. For applying genetic algorithm on given data, first it should be represented in some data representation forms. The most suitable representation is Binary representation of data. Generally genetic algorithm is given as under.

- Select the Initial Population.
- Check for termination criteria (fitness function).
- If termination criteria matched then stop, else go to next statement.
- Select the individuals using selection operation.
- Apply the genetic algorithm crossover and mutation operation to generate new population, go to second statement [5–7].

3 Cuckoo Search

In Cuckoo Search technique, we will adopt the strategy that Cuckoo used to follow in hatching its eggs. Cuckoo used to lay its eggs in the nest of other birds, to increase the hatching probability of their own eggs, Cuckoos remove the eggs of other birds from their native nest and replace its own eggs on the place of other bird's eggs. Some Cuckoos are good specialty in imitating the color and pattern of eggs like the eggs of other birds, so that one cannot easily recognize Cuckoo's eggs. There are some ideal rules about the Cuckoo search.

- Each Cuckoo randomly chose a nest and lays its one egg at a time.
- The best nests, with high quality of eggs (solutions) will carry over to the next generation.
- The host bird can discover Cuckoo's egg with a probability $p = [0, 1]$.
- If host bird finds that eggs does not belongs to her, the host bird throws the egg of Cuckoo or abandon the nest and build a new nest on a new location [8].

Cuckoo search Algorithm defined by Xin-She Yang is discussed below
Objective function $f(y) = (y_1,...,y_4)^T$;
Initial a population of n host nests $xi = (i = 1, 2, ..., n)$;

```
while(t<MaxGeneration) or (stop criterion)
        Get a Cuckoo (say i) randomly by Levy-fights;
        Evaluate its fitness function Fi;
        Choose a nest among n (say j) randomly;
        if (Fi>Fj)
                replace j by new solution;
        end
        Abandon a fraction (pa) of worse nests;
        Keep the best solutions;
        Rank the solutions and find the current best;
end while
```

4 Genetic Algorithm in Software Testing

Among all the search-based techniques genetic algorithm is more efficient than other optimization Algorithms [9]. The modified genetic algorithm improves the fitness function [10]. A concept of dominance relations between the nodes of control flow graph to reduce the software testing cost. Software testing data is relating the production of quality Software production indirectly. Improved genetic algorithm is superior to the basic genetic algorithm on effectiveness and efficiency of automatic test case generation [11].

5 Proposed Method

Here proposed a method given in which we merged two optimization techniques, One is Cuckoo search optimization and second is genetic algorithm.
    Algorithm is given below.

Step 1: Randomly generate the test cases for the given program with in a range.
Step 2: Refine the test cases using Cuckoo search optimization.
Step 3: Apply genetic algorithm's fitness function. If satisfied then stop otherwise go to next step.
Step 4: Select the initial population using tournament selection process.
Step 5: Apply genetic algorithm operations (crossover and mutation).
Step 6: Go to step 3.

The fitness function is maximum path coverage in our proposed method.

6 Experimental Setup

Algorithm to find X^Y for two positive numbers X and Y. For the above proposed method example is given below.

a. Power(X,Y)
b. if(X==1)
c. return 1;
d. if(Y==1)
e. return X;
f. i=1, t=1;
g. while(i<=Y)
h. { t=t*X;
i. i=i+1;}
j. return t;

First we have to take values of X and Y in binary. Basically, our algorithm will work on path coverage. If it is 100% path coverage then algorithm will stop. In proposed method, we first generate the test cases randomly and then apply Cuckoo search optimization for selecting the test cases which cover path in maximum. We then select other optimization technique, i.e., genetic algorithm for generating test cases from these first time random generated test cases.

In this example we have four paths.

Path 1: 1—2—3—10
Path 2: 1—2—4—5—10
Path 3: 1—2—4—6—7—10
Path 4: 1—2—4—6—7—8—9—7—10

Now, five test suits are taken and in each test suit some data is taken randomly. Below test suits are given with path covered across ponding to each data (Tables 1, 2, 3, 4, 5).

When we applied our proposed method on the given suites we found some of the test cases in a given suites useless. Our Algorithm refines these test suites using Cuckoo search optimization and generates the new test cases using genetic algorithm. After applying our method we find that these test cases are useful in each test suites (Tables 6, 7, 8, 9, 10).

We implemented our proposed method in C programming. After applying this method def use covered 100%. The results are given in Fig. 3 (Fig. 4).

Table 1 Test suit table

S. no	X	Y	X^Y	Path
1	1	2	1	1
2	4	2	16	4
3	3	1	3	2
4	4	1	4	2
5	9	2	81	4
6	6	2	36	4
7	1	4	1	1
8	5	2	25	4
9	8	3	512	4
10	3	6	324	4
11	4	1	4	2
12	1	6	1	1
13	1	9	1	1
14	2	3	8	4
15	8	1	8	2
16	6	2	36	4
17	3	5	243	4
18	1	4	1	1
19	4	1	4	2
20	7	2	49	4

Table 2 Test suit table

S. no	X	Y	X^Y	Path
1	2	3	8	4
2	9	2	81	4
3	1	10	1	1
4	3	6	729	4
5	4	2	16	4
6	2	6	64	4
7	4	4	256	4
8	3	2	9	4
9	4	3	64	4
10	2	2	4	4

Table 3 Test suit table

S. no	X	Y	X^Y	Path
1	2	1	1	2
2	1	2	2	1
3	2	2	4	4
4	3	3	27	4
5	4	2	16	4
6	5	3	125	4
7	6	2	36	4
8	2	3	8	4
9	1	8	1	1
10	3	2	9	4
11	4	3	64	4
12	3	3	81	4
13	4	1	4	2
14	2	4	16	4
15	6	2	36	4

Table 4 Test suit table

S. no	X	Y	X^Y	Path
1	4	2	16	4
2	2	5	32	4
3	1	8	1	1
4	6	1	6	2
5	9	2	81	4
6	4	3	64	4

Table 5 Test suit table

S. no	X	Y	X^Y	Path
1	8	2	64	4
2	10	2	100	4
3	4	3	64	4
4	1	9	1	1
5	1	4	1	1

Table 6 Test suit table

S. no	X	Y	X^Y	Path
1	1	2	1	1
2	4	1	4	2
3	5	2	25	4
4	6	2	36	4

Table 7 Test suit table

S. no	X	Y	X^Y	Path
1	1	10	1	1
2	9	1	9	2
3	3	2	9	4
4	2	2	4	4

Table 8 Test suit table

S. no	X	Y	X^Y	Path
1	1	2	1	1
2	2	1	2	2
3	3	2	9	4
4	2	4	16	4

Table 9 Test suit table

S. no	X	Y	X^Y	Path
1	1	8	1	1
2	6	1	6	2
3	9	2	81	4
4	4	3	64	4

Table 10 Test suit table

S. no	X	Y	X^Y	Path
1	1	9	1	1
2	4	1	4	2
3	8	2	64	4
4	10	2	100	4

Fig. 3 Data flow diagram for program X^Y

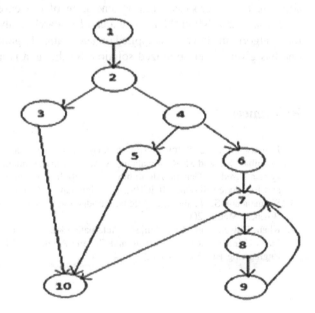

```
path fitness0.000000
5,6,7,8,9,10,11,12,13,18,19,20,21,25,26,27,28,29,30,19,20,21,25,26,27,28,29,30,1
9,31,32,

covered def use pairs:

path fitness0.000000
5,6,7,8,9,10,11,12,13,18,19,20,21,22,23,24,29,30,19,31,32,

covered def use pairs:

path fitness0.000000
5,6,7,8,9,10,14,15,16,17,18,19,31,32,

covered def use pairs:

path fitness0.000000
selected output paths are::
110111110111011 ,fitness of this path=0.652174
110111101111011 ,fitness of this path=0.130435
111011000000001 ,fitness of this path=0.130435
111011110100001 ,fitness of this path=0.086957
covered def use pairs=100.000000%
```

Fig. 4 Output of the GA functions for def use covered

7 Conclusion

In this paper, we proposed a method that combines two optimization techniques Cuckoo search optimization and genetic algorithm, for the purpose of optimization of test cases. In our paper, Cuckoo search first refine randomly generated test cases and genetic algorithm generate the new test cases from these remaining test cases. Our paper optimized the test cases and gives us those test cases which are useful in software testing process. Simply generation of test cases automatically is not an easy task, it is a NP problem. The area of research is always open in this field. So many algorithms have been applied on automatically generation of test cases but no one has given us an optimized solution. So the area is always open for research.

References

1. Jorgensen, Paul C. Software Testing: a craftsman's approach. CRC press, 2013.
2. Parthiban, M., and M. R. Sumalatha. "GASE-an input domain reduction and branch coverage system based on Genetic Algorithm and Symbolic Execution." Information Communication and Embedded Systems (ICICES), 2013 International Conference on. IEEE, 2013.
3. Sivanandam, S. N., and S. N. Deepa. Introduction to Genetic Algorithms. Springer Science & Business Media, 2007.
4. Khan, Rijwan, and Mohd Amjad. "Automatic Generation of Test Cases for Data Flow Test Paths Using K-Means Clustering and Generic Algorithm." International Journal of Applied Engineering Research 11.1 (2016): 473–478.

5. Mahajan, Manish, Sumit Kumar, and Rabins Porwal. "Applying Genetic Algorithm to increase the efficiency of a data flow-based test data generation approach." ACM SIGSOFT Software Engineering Notes 37.5 (2012): 1–5.
6. Srivastava, Praveen Ranjan, and Tai-hoon Kim. "Application of Genetic Algorithm in Software Testing." International Journal of Software Engineering and its Applications 3.4 (2009): 87–96.
7. Ghiduk, Ahmed S., and Moheb R. Girgis. "Using Genetic Algorithms and dominance concepts for generating reduced test data." Informatica 34.3 (2010).
8. Yang, Xin-She, and Suash Deb. "Engineering optimisation by Cuckoo search." International Journal of Mathematical Modelling and Numerical Optimisation 1.4 (2010): 330–343.
9. Andreou, Andreas S., Kypros Economides, and Anastasis Sofokleous. "An automatic Software test-data generation scheme based on data flow criteria and Genetic Algorithms." Computer and Information Technology, 2007. CIT 2007. 7th IEEE International Conference on. IEEE, 2007.
10. Liu, Dan, X. U. E. J. U. N. Wang, and J. I. A. N. M. I. N. Wang. "Automatic test case generation based on Genetic Algorithm‖." Journal of Theoretical and Applied Information Technology 48.1 (2013): 411–416.
11. Dong, Yuehua, and Jidong Peng. "Automatic generation of Software test cases based on improved Genetic Algorithm." Multimedia Technology (ICMT), 2011 International Conference on. IEEE, 2011.

Impact Analysis of Contributing Parameters in Audio Watermarking Using DWT and SVD

Ritu Jain, Munesh Chandra Trivedi and Shailesh Tiwari

Abstract This paper proposes a non-blind audio watermarking algorithm for copyright protection of audio files which has proved to satisfy the minimum requirements of optimal audio watermarking standards set by International Federation of Photographic Industry (IFPI). The algorithm is set to meet the IFPI requirements as it includes the two powerful mathematical tools: Discrete wavelet transform (DWT) and singular value decomposition (SVD). In this paper, we have also analyzed the contribution of parameters like watermark size and the embedding intensity factor on the algorithm.

Keywords DWT decomposition · SVD transformation · Imperceptibility · Embedding intensity factor

1 Introduction

With every new invention, there comes new challenges too. Similarly, as we observe the advancements in the technologies in terms of widened communication networks, cost worthy, and reliable storage devices, it has raised a threat to the same by making the replication and distribution of the crucial data over these networks quite easily. A considerable solution to this problem which has emerged over a few decades is technique of audio watermarking, which is basically the concept of infusing the copyright information into the valuable media files in a secret manner so that the

R. Jain (✉) · M.C. Trivedi · S. Tiwari
ABES Engineering College, Ghaziabad, India
e-mail: rjritu9@gmail.com

M.C. Trivedi
e-mail: munesh.trivedi@gmail.com

S. Tiwari
e-mail: hodcs@abes.ac.in

© Springer Nature Singapore Pte Ltd. 2018
S.K. Bhatia et al. (eds.), *Advances in Computer and Computational Sciences*,
Advances in Intelligent Systems and Computing 554,
https://doi.org/10.1007/978-981-10-3773-3_41

imperceptibility of the algorithm is maintained [1–3]. The inserted copyright information truly identifies the originator's and the receiver's and identity [4].

In order to achieve effective copyright protection, the audio watermarking algorithm must possess three basic properties: robustness, high data payload, and imperceptibility which are known to be the pillars of the watermarking algorithms [2, 5]. Imperceptibility is required to ensure that the watermark is perceptually unidentified and degrades the audio in the least possible manner. The attribute data payload significantly identifies the embedding capacity and is measured in bits per second (bps). The three properties are closely interrelated in the manner that imperceptibility is inversely proportional to data payload and in turn increase in data payload affects the robustness: hence, there is a trade-off between these properties depending on each application [6, 7].

PSNR stands for peak signal-to-noise ratio. It is calculated using the following formula.

$$PSNR = 10 \, log \, 10 (R^2 / MSE),$$

where R is the maximum fluctuation and MSE is the mean square error [8].

In this experiment, we are mainly considering WAVE (.wav) audio files, requiring *64* bits per sample of data type *double* that are in the data range $-1.0 \leq y \leq +1.0$, so R corresponds to +1.0 in this case [9].

SNR is signal-to-noise ratio and calculated as:

$$SNR = 10 \, log \, 10 (\textstyle\sum_n A_n^2 / \sum_n (A_n - A_n'^2)),$$

where A_n is the original audio signal and A_n' is the watermarked signal [10].

2 Related Work and Contribution

In this section, a gist of DWT-and SVD-based algorithms is discussed. The special contribution of these tools in audio watermarking algorithms is also described.

2.1 DWT Based Algorithms

Discrete wavelet transform (DWT) is any wavelet transform for which the wavelets are discretely sampled. As with other wavelet transforms, a key advantage it has over Fourier transforms is temporal resolution: it captures both frequency and location information (location in time). The input signal S when applied to the DWT operator produces two sets of coefficients: approximation coefficients (A) and detail coefficients (D) as shown in Fig. 1 [10–12]. The level of DWT decomposition is application dependent and varies accordingly.

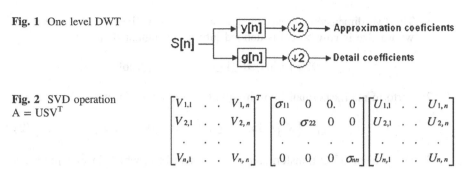

Fig. 1 One level DWT

Fig. 2 SVD operation
A = USV^T

$$\begin{bmatrix} V_{1,1} & . & . & V_{1,n} \\ V_{2,1} & . & . & V_{2,n} \\ . & . & . & . \\ V_{n,1} & . & . & V_{n,n} \end{bmatrix}^T \begin{bmatrix} \sigma_{11} & 0 & 0. & 0 \\ 0 & \sigma_{22} & 0 & 0 \\ . & . & . & . \\ 0 & 0 & 0 & \sigma_{nn} \end{bmatrix} \begin{bmatrix} U_{1,1} & . & . & U_{1,n} \\ U_{2,1} & . & . & U_{2,n} \\ . & . & . & . \\ U_{n,1} & . & . & U_{n,n} \end{bmatrix}$$

2.2 SVD-Based Algorithms

SVD is a numerical technique for diagonalizing matrices [13]. The SVD of an N × N matrix A is defined by the operation A = USV^T as shown in Fig. 2. The diagonal entries of S are called the singular values of A and are assumed to be arranged in decreasing order $\sigma_i > \sigma_{i+1}$ [10, 13].

3 Proposed Algorithm

Our proposed algorithm is an attempt to improvise the PSNR algorithm prescribed in [10]. In this paper, we have described an audio watermarking algorithm that employs two cascaded operators: DWT and SVD. Many such algorithms have been discussed in [14, 15]. The algorithm exploits unique features as: the robustness is ensured by scattering the bits of watermark throughout the audio signal by partitioning into frames, decomposing using DWT and SVD transformation. Another important property of imperceptibility is looked after by slightly varying the singular values of the diagonal elements of the matrix S for watermark embedding so that the perceptual degradation is least possible at hand. Likewise, the quality of the audio signal is also maintained. Hence, the name of the algorithm is said to be *dual transform algorithm*.

The algorithm comprises of two phases: *watermark embedding* and *watermark extraction*.

3.1 Watermark Embedding Procedure

A. Preprocessing of the watermark image and its SVD transformation is as follows:

1. The two-dimensional image in grayscale color to be used as watermark whose size is **row** × **col** is defined into a rectangular matrix **X**.

$$X = \{X(i,j); 0 \le i \le row; \quad 0 \le j \le col\} \qquad (1)$$

2. Perform matrix normalization by dividing the image in grayscale by 256.

$$X_{i,j} = \{X_{i,j}/256; \quad 0 \le i \le row; \quad 0 \le j \le col\} \qquad (2)$$

3. Perform the SVD transformation on the matrix X which in turn produces three matrices U, D and V, out of which U is a square matrix of order **i** × **i**.

$$X = U \cdot D \cdot V \qquad (3)$$

4. Multiply the matrix D with the embedding intensity denoted by **alpha**, such that

$$D = alpha * D \qquad (4)$$

B. Audio framing, DWT decomposition and SVD transformation on the audio data is as follows:

1. Read the audio particularly in .wav format into *filedata*.
2. The read audio signal is partitioned into frames of predefined number such that **no. of frames** × **frame size** resulting into the sampled audio signal, *audio*.
3. Reshape the matrix audio into *audio_mat* such that each row contains a single frame.
4. Perform 3-level DWT decomposition using the Daubechies wavelet (db1) on each frame such that it produces four multi-resolution subbands A3, D3, D2, D1, where A3 represents the low frequency components of the audio frame known as approximation subband while D3, D2, and D1 represents the detailed coefficients.
5. Perform SVD on matrix DwD3 which is 1/4th of the detailed sub-band D1.
6. The SVD operation on DwD3 produces SU, S, and SV matrices such that

$$DwD3 = SU \cdot S \cdot SV \qquad (5)$$

C. Embedding of watermark on audio

1. The embedding part is performed using the S matrix in Eq. (5) and D matrix used in Eq. (4). The 'k' indicates the image size. The embedding formula is as follows:

$$S\_mod = S(i, i) + D(i, i); \ 1 \leq i \leq k$$
$$S\_mod = S(i, i); \ (k+1) \leq i \leq N \tag{6}$$

2. Obtain the watermarked audio media file by performing the inverse of mathematical operators DWT and SVD.

3.2 Watermark Extracting Procedure

The extraction procedure is exactly a reverse manner of the steps described in the embedding procedure. To proceed, we need the original audio signal, the embedding intensity factor alpha, and matrices U and V. This is why, the algorithm is categorized as *non-blind* audio watermarking algorithm. The steps of watermark extraction are illustrated below:

1. Find the difference between the S_new and the matrix S.

$$D\_recn = S\_new(i, i) - S(i, i); 1 \leq N \tag{7}$$

2. Divide the difference b/w the matrices by the embedding intensity

$$D\_recn = (1/alpha) * D\_recn \tag{8}$$

3. Apply inverse SVD to produce normalized watermark image

$$img\_recn = U*D\_recn*V' \tag{9}$$

4. Finally, retrieve the row × col matrix of the extracted watermark as follows.

$$X_{i,j} = \{X_{i,j} \cdot 256; \ 0 \leq row, \ 0 \leq col\} \tag{10}$$

4 Experimental Results and Contribution

The N × N image, shown in Fig. 3, where N is taken as 128 is used as the watermark belonging to class unit 8, giving a total of 131072 watermark bits unlike [10] where the author has considered a gray scale image of varied specifications. In this experiment, we have tried to improve the performance of the algorithm described in [10] by tuning the parameters N and alpha. The value of alpha is interestingly varied and its effect on the values of SNR is observed thoroughly as shown in the Fig. 5 and Fig. 6, respectively.

'cameraman.tif' image for watermarking

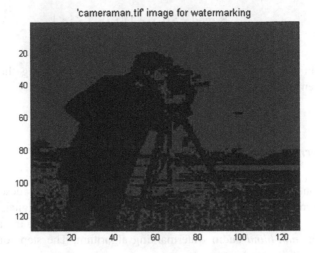

Fig. 3 Image used for watermarking

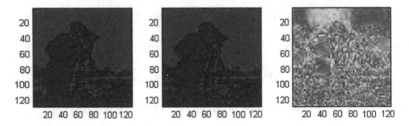

Fig. 4 Original image, reconstructed image, and the difference b/w the images

With increase in the value of alpha, the difference becomes negligible as shown in Fig. 4.

The Figs. 5 and 6 clearly illustrates the graph plot between the SNR values and the embedding intensity factor alpha and the differences in the watermark, respectively. With increase in watermark intensity, i.e., alpha, the quality of watermarked audio decreases because of decrease in its SNR; however, the quality of extracted watermark increases as watermark error decreases.

Table 1 represents the compared values between the proposed algorithm with modified parameters and the algorithm described in [10].

The correctness of the algorithm is ensured after testing it on various audio scripts like instrumental, jazz, animal speech, medley, etc. sampled at 44.1 kHz because at this frequency it gives out the accurate representation of the compressed signal [16].

Fig. 5 Plot of SNR versus embedding intensity factor

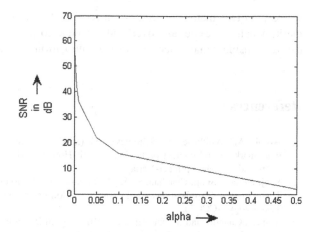

Fig. 6 Plot of SNR versus difference in the watermark image

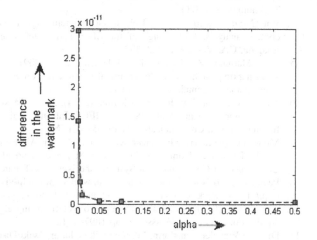

Table 1 SNR and PSNR values

Audio type	PSNR obtained in [10]	PSNR from proposed algorithm
Jazz audio	51.26	69.8
Instrumental audio	42.92	61.9
Speech audio	40.14	67.2
Average	44.77	66.3

5 Conclusion and Future Work

In this paper, a thorough study on two parameters, i.e., N and alpha is done which leads to a considerable improvement in the values of SNR for audio files. Another loophole in this algorithm is in the application of SVD transformation due to which we rely on two matrices U and V on receivers' end in order to extract the watermark, which comes out as a overhead for the algorithm. Our future works would be based on using some other type of decomposition methods during embedding.

References

1. Arnold M, Wolthusen S, Schmucker M (2003) Techniques and applications of digital watermarking and content protection. Artech House. Zwicker E, Fastl H. Psychoacoustics: Facts and models. Springer-Verlag.
2. Sanjay Pratap Singh Chauhan, S. A. M. Rizvi (2014) A survey: Digital Audio Watermarking Techniques and Applications: 4th International Conference on Computer and Communication Technology (ICCCT).
3. Munesh Chandra and Shikha Pandey (2010): Digital Watermarking Technique for Protecting Digital Images, Proc. IEEE 3rd Intl. Conference on Computer Science & Information Technology (ICCSIT).
4. Wu C, Su P, Kuo J (2000) Robust and efficient digital audio watermarking using audio content analysis. Proceedings of the SPIE 12th International Symposium on Electronic Imaging, CA. Vol. 3971, 382–392.
5. E.K. Marnani, Z. Karami, E. Molavian Jazi (2009) A Comparison of some audio watermarking methods 6th International Conference on Electrical Engineering, Computing Science and Automatic Control, CCE.
6. Munesh Chandra Trivedi, Naresh Kumar (2014): Audio Masking for Watermark Embedding under Time Domain Audio Signals, IEEE International Conference on Computational Intelligence and Communication Networks (CICN).
7. Munesh Chandra Trivedi, Naresh Kumar (2014): An Algorithmic Digital Audio Watermarking in Perceptual Domain Using Direct Sequence Spread Spectrum IEEE International Conference on Communication Systems and Network Technologies (CSNT).
8. Peak Signal to Noise Ratio: http://in.mathworks.com/help/vision/ref/psnr.
9. http://in.mathworks.com/help/matlab/ref/audioread.html#input_argument_samples.
10. Ali Al-Haj (2014): A dual transform audio watermarking algorithm, Multimedia Tools and Applications: Volume 73, Issue 3, pp 1897–1912.
11. Discrete Wavelet Transform: http://en.wikipedia.org/wiki/Discrete_wavelet_transform.
12. Munesh Chandra and Shikha Pandey (2010): A DWT Domain Visible Watermarking Techniques for Digital Images, Proc. IEEE Intl. Conference on Electronics and Information Engineering (ICEIE).
13. Andrews H, Patterson C (1976) Singular value decomposition (SVD) image coding. IEEE T Commun 42(4):425–432.
14. Vongpraphip S, Ketcham M (2009) An intelligence audio watermarking based on DWT-SVD using ATS, Global Congress on Intelligent Systems, 150–154.
15. B.Y. Lei, I.Y. Soon, Zhen Li (2011) Blind and robust audio watermarking scheme based on SVD-DCT, Elsevier: Volume 91, Issue 8, pp 1973–1984.
16. Harsh Verma, Ramanpreet Kaur, Raman Kumar (2009): Random Sample Audio Watermarking Algorithm for Compressed Wave Files, International Journal of Computer Science and Network Security (IJCSNS), Volume.9, Issue.11.

Digital Audio Watermarking: A Survey

Ritu Jain, Munesh Chandra Trivedi and Shailesh Tiwari

Abstract Counterfeiting and piracy of the intellectual property such as patents, copyrights, etc. is known to be a serious problem worldwide though it is more intense in some regions owing to the usage. Presently, millions of digital audio data is being copied over networks resulting into the loss of revenue to music industries at a big scale. Right owners and creators have been looking forward to find out ways to inhibit this process that ultimately prompted the research in digital watermarking to enable copyright protection in order to prevent illegal acts of forgery and pirate distribution of data. In this paper, we have surveyed the various existing methodologies of digital audio watermarking for preserving the copyright laws and highlighting the related issues.

Keywords Digital audio watermarking · Signal-to-noise ratio · Robustness

1 Introduction

During the last decade, with the advancement in Internet technologies in the form of multimedia, widespread availability of broadband communication networks, reliable and low cost storage devices, it is quite possible to replicate, transmit, and store the audio files anywhere unethically. These conveniences have significantly increased the risks of copyright violation and have challenged the preservation of intellectual property rights (IPR) [1, 2].

R. Jain (✉) · M.C. Trivedi · S. Tiwari
ABES Engineering College, Ghaziabad, India
e-mail: rjritu9@gmail.com

M.C. Trivedi
e-mail: munesh.trivedi@gmail.com

S. Tiwari
e-mail: hodcs@abes.ac.in

© Springer Nature Singapore Pte Ltd. 2018   433
S.K. Bhatia et al. (eds.), *Advances in Computer and Computational Sciences*,
Advances in Intelligent Systems and Computing 554,
https://doi.org/10.1007/978-981-10-3773-3_42

Initially, conventional encryption algorithms were adopted for this purpose but these restricts the permit of access to only authorized users and once decrypted, there is no way to prohibit further illegal distribution of the data. Thus, *digital watermarking* emerged out to be a promising solution to this problem. It has wide applications in copyright protection, fingerprinting, tamper proofing, broadcast monitoring, etc. [3, 4]. Until now, significant research has been done with the image but not much growth has been observed in the fields of video and audio watermarking owing to the complex and sensitive human auditory system (HAS).

The remainder of this paper contains the content as follows, Sect. 2 describes the classification of digital watermarking, introduction to audio watermarking is given in Sect. 3. Section 4 contains the techniques of audio watermarking pertaining to various domain. Section 5 gives the detailed survey on the existing methodologies followed by conclusion in Sect. 6.

2 Classification of Digital Watermarking

The process of digital watermarking can be categorized into different domains on various basis [3].

2.1 Based on Working Domain

1. **Spatial Domain**: Watermark is embedded in actual pixel value of host signal.
2. **Transform Domain**: Here, the initial step is to transform the original signal from time domain into other domains using DCT, DWT, etc.

2.2 Based on Extraction

I. **Blind Detection**: It is also known as public watermarking. The host signal as well as the watermark is not required for the purpose of extraction.
II. **Non-Blind Detection**: The original (host) signal on which the embedding is performed is needed at the time of extraction. It is not in use practically since it requires twice storage capacities and bandwidth required for communication thereby increasing the intervening overheads.

2.3 Based on Human Perception

I. **Robust Watermarking**: Embedded watermark can sustain various attacks like cropping, compression, D/A conversion, etc. It is applicable in copyright protection.
II. **Fragile Watermarking**: The fragile watermark gets tampered quite easily if encountered to slight modifications. It is applicable in tamper proofing, authentication, etc.

2.4 Based on Data

I. **Image**: Watermark is to be embedded in image, known as image watermarking.
II. **Text**: Watermark information is to be embedded in text data like organization's confidential documents.
III. **Audio**: Watermark is embedded into audio files like mp3, .wav, .au files, etc.
IV. **Video**: Watermarking is applied on video files for preserving their integrity.

2.5 Based on Key

I. **Symmetric Key**: As the name indicates, similar key is used at either sites; sender's and receiver's site. It is also known as public key.
II. **Asymmetric Key**: In asymmetric key method, different key is used at both the ends.

3 Audio Watermarking

Audio Watermarking is a practical methodology in which we embed the secret, robust, and imperceptible information into the audio signal to preserve the *integrity* of the original audio data [1, 2]. The embedded information may be an image, some other audio clip, or a randomly generated pseudo sequence. The watermark should distinctively identify the owner; and embedded in such a way that is imperceptible to the third-party. Embedding must tolerate the presence of common signal processing and compression techniques and various other deliberate attacks like cropping, requantization, sampling, etc. [4–6]. A good audio watermarking algorithm possesses the following special characteristics: robustness, imperceptibility, high data payload, security, verification, and reliability [1, 2, 4, 7].

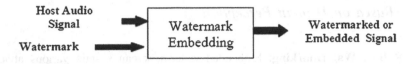

Fig. 1 Watermark embedding procedure [2]

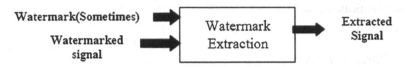

Fig. 2 Watermark extraction procedure [2]

Audio Watermarking algorithm can be divided into two main processes:

- Watermark Embedding
- Watermark Extraction

The *Embedding* procedure consists of original data and the watermark as inputs to the embedding block under which the embedding is performed [2] (Figs. 1 and 2).

The *Extraction* procedure consists of watermarked data and secret key as inputs and sometimes the original data for extracting the original signal. *Blind Detection* is often preferred for copyright verification owing to its practical feasibility [2, 3, 7].

4 Techniques of Audio Watermarking

The broad classification of audio watermarking techniques can be made as discussed below—

4.1 Time Domain Methods

Time domain watermarking or *spatial domain* watermarking is the oldest known watermarking techniques in which we add a watermark data or a pseudorandom sequence directly into the host signal; the processing overhead is much less than other methods. It takes much less time in the extraction process [8]. As compared to transform domain methods, it lacks in terms of robustness against various attacks.

Least Significant Bit (LSB) Modification: In [9], LSB technique for audio watermarking is suggested, where LSB is used to modify the original audio sample's amplitude producing an unnoticed perceptual difference. The watermarked signal y(i) is produced as follows:

$$y(i) = x(i) + f((x(i), w(i)), \tag{1}$$

where *x(i)* is the host signal and *w(i)* is the watermark signal in the range $[-\alpha, +\alpha]$ where α being a constant.

Spread Spectrum (SS) Method: SS is a type of correlation method where detection of watermark is done by calculating correlation between the embedded pseudorandom sequence and watermarked audio signal [10]. It, includes overheads like watermark shaping, sinusoidal modulation, etc. [10–13].

Perceptual Masking: In this technique, perceptually shaped pseudorandom sequence is embedded into the different frames of the host audio signal [13]. The time and frequency distribution of watermark is controlled due to which the amplitude of the watermark signal varies in accordance with the host audio signal.

Echo Hiding Method: This method embeds data into the host audio signal by introducing an *echo* which is not perceived by human ear such that

$$x(n) = s(n) + \alpha s(n - d), \tag{2}$$

where 'd' is the *echo delay* [14] and is detected by cepstrum or autocepstrum correlation methods at the time of extraction.

4.2 Transform Domain Methods

Transform domain methods are practically more adopted, as they are robust against the popular attacks like filtering operations, compression, and noise attacks. These methods derive benefit of HAS to embed a watermark into those components that are perceptually significant and results into high robustness and imperceptibility of noise in the signal [8]. The techniques are as follows:

Phase Modulation (PM) Method: PM is an improvement over the phase coding techniques as it allows higher data payload and individually alters each coefficient after modulation [14, 15].

Replica Modulation: Frequency shifts, phase shift and amplitude shift schemes come under Replica Modulation. The frequency shift method does transformation of signal into frequency domain and modulates the signal of a few ranges before inverse transformation into the time domain [16]. Other transform domain methods include discrete cosine transform (DCT), discrete wavelet transform (DWT), discrete fourier transform (DFT), and modified patchwork algorithm (MPA), etc. [7, 17–19].

5 Literature Survey

This section represents the review of various audio watermarking techniques proposed by various authors emphasizing on the techniques involved and their merits and demerits.

Bassia et al. [9] proposed the robust audio watermarking technique in the time domain using blind watermark detection.

- It directly modifies the amplitude values such that it does not create any perceptual difference.
- The watermarking key used for embedding was known only to copyright owner.
- Robustness could be increased with higher amplitude values but introduced distortion in the signal.
- This method was statistically imperceptible and was able to resist attacks like MPEG compression, rescaling, filtering, and quantization.

Kirovski and Malvar [10] proposed a SS technique in which each watermark bit is spread over a MCLT (Modulated Complex Lapped Transform) coefficients. SS techniques can be detected using correlation mechanisms.

- The MCLT coefficients within the range 2–7 kHz subband are considered for use in the detection process in order to minimize the carrier noise effects, sensitivity to down sampling, and cropping.
- It suggests a non linear cepstrum pre-processing to reduce the carrier noise.
- Enables effective watermarking with prevention against detection desynchronization, cepstrum filtering, and chess watermarks.

Huang et al. [20] proposed a blind, audio information bit-hiding algorithm with effective synchronization.

- The algorithm embeds synchronization signals in the time domain so that the computation for resynchronization can be kept lower.
- HAS features are exploited so that the watermark is placed in blockwise DCT coefficients of the original audio.
- Error-correcting codes are applied to lower the bit error rate.
- The hidden, imperceptible watermark is robust against attacks due to additive noise, MP3 coding, and cropping.

Wei Li et al. [21] proposed a novel localized and content-dependent audio watermarking algorithm.

- Emphasizes on determining the embedding regions with a steady-state high-energy using different strategies.
- Such regions have the potential to maintain audio quality without introducing much distortion.
- The limitation that is observed that it works well for modern music but not on jazz, classical music due to absence of obvious peaks on original waveforms.

Shijun Xiang and Jiwu Huang [22] proposed a histogram-based audio watermarking against time scale modification and cropping attacks.

- The paper proposed a multi-bit robust audio watermarking solution against the most challenging attacks like TSM, cropping attacks.
- Studied the insensitivity on audio histogram shape and the modified mean to TSM and cropping operations.
- Embedding process is done such that the histogram is extracted from a selected amplitude range where the watermark is resistant to amplitude scaling.
- It avoids exhaustive search in the extraction process.

Ali-Al Haj [23] proposed a technique of audio watermarking using DWT and Singular value decomposition (SVD) in the transform domain.

- When DWT is applied on host signal using Daubechies wavelet (dB1), Approximate coefficients (A) are obtained through low pass filtering while detailed coefficients (D) are produced by high pass filtering.
- SVD is a numerical technique based on which the matrices are diagonalized allowing imperceptibility and robustness to audio watermarking algorithms.
- Different levels of decomposition can be used for DWT but the author suggested 3-level decomposition.

Zamani and A. Manaf [24] proposed a new optimization technique in audio watermarking using genetic algorithms.

- Fragile audio watermarking emphasizing on genetic concept to reduce the distortion taking place because of LSB substitution.
- The PSNR is significantly improved using generic substitution-based audio watermarking (GSBAW).

6 Conclusion

In this paper, we have done an extensive literature survey of different existing digital audio watermarking techniques involving different methodologies. Each technique has its own pros and cons but the key pillars on which these algorithms stand are imperceptibility, robustness and data payload that share a trade off among them [3, 6]. If the payload is high, imperceptibility is affected and if payload is kept low then the robustness of the algorithm is affected. Thus, in future works there is a need to select such an algorithm which must be application specific and may be modulated according to the requirements beforehand. The summary of various audio watermarking techniques in given in Table 1.

Table 1 Taxonomy of different audio watermarking techniques

Authors	Technique of watermark embedding	Observations	Pros and cons	Comparison with previous technologies
Bassia et al.	Amplitude modification in time domain	Statistically Imperceptible Watermark	• Signal Distortions • Easy due to its presence in time domain	• Resist many attacks like MPEG2 compression, resampling, etc. • Not much robustifying against sophisticated attacks
Kirovski and Malavar	Spread spectrum (SS) and correlation mechanism	MCLT coefficients enabled to minimize the noise and cropping attacks	• Non linear cepstrum preprocessing required to reduce noise • Prevention against detection desynch	• Overcomes the major weakness of SS technique. • Reliably detects watermark of audio clips with degradation beyond acceptable limits
Huang et al.	Bit hiding algorithm with effective synchronization	Embedding is performed block wise in DCT coefficients	• ECC lowers bit error rate • Considerably robust against attacks like MP3 coding, cropping, etc.	• It effectively exploits the well-known cryptographic method: ECC for lowering the computation for extraction
Wei Li et al.	Content dependent algorithm	Few regions of interest are chosen with different strategies	• Much less distortion in the signal • Effective only for few kinds of audio type like jazz, etc.	• Embedding is done in translation-invariant FFT domain to resist small distortions • Stands out as it is content dependent and based on selecting the embedding regions accurately
Xiang and Huang	Histogram-based audio watermarking	Analyses the insensitivity on audio histogram shape	• Avoids exhaustive search during extraction as selected amplitude range is preferred • Effective against popular attacks like TSM and cropping	• Histogram shape and modified mean value provides improved results of resistance against significant attacks • Future scope focus on securing the watermarking scheme

(continued)

Table 1 (continued)

Authors	Technique of watermark embedding	Observations	Pros and cons	Comparison with previous technologies
Ali Al Haj	DWT-SVD-based transform domain algorithm	3-level decomposition using dB1 wavelets performed	• Data payload increases with clip length • IFPI requirements met	• Using mathematical tools, effectively improved the robustness of the algorithm • shortcomings include high data payload being a non-blind algorithm
Zamani and Manaf	Genetic optimization and GSBAW	Aimed to improvise the loopholes in other techniques	• genetic optimization figures out the significant regions efficiently • Requires skill in coding as genetic concept requires sampling to be done wisely	• Payload permissible through GSBAW was much higher while maintaining the imperceptibility • Listening tests successfully identified the noise imperceptibility threshold at 4 bps

References

1. R. F.Olanrewaju, Othman Khalifa: Digital Audio Watermarking; Techniques and Applications: International Conference on Computer and Communication Engineering (ICCCE 2012), pp 3–5 J (2012).
2. Sanjay Pratap Singh Chauhan, S. A. M. Rizvi: A survey: Digital Audio Watermarking Techniques and Applications: 4th International Conference on Computer and Communication Technology (ICCCT) (2014).
3. Siddharth Gupta, Vogesh Porwal: Recent Digital Watermarking Approaches, Protecting Multimedia Data Ownership: ACSIJ Advances in Computer Science: an International Journal, Vol. 4, Issue 2, No.14, ISSN: 2322-5157 (2015).
4. Digital Watermarking: https://en.wikipedia.org/wiki/Digital_watermarking.
5. Fabien, A. P. Petitcolas, Ross J. Anderson and Markus G. Kuhn: Information Hiding: A Survey: Proceedings of the IEEE, special issue on protection of multimedia content, 87(7): 1062{1078 (1999).
6. W. Bender, D. Gruhl, N. Morimoto, A. Lu: Techniques for data hiding: IBM Syst. J. 35 (3–4) pp. 313–336 (1996).
7. M. Arnold: Audio watermarking: Features, applications and algorithms: Proceeding of the IEEE International Conference on Multimedia and Expo, pp. 1013–1016 (2000).
8. J. S. Pan, H. C. Huang, L. C. Jain,: Intelligent Watermarking Techniques: World Scientific Pub Co Inc. (2004).
9. P. Bassia, I. Pitas, and N. Nikolaidis: Robust audio watermarking in the time domain: IEEE Trans. Multimedia, vol. 3, no. 2, pp. 232–241 (2001).
10. D. Kirovski, & H. Malvar: Robust Spread Spectrum Audio Watermarking: Proceeding ICASSP '01 Proceedings of the Acoustics, Speech, and Signal Processing, on IEEE International Conference Volume 03, pp. 1345–1348 (2001).
11. D. Kirovski, & H. Malvar: Spread Spectrum watermarking of Audio Signals: IEEE Transactions on Signal Processing; vol. 51, no.4, pp. 1020–1033 (2003).
12. I.J. Cox, J. Kilian, F.T. Leighton, T. Shamoon: Secure spread spectrum watermarking for multimedia: IEEE Trans. Image Process. 6 (12) pp 1673–1687 (1997).
13. M.D. Swanson, Bin Zhu, A.H. Tewfik, Lawrence Boney: Robust Audio Watermarking using perceptual masking: Elsevier Signal Processing 66, pp. 337–355 (1998).
14. Vivekananda Bhat K, Indranil Sengupta, Abhijit Das: An adaptive audio watermarking based on the singular value decomposition in the wavelet domain: Digital Signal Processing, Vol. 20, Issue 6, pp. 1547–1558 (2010).
15. X. Dong, M. F. Bocko, Z. Ignjatovic: Data hiding via phase manipulation of audio signals: IEEE Inter. Conf. on. Acoustics, Speech, and Signal Processing, Proceedings. (ICASSP '04). vol. 5, no., pp. V-377-80 (2004).
16. R. Petrovic, Audio Signal Watermarking Based on replica modulation, International Conference on Telecommunications in Modern Satellite, Cable and Broadcasting Service, Vol. 1, pp. 227–234 (2001).
17. Ali Al-Haj, Ahmad Mohammad: Digital Audio Watermarking Based on the Discrete Wavelets Transform and Singular Value Decomposition: European Journal of Scientific Research ISSN 1450–216X., vol. 39, no. 1, pp. 6–21 (2010).
18. IK. Yeo, & HJ. Kim: Modified patchwork algorithm: A novel audio watermarking scheme: IEEE Transactions on Speech and Audio Processing; vol. 11, no. 4, pp. 381–6 (2003).
19. X. Wang, & H. Zhao: A novel synchronization invariant audio watermarking scheme based on DWT and DCT: IEEE Transactions on Signal Processing; vol. 54, no. 12, pp. 4835–40 (2006).
20. Jiwu Huang, Yong Wang, Yun Q. Shi: A Blind Audio Watermarking Algorithm With Self-Synchronization: IEEE International Symposium on Circuits and Systems, ISCAS, Vol. 3, pp. 627–630 (2002).

21. Wei Li, Xiangyang Xue, and Peizhong Lu: Localized Audio Watermarking Technique Robust Against Time-Scale Modification: IEEE TRANSACTIONS ON MULTIMEDIA, VOL. 8, NO. 1 (2006).
22. Shijun Xiang and Jiwu Huang: Histogram-Based Audio Watermarking Against Time-Scale Modification and Cropping Attacks: IEEE TRANSACTIONS ON MULTIMEDIA, VOL. 9, NO. 7 (2007).
23. Ali Al-Haj: A dual transform audio watermarking algorithm, Multimedia Tools and Applications: Volume 73, Issue 3, pp 1897–1912 (2014).
24. Mazdak Zamani, Azizah Bt Abdul Manaf: Genetic Algorithm for fragile audio watermarking: Telecommunication Systems, vol. 59, Issue 3, pp. 291–304 (2015).

Brain CT and MR Image Fusion Framework Based on Stationary Wavelet Transform

Sharma DileepKumar Ramlal, Jainy Sachdeva, Chirag Kamal Ahuja
and Niranjan Khandelwal

Abstract In this paper, a technique is proposed for merging of complementary information of two images to obtain a fused image for more suitable human visual perception and for clinical diagnosis. The stationary wavelet transform (SWT) is one of the methods which has proved to be valuable and efficient in image fusion. In this paper, Entropy of square of low frequency subband coefficients is applied. Further, weighted sum modified Laplacian for high frequency subband coefficients is implemented. These two fusion rules are used as activity measure to fuse Computed Tomography (CT) and Magnetic Resonance (MR) images. The proposed method is compared with existing fusion method of averaging low frequency subband and maximum coefficient selection using SWT. The visual and quantitative analysis is done using cross correlation and root mean square error. Both the analysis showed that the proposed method is superior in fusing CT images with T2-weighted MR images.

Keywords Stationary wavelet transform (SWT) · Fusion · Weighted sum modified Laplacian · Entropy · Activity measure

S.D. Ramlal (✉) · J. Sachdeva
Electrical and Instrumentation Engineering Department,
Thapar Institute of Engineering and Technology University, Patiala, India
e-mail: dileepsharma28@yahoo.com

J. Sachdeva
e-mail: jainysachdeva@gmail.com

C.K. Ahuja · N. Khandelwal
Department of Radio Diagnosis, Postgraduate Institute of Medical Education
and Research, Chandigarh, India
e-mail: chiragkahuja@rediffmail.com

N. Khandelwal
e-mail: khandelwaln@hotmail.com

© Springer Nature Singapore Pte Ltd. 2018   445
S.K. Bhatia et al. (eds.), *Advances in Computer and Computational Sciences*,
Advances in Intelligent Systems and Computing 554,
https://doi.org/10.1007/978-981-10-3773-3_43

1 Introduction

Medical imaging implies to a number of diverse technologies such as Computer Tomography (CT) or Magnetic Resonance Imaging (MRI) that are used to examine the human body in order to identify, supervise or heal medical conditions such as a possible disease or injury [1–3]. X-ray and CT images clearly depict bony structures allowing assessment of damaged bones [1, 4]. MRI images better visualize soft tissues like nerves, cerebrospinal fluid, white matter, gray matter, etc. [1, 3]. The images acquired are combined to have more detailed information of medical conditions generally termed as fusion. For example, a CT and MRI image is fused (combined) which visualizes both bones and soft tissues in a single combined image. Fusion process can be categorized into pixel level, feature level, and decision level [5–7]. Pixel level fusion is done by combining pixels of two images directly. Feature level fusion is done by extracting and combining representative features of source images, such as edges, textures, etc. [5, 8]. Decision level fusion is done by fusing decisions of several expert systems together to form fused image [3, 9]. As the fusion process is completed its performance is evaluated using visual analysis and quantitative parameters, such as entropy, correlation coefficient, root mean square error (RMSE), etc. [9].

Many researchers such as Ellmauthaler et al. [10], S. Li and B. Yang [7], Y. Yang [11], etc. have been doing research in the fusion area of brain MR and CT images using wavelet transforms, such as undecimated, stationary wavelet, translation invariant wavelet, etc. Few of the researches are discussed below.

Ellmauthaler et al. [10] used undecimated wavelet transform. They used choose maximum (CM), choose maximum with intra-scale grouping (CM-IS) with and without window related activity. Mutual information, $Q^{AB/F}$ and Q^P parameters were used for comparison with NSCT, dual tree complex wavelet transform and undecimated wavelet transform without spectral factorization.

F. Xiao [6] worked on translation invariant wavelet transform to fuse multifocus and remote sensing images and used weighted combination of coefficients as fusion rule where weights were computed using local weighted energy with consistency verification. Experimental results were compared with DWT and SWT using RMSE, mutual information, and information Entropy as parameters. Visual and quantitative analysis showed superiority of their methods.

S. Li and B. Yang [7] used stationary wavelets with NSCT for hybrid fusion of CT and MR image. RMSE and image fusion quality matrices were used to compare the results with stationary wavelet transform, complex wavelet transform (CWT), Curvelets, WBCT (contourlet transform based on wavelets).

Y. Yang [11] used edge-based scheme and variance based scheme for fusion. The entropy, overall cross entropy and gradient (avg.) were used as parameters for comparison with pixel averaging method.

Zheng et al. [12] fused multisource images using support value transform. They used support value using support value transform to denote important features. In support vector machines (SVM) the pixels with large support values are more important for SVM model, this feature was used for their research. Multiscale support value filters were used to do support value analysis. Quality of visual information ($Q^{AB/F}$), conditional entropy (CE), and mutual information (MI) were used to compare their method with Laplacian pyramid, discrete wavelets, and undecimated "atrous" wavelets.

R. Li and Y. J. Zhang [13] performed experiment to find out the appropriate no. of decomposition levels for fusion of out of focus images. They used the difference in energy levels from higher bands for decomposition. The level selection was done by minimizing the mean square error (MSE) among fused image and input image. It was observed that MSE decreases with the rise in decomposition level. It reached minimum point at third level of decomposition. The MSE remained low with a rise in decomposition level. The actual best levels and computed best levels by the proposed method were found nearly equal maximum difference being equal to 1. All the results showed that the estimated levels provide better results.

The wavelet-based methodologies are used in previous researches. It is observed that the main limitations while fusing CT and MR images are low directional selectivity and inability to resolve long curves. Therefore, in this paper, the method using stationary wavelet transform which falls in pixel level fusion is proposed. Further, a fusion rule based on Entropy of square of low frequency subband coefficients and weighted sum modified Laplacian (WSML) for high frequency subband coefficients is implemented which distinguishes our approach from existing approaches. High entropy represents high information content and WSML produces sharp edges to signify HF information. This method provides improved directional selectivity and long curves.

2 Methodology of the Proposed Fusion Scheme

The stationary wavelet transform (SWT) is related to discrete wavelet transform with the difference that its coefficients are not decimated at each stage. At each decomposition level SWT coefficients contain the same number of samples as the input image. It finely represents angles and textures of the image and is shift invariant (approximately) [7]. In the proposed approach SWT is used to decompose the input images into approximation and detail coefficients as represented by Eqs. (1), (2), (3), and (4). Detail coefficients and approximation coefficients compute activity measures according to Eqs. (6), (7), (8), (10), and (11), respectively. Fused image reconstruction is done using Eq. (15). The block diagram of the fusion process is shown in Fig. 1. The detailed methodology is described in the following steps.

Step 1: Decompose source image X and Y using SWT into one low frequency (LF) subband and a sequence of high frequency (HF) subbands (horizontal, vertical, and diagonal components).

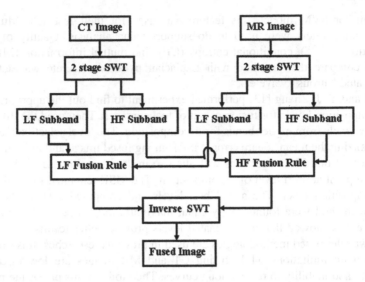

Fig. 1 Block diagram of the proposed fusion scheme

$$\left[L1F1_X^{SWT}, H1F1_X^{SWT}\right] = SWT(X) \tag{1}$$

and

$$\left[L1F1_Y^{SWT}, H1F1_Y^{SWT}\right] = SWT(Y) \tag{2}$$

Step 2: Decompose Low frequency subband further into one LF subband and a sequence of HF subbands, where HF subband represent Horizontal, vertical, and diagonal components.

$$\left[L2F2_Y^{SWT}, H2F2_Y^{SWT}\right] = SWT(L1F1_Y^{SWT}) \tag{3}$$

and

$$\left[L2F2_X^{SWT}, H2F2_X^{SWT}\right] = SWT(L1F1_X^{SWT}) \tag{4}$$

Step 3: Use an activity level measurement such as weighted sum modified Laplacian (WSML) of coefficients for high frequency subband fusion. A window size of 3×3 is used for neighborhood operation. WSML is evaluated using following steps:

Step 3.1: Modified Laplacian is

$$ML[f(m,n)] = |2f(m,n) - f(m-1,n) - f(m+1,n)| + |2f(m,n) - f(m,n-1) - f(m,n+1)| \tag{5}$$

Step 3.2: WSML of f(m, n) is computed using following expression:

$$WSML[f(m,n)] = \sum_{i=-1}^{i=1} \sum_{j=-1}^{j=1} w(i+1,j+1)ML[f(m+i,n+j)], \tag{6}$$

where w is city block distance.

Step 3.3: Compute WSML as activity measure at each high frequency subband coefficient of image A and B

$$a^A(m,n) = WSML[C^A(m,n)] \tag{7}$$

$$a^B(m,n) = WSML[C^B(m,n)] \tag{8}$$

where $C^A(m,n)$ and $C^B(m,n)$ are either horizontal, vertical, or diagonal coefficients of image A and B at location (m, n).

Step 4: Final fusion decision is achieved by choose maximum fusion rule.

$$d_i(m,n) = \left\{ \begin{array}{ll} 1 & \text{if } a^A(m,n) \geq a^B(m,n) \\ 0 & \text{if } a^A(m,n) < a^B(m,n) \end{array} \right\} \tag{9}$$

Step 5: Low frequency subband of SWT is fused using Entropy of square of the coefficients as activity measure. A window size of 3×3 is used to compute Entropy. It is given by following equation:

$$a_L^A(m,n) = \sum_{i=-1}^{i=1} \sum_{j=-1}^{j=1} C_L^{A^2}(m+i,n+j) \log\left(C_L^{A^2}(m+i,n+j)\right)/9 \tag{10}$$

$$a_L^B(m,n) = \sum_{i=-1}^{i=1} \sum_{j=-1}^{j=1} C_L^{B^2}(m+i,n+j) \log\left(C_L^{B^2}(m+i,n+j)\right)/9 \tag{11}$$

Step 6: Final fusion decision map is evaluated as

$$d_i(m,n) = \left\{ \begin{array}{ll} 1 & \text{if } a_L^A(m,n) \geq a_L^B(m,n) \\ 0 & \text{if } a_L^A(m,n) < a_L^B(m,n) \end{array} \right\} \tag{12}$$

Step 7: Fused low frequency coefficients are computed using following equation

$$LF_F^{SWT} = d_i * C_L^A + (\sim d_i) * C_L^B \tag{13}$$

Fused high frequency coefficients are computed using following equation

$$HF_F^{SWT} = d_i * C^A + (\sim d_i) * C^B \tag{14}$$

Step 8: Perform inverse SWT($^{SWT-1}$) on the fused LF and HF subband coefficients of SWT to get final fused image.

$$F = SWT^{-1}(LF_F^{SWT}, HF_F^{SWT}) \tag{15}$$

3 Experimental Setup

In the present study, 10 dataset pairs of CT and MRI are obtained from [14]. Other 15 dataset pairs of CT and MRI of various patients are taken from Department of Radiodiagnosis, Postgraduate Institute of Medical Education and Research (PGI-MER), Chandigarh, India.

The proposed method is coded in Matrix lab R2014a, is tested on various CT and MR images of size 256 × 256. The conducting tests are done with a Intel core i5 computer and 8 GB memory.

The fusion experiment is done on 25 dataset pairs of CT and MR images. The fusion results of five pairs of CT and MR images are shown in Fig. 2. The quantitative results are shown in Tables 1 and 2. The proposed method is compared with existing method of averaging scheme of fusion for LF subband coefficients [5, 6] and maximum selection scheme for fusing HF subband coefficients [5, 10]. This method is named as method 1. The proposed fusion scheme, i.e., entropy of square of low frequency subband coefficients and WSML for high frequency subband coefficients is implemented. This is named as method 2. Table 1 shows root mean square error (RMSE) values obtained by averaging and maximum selection-based rule (method 1) and Entropy of square of the low frequency subband coefficients and WSML scheme (method 2). Table 2 shows cross correlation (CC) values obtained by the two fusion methods. These methods are applied on 25 pairs of datasets.

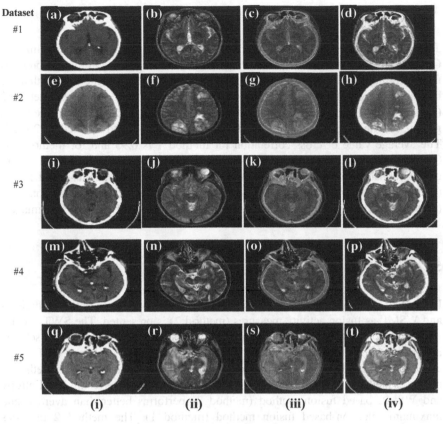

Fig. 2 (Column wise) Comparative analysis of fusion methods (*i*) CT images (*ii*) MR images (*iii*) Fused images by method 1 (*iv*) Fused images by method 2

Table 1 Root mean square error values obtained on two fusion schemes on dataset shown in Fig. 2

Dataset	Fusion method 1	Fusion method 2
#1	37	20
#2	40	19
#3	36	33
#4	42	38
#5	42	35

Table 2 Cross correlation values obtained on two fusion schemes on dataset shown in Fig. 2

Dataset	Fusion method 1	Fusion method 2
#1	0.93	0.98
#2	0.95	0.98
#3	0.94	0.95
#4	0.90	0.92
#5	0.94	0.95

4 Results and Discussions

The image results are shown in Fig. 2. Column 1 represents brain CT images. Column 2 represents brain MR images whereas column 3 and column 4 represents fusion results by method 1 and method 2. It is observed from the results that the both the bony structure and soft tissues are better seen in fused images by method 2 (column 4-image no.-d, h, l, p, t) as compared to method 1 (column 3-image no.-c, g, k, o, s). The average value of RMSE for method 1 is 40 and for method 2 is 28. The average value of cross correlation for method 1 is 0.93 and for method 2 is 0.97. From the average value of the two results it is noticed that Entropy of square of LF subband coefficients and WSML of HF subband coefficients-based fusion (method 2) rules are better than average and maximum selection-based fusion rules (method 1). This method provides good results for fusion of CT and MR images.

5 Conclusion

In this paper, a fusion scheme based on SWT and Entropy of square of the coefficients and WSML as fusion activity measures (method 2) is presented. The SWT divides input images into multiscale and multiple direction subimages. The proposed scheme is compared with existing method of averaging fusion for LF subband coefficients and maximum selection scheme for fusing HF subband coefficients. This method is named as method 1. It is evident from qualitative and quantitative results that Entropy and WSML based fusion method (method 2) performs better than average and maximum selection-based fusion method (method 1). The method 2 provides improved directional selectivity and long curves.

References

1. Ali F. E., El-Dokany I. M., Saad, A. El-Samie F. E.: Curvelet Fusion of MR and CT images. Progress in Electromagnetics Research. 3, 215–224 (2008)
2. Ganasala P. and V. Kumar V.: Multimodality medical image fusion based on new features in NSST domain. Biomedical Engineering Letter. 4, 414–424 (2014).
3. James A. P., Dasarathy B. V.: Medical image fusion, A survey of the state of the art. Information Fusion. 19, 4–19 (2014)
4. Ganasala P. and Kumar V.: Feature motivated simplified adaptive PCNN-based medical image fusion algorithm in NSST domain. Journal of Digital Imaging. 29, 73–85 (2015)
5. Ganasala P. and Kumar V. CT and MR image fusion scheme in nonsubsampled contourlet transform domain. Journal of Digital Imaging, 27, 407–418 (2014)
6. Xiao F.: A novel image fusion method based on translation invariant wavelet transforms. Journal of Software. 9, 859–866 (2014)
7. Li S. and Yang B., Hybrid multiresolution method for multisensor multimodal image fusion. IEEE Sensors. 10, 1519–1526 (2010)

8. Kumar M. and Dass S.: A total variation based algorithm for pixel level image fusion. IEEE Transaction on Image Processing. 18, 2137–2143 (2009)
9. Schild H. H.: MRI made easy. Schering, Made Easy, Canada (2007)
10. Ellmauthaler A., Pagliari C. L., E. Da-Silva A.B.: Multiscale image fusion using the undecimated wavelet transform with spectral factorization and non orthogonal filter banks. IEEE Transactions on Image Processing. 22, 1005–1017 (2013)
11. Yang Y.: Multiresolution image fusion based on wavelet transform by using a novel technique for selection coefficients. Journal of Multimedia. 6, 91–98 (2011)
12. Zheng H S., Shi W. Z., Liu J., Zhu G. X., Tian J. W.: Multisource image fusion method using support value transform. IEEE Transaction on Image Processing, 16, 1831–1839 (2007)
13. Li R. and Zhang Y. J.: Level selection for multiscale fusion of out of focus image. IEEE Signal Processing Letters. 12, 617–620 (2005)
14. Harvard University site, http://www.med.harvard.edu/AANLIB/home.html

A Feature-Based Semi-fragile Watermarking Algorithm for Digital Color Image Authentication Using Hybrid Transform

Hiral A. Patel and Nidhi H. Divecha

Abstract Digital watermarking plays vital role to preserve content of digital color image authentication. This paper proposed semi-fragile watermarking technique using feature extraction. The features of original image are extracted through canny edge detection and utilize it as watermark. Watermark is scrambled for providing more security. To achieve higher level of imperceptibility and robustness against attacks, amalgamation of DCT, DWT, and SVD are applied. Experimental results prove that, proposed technique is robust against group of attacks with adequate hiding capacity and imperceptibility as compare to existing systems.

Keywords Image authentication · DCT-DWT-SVD · Semi-fragile watermarking · Scrambling · Digital watermarking

1 Introduction

With a frequent use of data communication through internet, there is need to provide security to the digital assets. Authentication and integrity issues are increased day by day for digital assets especially digital images. Digital watermarking is the electronic security mechanism, which helps to knob these issues. Digital watermarking is one of the data hiding techniques, which hides the secret messages like logo of company, owner's detail, and features of image behind the original image or the cover image. These secret messages are considered as watermark. At the time of image verification, watermark helps to know the authentication and integrity of that image.

H.A. Patel (✉)
KSV, Gandhinagar, Gujarat, India
e-mail: hiral_shreya2003@yahoo.com

N.H. Divecha
SKMIPCS, Gandhinagar, Gujarat, India
e-mail: nidhidivechavalu@gmail.com

© Springer Nature Singapore Pte Ltd. 2018
S.K. Bhatia et al. (eds.), *Advances in Computer and Computational Sciences*,
Advances in Intelligent Systems and Computing 554,
https://doi.org/10.1007/978-981-10-3773-3_44

Watermarking technique is classified into fragile, semi-fragile, and robust. With robust watermarking technique, the extraction process is able to extract the watermark even if the malicious attacks are applied. With fragile watermarking technique, the watermark cannot survive for malicious as well as non-malicious modifications. With semi-fragile watermarking technique, the extraction process has the ability to differentiate malicious and non-malicious modifications [1]. Malicious attacks are those attacks which are intentionally manipulates to modify the image. Non-malicious attacks are those attacks which are not intentionally manipulated to destruct the image, but these are required during the transmission. Semi-fragile watermarking technique is robust with non malicious attacks where as fragile with malicious attacks [2].

Watermark can also be generated by using the features of original image. It is especially helpful for the authentication and integrity. The features of image are extracted from by using any of the image processing techniques. This watermark is embedded into cover image or original image and generates the watermarked image. This watermarked image travels through communication channel where, multiple attacks may be affected the watermark image. Whenever, the image needs verification, the hidden watermark is extracted from the watermarked image and it is compared with the original watermark. If both the watermarks are matched then the cover image is considered as authentic image [3].

The proposed semi-fragile watermarking system is based on the feature extraction technique. The features of cover image are extracted using canny edge detection method and it is used as watermark. The proposed system also scrambles the watermark by column shifting and row shifting techniques sequentially to increase the security. These techniques shift pixels of watermark image column wise and then row wise. After applying scrambling, watermark becomes difficult to interpret by unauthorized users. This technique provides more security as compare to existing systems. The rest of the sections discuss about the related work, proposed algorithm, experimental results, and conclusion.

2 Related Work

Number of researchers work under the authentication issues and some of them achieved their target where as others are trying to improve it. Some of the research works are listed below:

Chitra Arathi [4] had embedded the watermark using SVD method in grayscale image but the system was less imperceptible and not robust. Tanmay Bhattacharya et al. [5] had worked with medical images and embedded the watermark of size 64×64 using DWT in R, B, and G channels of image with size 180×180. They achieved the PSNR up to 34.5409. Buddhika Madduma et al. [6] had worked with feature-based watermark and generated watermark from original image using Sobel edge detection and Zernike Movement, embedded the watermark using DCT and DWT. This method was tolerated JPEG Compression up to 40% and Gaussian noise up to 0.3 variance. Khaled Loulhaoukha et al. [7] had given blind watermarking in which RGB color image was converted to YCbCr image and embedded watermark using DWT, DCT, and SVD transforms. D. Vaishnavi et al. [8] had embedded watermark in Blue channel using SVD transform also they scrambled watermark using Arnold transform. They had achieved robustness for JPEG Compression up to 75%, Gaussian 0.01 variance, and salt and pepper 0.002 density with 0.9809, 0.9932, and 0.9809 NCC values. Priti Sharma et al. [9] had used Red channel to embed the watermark. Their system embedded watermark using DWT and SVD transform and achieved imperceptibility up to 52.92. Jenil Geeorge et al. [10] had worked with semi-blind watermarking using secret key which was generated using chaotic map. They had selected Blue channel of color image, scrambled watermark of size 64×64 using Arnold transform and embedded watermark using DWT, SVD with secret key. Laur et al. [11] had embedded watermark of size 128×128 in all R, G, and B channels using DWT, SVD, CTZ, entropy, and DR decomposition methods. Robustness of method was tested against attacks like JPEG compression, Gaussian noise, salt and Pepper, and achieved results 0.8707, 0.9252, 0.9571, respectively. Shaila R. Hallur et al. [12] had embedded watermark of size 256×256 in one of the color channel of image using DWT, DCT, SVD, and got higher resistance value against different attacks. S.S. Sujatha et al. [13] had worked with grayscale image and had used DWT. They scrambled watermark using Arnold transform and had achieved PSNR 59.1168 and tolerated JPEG Compression up to 10% with Similarity ratio 0.8158. Woo Chaw Seng et al. [14] had embedded the watermark using DWT method in grayscale image, got 85.09 PSNR and tolerated JPEG Compression up to 85% and noise with density 0.03.

Challenges observed from the existing systems are:

- Semi-fragile watermarking needs robustness improvement with non-malicious attacks.
- With color image, robustness against attacks like JPEG compression, Gaussian noise, salt, and peppers needs improvement.
- There is a need to improve the capacity of watermark as well as imperceptibility of watermarked image.

Authors are trying to solve the above mentioned challenges.

3 Proposed Algorithm

The proposed algorithm is divided into four sections: Generating watermark and scrambling process, embedment process, extraction process, and reverse scrambling process. The general framework for proposed system is demonstrated in Fig. 1.

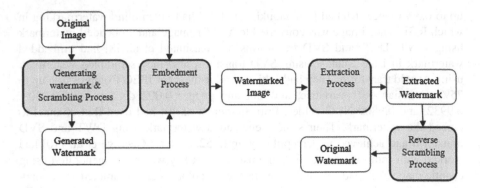

Fig. 1 Framework for proposed system

3.1 Generating Watermark and Scrambling Process

The features of original image are extracted from image and it is used as a watermark image. For providing more security, scrambling is applied to watermark. Here input is original Image and outputs are generated watermark and scrambled watermark.

1. Let IM is the original color image of M × N size. Select Blue Channel of color image (BIM).
2. Apply 1st level Haar Wavelet transform to decompose the BIM image into four sub bands LL1, LH1, HL1, and HH1 which are of M/2 × N/2 size.
3. Select LL1 sub-band of IM and apply the canny edge detection method to obtain M/2 × N/2 size logical image (G_WM).
4. Consider G_WM as watermark image which is of M/2 × N/2 size.
5. Again apply first level Haar wavelet transform on watermark image (G_WM) which decomposes the watermark in wLL1, wLH1, wHL1, and wHH1 which are of M/4 × M/4 size.
6. Apply scrambling to wLL1 using column shifting and row shifting. Consider this image as S_WM.
7. Use S_WM image for embedding which is of M/4 × M/4 size.

3.2 Embedment Process

The embedment process is described below where inputs are original and scrambled watermark Images where as output is Watermarked Image.

1. Let IM is the original color image of M × N size. Select Blue Channel of color image (BIM).

2. Apply first level Haar DWT on BIM which decompose it into LL1, LH1, HL1, and HH1 subbands of size M/2 × N/2
3. Apply second level DWT on LL1 which decompose it and get LL2, LH2, HL2, and HH2 of size M/4 × N/4.
4. Select the LL2 subband of the image and apply DCT on it. It gives DCT matrix B of M/4 × N/4 size.
5. Apply SVD on DCT matrix B to get U, S, and V.
6. To embed the scrambled watermark S_WM, first apply SVD on it to get U_W, S_W, and V_W.
7. Modify the singular values S with S_W using scaling factor (α) and following formula: S_NEW = S + α * S_W
8. Apply inverse SVD using iB = U * S_NEW * V'.
9. Apply inverse DCT on iB to produce iLL2.
10. Apply inverse DWT to iLL2, LH2, HL2 and HH2 to get iLL1.
11. Apply inverse DWT to iLL1, LH1, HL1 and HH1 to get watermarked image (B_WMD).
12. Substitute B_WMD as Blue color channel and get the color watermarked image (WMD)

3.3 Extraction Process

Extraction process is based on non-blind watermarking technique which is used to extract the scrambled watermark and steps are described here where inputs are watermarked and original images, where as output is Extracted scrambled watermark.

1. Select the Blue channel of WMD color image. Let it be B_WMD.
2. Apply first level DWT to B_WMD to get LL1, LH1, HL1 and HH1 sub bands.
3. Apply second level DWT to LL1 to get LL2, LH2, HL2 and HH2 sub bands.
4. Calculate DCT of LL2 subband to get DCT matrix A.
5. Apply SVD on A to acquire WU, WS and WV.
6. Obtain the singular values of watermark using formula SR = (WS − S)/α.
7. Apply inverse SVD using formula S_WM_E = U_W * SR * V_W' to extract the embedded scrambled watermark.

3.4 Reverse Scrambling Process

Scrambling which was applied earlier, needs to be converted back to the original watermark. This process converts this extracted scrambled watermark into the original watermark. Here input is extracted scrambled watermark and output is original watermark.

1. Select S_WM_E and again apply scrambling on it in reverse order. It means first apply reverse row shifting and then apply reverse column shifting. Let it be iwLL1.
2. Now apply inverse DWT on iwLL1, wLH1, wHL1 and wHH1 to get the original watermark, i.e., WM_E.

4 Experimental Results

The proposed algorithm is implemented in MATLAB 7.10 (2010a). The original image is of 512×512 size where as the watermark is of 256×256 size. The algorithm is tested with color images which are shown in Fig. 2. Also the generated watermarks are displayed in Fig. 2.

The original image, features of cover image as watermark, scrambled watermark, watermarked image, extracted watermark, and converted original watermark, all these images are displayed in Fig. 3, respectively.

The imperceptibility is measured using PSNR values and the robustness is measured using NCC values. PSNR value is calculated based on Eqs. 1 and 2. NCC value is calculated based on Eq. 3.

$$PSNR = 10 \times lg\left(\frac{255^2}{MSE}\right) \tag{1}$$

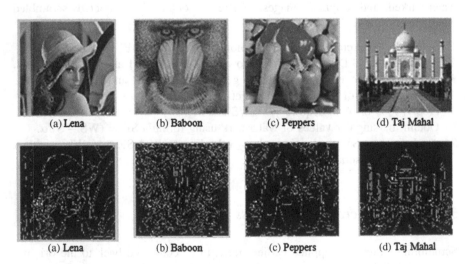

(a) Lena   (b) Baboon   (c) Peppers   (d) Taj Mahal

(a) Lena   (b) Baboon   (c) Peppers   (d) Taj Mahal

Fig. 2 Color image database and their features as watermark

Original Image   Generated Watermark   Scrambled Watermark   Watermarked Image   Extracted Watermark   Reverse Scrambling

Fig. 3 Output of proposed system without applying attacks

Table 1 PSNR and NCC values for database images

Images	PSNR	NCC
Lena	93.2021	1
Baboon	90.5951	1
Peppers	93.4931	1
Taj Mahal	93.0783	1

$$MSE = \frac{1}{M \times N} \sum_{i=1}^{N} \sum_{j=1}^{M} \left[I(i,j) - I'(i,j) \right]^2 \tag{2}$$

$$NC = \left(\sum_{i=0}^{M} \sum_{j=0}^{N} OW * EW \right) \bigg/ \left(\sum_{i=0}^{M} \sum_{j=0}^{N} OW * OW \right) \tag{3}$$

Here I and I' are the original image and watermarked image pixel values. OW and EW are the original watermark and the extracted watermark.

PSNR and NCC values with blue color channel without attacks are displayed in Table 1.

To check the feasibility of proposed system, different attacks are applied on it and robustness is measured. The Table 2 displays the NCC values with different attacks.

Comparison of proposed system and existing systems given in Table 3 is for Lena image. Vaishnavi et al. [8] proposed two methods; out of them, first method is taken for comparison with proposed system.

Imperceptibility comparison of proposed system with existing system is demonstrated through the chart shown in Fig. 4.

From the comparison, it is concluded that the proposed system is more suitable for content-based watermarking as it provides more robustness and imperceptibility.

Some related outputs of attacked images and extracted watermarks are shown in Fig. 5.

Table 2 Robustness test with different attacks

Attacks	Lena	Baboon	Pepper	Taj Mahal
JPEG Comp. (90)	1	1	1	1
JPEG Comp. (50)	1	1	1	0.9995
JPEG Comp. (10)	1	1	1	0.9991
Gaussian 0.001	1	1	1	1
Gaussian 0.009	1	1	1	0.9813
Gaussian 0.01	1	1	1	0.9670
Gaussian 0.03	0.9934	0.9844	0.9785	0.5038
Low pass filter	0.9945	0.9931	0.9990	0.9666
Rotation and crop	0.8241	0.6930	0.6303	0.8153
Salt and pepper 0.01	1	1	1	1
Salt and pepper 0.03	1	1	1	0.8612
Salt and pepper 0.04	0.9967	0.9996	0.9992	0.7389
Blur (1%)	1	1	1	1
Blur (2%)	0.9043	0.9247	0.9699	0.7982
Sharpen	0.9987	0.9986	1	0.9903
Motion blur (1,180)	1	1	1	1
Motion blur (3,180)	1	1	1	1
Motion blur (4,180)	0.9868	0.9972	0.9994	0.9801

Table 3 PSNR and NCC comparison with attacks

	Hallur's system (2015) [12]		Vaishnavi's system (2015) [8]		Proposed system	
Attacks	PSNR	NCC	PSNR	NCC	PSNR	NCC
Without attack	57.4913	1	28.3390	–	93.2021	1
JPEG Comp. (75%)	–	–	41.3565	0.9809	86.5679	1
JPEG Comp. (50%)	50.5384	0.9999	–	–	84.5944	1
Gaussian filter (0.01)	57.0059	1	25.2505	0.9932	72.8946	1
Salt and pepper (0.002)	50.4249	0.9990	43.5849	0.9809	84.9049	1
Sharpen image	46.5720	1	–	–	75.6461	0.9987
Motion blur (10, 20)	–	–	36.4774	0.9978	79.0741	0.4101

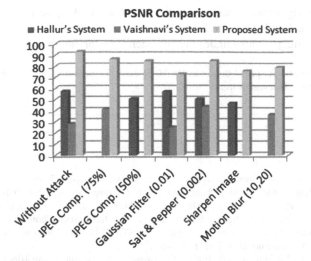

Fig. 4 PSNR comparison with existing systems

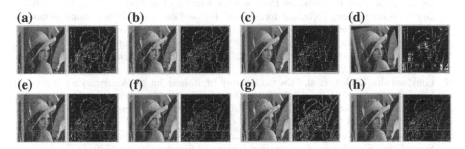

Fig. 5 Outputs with different attacks: **a** JPEG 90% compressed, **b** Gaussian filter (0.01 variance), **c** Low pass filter, **d** Rotation and crop, **e** Salt and pepper (0.03 density), **f** Blur (1%), **g** Sharpen image, **h** Motion blur (3,180)

5 Conclusion

Based on the above results, it is concluded that the proposed semi-fragile watermarking system is acceptable for image authentication. The amalgamation of DWT, DCT, and SVD is done in such a way that the system gives imperceptible result near to 93 dB as PSNR value and robustness result with 1 as NCC value. The given system is more robust with JPEG Compression up to 90% compressed, Gaussian Noise up to 0.03 variance, salt, and pepper up to 0.03 density, sharpen image and motion blur. These are considered as different non-malicious attacks.

The given algorithm is based on non-blind watermarking technique, which can be improved as blind system in future. In the same manner, Tamper detection from watermarked image is also under the consideration for future work.

Acknowledgements A special thanks to Dr. N. N. Jani, for providing his valuable suggestions and guidance for this research work.

References

1. Saha Shilpi Debnath Bhattacharyya, and Samir Kumar Bandyopadhyay. "Security on fragile and semi-fragile watermarks authentication." *Int. J. Comp. Applicat* 3.4 (2010): 23–27.
2. Vartak Reshma, et al. "Survey of Digital Image Authentication Techniques." (2014).
3. Kommini Chaitanya, Kamalesh Ellanti, and E. Harshavardhan Chowdary. "Semi-Fragile Watermarking Scheme based on Feature in DWT Domain."*International Journal of Computer Applications* 28.3 (2011): 42–46.
4. Arathi Chitla. "A Semi Fragile Image Watermarking Technique Using Block Based SVD." *International Journal of Computer Science and Information Technologies* 3.2 (2012): 3644–3647.
5. Bhattacharya Tanmay, Sirshendu Hore, and SR Bhadra Chaudhuri. "A Semi-Fragile Blind Digital Watermarking Technique for Medical Image File Authentication using Stationary Wavelet Transformation." *International Journal of Computer Applications* 104.11 (2014).
6. Madduma Buddhika, and Sheela Ramanna. "Content-based image authentication framework with semi-fragile hybrid watermark scheme." *Man-Machine Interactions 2*. Springer Berlin Heidelberg, 2011. 239–247.
7. Loukhaoukha Khaled, et al. "On the Security of Robust Image Watermarking Algorithm based on Discrete Wavelet Transform, Discrete Cosine Transform and Singular Value Decomposition." *International Journal of Applied Mathematics and Information Sciences* (2015): 1159–1166.
8. Vaishnavi D. and T. S. Subashini. "Robust and Invisible Image Watermarking in RGB Color Space Using SVD." *Procedia Computer Science* 46 Elsevier (2015): 1770–1777.
9. Sharma Parmanand, and Trapti Jain. "Robust digital watermarking for coloured images using SVD and DWT technique." *Advance Computing Conference (IACC), 2014 IEEE International*. IEEE, 2014.
10. George Jeril, Satishkumar Varma, and Madhumita Chatterjee. "Color image watermarking using DWT-SVD and Arnold transform." *India Conference (INDICON), 2014 Annual IEEE*. IEEE, 2014.
11. Anbarjafari G., et al. "A Robust Color Image Watermarking Scheme Using Entropy and QR Decomposition." *Radioengineering* (2015).
12. Hallur Shaila R., and Suresh Kuri. "Robust Digital Watermarking using DWT-DCT-SVD Algorithms for Color Image." (2015).
13. Sathik M. M., and S. S. Sujatha. "Authentication of digital images by using a semi-fragile watermarking technique." *Int. J. Adv. Res. Computer.Sci. Softw. Eng* 2.11 (2012): 39–44.
14. Seng Woo Chaw, Saied Ali Hosseini, and Leong Lai Fong. "Semi-fragile watermarking for gesture authentication." *Open Systems (ICOS), 2011 IEEE Conference on*. IEEE, 2011.
15. Divecha Nidhi, and N. N. Jani. "Implementation and performance analysis of DCT-DWT-SVD based watermarking algorithms for color images." *Intelligent Systems and Signal Processing (ISSP), 2013 International Conference on*. IEEE, 2013.

Data Set

16. Lena Image: http://graphics.stanford.edu/~jowens/223b/examples.html.
17. Babbon Image: http://old.vision.ece.ucsb.edu/segmentation/jseg/jsegcolor.html.
18. Peppers Image: http://www.math.tau.ac.il/~turkel/images.html.
19. Taj Mahal Image: https://en.wikipedia.org/wiki/File:Taj_Mahal_in_March_2004.jpg.

Inventory Control Using Fuzzy-Aided Decision Support System

Mahuya Deb, Prabjot Kaur and Kandarpa Kumar Sarma

Abstract Virtually every segment of the economy relies on some form of inventory for their operations. The ubiquitous nature of inventory has motivated to carry out this study wherein a decision support system (DSS) could be formulated to assist the decision-makers for effective monitoring of inventory levels and to ensure continuous availability of goods. The DSS needs be designed in a manner which can communicate its information to its user that is comprehensible and useful within the context of the decision situation. However, while dealing with the parameters of the system it is often seen that they are uncertain, imprecise and vague. Fuzzy-based approaches are best suited for such situations but these cannot provide automated decision support unless combined with learning systems like artificial neural network (ANN). When ANN and fuzzy are combined, fuzzy neural system (FNS) and neuro-fuzzy system (NFS) are created. The model of DSS developed in this study is based on a new framework using a system called adaptive neuro-fuzzy inference system. The model established has the advantage of the ANFIS for the DSS for use as part of inventory control.

Keywords Inventory control · DSS framework · ANFIS · Decision-making

M. Deb (✉) · P. Kaur
Department of Mathematics, Birla Institute of Technology, Mesra, Ranchi,
Jharkhand, India
e-mail: mahuya8@gmail.com

P. Kaur
e-mail: tinderbox_fuzzy@yahoo.com

K.K. Sarma
Department of Electronics and Communication Technology,
Gauhati University, Guwahati, India
e-mail: kandarpaks@gmail.com

© Springer Nature Singapore Pte Ltd. 2018   467
S.K. Bhatia et al. (eds.), *Advances in Computer and Computational Sciences*,
Advances in Intelligent Systems and Computing 554,
https://doi.org/10.1007/978-981-10-3773-3_45

1   Introduction

The traditional inventory models are built considering the parameters as crisp. But in practical situations inventory control involves a range of uncertain situations which include demand and supply on one hand and control on the cost components on the other. The various elements associated with how much to order are normally concerned with inventory costs and inventory lot sizing models. The inventory cost includes the ordering cost, procurement cost, holding cost and shortage cost. The basic objective in inventory control model is therefore to minimize inventory cost by establishing a trade-off between inventory investment and customer satisfaction. This objective can be simplified with the use of a decision support system, first coined by P.G. W Keen, a British academician wherein he defined DSS as a computer system essential in decision-making wherein the computer along with its analytical tools can be of value but the manager's judgement is essential. Turban et al. [1] broadly define a DSS as "a computer based information system that combines models and data in an attempt to solve semi—structured and some unstructured problems with extensive user involvement." The construction of a DSS is essential in an inventory control due to the ubiquitous nature of inventory and therefore it demands the design of appropriate decision support systems (DSS) if inventory management is to be made automated and efficient. Decision support system which consists of components such as inputs, user knowledge and expertise acts as a viable support for effective decision-making with regard to inventory policy decisions [2]. However, the parameters that define the inventory control models are not always crisp they are often uncertain, ambiguous and vague. Further, uncertainties revolve around the fluctuating buyer seller interaction. Against this backdrop, fuzzy-based approaches formulated by Zadeh [3] are best suited to tackle with the imprecise and uncertain parameters. It is almost over 30 years since the first article of fuzzy appeared in inventory. Since it is decided to create a decision support system that allows for an easy and smooth interaction with the human operators, with a decision process that is both intuitive and can be explained in linguistic terms, a decision support system based on a fuzzy knowledge base is discussed here. In addition, a fuzzy inference system is applied wherein a connection between the input and output variables of the system is established. The objective here is to convert the crisp inputs into fuzzy inputs by using fuzzification interface. After which the rule bases are developed. Fuzzy logic toolbox of MATLAB is applied to the fuzzy inventory system (FIS) model to compute the cost coefficients in any time period. The fuzzy input variable is demand and the output variable is the cost component. FIS has wide application in manufacturing and therefore can be an ideal tool in inventory control. But FIS cannot alone provide automated decision unless combined with learning tools like artificial neural networks ANN, which consist of a number of interlinked cells as neurons with weights running coincidently to initiate artificial intelligence. Recently, learning systems like Artificial Neural Networks (ANN) have emerged as a suitable option because of the fact that these can capture the intricate linkages between input and output data without bothering at all about the model involved, retain the know-how and use it subsequently. ANNs attempts to simulate the structure of the human brain, their thinking capabilities and learning in a machine. This gives an advantage for the modelling of complex and nonlinear data. It is neuroscience which has inspired neural networks and not because they are considered good models of biological phenomenon. These networks are "neural" in the sense that they have been inspired by neuroscience but not necessarily because they are faithful models of biological cognitive phenomena. This ability however is quantitative but to have qualitative attributes, the system must be provided with mechanisms that could track finite variations. ANFIS unlike FIS creates sufficient input–output data using the benefit of learning capability of the neural networks. It has received wide recognition due to its application in areas of pattern recognition, system identification, etc. The advantage of this system is that it hardly needs expert opinion for modelling and training a system.

Various researchers time and again have incorporated fuzziness in inventory control models like Chang [4], Park [5], Kacpryzk and Staniew [6], Wee [7], Maiti [8] to name a few. Even Paul [9] used fuzzy logic for selecting optimal number of shifts using inventory information, reliability of machines and customer requirements. Zeng [10] formulated a web-based fuzzy decision support system for spare parts inventory control in a nuclear plant where he was successful in decreasing the inventory holding cot. Lo [11] formulated a paper wherein he designed a decision support system for an integrated inventory model wherein the objective was to minimize the joint total expected cost by taking parameters such as ordering cost, backorder discount, lead time, ordering quantity, safety factor, and the lot size. Fuzzy-based approaches are the best befitting tool in such circumstances but these cannot provide automated decision support unless combined with learning systems like ANN. In his paper, Paul and Azeem [12] determined the optimal level of inventory using an artificial neural network (ANN) which was supposed to be a function of product demand, material costs, setup and holding cost. In the study conducted by [13], MRP problem of lot sizing was solved by a neural network model. Then he compared the performance of the model taking different scenarios and compared the same to common heuristics. Thereafter, the results proved that artificial neural network model was far better to solve lot sizing problems with reasonable accuracy. Lin [14] in his study tried to find an optimal WIP value of wafer fabrication processes. The algorithm was developed by integrating an ANN and sequential quadratic programming (SQP) method. It proved to be quite effective to find the solution with this approach. When ANN and fuzzy are combined, fuzzy neural system (FNS) and neuro-fuzzy system (NFS) are created. These two hybrid blocks combine the advantages of both the two mechanisms hence acquire quantitative and qualitative abilities with learning which makes it an efficient DSS. Önüt et al. [15] in his paper forecasted the incomplete demand information by artificial intelligence approaches wherein he compared both artificial neural networks and adaptive network-based fuzzy inference system techniques. Aengchuanin [16] proposed the fuzzy inference system (FIS) model with artificial neural network (ANN) model and FIS with adaptive neuro-fuzzy inference system (ANFIS) model for an inventory system where both supply and demand are uncertain. Fuzzy rules were generated so as to obtain the fuzzy order quantity continuously. The order quantity was thereby adjusted with the evaluation algorithm to fit the FIS inventory model. This model thus gave the best performance of total inventory cost saving by more than 75% as compared to stochastic EOQ model. Sarwar et al. [17] in their paper had successfully implemented Adaptive Neuro-Fuzzy Inference System (ANFIS) through an algorithm which was useful; in predicting the long term demand of Natural Gas Consumption.

In this work, we report the design of an ANFIS-based DSS configured to work as DSS for inventory management. The system accepts demand as input and generates procurement, ordering and holding cost to control production and supply. How effective the system was checked in terms of number and types of membership (MF) used, accuracy generated and computational efficiency accounted by computation cycles required. The experiments are carried out for certain data acquired from a shopping complex distributed over 6 months. The results are also compared with that obtained from ANN formulated in two different forms. The organization of the paper is as follows: In Sect. 2, the formulated model of ANFIS-based DSS for inventory control finds a place. Section 3 discusses the experimental details and results. Section 4 exclusively mentions the conclusion of the paper.

2 Proposed Model of ANFIS-Based DSS for Inventory Control

The proposed system has a block diagram as shown in Fig. 1. It is formed by an ANFIS which has a number of MFs and its configuration is fixed after performing a range of experiments in terms of number and types of membership (MF) used, accuracy generated and computational efficiency accounted by computation cycles required. The process logic involved in linking demand which is the input and the outputs, i.e. procurement, ordering and holding costs are outlined in Fig. 1.

The mechanism involves learning of the variations in demand of certain goods and establishing their association with the related costs observed in the data set used for the purpose. The work has two phases. First, the learning takes place. Next, the system is tested. During the learning phase, the input–output mapping is acquired

Fig. 1 Process logic of the proposed decision support system for inventory control using fuzzy-based approach

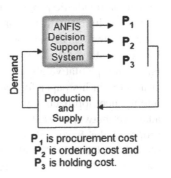

P_1 is procurement cost
P_2 is ordering cost and
P_3 is holding cost.

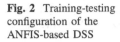

Fig. 2 Training-testing configuration of the ANFIS-based DSS

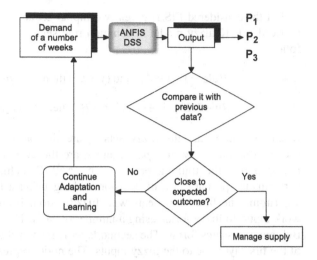

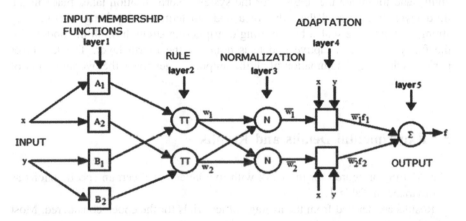

Fig. 3 Different layers of the ANFIS

by the ANFIS which it tracks adaptively and identifies the discrimination boundaries despite the data having finer variations. The training-testing cycle is shown in Fig. 2. The learning is sustained till the objectives are achieved. To ascertain the state of learning, two variables are used. First, number of training cycles or epochs the ANFIS takes to attain the convergence goal and next the accuracy generated in making the decision. These two aspects are linked with the numbers and types of MFs used with the ANFIS. This aspect is discussed in the next section. A major portion of the work is the ANFIS. It is depicted in Fig. 3.

ANFIS has been preferred as DSS in a range of applications. The fuzzy model involved can be configured either in terms of the Mamdani or Takagi, Sugeno, and Kang (TSK) representation. However, TSK model is computationally efficient than the Mamdani's model hence is popular in many applications. The selection of the model is based on the fuzzy inference system (FIS).

Let the considered FIS is given two inputs 'x' and 'y' and one output 'f'. For a first-order TSK fuzzy model, a common rule set with two fuzzy if-then rules is as follows:

$$\text{Rule 1: If } (x \text{ is } A_1) \text{ and } (y \text{ is } B_1) \text{ then } (f_1 = p_1 x + q_1 y + r_1) \tag{1}$$

$$\text{Rule 2: If } (x \text{ is } A_2) \text{ and } (y \text{ is } B_2) \text{ then } (f_2 = p_2 x + q_2 y + r_2), \tag{2}$$

where A_i and B_i are the fuzzy sets, f_i are the outputs within the fuzzy region specified by the fuzzy rule, p_i, q_i, and r_i are the design parameters that are determined during the training process. The ANFIS architecture having five layers performs the primary task of decision making using a fuzzified back propagation algorithm. The first layer deals with fuzzification. It adaptively converts the real work inputs to fuzzy forms using a number of MFs. The common MF types include Bell and Gaussians forms. The second layer is responsible for generating the output of the first layer due to the fuzzy inputs. The nodes represent the firing or triggering ability and formulate the rule set for the system. Normalization takes place in the third layer. It expressed the value of a node in terms of the firing or triggering strength of the node scaled by the firing or triggering strength of all the nodes. For the fuzzy responses adaptive update is done by the fourth layer. The last layer performs the task of summation of all the outputs. It generates the overall output of the network.

3   Experimental Details and Results

The ANFIS configured for the work with five layers have certain specifications as summarized in Table 1.

Results are derived from the average of ten trials for the epoch considered. Most reliable real-world experiences are provided by the Gaussian curve membership functions and as such it provides the most reliable real world to fuzzy conversion.

Table 1 ANFIS specifications

Training, testing and validation data	Data of a shopping complex for 6 months
Variations	50%
ANFIS type	ANFIS with five layers
ANFIS training method	Fuzzified back propagation gradient descent method
Training qualitative measure	Least squares distance measure
Average training epochs	10–200
Total no. of membership functions	10
Type of membership function	Generalized Gaussian and Bell MFs
RMSE with 9 GaussMFs at 200 epochs	1.97×10^{-3}

It is therefore used for the fuzzification in the ANFIS layer. Ten such MFs give optimum results, hence form the basis of the ANFIS for training and testing. The cost functional considered for training the ANFIS is the root mean squared error (RMSE). The system is trained using a data set for a commodity for several weeks. One such example is elaborated in Table 2.

If the actual demand is more than the quantity available, the system triggers a decision state for supply and production otherwise the inventory is maintained and no order is placed for restocking. The training considerations also involve certain profitability rating which is combined with the day to day inventory records. Table 3 outlines the formulation of data relating profitability rating with the demand available cycle. A demand–cost profile with profitability rating is formed to train the ANFIS. This is shown in Table 4. The states shown in Table 4 are codified and training executed for a number of weeks. Some of the testing results are shown in Table 5.

Table 2 Training data for a commodity for a week

Day	Quantity available	Actual demand
Sunday	100	112
Monday	50	48
Tuesday	60	60
Wednesday	40	38
Thursday	60	58
Friday	65	62
Saturday	80	83

Table 3 Profitability rating considered with the training data

State	Rating	Binary code
High profit	P3+	111
Medium profit	P2+	110
Low profit	P+	101
Profit	P	100
Loss	P–	11
Medium loss	P2–	10
Heavy loss	P3–	1

Table 4 Demand-cost profile with profitability rating for training ANFIS

Day	Profitability rating		
	P_1	P_2	P_3
Sunday	P3+	P+	P3–
Monday	P3+	P+	P3–
Tuesday	P2+	P2+	P+
Wednesday	P2+	P+	P2+
Thursday	P2+	P+	P2+
Friday	P2+	P+	P–
Saturday	P3+	P–	P–

Table 5 Calculation of time and average RMSE for 200 epochs for inventory

Input	Type of membership function	No. of membership functions (MFs)	Time required to reach 200 epochs by (in sec.)	Minimum RMSE attained by 200 epochs
Data of four weeks	Bell MF	3	1.95	2.72×10^{-3}
		6	2.1	2.15×10^{-3}
		9	2.2	1.8×10^{-3}
	Gauss MF	3	1.8	2.6×10^{-3}
		6	1.9	2.2×10^{-3}
		9	2.0	1.8×10^{-3}

Fig. 4 RMSE plot for 50 epochs with 10 Gaussian MFs

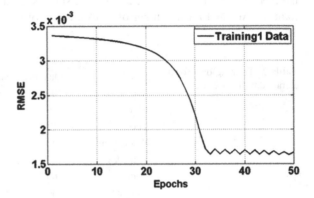

During execution, the training time recorded is confined to 1.8–2.2 s for the sample sets. Ten trials are performed for the sample set and average results considered. The RMSE is calculated using the expression:

$$\text{RMSE} = \sqrt{\frac{1}{N} \sum_{t=1}^{N} (A_t - F_t)^2}, \tag{3}$$

where A_t and F_t are the observed and fitted values respectively and N is the number of training or testing sample.

The error curve for a 50 epochs training with 10 numbers of Gaussian MFs is shown ion Fig. 4. The RMSE convergence is fast which ensures an efficient decision-making. For a four week data set the readings have been found to be consistent. The above results are also compared with that of an ANN formulated in feed forward form with two configurations of three hidden layers (15, 25 and 30 and 5, 12, 16 numbers of log-sigmoidal, tan-sigmoidal and log-sigmoidal activation functions). The performance is measured in terms of epochs of training and accuracy achieved while making decision. Summary of the results are depicted in Table 6.

Table 6 Comparison of performance of ANFIS with ANN

Slno.	Minimum Epochs by ANN 1 (hidden layers of 15, 25 and 30 neurons)	Minimum Epochs by ANN 2 (hidden layers of 5, 12, 16 neurons)	Epochs by ANFIS	% Success rate with ANN 1	% Success rate with ANN 2	% Success rate with ANFIS
1	15	18	9	87	87	89
2	30	31	19	89	88	91
3	45	43	31	92	91	92
4	60	56	36	93	92	95
5	74	71	48	93	93	95

4 Conclusion

Here, we have discussed the design of an ANFIS-based DSS for use as part of an inventory management system. The primary focus has been to configure the ANFIS for generating decision for 7 days of a week by including demand and supply on one hand and several other factors like procurement, ordering and holding costs to name a few. For the design of ANFIS, Gaussian MFs have been used. Number of MFs used is also an important aspect. For ascertaining the training, the RMSE is considered to be an important factor. The performance is measured in terms of epochs of training and accuracy achieved while making decision. The results are compared to that obtained from an ANN as well which also has learning-based decision-making capability. Experimental results establish the advantage of use of the ANFIS for the DSS for use as part of an inventory management system.

References

1. Turban, E, Aronson J, Liang T, Shard R. (2006). Decision Support and Business Intelligence Systems (Eight Edition). Pearson
2. Pitchipoo, P., Venkumar, P. and Rajakarunakaran, S. (2013). Modeling and development of a decision support system for supplier selection in the process industry. *Journal of Industrial Engineering International*, 9(1), pp. 1–15.
3. Zadeh, L. A. (1965). Fuzzy sets. *Information and control*, 8(3), pp. 338–353.
4. Chang, H. C., (2004). An application of fuzzy sets theory to the EOQ model with imperfect qualityitems. *Computers & Operations Research*, 31, pp. 2079–2092.
5. Park, K. S. (1987). Fuzzy-set theoretic interpretation of economic order quantity. *Systems, Man and Cybernetics, IEEE Transactions on*, 17(6), pp. 1082–1084.
6. Kacprzyk, J. and Stanieski, P. (1982). Long-term inventory policy-making through fuzzy decision-making models. *Fuzzy sets and systems*, 8(2), pp. 117–132.
7. Wee, H. M., Yu, J. and Chen, M. C., (2007). Optimal inventory model for items with imperfect quality and shortage backordering. *Omega*, 35, 7–11.
8. Maiti, M. K., and Maiti, M., (2006). Fuzzy inventory model with two warehouses under possibilityconstraints. *Fuzzy Sets and Systems*, 157, pp. 52–73.

9. Paul, S. K. and Azeem, A. (2010). Selection of the optimal number of shifts in fuzzy environment: manufacturing company's facility application. *Journal of Industrial Engineering and Management*, 3(1), pp. 54–67.
10. Zeng, Y., Wang, L., & Zhang, J. (2007). A web-based fuzzy decision support system for spare parts inventory control. In *Fuzzy Information and Engineering* (pp. 601–609). Springer Berlin Heidelberg.
11. Lo, M. C. (2007). Decision support system for the integrated inventory model with general distribution demand. *Information Technology Journal*, 6(7), pp. 1069–1074.
12. Paul, S. and Azeem, A. (2011). An artificial neural network model for optimization of finished goods inventory. *International Journal of Industrial Engineering Computations*, 2(2), pp. 431–438.
13. Gaafar, L. K. and Choueiki, M. H. (2000). A neural network model for solving the lot-sizing problem. *Omega*, 28(2), pp. 175–184.
14. Lin, Y. H., Shie, J. R. and Tsai, C. H. (2009). Using an artificial neural network prediction model to optimize work-in-process inventory level for wafer fabrication. *Expert Systems with Applications*, 36(2), pp. 3421–3427.
15. Efendigil, T., Önüt, S. And Kahraman, C. (2009). A decision support system for demand forecasting with artificial neural networks and neuro-fuzzy models: A comparative analysis. *Expert Systems with Applications*, 36(3), pp. 6697–6707.
16. Aengchuan, P. and Phruksaphanrat, B. (2015). Comparison of fuzzy inference system (FIS), FIS with artificial neural networks (FIS + ANN) and FIS with adaptive neuro-fuzzy inference system (FIS + ANFIS) for inventory control. *Journal of Intelligent Manufacturing*, pp. 1–19.
17. Ferdous Sarwar, M., Rashid, M. and Ghosh, D. (2014). An Adaptive Neuro-Fuzzy Inference System based Algorithm for Long Term Demand Forecasting of Natural Gas Consumption, Fourth International Conference on Industrial Engineering and Operations Management (IEOM, 2014) Bali, Indonesia, January 7–9

Assessment of Examination Paper Quality Using Soft Computing Technique

Shruti Mangla and Abhishek Singhal

Abstract The level of the education is an important aspect in the career of a student. Quality of education is the root parameter to enhance the standard of teaching. In order to achieve superior quality of teaching examination paper's quality plays a vital role. Student's knowledge, potential, and skills are all judged with the help of exam papers. Thus, examination papers play a very crucial role in the phase of student's life and also help in judging the quality of teaching. So, the foremost aim of this paper is to assess the quality of examination paper using fuzzy logic.

Keywords Examination paper · Fuzzy logic · Membership function · Quality

1 Introduction

In India, education has shown several miracles. Statistics shows that there are over 20 million enrollments of students for higher education. The level of education is an important aspect for the growth and career of a student [1]. Standard of the education is totally dependent on the teaching quality. Teaching quality is usually judged by feedback questionnaire, which is conducted during mid semesters and end semesters. This helps in raising the levels of teaching. Thus, quality of teaching is an essential parameter in enhancing the standard of education.

In order to achieve excellent quality of teaching examination papers' quality plays a vital role. Conducting exams is mandatory for judging the capabilities, skills, and knowledge of a student. Examination not only helps in learning and enhancing knowledge but also helps in achieving the career goals.

S. Mangla (✉) · A. Singhal
Department of Computer Science & Engineering,
ASET, Amity University, Uttar Pradesh, India
e-mail: shrutimangla55@gmail.com

A. Singhal
e-mail: asinghal1@amity.edu

© Springer Nature Singapore Pte Ltd. 2018
S.K. Bhatia et al. (eds.), *Advances in Computer and Computational Sciences*,
Advances in Intelligent Systems and Computing 554,
https://doi.org/10.1007/978-981-10-3773-3_46

Regular conduction of exams is necessary for determining the performance of the students and also it helps the teacher to determine the position and grade of the student. The final result of the exam papers helps students to identify their weak points. Hence, the exam papers need to be designed with full concentration and its quality must be checked and verified properly.

The examination paper should be set in such a way that it covers all the main points. However, checking and judging the quality of the examination paper is too difficult for the setter. Due to the lack of awareness, the moderator may raise the difficulty level of examination paper, which can adversely affect the student's performance. So, there is a need for powerful tool to check the standard of exam paper for the sake of students' future.

High-quality examination paper must deal with the following issues:

- Is it possible to make examination paper of different difficulty levels?
- Does the exam paper meet marks distribution and time limitation?
- Is it feasible to avoid similarity among the questions?
- Are the contents of the examination paper covering all the important topics?

The main motivation of this paper is to assess the quality of examination papers conducted in different schools and colleges in various courses. Using fuzzy logic, an attempt has been made to categorize the level of examination paper using various parameters.

This paper comprises five sections: Sect. 2 provides literature review. Fuzzy logic is described in Sect. 3. Section 4 provides the proposed model. Results are discussed in Sect. 5. Finally, Sect. 6 summarizes the paper.

2 Literature Review

In past, a large amount of work has been done in order to evaluate the quality of the examination paper using various algorithms and techniques.

Guo and Wang [2] presented an evaluation model for determining the quality of examination paper by effectively utilizing the Delphi method and Fuzzy theory that was applied in practicing the characteristics of comprehensive operations. This model helps in expressing the result in scores, which decrements the randomness and subjectivity regarding the evaluation pertaining to quality of paper.

Cheng and Zheng [3] gave an approach for auto generation of exam papers in order to achieve reliable results. In this paper questions were selected from database using fuzzy similarity. This approach helped in removing the identical questions.

Yaakob et al. [1] used Bloom's taxonomy for the purpose of evaluation of paper. A representation model was build to calculate the accuracy percentage between the test specification table and the examination paper specification. The model proposed helped instructor to reduce the time in making the question paper.

Paul et al. [4] used evolutionary algorithm to generate dynamic question paper template. Different requirements of paper setter like each module section, degree of difficulty, kinds of different questions, etc. were considered in the paper. Proposed template was beneficial to generate examination paper manually by selecting the relevant questions from the question bank.

Examination paper automatic construction scheme (EPAC) was provided by Song and Hong [5] using the K-means clustering. EPAC was build with the help of Java Servlets and JSP. With the help of EPAC all the knowledge points can be covered in paper easily.

Exam paper components have been analyzed by Han and Li [6] using genetic algorithm. Certain parameters like discrimination, difficulty coefficient, knowledge, quantity, questions types, etc. were considered. Test cases were built. As a result, this helped in examining the difficulty level of paper in less time.

Fang et al. [7] proposed hierarchical clustering-based approach in order to process reform of practical teaching whose result became basis for the open education management.

Jing [8] presented algorithm of volume, which helped in deciding the total time, distribution of questions, and degree of difficulty during the construction of online examination paper.

Using genetic algorithm Huang and Wang [9] introduced the design of a universal automatic examination paper generation system. The design helped in increasing the generating speed and success rate.

Xu and Kong [10] gave a method that considered teacher's weigh as well as the scores given by the teacher in marking examination papers. This method helped in making the score of student fair and transparent.

Stergiopoulos et al. [11] made an application which helped in performing computerized tests and proved electronically exams better than pen–paper examination.

Swart [12] using Bloom's taxonomy distinguish between questions based on application, analysis, synthesis, evaluation, knowledge, and comprehension. This comparison proved beneficial for the evaluation of final examination papers.

Shiyan [13] with the help of questionnaire proved that training and group discussions are important part of teaching.

Wang and Wen [14] analyzed the problems which occur by auto-generating examination papers and present some algorithms to solve the problems. They used Monte Carlo method for the selection of questions of various difficulty levels.

Jiang and Zhang [15] discussed problems of systems generating intelligent examination papers. Using genetic algorithm structure of examination paper is analyzed.

Raus et al. [16] focused on the evaluation system of teaching and web-based question bank. Feasibility study of i-QuBES in detail is done.

Mogale et al. [17] presented security procedures for ensuring that the examination papers are securely handled. They provided a model which helped in protecting the papers from unauthorized access.

Fig. 1 Fuzzy logic system [18]

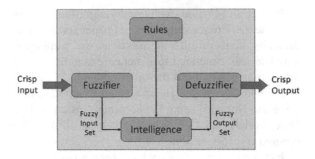

3 Fuzzy Logic

Fuzzy logic (FL) was introduced by Lotfi A. Zadeh in the year 1965. It is problem-solving system, which tackles with uncertainty and imprecision. FL can be executed in software, hardware, or with both as well. It mainly provides a decision as output by taking vague, noisy, and imprecise data as inputs. Fuzzy logic system is described in Fig. 1 [18].

Fuzzy set has set of elements and the elements have the membership degrees. This can be described as

- Fuzzy subset F for a set S can describe a set of ordered pairs. First member of ordered pair will be of set S and second member will be from the interval [0, 1].
- Zero value will denote non-membership, one value will depict complete membership and values between zero and one will denote degrees of membership.

There are various fuzzy set operations like Union, Intersection, Complement, etc. Various properties like commutativity, associativity, distributivity, idempotency, identity, involution, and transitivity are also used.

A mathematical function namely membership function is used to define the membership of an element.

Various built-in membership functions are Gaussian distribution function, sigmoid curve, piecewise linear function, etc.

Fuzzy IF-THEN rules are used in which IF part is known as antecedent and THEN is known as consequent. These rules are basically used for converting the input conditions into the output conditions. With the help of output, appropriate decision can be obtained.

4 Proposed Model

In this paper, a model has been proposed for calculating the quality of all the examination papers conducted for various courses. This model will help in giving accurate results in less time.

In execution of fuzzy logic various inputs and membership functions are used. The range of the membership function is in between 0 and 1. Triangular membership function is used in calculating the examination paper quality.

In this model five inputs have been used to calculate the examination paper quality as the output. Using inputs and triangular membership function the output is computed as shown in Fig. 2.

Different input parameters used for the assessment of examination paper quality are as follows:

(1) Question Hardship: Low, Medium, High
    Hardship of the questions should be set high in order to achieve the high quality of the examination paper.
(2) Case Study: Easy, Medium, Hard
    Case study is an analytical tool, which helps in determining the problem-solving capability of a student. This is one of the important parameters in judging the quality of examination paper.
(3) Question Type: Direct, Not Direct
    In examination paper diversities of questions are involved. Type of question can be straight forward or not.

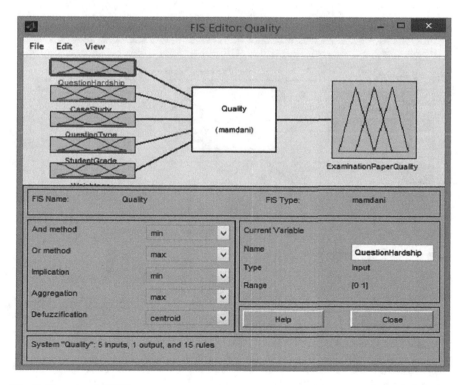

Fig. 2 Fuzzy model [19]

(4) Student Grade: Poor, Average, Good, Excellent
    Grade of the student imparts direct impact on the quality of paper.
(5) Weightage: Less, Accurate, More
    Bifurcation of marks for various questions should be done seriously. Accurate
    weightage will help in achieving high quality of examination paper.

Table 1 Rules

Rules	Inputs					Output
	QH	CS	QT	SG	WT	Examination paper quality
1	Low	–	Direct	Excellent	Less	Very low
2	Low	Easy	Direct	Excellent	Less	Very low
3	Low	–	Direct	–	–	Very low
4	Low	Easy	Direct	–	–	Very low
5	Medium	Medium	Direct	Good	Accurate	Average
6	High	–	Not direct	Poor	More	Very high
7	High	Medium	Not direct	Poor	More	Very high
8	High	Hard	Not direct	–	More	Very high

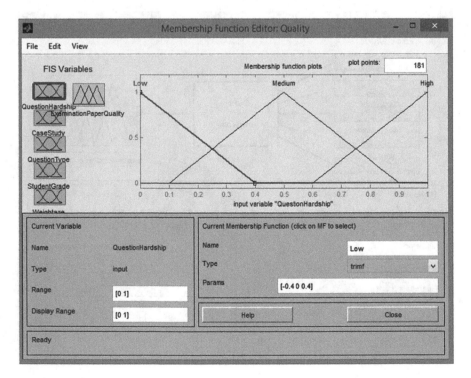

Fig. 3 Membership function

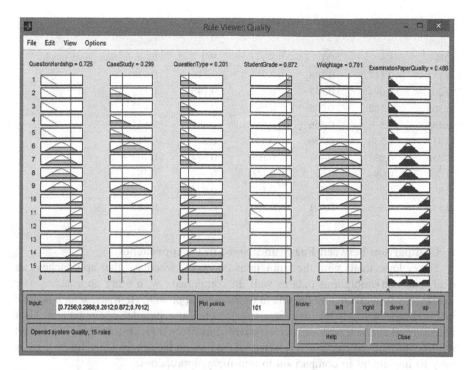

Fig. 4 Rule viewer with different input values

The value of fuzzy set for output parameter is as follows:

Examination Paper Quality: Very low, Low, Average, High, Very High.

Many rules have been applied in this model. A sample of rules has been shown in Table 1.

Membership function for the Question Hardship input is depicted in Fig. 3.

With the help of rule viewer, paper quality is observed using all the five inputs for the different sets of the input values as shown in Fig. 4.

5 Result

With the help of fuzzy logic an evaluation model for assessing the examination paper quality is proposed in this paper. Table 2 presents the result of assessment of examination paper quality produced by the proposed method.

Different input values are given to assess the exam paper quality. It is observed that very high examination paper quality can be obtained with high value of Question Hardship, Case Study, and Weightage.

Table 2 Assessment of examination paper quality

Inputs					Output
QH	CS	QT	SG	WT	Examination paper quality
0.117	0.117	0.108	0.966	0.116	0.428
0.207	0.094	0.153	0.937	0.146	0.451
0.305	0.102	0.205	0.907	0.728	0.472
0.485	0.5	0.556	0.504	0.481	0.631
0.53	0.538	0.556	0.444	0.5	0.644
0.838	0.831	0.407	0.108	0.884	0.702
0.084	0.994	0.511	0.146	0.994	0.783
0.898	0.951	0.578	0.108	0.974	0.81
0.898	0.996	0.601	0.116	0.996	0.908

Comparison between Fuzzy and Non-Fuzzy Approaches

Fuzzy logic deals with the ambiguous concepts, whereas other approaches can deal with non-ambiguous data as well. In comparison with other approaches user interface is convenient and flexible in fuzzy logic approach. Heuristic search is done in fuzzy systems, whereas non-fuzzy approaches use parallel computations. User friendly and efficient performance is provided in fuzzy logic. Fuzzy logic can model nonlinear functions easily and is based on the natural language. Thus, fuzzy logic is easy to implement in comparison to non-fuzzy approaches.

6 Conclusion

In this paper assessment of examination paper quality is done using the fuzzy approach in order to ease the work of teacher so that appropriate modifications can be done before giving the hard copies of the exam paper to the students. Better understanding can be achieved easily with the help of fuzzy logic. Paper quality is the most important aspect for the future of the students and for the quality of teaching as well. Different membership functions can also be used like Gaussian Membership Function and Trapezoidal Membership Function in order to compare the result obtained. Different parameters like Question Hardship, Case Study, Question Type, Student Grade, and Weightage helped a lot in determining the good results.

References

1. Farhana Yakoob, Noor Hasimah Ibrahim Teo and Nur Atiqah Sia Abdullah, "Cognitive Knowledge Representation for Examination Questions Specification Analysis", International Conference on Advanced Computer Science Applications and Technologies, 2013, pp. 530–533.

2. Yemin Guo and Lanmei Wang, "Preliminary Study on Evaluation of the Quality of Examination paper Based on Fuzzy theory", 7th International Conference on Fuzzy Systems and Knowledge Discovery, 2010, pp. 1328–1331.

3. Xien Cheng and Jinghua Zheng, "Novel Approach of Auto-generating Examination Papers", International Conference on e-Education, e-Business, e-Management and e-Learning, 2010, pp. 411–414.

4. Dimple V. Paul, Shankar B. Naik, Priyanka Rane and Jyoti D. Pawar, "Use of an Evolutionary Approach for Question Paper Template Generation", 4th International Conference on Technology for Education, 2012, pp. 144–148.

5. Yuli Song and Feng Hong, "EPAC: Examination Paper Automatic Construction Based on K-Means Algorithm", International Conference on Information Technology and Applications, 2013, pp. 232–235.

6. Ling Ling Han and Xiao Dong Li, "The Analysis of Exam Paper Component based on genetic algorithm", 4th International Conference on Communication Systems and Network Technologies, 2014, pp. 561–564.

7. Liu Fang, Yang Ting-Ting, Chen Shou-Gang, Liu Jing-Duo, Zhang Shao-Gang and Chen Pu, He Jie-Tao, He Bin-Sheng, "Hierarchical clustering based teaching reform courses examination data analysis approach applied in China Open University system", Seventh International Symposium on Computational Intelligence and Design, 2014, pp. 377–381.

8. Wang Jing jing, "Research On Intelligent Test Paper of WEB-Based", International Conference on Computer Science and Information Processing (CSIP), 2012, pp. 369–371.

9. Wei Huang and Zhao-hui Wang, "Design of Examination Paper Generating System from Item Bank by Using Genetic Algorithm", International Conference on Computer Science and Software Engineering, 2008, pp. 1323–1325.

10. Qingzhen Xu and Jing Kong, "Fuzzy Mathematics application in marking examination papers", ISECS International Colloquium on Computing, Communication, Control, and Management, 2009, pp. 67–70.

11. Stergiopoulos, Charalampos, Panagiotis Tsiakas, Dimos Triantis, and Maria Kaitsa. "Evaluating Electronic Examination Methods Applied to Students of Electronics. Effectiveness and Comparison to the Paper-and-Pencil Method." In Sensor Networks, Ubiquitous, and Trustworthy Computing, 2006. IEEE International Conference, 2006, pp. 143–151.

12. Swart, Arthur James. "Evaluation of final examination papers in engineering: A case study using Bloom's Taxonomy." Education, IEEE Transactions on 53, no. 2 (2010): 257–264.

13. Shiyan, Wen. "Teaching methods for a school-based curriculum." In Networking and Digital Society (ICNDS), 2nd International Conference, 2010, pp. 508–511.

14. Wang, Xiuhui, and Junqin Wen. "Architectures and algorithms for auto-generating examination paper system." In Biomedical Engineering and Informatics (BMEI), 3rd International Conference, 2010, pp. 2893–2895.

15. Jiang, Lanling, and Fang Zhang. "Research and implementation of intelligent paper generating system." In 2011 International Conference on Computer Science and Service System (CSSS). 2011.

16. Raus, Mohd Ikhsan M., Roziah Mohd Janor, Roslan Sadjirin, and Zahriah Sahri. "The development of i-QuBES for UiTM: From feasibility study to the design phase." In Control and System Graduate Research Colloquium (ICSGRC), 2014, pp. 96–101.

17. Mogale, Miemie, Mariana Gerber, Mariana Carroll, and Rossouw Von Solms. "Information Security Assurance Model (ISAM) for an Examination Paper Preparation Process." In Information Security for South Africa (ISSA), 2014, pp. 1–10.

18. Artificial Intelligence-Fuzzy Logic Systems. http://www.tutorialspoint.com/artificial_intelligence/artificial_intelligence_fuzzy_logic_systems.htm.

19. MATLAB Toolbox, http://www.mathworks.com.

Moving Shadow Detection Using Fusion of Multiple Features

Yajing Lin, Bingshu Wang and Yong Zhao

Abstract Moving shadow detection is an important technique for the integrity of object detection. This paper is based on the assumption that the shadow area is darker than the corresponding background area but keeps color constancy and texture consistency. The main contributions of this paper include two parts. First, an adaptive mechanism for shadow detection is proposed using texture of improved local ternary pattern. The main idea is to detect partial real shadows to estimate relative accurate threshold parameters for shadow detector. Second, we utilize a model of genetic programming model to fuse multiple features: texture, color, and gradient information. Experimental results on indoor and outdoor sequences demonstrated that the proposed method outperforms some state-of-the-art methods.

Keywords Moving shadow · Improved local ternary pattern · Genetic programming model · Multiple features

1 Introduction

Shadows are always divided into two categories: static shadows that are casted by static objects such as buildings and trees; and moving shadows that are casted by moving objects. Shadow detection plays an important role in the field of motion detection. This paper focuses on the detection of moving shadows, so the static shadows detection goes beyond scope of this paper. There are two kinds of important properties for shadow detection: first, shadows share similar movement

Y. Lin · B. Wang · Y. Zhao (✉)
School of Electronic and Computer Engineering,
Shenzhen Graduate School of Peking University, Shenzhen, China
e-mail: zhaoyong@pkusz.edu.cn

Y. Lin
e-mail: yjlin@sz.pku.edu.cn

B. Wang
e-mail: wangbingshu@sz.pku.edu.cn

© Springer Nature Singapore Pte Ltd. 2018   487
S.K. Bhatia et al. (eds.), *Advances in Computer and Computational Sciences*,
Advances in Intelligent Systems and Computing 554,
https://doi.org/10.1007/978-981-10-3773-3_47

pattern with moving objects; second, the shadow regions are darker than the corresponding background regions but remain texture consistency and color.

In recent years, many approaches were developed and applied for moving shadow detection. All these methods [1–5] can be categorized into four groups [6]: Chromaticity-based [7] methods, physical-based [8] methods, geometry-based [9] methods, and texture-based [10] methods [11]. Discussing all the methods in detail, it proposes that all the methods can be easily performed and implemented. However, for changing illumination condition, most of them cannot perform ideally [11]. Chromaticity-based methods still is the chrominance. The central ideal of chromaticity-based methods is assumption that shadow regions are darker than corresponding background regions. One of the most typical methods is using HSV color information. Geometry-based methods can work straightforward on input frame with the prior knowledge such as illumination source, object shape, etc. However, obtaining the prior knowledge is a difficult task. This is the main limitation of these methods. Compared with chromaticity-based methods, the physical-based [8] methods show better performance but have limitation in spectral properties. The geometry-based [9] methods are using the information of shadow itself to remove them. All the information required is based on geometrical setting. In other words, it is not effective when the geometrical relationship changed. The texture is a more stable property, which means it does not depend on illumination conditions [6]. Texture-based methods are mainly based on the assumption that shadows remain the most of their textures. Jiang [12] proposes an adaptive method to detect shadow. He employs global texture and sampling deduction. However, the processing time tends to be long which cause low efficiency of shadow detection. Each of these methods utilized one or two features. Moreover, multiple features were used for moving shadow detection in [13] and performed large region texture-based method.

Inspired by the application of multiple features, this paper proposed an adaptive shadow detection method using improved local ternary pattern. Chromaticity and texture information are also used and combined in our method. Experimental results on indoor and outdoor scenes demonstrated that the proposed method outperformed several state-of-the-art methods.

2 The Proposed Shadow Detection Method

2.1 The Description of Flowchart

It can be seen from Fig. 1 that three shadow detectors are utilized: Color-based shadow detector, gradient-based shadow detector, and adaptive ILTP shadow detector. The gradient-based shadow detector is from large region of texture-based method [10]. The color-based shadow detector is from HSV color space-based method [7]. Specially, the adaptive ILTP shadow detector is proposed by this paper

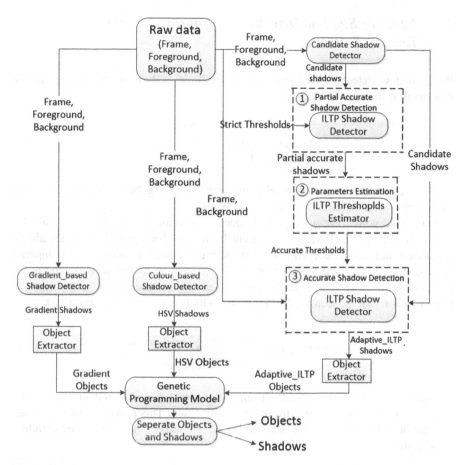

Fig. 1 The framework of the proposed method

using an improved local ternary pattern [14]. The proposed method of adaptive ILTP shadow starts by candidate shadow detection. It refers to the assumption that shadow regions are darker than the corresponding background region. The adaptive ILTP shadow detector mainly includes three steps, which will be detailed introduced in Sect. 2.2.

Three shadow components are gradient shadows, HVS shadows, and adaptive ILTP shadows are generated by three shadow detectors. Then they are used by object extractor to obtain three objects components: Gradient objects, HSV objects, and adaptive ILTP objects. These three objects components are sent to genetic programming model to obtain relatively accurate objects. Then the objects and shadows are separated successfully.

2.2 Adaptive Shadow Detection by Improved Local Ternary Pattern

ILTP shadow detector that mentioned in multiple features method [13] can be described in the equation below:

$$v(x, i_c, t) = \begin{cases} 10, & i_c \geq x+t \\ 01, & |x - i_c| < t, \\ 00, & i_c \leq x - t \end{cases} \quad (2.1)$$

where x represents the present pixel point, represents neighborhood pixel point and t represents the tolerance of noise, respectively.

Compared with LTP, see Fig. 2, it added four values in neighborhood calculation in ILTP. The ILTP values of present frame and background frame should be calculated and compared. The array ar should be used to record the compared results. If they are same, then record as 1; Otherwise, record it as 0. The accumulated result of ar values can be regarded as the matched level. If the result is higher then it can be seen as shadow point, which is explained in Eq. (2.2):

$$ILTP \, Shadow = \begin{cases} 1, & \text{if } \sum_{i=1}^{n} ar(i)/n > \delta \\ 0, & \text{otherwise} \end{cases} . \quad (2.2)$$

The kernel of the adaptive ILTP algorithm was shown in red in Fig. 1. The candidate shadow should be extracted in the first step. This process is mainly based on the assumption that the shadow area is darker than the corresponding background:

$$\Psi_r = \frac{C_b/C_g}{B_b/B_g}, \quad \Psi_g = \frac{C_b/C_r}{B_b/B_r}, \quad \Psi_b = \frac{C_g/C_r}{B_g/B_r} \quad (2.3)$$

$$Candidate \, Shadow = \begin{cases} 1, & \text{if} |\Psi_i - \mu| < \lambda, i \in \{b, g, r\} \\ 0, & \text{otherwise} \end{cases} \quad (2.4)$$

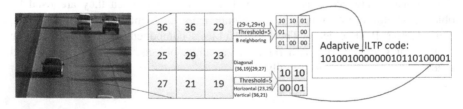

Fig. 2 The coding schematic of ILTP

Equation (2.3) represents the detected channel value of present frame and background frame of corresponding foreground point.

When the value of candidate shadow is 1, it represents the detected point as shadow point, which is shown in Eq. (2.4).

The first step is using the ILTP shadow detector to process the candidate shadow, which the partial accurate shadows can be got. The second step is to use the output of partial accurate shadows to estimate the parameter analysis of ILTP. The characteristic detector has parameter threshold. But the implementation of algorithm that set parameters for each scene was limited by changing scenes. Hence, a strategy that could predict more accurate parameters is needed. Using setting of the partial valid shadow area, which is shown in Eq. (2.5), the initial values of main parameters can be decided and the related value can be calculated from shadow area. The parameters of shadow detector are determined by linear combination of initial value and correlated value. The linear correlation factor is

$$\delta = \delta^{ini} - \alpha * \left| \delta^{ini} - \delta^{cor} \right| \qquad (2.5)$$

The third step is to use the candidate shadow area in step 1 and the parameters in step 2 to do the accurate detection. The adaptive ILTP shadow can be got through this step. This result will be combined with HSV shadows, and gradient shadows to be processed through object extractor. Then it will be processed in genetic programming model. The final result then can be got.

2.3 The Fusion of Multiple Features by Genetic Programming Model

The three-detection operator will detect three results separately. In fact, the test results are not very accurate. They need to be combined together, and combining all the advantages of all methods. Genetic algorithm [15] is automatically selected from existing multiple results in optimal combination algorithm results in different ways, and selecting the most stable post-treatment method, so as to achieve optimum test results, and thus enhance the detection rate and reduce the false alarm rate. The algorithm has been verified in the field of motion change detection. By applying genetic algorithm, it shows an excellent result compared to any single algorithm in CDnet 2014 testing. The experimental results of three methods, Chr, LR, and adaptive ILTP, were combined and put into the solution of IUTIS-3 solution to do the integrated treatment.

3 Experimental Results

In this section, our approach is compared with the recent methods which are chromaticity method [7], geometry method [9], physical method [8], SR method [10], and LR method [16]. We applied all these methods to six standard sequences which are used for shadow detection. We used the shadow detection η and shadow discrimination ξ, to evaluate the performance of prior methods. In order to analyze the comprehensive performance, *F-measure* is the cogent evaluate metric. They are defined as

$$\eta = \frac{TP_S}{TP_S + FN_S}, \quad \xi = \frac{TP_F}{TP_F + FN_F} \quad F - measure = \frac{2\eta\xi}{\eta + \xi}, \quad (3.1)$$

where TP_S and FN_S are the true positives and false negative for shadows, and TP_F and FN_F are the true positives and false negatives for objects.

3.1 Dataset and Evaluation Metric

As it can be seen from Table 1 and experimental data, it is an effective shadow detection strategy by extracting the candidate shadow first and then using the accurate detection. LR is a strategy that uses HSV first then using gradient information to use accurate detection. The experiment result shows that the average result of LR is higher than Chr, Geo, Phy, and SR by 8%. The adaptive ILTP method introduced in this article shows 5% higher than LR by comparing the average result. It can be concluded that the adaptive ILTP method not only to achieve enhanced adaptive parameter generalization capability, and improved test results.

Table 1 The F-measure results of all methods

	Campus	Hallway	Highway1	Highway3	Room	Caviar	Average
Chromaticity	0.532	0.864	0.704	0.534	0.782	0.687	0.684
Geography	0.589	0.592	0.706	0.546	0.613	0.573	0.603
Physical	0.553	0.689	0.568	0.488	0.716	0.541	0.592
SR	0.674	0.779	0.287	0.109	0.751	0.798	0.566
LR	0.668	0.956	0.740	0.536	0.886	0.814	0.767
Adaptive ILTP	0.796	0.956	0.711	0.648	0.919	0.866	0.816
Multiple feature	0.790	**0.963**	0.779	0.581	0.917	**0.914**	0.824
Ours	**0.812**	0.955	**0.788**	**0.777**	**0.930**	0.899	**0.860**

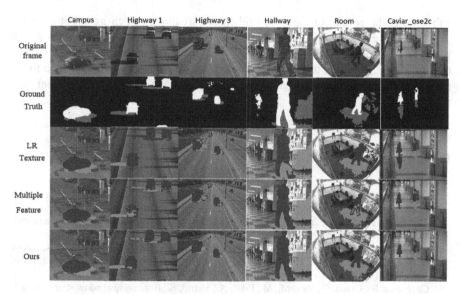

Fig. 3 The qualitative results of six test sequences by seven methods

3.2 Qualitative and Quantitative Results

Figure 3 shows the compared results of qualitative test of seven different methods according to six scenes. In the figure, the blue area represents the target area detected. The green area represents shadow area detected. In can be seen from the figure that the result is better compared with other six methods by combing the Chr, LR, adaptive ILTP genetic algorithms. The most of shadows can be detected by Chr. However, it cannot detect all and sometimes misdetection occurred. There were also too much misdetection by using Geo, especially the shape was distorted when cutting. SR could only detect a little shadow. For most of conditions, the shadow cannot be detected by applying LR.

From Table 1, it can be concluded that the methods used in this article have more advantages compared with other six methods in comprehensive assessment. It also shows better results and more robust. It can be approved from the experiment that the result is much better by combining all the methods rather than only one of them. And the result is better than the method mentioned in multiple features using ballot as the post-processing.

4 Conclusion

This paper presents a novel method for moving cast shadow detection using color and texture information. The result is much better by combining all the methods rather than only one of them. For shadow detection, detection using a plurality of sub-testing, the results of their comprehensive treatment (such as ballot or using genetic algorithms) will be more robust than the results of using single operator. It is mainly because of the following: 1. The adaptive parameter estimation can be estimated for the current threshold parameters; 2. The robustness of algorithm to the complex scene can be improved by combining information of color, gradient, and texture.

References

1. Sanin, A., Sanderson, C., Lovell, B.C.: Shadow detection: a survey and comparative evaluation of recent methods. J. Pattern Recogn. 45(4), 1684–1695 (2012)
2. Zhang, W., Fang, Z.Z., Yang, X.K., Wu, Q.M.J.: Moving cast shadows detection using ratio edge. J. IEEE Trans. Multimedia 9(6), 1202–1214 (2007)
3. Qin, R., Liao, S.C., Lei, Z., Li, S.Z.: Moving cast shadow removal based on local descriptors. In: IEEE 20th International Conference on Pattern Recognition, pp. 1377–1380. IEEE Press, New York (2010)
4. Cucchiara, R., Grana, C., Piccardi, M., Prati, A., Sirotti, S.: Improving shadow suppression in moving object detection with HSV color information. In: 2001 IEEE Conference on Intelligent Transportation Systems, pp. 334–339. IEEE Press, New York (2001)
5. Martel-Brisson, N., Zaccarin, A.: Kernel-based learning of cast shadows from a physical model of light sources and surfaces for low-level segmentation. In: 2010 IEEE Conference on Computer Vision and Pattern Recognition, pp. 1–8. IEEE Press, New York (2010)
6. Al-Najdawi, N., Bez, H.E., Singhai, J., Edirisinghe, E.A.: A Survey of cast shadow detection algorithms. J. Pattern Recogn. Lett. 33(6), 752–764 (2012)
7. Cucchiara, R., Grana, C., Piccardi, M., Prati, A.: Detecting Moving Objects, Ghosts, and Shadows in Video Streams. J. IEEE Trans. Pattern Analysis and Machine Intelligence. 25(10), 1337–1342(2003)
8. Huang, J.B., Chen, C.S.: Moving Cast Shadow Detection Using Physics-Based Features. In: 2009 IEEE Conference Computer on Vision and Pattern Recognition, pp. 2310–2317. IEEE Press, New York(2009)
9. Hsieh, J.W., Hu, W.F., Chang, C.J., Chen, Y.S.: Shadow Elimination for Effective Moving Object Detection by Gaussian Shadow Modeling. J. Image and Vision Computing. 21(3), 505–516(2003)
10. Leone, A., Distante, C., Buccolieri, F.: Shadow Detection for Moving Objects Based on Texture Analysis. J. Pattern Recognition. 40(4), 1222–1233(2007)
11. Sanin, A., Sanderson, C., Lovell, B.C.: Shadow detection: a survey and comparative evaluation of recent methods. J. Pattern Recogn. 45(4), 1684–1695 (2012)
12. Jiang, K., Li, A.H., Cui, Z.G., Wang, T., Su, Y.Z.: Adaptive shadow detection using global texture and sampling deduction. J. IET Comput. Vis. 7(2), 115–122 (2013)
13. B Wang,W Zhu,Y Zhao,Y Zhang: Moving Cast Shadow Detection Using Joint Color and Texture Features with Neighboring Information. Springer International Publishing, 2015
14. Tan, X., Triggs, B.: Enhanced Local Texture Feature Sets for Face Recognition under Difficult Lighting Conditions. J. IEEE Trans. Image Processing. 19(6), 1635–1650(2010)
15. S Bianco,G Ciocca,R Schettini: How Far Can You Get By Combining Change Detection Algorithms? Computer Science, 2015
16. Sanin, A., Sanderson, C., Lovell, B.C.: Improved Shadow Removal for Robust Person Tracking in Surveillance Scenarios. In: IEEE 20th International Conference on Pattern Recognition, pp. 141–144. IEEE Press, New York (2010)

Caption Text Extraction from Color Image Based on Differential Operation and Morphological Processing

Li-qin Ji

Abstract With the continuous progress and development of multimedia technology, it becomes more and more valuable to extract text from the image. The text image can be divided into document text image, caption text image, and scene text image due to the different characteristics of the text in the image; so many researchers have proposed different methods to extract text. This paper puts forward a new method which is based on differential operation and morphological for caption text extraction. First, using three differential operator of vertical, horizontal, and diagonal direction to detect caption text information, and then using morphological processing to further determine caption regions. Finally, Using mathematical logic "and" operation to process three pieces of different direction of caption region image, combined with recursive statistics method to delete noise and extract the final caption region. Results in this paper show that this method can effectively extract the caption information in the image.

Keywords Text extraction · Edge detection · Morphology · Binary · Image processing

1 Introduction

With the continuous progress and development of modern multimedia technology, image has become one of the important medium of mutual communication, and the texts in the image can illustrate the meaning of the image, therefore, text extraction from complex background image is of great significance [1–3]. The text image can be divided into document text image, caption text image, and scene text image according to the characteristics of the text in the image. There are many experts from home and abroad to study the text extraction technology; they use different

L. Ji (✉)
Electrical Engineering College in SuZhou Chien-Shiung Institute of Technology,
Taicang, Jiangsu Province, China
e-mail: Jiliqin2003@163.com

© Springer Nature Singapore Pte Ltd. 2018
S.K. Bhatia et al. (eds.), *Advances in Computer and Computational Sciences*,
Advances in Intelligent Systems and Computing 554,
https://doi.org/10.1007/978-981-10-3773-3_48

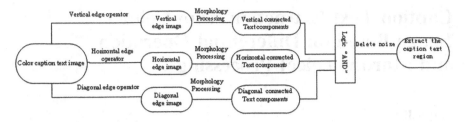

Fig. 1 The flowchart of algorithm for caption text extraction

processing methods for different text images. Alvaro et al. [4] used text recognition to determine the text area. This method included three steps: First, find out the character candidates by segmentation, then analyze further the connected component; finally, using gradient features of texts and the method of support vector machines to classify the text line. H. Chen et al. [5] presented a new method for text detection. First, using the maximum and Stable Extreme Regions as text candidates, then eliminate nontext regions by geometric and stroke width of candidates. Sumathi et al. [6] used gamma correction method (GCM) to extract text from image. This approach used texture analysis and measurement to evaluate the gamma value, which is applied to an original input image to obtain the background image. Keechul and Jung [7] puts forward a kind of method which uses the texture and connected component analysis to locate the text, but the whole algorithm is complex and requires enough training samples. Kim et al. [8] used texture classifier which supports vector machine to detect text, the results of this method are good, but the amount of computation is too large.

About caption text image, this paper puts forward a novel method, which uses three vertical, horizontal, and diagonal direction of the differential operator to detect caption information, and then uses morphological opening and closing, dilation, erosion to determine caption regions. Finally, using logic "and" arithmetic to process three pieces of different direction of caption region images, combined with recursive statistics method to extract the final caption region. Experimental results show that this method can effectively extract the caption information in the image. The flowchart of algorithm for caption text extraction is shown in Fig. 1.

2 Edge Detection for Caption Text

Usually, edge detection operators for image enhancing are not only Roberts, Sobel, Prewitt, Krisch, and Laplass Gauss, but also the gradient and differential operator [9]. Because the caption edge image usually consisted of different gray level pixel, we use the differential operation of three vertical, horizontal, and diagonal directions to detect the edge of the caption. The processing object is 24 bit true color image, where each pixel is composed of three components R, G, B. Therefore, each component is to be detected by differential operation of three directions. Take the pixel point f(i, j) as an example, its three component are R(i, j), G(i, j), B(i, j), according to the Formula 1, the value of the red component in the vertical direction can be obtained. According to the Formula 2, the differential value of the red component in the horizontal direction can be calculated, also the differential value of the red component in the diagonal direction can be obtained by the Formula 3. In the same way, it can be concluded that the differential values in different directions of other color components

$$R_V(i, j) = 3*(R(i, j) - R(i, j - 1)) \tag{1}$$

$$R_H(i, j) = 3 * (R(i, j) - R(i - 1, j)), \tag{2}$$

$$R_D(i, j) = 3*(R(i, j) - R(i - 1, j - 1)) \tag{3}$$

The specific code implementation of vertical edge detection are as follows.

```
void Convolute(int *R, int *G, int *B, int i, int j, WORD
BytesPerLine, LPBYTE lpDIB, KERNEL1 *lpK)
{
   BYTE bb[9], gg[9], rr[9];
   LONG Offset;
   Offset= P_OFFSET(i-1,j-1, BytesPerLine);// Get the up-
per-left pixel address of pixel(i,j)
   bb[0] = *(lpDIB + Offset++);
   gg[0] = *(lpDIB + Offset++);
   rr[0] = *(lpDIB + Offset);
   *R = *G = *B = 0;
   for (int k=0; k<9; ++k)
   {
      *R+= lpK->Element[k]*rr[k];// The red component value
is calculated by the formula.1
      *G+= lpK->Element[k]*gg[k];// The green component
value is calculated by the formula.1
      *B+= lpK->Element[k]*bb[k];// The blue component val-
ue is calculated by the formula.1
   }
}
```

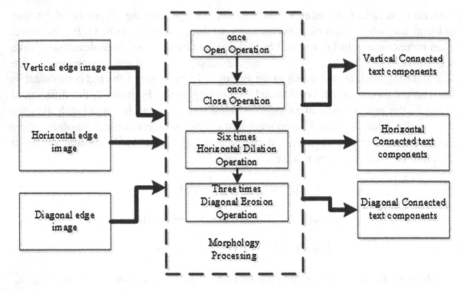

Fig. 2 Flowchart of morphology processing

3 Morphological Processing

The main function of morphology is to obtain the topology and structure information of objects [10, 11]. This paper uses the basic operations of morphology
including opening, closing, erosion, and dilation to obtain the unknown caption
connected regions. The flowchart of the process is shown in Fig. 2. When the
dilation operation is realized, the horizontal structure element B = (1, 1, 1, 1, 1) is
adopted. The experimental results show that this kind of structure element can form
the caption connected regions.

4 Caption Text Location

- Logical "AND" Operation

   After getting the three directions of the caption connected regional image, it is
easy to exclude most of noncaption regions, just computing the three images by
logic "and" operation. However, there are still pseudo caption regions. In this paper,
we use the recursive algorithm [12] to calculate the white pixel number of unde-
termined caption regions. If the total number of white pixels is less than the
specified value (here the specified value = image height * width of the image/150),
then remove the pseudo region (Fig. 3h).

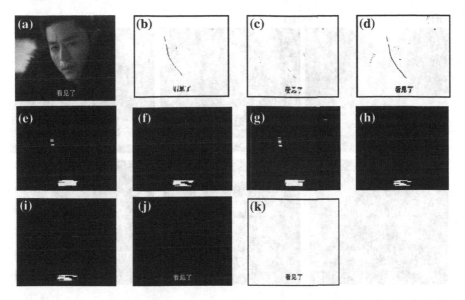

Fig. 3 The processing results of each step. **a** Original input image. **b** Vertical edge detection. **c** Horizontal edge detection. **d** Diagonal edge detection. **e** Vertical connected text components. **f** Horizontal connected components. **g** Diagonal connected components. **h** The final text region after logic "AND" and recursive algorithm. **i** Locate the coordinate of caption text region. **j** Extraction of caption text region. **k** Binarization of caption text region

- Locate the Caption Text

    After "AND" operation, we can use the following algorithm to locate the coordinate of the caption region.

- STEP1: Take the pixel point f(i, j) as an example, if it is black, then exit the algorithm. Else scan the 4 neighbor fields.
- STEP2: if there are no white pixels in the 4 neighbor fields, exit the algorithm, return the coordinates of the upper-left corner and the coordinates of the caption region.
- STEP3: if there are white pixels in the 4 neighbor fields, then adjust the coordinates of caption region.

- Results of Extraction for Caption Text

    To get the 24 bit binary caption text image, we still need to process the extracted caption regions by gray transform, binarization processing, and color inversing transform. The whole process of caption text extraction is shown in Fig. 3.

- Comparison of Results

    It can be seen from Fig. 4 that the caption text extraction method proposed in this paper is simple and the location of the text area is very accurate. However, there are many noncaption text regions using the method of literature 7. So our method is better than that method of literature 7.

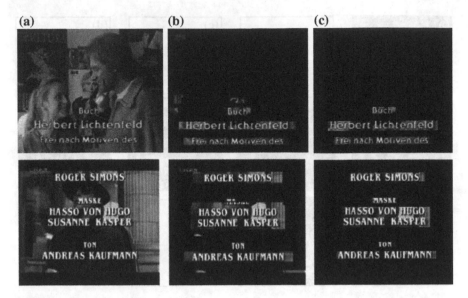

Fig. 4 Comparison of the extracted text regions. **a** Original image. **b** Extracted caption text regions by the method of Ref. [7]. **c** Extracted caption text regions by our method

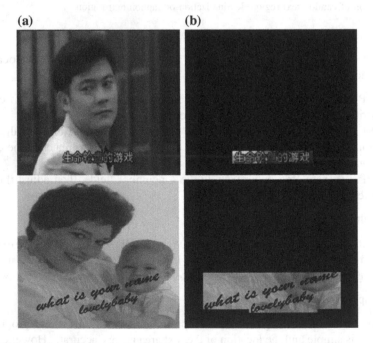

Fig. 5 Robustness verification. **a** Original image. **b** Extracted caption text regions

- Robustness Verification

The caption text regions in the above images are always horizontal, and the color of characters are single. To verify the robustness of the algorithm in this paper, we tested a variety of color images which have the tilt and colorful text regions (Fig. 5).

5 Conclusion

There are many changes and differences in text size, texture, color, and other information in the caption text image, so it is some difficult to accurately extract this type of text. This paper puts forward a novel method which uses vertical, horizontal, and diagonal differential operator to detect caption information, and then uses morphological processing to determine caption regions. Finally, using logic "and" arithmetic to delete the noise, combined with recursive statistics method to extract the final caption region. The method based on differential operator and morphological processing is simple and effective for caption extraction. Also this method is robust for tilt and colorful caption texts.

The next work can be carried out from the following aspects: (1) The extracted caption region can be continued processing by segmentation, feature extraction, and recognition, or removing the caption text and restoring the background of image. (2) This system is completed in visual c++6.0 [13, 14] environment, the achievement of some algorithms is very complex, it can be simplified by using GDI+ [15, 16] technology or OpenCV and other open source library. (3) The extracted caption texts are sometimes incomplete, and it needs to further improve the algorithm.

References

1. Wang Qi, Chen Linqiang, Liang Xu. Caption extraction in video image [J], computer engineering and application, 2012, 48 (5): 177–216.
2. C.P. Sumathi, T. Santhanam, and G. Gayathri Devi, "Survey on various approaches of text extraction in im-ages," International Journal of Computer Science & En-gineering Survey, vol. 3, pp. 803–806, August 2012.
3. Cao Xixin, Liu Jing, Yang Xudong, Wu handsome, Zhang Qihui. A new video caption extraction algorithm [J]. Journal of Peking University (NATURAL SCIENCE EDITION), 2013,49 (2): 197–202.
4. Alvaro Gonzalez, Luis M. Bergasa, J. Javier Yebes, Sebastian Bronte, "Text Location in Complex Images", 21st International Conference on Pattern Recognition (ICPR2012), November 11–15, 2012. Tsukuba, Japan, pp. 617–620.
5. H. Chen, S. Tsai, G. Schroth, D. Chen, R. Grzeszczuk, and B. Girod. Robust text detection in natural images with edge-enhanced maximally stable extremal regions. In ICIP, 2011.
6. C.P. Sumathi and G. Gayathri Devi, "Automatic text extraction from complex colored images using gamma correction method," Journal of Computer Science, vol. 4, pp. 705–715, October 2014.
7. Keechul J, Jung H H. Hybrid approach to efficient text Extraction in complex color image[J]. Pattern Recognition Letters, 2004, 25(6):679–699.

8. Kim, K.I., Jung, K., Park, S.H., etc. Support vector machines for texture classification [J]. In: IEEE Trans Image Processing, 2002, 24(11):1542–1550.
9. Yang Shuying. VC++ image processing program design (Second Edition) [M], Beijing: Tsinghua University press, Beijing Jiaotong University press, 2005.
10. Yu Zhaohui, Pang Yechi et al. Visual C++ digital image processing and engineering application practice [M]. Beijing: China Railway Publishing House, 2012.
11. Lang Rui. digital image processing. Beijing: hope electronic publishing house, 2003.
12. Xu Hui. Visual c++ digital image practical engineering case selection [M], Beijing: People's Posts and Telecommunications, 2004.
13. Liu Haibo, Shen Jing, Guo song. Visual C++ digital image technology [M]. Beijing: Mechanical Industry Press, 2010.
14. Zhou Changfa. First Edition: The Mastery of image processing programme with Visual C++ [M]. Beijing: Publishing House of Electronics Industry, 2004.
15. Liu Ruining, Liang Shui, Li Weiming et al. Visual C++ full record about project development case (Second Edition) [M]. Beijing: Mechanical Industry Press, 2012.
16. Sun Xiumei, Li Xin et al. Actual combat assault Visual c++ project development case integration [M], Beijing: The Electronics Industry Press, 2011.

Reversible Data Hiding Based on Dynamic Image Partition and Multilevel Histogram Modification

Wenguang He, Gangqiang Xiong and Yaomin Wang

Abstract To better exploit block spatial redundancy, dynamic image partition method is proposed to construct a sharper difference histogram. Instead of figuring the optimal block size, we scan the whole image in specified direction and partition the host image into blocks with strategy as follow: the first three pixels construct the original block which may grows depending on the difference between the coming pixel and the median of the first three pixels. In this way, the image content and capacity requirement can be taken into account to achieve efficient image partition. In addition, EMD mechanism can be adopted for the extra part of block to achieve extensive embedding capacity promotion.

Keywords Reversible data hiding · Dynamic image partition · Multilevel histogram modification

1 Introduction

Data hiding is a technique that embeds secret message into host media to ensure copyright protection, authentication, and so on. As an important branch of data hiding, reversible data hiding (RDH) for image develops very fast in recent years and has been applied to many quality sensitive fields such as military, medical imaging and remote sensing for the feature that marked image can be precisely recovered after secret message extraction.

Many RDH schemes have been proposed so far. Among them, Tian's difference expansion (DE)-based scheme [1] has attracted much attention from researchers with the idea of embedding the secret message by expanding the difference between adjacent pixels. Afterwards, the DE technique has been widely investigated in many aspects such as Alattar et al. [2] extended the DE method by embedding secret

W. He (✉) · G. Xiong · Y. Wang
School of Information Engineering, Guangdong Medical University,
Guangdong 524023, China
e-mail: 56207403@qq.com

© Springer Nature Singapore Pte Ltd. 2018
S.K. Bhatia et al. (eds.), *Advances in Computer and Computational Sciences*,
Advances in Intelligent Systems and Computing 554,
https://doi.org/10.1007/978-981-10-3773-3_49

message into the k − 1 differences of k adjacent pixels, which provides larger embedding capacity (EC). In [3], Thodi et al. proposed a major extension of DE called prediction error expansion (PEE), where the difference between pixel intensity and its predicted value is utilized for expansion embedding. In this approach, prediction accuracy is close related to embedding performance. So, many prediction methods are investigated, for examples, median edge detector (Thodi et al. [3]), rhombus prediction (Sachnev et al. [4]), gradient adjusted prediction (Pei et al. [5]). Moreover, the PEE technique can also be developed in other aspects such as location map reduction [6], adaptive embedding [7].

Besides DE, Ni et al. proposed a remarkable RDH scheme based on histogram modification [8]. The histogram bins between the peak point and zero point are shifted before the peak points are employed for data embedding. However, its EC is limited despite high marked image quality and low computational complexity. To construct a sharper histogram, Lee et al. [9] proposed to utilize the difference histogram instead and Tsai et al. [10] proposed to utilize the prediction error histogram. Afterwards, difference histogram modification technique is also adopted in many works [11–13] while the correlation of two adjacent pixels is considered in [14, 15] and the correlation of pixels within block is considered in [11–13]. To better exploit block spatial redundancy, we present a new RDH scheme in this paper by incorporating dynamic image partition into multilevel histogram modification.

2 Proposed Scheme

2.1 Dynamic Image Partition

In [12], Luo et al. improved Kim et al.'s scheme [11] by selecting the median of each block to construct the reference image, which leads to a sharper histogram. However, uniform block size is not conducive to exploit block spatial redundancy. To solve this problem, we introduce the region growing technique and design a dynamic image partition (DIP) method with steps described as follows.

Step1. Scan all pixels in specified direction shown in Fig. 1 to form a pixel array.

Fig. 1 Direction for DIP

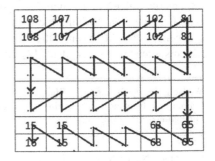

Step2. Sort the next three pixels which constructs the original block in ascending order such that $P_{s1} \leq P_{s2} \leq P_{s3}$ and select the median as the reference pixel. If any of the differences between reference pixel and other pixels is not larger than T, block will grow. Otherwise, turn to Step 2 to construct the next block.

Step3. Integrate the coming pixel with block. If the difference between reference pixel and the coming pixel is larger than T, the coming pixel will be the last element of block and turn to Step 2 to construct the next block. Otherwise, block keeps on growing.

In the above partition procedure, the parameter T is taken as the embedding level (EL) from [12]. After multilevel histogram modification based data embedding, the last element of block is surely enlarged by EL + 1. Then the decoder can achieve the same image partition by taking $T = 2 \times EL + 1$. This guarantees the reversibility.

After processing all pixels, the host image is partition into a set of size non-predictable blocks. In smooth region, larger block would produce more expandable differences. In rough region, smaller block helps to reduce the number of to-be-shifted differences. As lager EL would always guarantees more differences and larger maximum modification, EL is taken as the smallest positive integer such that the secret data can be successfully embedded.

By taking into account the fact that multilevel histogram modification is also adopted in [12, 15], Table 1 compares the numbers of expandable difference and to-be-shifted difference obtained without considering other factors such as overflow/underflow prevention. We can see that DIP always provides approximate expandable differences and much fewer to-be-shifted differences on all test images.

Table 1 Comparison of expandable differences and to-be-shifted differences

Image	Scheme	Expandable difference (E + 4)			To-be-shifted difference (E + 4)		
		EL = 2	EL = 3	EL = 4	EL − 2	EL − 3	EL = 4
Lena	Proposed	12.1	15.2	17.3	**8.9**	**6.8**	**5.4**
	Zhao et al.	12.1	15.3	**17.7**	14.1	10.9	8.5
	Luo et al.	**12.6**	**15.4**	17.3	11.9	9.1	7.3
Boat	Proposed	6.8	9.1	11.2	**12.7**	**11.0**	**9.5**
	Zhao et al.	6.9	9.2	11.4	19.3	17.0	14.8
	Luo et al.	**7.6**	**9.9**	**12.0**	17.0	14.6	12.6
Pepper	Proposed	8.5	11.3	13.7	**11.5**	**9.4**	**7.7**
	Zhao et al.	8.4	11.3	13.9	17.8	14.9	12.3
	Luo et al.	**9.6**	**12.6**	**14.9**	14.9	12.0	9.7
Barbara	Proposed	7.4	9.7	11.4	**12.2**	**10.7**	**9.5**
	Zhao et al.	7.5	9.8	**11.7**	18.7	16.4	14.5
	Luo et al.	**7.7**	**9.8**	11.4	16.9	14.8	13.2

2.2   EMD Mechanism

We define the first three pixels as the original part of block. For differences obtained from the extra part of grown block, differences with magnitude zero among them are utilized by using EMD mechanism.

First, transform seven continuous secret bits into a decimal number S. If S is not in range of [64, 79], turn to transform the former six continuous bits into S. Second, transforms into two nonary numbers and embed each of them by modifying a difference pair (0, 0) according to the nonary EMD table shown in Fig. 2. For example, because the decimal number of seven bits "1001010" is 74 and it is in range of [64, 79], we can directly transform 74 into two nonary numbers: $74 = 8 * 9 + 2$. Finally, we embed 8 and 2 by modifying (0, 0), (0, 0) into $(-1, 0)$ and $(-1, 1)$.

As two differences with magnitude zero are exploited to embed a nonary data, more than 1.5 bit data is embedded at each difference with magnitude zero. Table 2 shows the number of those differences with magnitude zero obtained from all test images.

2.3   Overflow and Underflow Prevention

We identify the exceptional block during the process of DIP. For each block, we suppose the data bits to be embedded are all "1"s and calculate the marked value of all pixels. Once overflow/underflow occurs, block will be resized to original and marked exceptional using the location map. Consequently, the location map is a

Fig. 2 Nonary EMD table

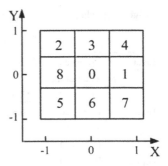

Table 2 Differences utilized for EMD mechanism

Image	EL = 2	EL = 3	EL = 4
Lena	16792	20001	21434
Boat	5321	6800	7927
Pepper	6961	9146	10991
Barbara	7231	9191	10398
Lena	16792	20001	21434

binary sequence with the same length of block set. Then losslessly compress it using arithmetic coding. Denote the length of the compressed location map as *Lclm*.

2.4 Data Embedding

The data embedding procedure is described as follows step by step.

Step 1 (Image partition and Location map construction): Partition the host image into a set of non-overlapped blocks and meanwhile obtain the location map.

Step 2 (Data embedding): We do nothing with exceptional blocks. For others, calculate the difference as

$$d_k = p_k - p_{s2} \tag{1}$$

Suppose w is a data bit to be embedded. For d_{s1} and d_{s3}, implement

$$d_{s1}^w = \begin{cases} 2 \times d_{s1} - w & \text{if } d_{s1} \geq -EL \\ d_{s1} - (EL+1) & \text{if } d_{s1} < -EL \end{cases} \tag{2}$$

$$d_{s3}^w = \begin{cases} 2 \times d_{s3} + w & \text{if } d_{s3} \leq EL \\ d_{s3} + (EL+1) & \text{if } d_{s3} > EL \end{cases} \tag{3}$$

For the rest except those which have been utilized for EMD embedding, implement

$$d_k^w = \begin{cases} 2 \times d_k + w & \text{if } d_k \in [1, EL] \\ 2 \times d_k - w & \text{if } d_k \in [-EL, -1] \\ d_k + (EL+1) & \text{if } d_k > EL \\ d_k - (EL+1) & \text{if } d_k < -EL \end{cases} \tag{4}$$

Finally, the marked value is obtained as

$$p_k^w = p_{s2} + d_k^w \tag{5}$$

This step will stop when all data bits are embedded. Then we denote *KB* as the index of last data-carrying block and *PB* the as the index of the first pixel of next block.

Step 3 (Auxiliary information and location map embedding): Record the least significant bits (LSB) of first $4 + 3\lceil \log_2(MN) \rceil + Lclm$ image pixels to obtain a binary sequence *SLSB*. Then replace these LSBs by the following auxiliary information and the compressed location map:

EL, *KB*, *PB*, *Lclm*. Finally, embed the sequence into the remaining blocks using the same method in Step 2 to obtain the marked image.

2.5 Data Extraction and Image Restoration

Step 1 (Auxiliary information and location map extraction): Read the LSB of first several pixels to get the values of *EL*, *KB*, *PB*, *Lclm* and the compressed location map.

Step 2 (Sequence *SLSB* extraction and image restoration): Starting from the *PB* one, retrieve all pixels to construct blocks by taking $T = 2 \times EL + 1$. Then extract *SLSB* and meanwhile realize the restoration.

Step 3 (Data extraction and image restoration): Replace LSB of first image pixels by *SLSB*. Then construct blocks and extract the hidden data and meanwhile realize restoration for those blocks.

3 Experimental Results

Our scheme is compared with three state-of-the-art schemes of [12, 13, 15, 16]. The reason to compare with schemes of [12, 13, 15] lies in that they all embed secret data by multilevel histogram modification. So the comparisons can reflect the advantage of DIP. Although multi-layer embedding is adopted in [16], they similarly aimed to exploit block spatial redundancy and multiple block sizes are employed meanwhile.

First, our scheme is compared with Luo et al.'s scheme [12] using 4×4 block size. According to the compassion results shown in Fig. 3, our scheme can provide higher PSNR with the same payload evidently. Moreover, with the increment of EL our scheme can provide larger EC and higher PSNR meanwhile with the same EL on images Lena, Boat and Barbara. The comparisons demonstrate the superiority of DIP.

Compared with Zhao et al.'s scheme [15], our scheme achieve better performance when EL > 0. When EL = 0, only differences with magnitude larger than 0 are shifted in Zhao et al.'s scheme and thus higher PSNR is achieved. Given EL > 0, we can see that our scheme always provides similar number of expandable differences compared with Zhao et al.'s. However, lots of reference pixels remain unchanged in our scheme. This helps to reduce the number of shifted pixels and thus higher PSNR is achieved.

Fu et al. [16] proposed to introduce EMD into prediction error expansion. From Fig. 3 we can see that DIP achieves similar efficiency in difference computation to PEE with moderate EC. However, the advantage of DIP shrinks when EL is large.

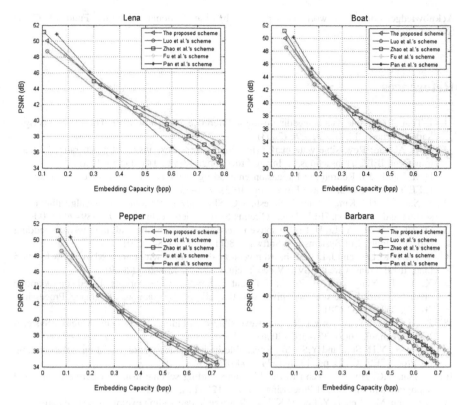

Fig. 3 Comparison of PSNR and EC between our scheme and related schemes

In Pan et al.'s scheme [13], the peak point of each block is selected as the reference pixel and only 1 and −1 are utilized for data embedding. This helps a lot to achieve high PSNR although maximum EC is limited. From Fig. 3 we can see that the maximum ECs brought by one-layer embedding is 0.39 dB, 0.24 dB, 0.29 dB, 0.25 dB for images respectively and highest PSNR is also achieved. However, Pan et al. failed to maintain high PSNR on the way to achieve larger EC through multi-layer embedding.

4 Conclusion

In this paper, we present a reversible data hiding scheme based on dynamic image partition and multilevel histogram modification. To better exploit block spatial redundancy, uniform block size is replaced by dynamic image partition and EMD mechanism is introduced. Performance comparisons of several available schemes demonstrate the effective improvement achieved by the proposed scheme.

Acknowledgements This work is supported by the National Scientific Fund of China (No. 61170320).

References

1. J. Tian.: Reversible data embedding using a difference expansion. IEEE Trans. Circuits Syst. Video Technol. 13(8), 890–896 (2003)
2. A.M. Alattar.: Reversible watermark using the difference expansion of a generalized integer transform. IEEE Transactions on Image Processing. 13(8), 1147–1156 (2004)
3. D.M. Thodi, J.J. Rodriguez.: Expansion embedding techniques for reversible watermarking. IEEE Transactions on Image Processing. 16(3), 721–730 (2007)
4. V. Sachnev, H.J. Kim, J. Nam, S. Suresh, Y.Q. Shi.: Reversible watermarking algorithm using sorting and prediction. IEEE Trans. Circuits Syst. Video Technol. 19(7), 989–999 (2009)
5. Q. Pei, X. Wang, Y. Li, H. Li.: Adaptive reversible watermarking with improved embedding capacity. Journal of Systems and Software. 86(11), 2841–2848 (2013)
6. Y. Hu, H.K. Lee, and J. Li.: DE-based reversible data hiding with improved overflow location map. IEEE Trans. Circuits Syst. Video Technol. 19(2), 250–260 (2009)
7. X. Li, B. Yang, and T. Zeng.: Efficient reversible watermarking based on adaptive prediction-error expansion and pixel selection. IEEE Transactions on Image Processing. 20(12), 3524–3533 (2011)
8. Z. Ni, Y.-Q. Shi, N. Ansari, W. Su.: Reversible data hiding. IEEE Trans. Circuits Systems Video Technol. 16(3), 354–362 (2006)
9. S.K. Lee, Y.H. Suh, and Y.S. Ho.: Reversible image authentication based on watermarking. In: Proc. IEEE ICME, pp. 1321–1324 (2006)
10. P. Tsai, Y.C. Hu, H.L. Yeh.: Reversible image hiding scheme using predictive coding and histogram shifting. Signal Processing. 89(6), 1129–1143 (2009)
11. K.S. Kim, M.J. Lee, H.Y. Lee, H.K. Lee.: Reversible data hiding exploiting spatial correlation between sub-sampled images. Pattern Recognition. 42(11), 3083–3096 (2009)
12. H. Luo, F.-X. Yu, H. Chen, Z.-L. Huang, H. Li, P.-H. Wang.: Reversible data hiding based on block median preservation. Information Sciences. 181(2), 308–328 (2011)
13. Z.-B. Pan, S. Hu, X.-X. Ma, L.F W.: Reversible data hiding based on local histogram shifting with multilayer embedding. J. Vis. Commun. Image R. 31, 64–74 (2015)
14. Y.C. Li, C.M. Yeh, C.C. Chang.: Data hiding based on the similarity between neighboring pixels with reversibility. Digital Signal Processing. 20(4), 1116–1128 (2010)
15. Z.-Z. Zhao, H. Luo, Z.-M. Lu, J.-S Pan.: Reversible data hiding based on multilevel histogram modification and sequential recovery. Journal of Electronics and Communications. 65, 814–826 (2011)
16. D.-S. Fu, Z.-J. Jing, S.-G. Zhao, J. Fan.: Reversible data hiding based on prediction-error histogram shifting and EMD mechanism. Journal of Electronics and Communications. 68, 933–943 (2014)

Part IV
ADBMS and Security

Threshold-Based Hierarchical Visual Cryptography Using Minimum Distance Association

Pallavi Vijay Chavan and Mohammad Atique

Abstract In this paper, we consider the novel type of visual cryptography scheme, which can decode concealed images without any cryptographic computations. The hierarchical visual cryptography scheme is perfectly secure and very easy to implement. It divides secret in number of pieces called shares. We extend this scheme with variant of threshold λ. Shares generated out of this scheme are expansionless and capable to reconstruct high-contrast secret. The approach associates the shares of hierarchical visual cryptography using Euclidean distance measure. This visual cryptographic scheme superimposes the shares to decode the secret using minimum distance association.

Keywords Authentication · Hierarchical visual cryptography · Secret sharing · Thresholding

1 Introduction

Visual cryptography is a technique of encrypting information such as handwritten text and images in such a way that the decryption is possible without any mathematical computations and human visual system is sufficient to decrypt the information. In order to encode the secret, visual cryptography split the original secret among *n* modified versions (referred as shares) such that each pixel in a share is theoretically subdivided into *n* black and white sub-pixels. To decode the secret, a subset *S* of those *n* shares are picked and copied on separate transparencies. Visual cryptography was independently introduced by Shamir in the form of secret sharing. Shamir divided secret information D into n number of pieces such that D is

P.V. Chavan (✉)
Department of Computer Engineering, BDCE, Sevagram, MH, India
e-mail: pallavichavan11@gmail.com

M. Atique
Department of Computer Science, SGBAU, Amravati, MH, India
e-mail: mohd.atique@gmail.com

© Springer Nature Singapore Pte Ltd. 2018   513
S.K. Bhatia et al. (eds.), *Advances in Computer and Computational Sciences*,
Advances in Intelligent Systems and Computing 554,
https://doi.org/10.1007/978-981-10-3773-3_50

reconstructable from any k number of shares. Knowledge of any (k−1) shares is not capable of reconstructing the secret information D [1]. The simplest form of visual cryptography separates a secret into two parts so that either part by itself conveys no information. In traditional visual cryptography system, encryption and decryption requirement result from the physical characteristics of the secret to be processed [2]. Some visual cryptography applications must meet external requirements too. For example, plain binary secret is encrypted by some visual cryptographic technique; the same technique has to handle the secret of biometric form or any other form of spatial domain. Most visual cryptography schemes implement single level secret encryption [3, 4]. Noticeably, they are degrading the quality of the reconstructed secret. Even though meeting such expected quality is a primary characteristic of the visual cryptography applications.

Visual cryptography schemes are mostly vulnerable to many security threats and give birth to new security challenges. Some of them are permutation and combination method applied on individual share to reveal the secret without existence of all the participants [5]. In particular, visual cryptography is subjected to compromise information integrity, authenticity, user privacy and system performance. Traditional visual cryptography schemes are limited to print and scan applications. These schemes lead to excessive use of transparencies over which every individual share is supposed to be printed. All the schemes stated in literature are subjected to existence of greying effect in reconstructed secret. Greying effect is some sort of noise observed in the background of the revealed secret. Reduction of greying effect is also a considerable challenge for proposing the new approach of hierarchical visual cryptography [6]. The pixel expansion ratio is another issue wherein every secret after reconstruction is observed in expanded form. Expansion does not affect textual secrets at all. For informative secrets in the form of images, expansion ratio may lead to wrong reconstruction of secret. Reduction of expansion ratio to considerable level is necessity of upcoming visual cryptographic techniques. There are several analytical studies on the overall performance of few visual cryptography schemes. However, these schemes are subject to the number of participants involved in secret reconstruction. These schemes fail to potentially exploit the shares in digital form. Necessarily, shares generated by them are printed on transparencies for staking operation. Staking operation is equivalent to the superimposition of the shares [7, 8]. Digital shares can be achieved when the encryption process is modeled differently with respect to traditional approaches. This leads to give birth to special issues about the reconstruction of secret. Obviously, such form of digital shares works differently. Another important factor in proposing the HVC approach is the elapse time required for reconstruction. In all traditional approaches, for reconstruction, shares are superimposed in the form of images. This approach is not feasible for n number of shares [9]. By considering all of the above standing issues, the authors designed hierarchical approach of encryption. In this paper we consider the encryption of biometric secrets in absolute secure way which can be decoded without any cryptographic computations. The human visual system is necessary and sufficient condition for decoding. The original secrets are reconstructed by superimposition of shares (the best way to visualize the art of visual cryptography is to print the shares on transparencies and to place them one above other. This is traditional approach of visual cryptography, although this approach is not feasible, and now a day's digital shares superimposition is widely adopted) [10, 11].

2 Hierarchical Visual Cryptography

Security is the most important aspect over the Internet. There has been an incredible amount of research conducted in effective encryption of secret using visual cryptography. Visual cryptographic schemes are implemented in several ways; more importantly number of levels is not introduced during encryption.

This section is concerned with the issues conceived in the encryption process, specifically, quality of reveled secret and time complexity of encryption. The issues concerned in design of hierarchical visual cryptography ranges within: the decision of a number of levels in encryption, quality of revealed secret, time complexity and space complexity. In Visual Cryptography context, hierarchy indicates the depth of encryption in terms of number of levels. Hierarchical visual cryptography evolved as a multilevel architecture of visual cryptography. The design of hierarchical visual cryptography is fundamentally determined by how the secret is handled in multiple levels without expansion. In general, the multilevel encryption constraint in visual cryptography has not been widely studied as a general visual cryptography problem. The proposed model of hierarchical visual cryptography attempts to integrate the concept of layered encryption with visual cryptography.

2.1 The Model

The simplest version of the hierarchical visual cryptography (HVC) assumes that the secret consists of collection of black and white pixels. Biometric secrets are dithered in order to satisfy this requirement. Each pixel in original secret appears in 5 modified versions. Each share is a collection of black and white pixels, which are in close proximity to each other. HVC shows the variations in the encryption process with respect to the threshold values. The secrets are encrypted using multiple threshold values. The variety of secrets that HVC model encrypts ranges from passwords, biometric information to handwritten texts and images.

HVC is the art of encrypting secret information in number of levels. Unlike visual cryptography, initially, HVC encrypts the secret among two shares in the first level. The resulting structure is referred as *share1* and *share2*. At second level of encryption, HVC encrypts *share1* and *share2* independently. The outcome of the second level is a collection of four shares viz. *share11, share12, share21* and *share22*. *Share11* and *share12* are the outcome of *share1*. Similarly, *share21* and *share22* are generated by encrypting *share2*. The architecture of HVC model is presented in Fig. 1. In this architecture, S denotes dithered secret while S' is

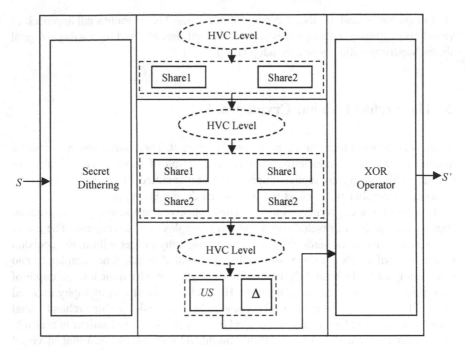

Fig. 1 Hierarchical visual cryptography architecture

revealed secret. This framework resembles the framework of the *(k, n)* thresholding scheme of visual cryptography with the important difference that encryption process is hierarchical rather than sequential. Level 3 is of great importance leading to the formation of key share Δ (It is considered as a combination of three successive shares of the second level viz. *share12*, *share21*, *share22*). Outcome of the second level of encryption helps third level for the formation of Δ. The remaining share (*share11*) is treated as user share *US*. The hierarchical encryption is based upon thresholding approach.

2.2 Thresholding Approach

Thresholding approach is used in many ways in the field of cryptography. Threshold is a standard and precise value uniquely determined prior to the encryption process. Based upon the domain of experimental inputs, the threshold value λ appears with the interval of [0 1]. Dithered secret *S* indicates black and white tendency. In this scheme threshold is denoted by λ. Value of λ determines the concentration of black and white pixels in the resultant shares. Following is the algorithm for share generation using HVC.

Input: secret S, Output: shares.

1. *Read S.*
2. *Normalization and dithering of S.*
3. *Generate random number* α.
4. *p* = α > λ
5. *Repeat steps 5 through 6 for all pixels in S.*
6. *For every white pixel in S*
   Share1 is assigned p and Share2 is assigned p.
7. *For every black pixel in S*
   Share1 is assigned ~p and Share2 is assigned p.
8. *Output Share1 and Share2.*

While generating shares using thresholding approach, 0.5 is the probability of occurrence of black pixel while 0.5 is the probability of occurrence of white pixel. In order to show the high contrast level in decrypted secrets, λ is experimented with the value of 0.5. For 0.5 > λ > 0 *share2* tends to reflect information of secret which is visible by necked eyes. Similarly, for 1 > λ > 0.5, *share1* tends to reflect more black information and *share2* tends to expose the information about secret visually. Suitable value of λ is 0.5 maintaining the balance of randomness in both of the shares. Thus λ forms the first level of visual cryptography in this model (first level of hierarchy). In the later stage of encryption, same model is applied and forming resultant four shares indicating second level of hierarchy in encryption.

During experimentation, we have analysed that 0.5 is the best suitable value of λ for random patterns in shares. Shares generated by λ = 0.5 are perfectly random and reveals no information about the secret. In general, λ ranges between 0 and 1 inclusive. The algorithm is experimented for different values of λ within specified range. Figure 2 shows the shares generated by different values of λ in the interval of

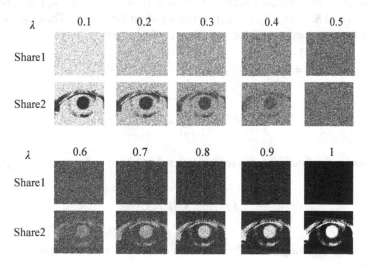

Fig. 2 Shares at different values of λ

[0 1]. Shares are observed from threshold 0.1 to 1 with unit step of 0.1. It is observed that λ close to 0 gives lighter shade in share while λ close to 1 gives darker shade of share. In both of the cases, share2 reflects secret information. The average case time complexity of HVC algorithm is $O(n^2)$.

3 Minimum Distance Association of Shares

Minimum distance association of shares (MDAS) is defined using Euclidean distance. Unlike traditional Euclidean distance, MDAS takes into account the relationship among the pixels in spatial domain. The Euclidean distance plays vital role in identifying the closeness between two entities. Therefore, this approach is robust for digital secrets in any form viz. handwritten text, biometric information (fingerprints and iris). Common problem associated with the secrets in form of digital images is to find the distance between pair of images. Euclidean distanced (ED) concept is used most often due to its reliability and simplicity. Euclidean distance (ED) or Euclidean norm is the distance between two patterns. Here a pattern refers to the shares. ED(U, S) is the Euclidean distance between the user share U and the secret information S. The pattern space to compute Euclidean distance is 2 dimensional here. Similarly, ED (K, S) is the distance between the key share and the secret information. In order to compute Euclidean distances, the patterns in 2 dimensional patter space are normalized. Normalization brought the information of each pattern in the interval [0 1]. Each secret is encrypted using hierarchical visual cryptography. After encryption in 3 successive levels, two shares are generated. Let S be the set of secrets such that S = {s_1, s_2, ..., s_n}. All the secrets are of the same size 280 by 320 considered for experimentation. However, secret of any size undergoes MDAS. Consider U = {u_1, u_2, ..., u_p} as a set of all user shares and K = {k_1, k_2, ..., k_p} be the set of all key shares. Then the Euclidean distance between any user share and the secret is given by,

$$ED = sqrt\left[(u_1 - s_1)^2 + (u_2 - s_2)^2 + \cdots + (u_p - s_p)^2\right] \tag{1}$$

$$ED = \sqrt{\sum_{i=1}^{2}(u_i - s_i)^2} \tag{2}$$

Similarly, the Euclidean distance between key share and the secret is given by,

$$ED = \sqrt{\sum_{i=1}^{2}(k_i - s_i)^2} \tag{3}$$

Following is the algorithm for minimum distance association among shares and the secret.

Inputs: S, U, K, Outputs: Euclidean norms E(U, S), E(U, K).

1. *Begin*
2. *Read S, U, K.*
3. *Normalize S, U, K.*
4. *Repeat for all U*
   Compute dist1 = ED(S, U).
5. *Repeat for all K*
   Compute dist2 = ED(S, K).
6. *Find minimum association of S with respect to U- min(dist1).*
7. *Find minimum association of S with respect U- min(dist2).*
8. *Output dist1, dist2.*
9. *Stop.*

While associating the share using MDAS, two different Euclidean distances are computed. First is the distance between user share u_i and all n number of secrets. This provides as array of Euclidean distances of size n meaning that for n number of secrets, the system computes n Euclidean distances between supplied user share u_i and all n secrets. Once the array of Euclidean distances is available, minimum Euclidean distance is observed among all. Let us assume that for ith user share u_i, we get ith secret s_i which is providing minimum distance between the user share and the secret. Secondly, ith secret is useful to compute further distances. Again n number of Euclidian distances is computed between the ith secret s_i and all the key shares available in the set (n in number). This is another array of Euclidean distances. Same principle of minimum distance identification is applied here. It is found during experimentation that ith secret is giving the minimum distance with respect to applied user share in first array of distances and for the same secret i, key share k_i is giving minimum Euclidean distance. Euclidean distances for all pairs of user share and secret are presented in Fig. 3. Euclidean distances for all pairs of key share and secret are also represented in Fig. 3.

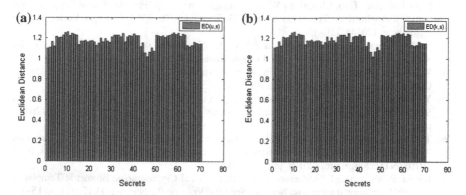

Fig. 3 **a** Euclidean distances *ED(u, s)* for all pairs of the user share and dithered secret. **b** Euclidean distances *ED(k, s)* for all pairs of the key share and dithered secret

However, Euclidean distance measure is highly sensitive to small changes in the dithered secret. Reasonable association of shares with secret should represent the minimum distance between user share and the secret. In parallel, minimum should be the distance between keys share and the same secret. If both of these two conditions are achieved by MDAS then it is inferred that the desire secret is revealed. Some key features of MDAS: It is applicable to any secret recognition in digital form. Small changes in the secret or in any share cause the distance factor, simplicity in computation of distance measure and it can be easily embedded with secret in any form like spacial or temporal data.

4 Conclusion

In this paper, we suggested hierarchical visual cryptography model for encryption of digital secrets. The suggested threshold based approach of encryption is capable of generating the expansionless shares at each level and also eliminates the greying effect from the background of the revealed secret. The major goal achieved in this contest is about the quality of reconstructed secret. The novel association scheme among the shares is conceived based on Euclidean distance. The advantage of Euclidean distance is that the shares belonging to the same secrets are determined on basis of minimum Euclidean distance. It has been demonstrated that the proposed minimum distance associator is capable of identifying the secret associated with the pair of share. This pair reflects the minimum Euclidean distance with respect to the same secret.

References

1. A. Shamir, How to Share a Secret, Communication in ACM, 22, (1979), pp 612–613.
2. Feng Liu, Teng Guo, ChaunKun Wu and Ching-Nung Yang, Flexible Visual Cryptography and It's Application, Transactions on DHMS IX, LNCS 8363, Springer-Verlag, Berlin Heidelberg, (2014), 110–130.
3. S. Arumugam, R. Lakshmanan and Atulya K. Nagar, On (k, n)*- Visual Cryptography Scheme, Journal of Design Code and Cryptography, 71(1), (2014), pp 153–162.
4. Wen-Guey Tzeng and Chi-Ming Hu, A New Approach for Visual Cryptography, Design Code and Cryptography, 27, (2002), pp 207–227.
5. Yan-Xiao Liu, Lein Harn, Ching-Nung Yang, Yu-Qing Zhanga, "Efficient (n, t, n) Secret Sharing Schemes", in Elsevier Journal of Systems and Software, Science Direct, 85, (2012), pp 1325–1332.
6. Thomas Monoth, Babu A. P., "Contrast-Enhanced Visual Cryptography Schemes based on Additional Pixel Patterns", in Proceedings of International Conference on Cyberworlds, (CW 2010), NTU, Singapore, (2010), pp 171–178.
7. Rahna. P. Muhammed, "A Secured Approach to Visual Cryptographic Biometric Template", ACEEE International Journal on Network Security, Volume-02, Number-03, (2011), pp 15–17.

8. Mohd. Junedul Haque, Mohd. Muntjir, Mohd. Rahul, "Image Encryption – An Intelligent Approach of Color Visual Cryptography", in International Journal of Computer Applications, 83(5), (2013), pp 7–9.

9. Arun Ross, Asem Othman, "Visual Cryptography for Biometric Privacy", in IEEE Transactions on Information Forensics and Security, 6(1), (2011), pp 70–81.

10. Feng Liu, Teng Guo and Ching-Nung Yang, "Flexible Visual Cryptography Scheme and It's Application", in Springer Transaction on DHMS IX LNCS 8363, (2014), pp 110–130.

11. Arun Ross, Asem Othman, "Visual Cryptography for Biometric Privacy", in IEEE Transactions on Information Forensics and Security, 6(1), (2011), pp 70–81.

Security in IoT-Based Smart Grid Through Quantum Key Distribution

Maninder Kaur and Sheetal Kalra

Abstract The Internet of Things (IoT) is made up of a large group of networks, which consists of many physical objects provided with the capabilities of making computation and communication is logically integrated at various distinctive levels of internet. The Smart Grid (SG) forms a system which consists of various operative and energy measures including smart objects, various smart appliances, smart meters, sensors, energy efficiency resources. The Smart Grid is considered as the largest IoT network where security is the prime factor that hinders the fast and extensive adoption of the IoT-based Smart Grid. In this paper, various security-related issues are described and a mutual authentication scheme for the Smart Grid Users and Smart Grid Server based on Quantum Key Distribution has been proposed. The proposed scheme achieves mutual authentication between Smart Grid Users and Smart Grid Server and provides security against various security attacks by generating secret key, which consists of qubits. The formal verification of the proposed scheme is performed using QKD simulator that confirms its security in the presence of an adversary.

Keywords Authentication · Internet of Things · Quantum Cryptography · Quantum Key Distribution · Security · Smart Grid

1 Introduction

The Internet of Things (IoT) consists of a network in which unique identifiers are provided to objects or people that is able to transmit the required data through the network unaccompanied by either human-to-human or human-to-computer form of

M. Kaur (✉) · S. Kalra
Department of Computer Science & Engineering, Guru Nanak Dev University,
Regional Campus, Jalandhar, India
e-mail: bhogalmani9547@gmail.com

S. Kalra
e-mail: sheetal.kalra@gmail.com

© Springer Nature Singapore Pte Ltd. 2018
S.K. Bhatia et al. (eds.), *Advances in Computer and Computational Sciences*,
Advances in Intelligent Systems and Computing 554,
https://doi.org/10.1007/978-981-10-3773-3_51

interaction. Thus, IoT communication is the development of machine-to-machine (M2M) communication.

1.1 Smart Grid

The Smart Grid is thought to be the first and biggest IoT example. Smart Grid is the combination of Internet technology and electric grid. These systems are used, from the various power plants generation on electricity networks to the buyers of electricity consumers. They provide various advantages to consumers and utilities that enhance betterment in energy resources efficiency and reliability on the electricity grid [1].

2 Related Work

Reddy et al. [2] discussed various techniques that are very beneficial for controlling functions of metering and communication in Smart Grids. Xia et al. [3] proposed a secure and efficient key distribution protocol for a Smart Grid network between smart meters and service providers, using a trusted third party which does not have issue on key revocation. Wei et al. [4] proposed a layered framework for the protection of Smart Grid from many internal and external security attacks and discussed the various challenges and strategies used to protect the Smart Grid. Eissa et al. [5] discussed the efficient data schemes for analysis of data in detection and prediction purposes in Smart Grid. Qinghua et al. [6] presented a One-Time Signature scheme for providing Multicast Authentication in the Smart Grid and offers various advantages, turntable signing and verification (TSV) generates very smaller signature which have lower storage requirement, hence reduces storage overhead. Hasen et al. [7] proposed an authentication Scheme and Key Management protocol for Home Area Network via Public/Private key pair technique based on identity-based cryptographic technique, which helps in reduces overhead. Hongwei et al. [8] presented Authentication Scheme for the Smart Grid by using symmetric cryptography Algorithm scheme and technique of Merkle tree, which reduces complexity and computational cost.

3 Security Issues of IoT-Based Smart Grid

1. **Authentication**: The ability to assure or certify the identity of every smart device that wants to communicate in the Smart Grid.

2. **Data Integrity**: It is prime security issue, an intruder can modify exchanged data, hence effect the integrity of data.
3. **Eavesdropping**: In IoT-based Smart Grid, various devices communicate with each other, through public communication channel, a third party can access their exchanged confidential information. This is termed as eavesdropping [9].

4 Preliminaries of Quantum Key Distribution (QKD)

QKD is based on the Heisenberg principle, according to which physical properties of certain pairs are described in a way, that observer cannot be able to simultaneously measure the value of two properties [10, 11]. This principle is very important to prevent attacks of an eavesdroppers on a quantum-based cryptosystem. Photon polarization principle states that a light photon is polarized in a particular direction, so an eavesdropper is not able to copy unknown qubits according to no-cloning theorem, which was first presented in 1982 [12].

4.1 How Quantum Key Distribution Helps to Solve Security Issues

1. **How QKDP's can help in authentication**: For the authentication purpose, both the sender and the receiver has to register themselves into the database that is separately maintained for them and a secure key is generated by the authentication server whenever each user login on it.
2. **How QKD helps in maintaining data integrity**: Only authenticated users are allowed to access the data in SG.
3. **How QKD can help in detection of eavesdropping**: QKD is employed to identify the presence of any adversary. If an adversary tries to record the quantum state, it will disturb the functioning of quantum system. This detectable disturbance enables to find the participation of the third party who tries to eavesdrop [13].

5 Proposed Protocol

The devices utilizing QKD provides authentication services over the optical (quantum) channel. An authentication protocol for keeping up quantum states must have the capacity to secure superposition of states. Authentication is the process in which the credentials provided, are compared to those on file in the database stored on local operating system within an Authentication Server. The following steps have to be followed to achieve authentication:

1. If Smart Grid User wants to communicate with Smart Grid Server, it sends the request message and it's UserId which is encrypted using with its quantum basis for verification to Server through classical channel.
2. Smart Grid Server receives its request. If it is interested for communication, it processes the requests.
3. Smart Grid Server verifies the UserId, which can be decrypted using public key cryptography, with centralized authority. (Assume Centralized authority is a trusted one).
4. It maintains Smart Grid User's Id random basis for further communication.
5. Smart Grid Server sends the reply message containing UserId and its authenticated message to Smart Grid User denoting acceptance of communication.
6. Smart Grid User verifies it and send the quantum basis as nibble through quantum channel.
7. Smart Grid Server verifies the quantum basis with stored basis. If it matches, it is ready to share the secret key. Send same quantum basis to Smart Grid User. Smart Grid User verifies it and ready to transfer the secret key.
8. In proposed scheme, after authentication of user, server generates the session key, which is required to encrypt communication to assure confidentiality of the message. A one way hash function and a random number is used to generate the session key (S2) (Fig. 1 and Table 1).

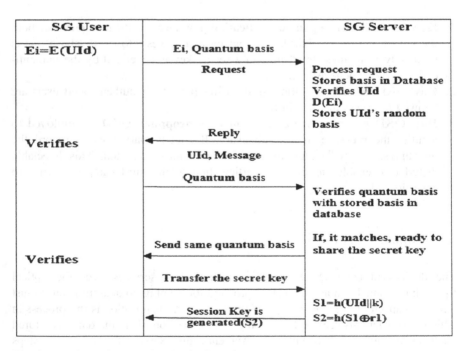

Fig. 1 Workflow of proposed authentication scheme

Table 1 Notations used in the protocol

UId	User Id
E	Symmetric encryption algorithm
D	Decryption algorithm
k	Unique secret value of SG server
r1	Random number generated by the SG server
$\oplus$	XOR operation
||	Concatenation

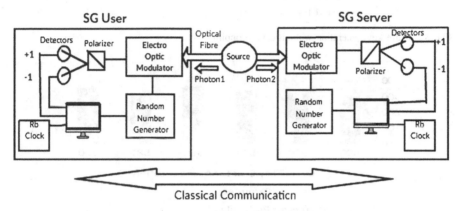

Fig. 2 Experimental setup of proposed scheme

6 Experimental Setup

The experimental realization of proposed scheme is sketched in the Fig. 2. The formal verification of the proposed scheme is performed using QKD simulator that conforms its security in the presence of an adversary. QKD simulator is a web application which analyses and simulates all aspects of quantum key distribution using a powerful engine which includes quantum key distribution toolkit. The tool has been implemented in Python and make use of different scientific libraries such as Quantum Information Toolkit (QIT), Scipy, BitVector. The experiment has been performed using different values of qubit count.

6.1 Key Generation Without Eavesdropping

Figure 3 shows implementation of Quantum Key Distribution using different values of qubits without enabling eavesdropping, size of qubits to be sent is taken as 500, 520, 540, 560, 580, 600 and is taken along the X-axis.

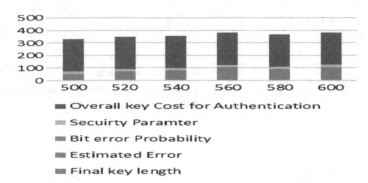

Fig. 3 Secure key generation without eavesdropping

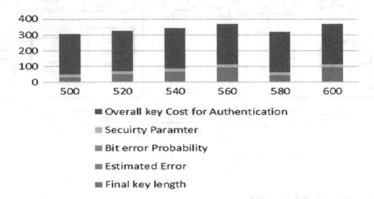

Fig. 4 Secure key generation with eavesdropping

6.2 Key Generation with Eavesdropping

Figure 4 shows values of different parameters on implementing Quantum Key Distribution with eavesdropping. With eavesdropping, final key length has reduced but eavesdropper is not successful to know the secret key. Thus Quantum Key Distribution is very secure.

6.3 Comparison of Different Authentication Schemes

Following Table 2 describes the comparison between different authentication schemes with above discussed proposed scheme.

Table 2 Comparison of different authentication schemes

Authentication schemes	One time signature scheme [6]	Authentication scheme and key management protocol for home area network [7]	Authentication scheme for the smart grid [8]	Proposed scheme
Method used	Utilizes HSLV (Heavy signing light verification) with LSHV (light signing heavy verification) and acquire the scheme turntable signing and verification (TSV), design a multicast authentication protocol	Public/Private key pair strategy based on identity based cryptographic technique	Follows symmetric cryptography algorithm scheme and strategy of Merkle tree	Quantum key distribution
Attacks	Replay attack	DOS -attack, Man-in-middle, replay attacks	Message injection attack, replay attack	Till now, no attacks are possible on quantum key distribution
Disadvantages	The security of the proposed scheme depends on PKI (public key infrastructure). PKI has disadvantages identified with authentications which will be enormous concern	It utilizes private key generator which deals with the private key so privacy is significant concern	Utilizing symmetric keys for whole smart grid is not adaptable because of the expansive number of gadgets and hubs	There is only one constraint, its high cost for business-oriented QKD systems and it is subjected to limited distance

7 Conclusion

This paper gives an overview of security issues of the IoT-based Smart Grid. The Smart Grid is considered as the largest IoT network where security is the major concern that hinders the fast and extensive adoption of the IoT-based Smart Grid. Thus, this paper proposes a scheme to authenticate Smart Grid server and Smart

Grid user using QKD. The proposed scheme mutually authenticates server and user guarantees security against different security attacks by creating secret key which comprises of qubits. The formal verification of the proposed scheme is performed using QKD simulator and the outcomes obtained affirms the security of proposed scheme in presence of an eavesdropper.

References

1. http://energy.gov/oe/services/technology-development/smart-grid.
2. Reddy KS, Kumar M, Mallick TK, Sharon H, Lokeswaran S. Are view of integration, control, communication and metering (iccm) of renewable energy based smart grid. Renew Sustain Energy Rev 2014; 38:180–92.
3. Xia J, Wang Y. Secure key distribution for the smart grid. IEEE Trans Smart Grid 2012; 3 (3):1437–43.
4. D. Wei, Y. Lu, M. Jafari, P. Skare, K. Rohde, Protecting smart grid automation systems against cyber-attacks, IEEE Trans. Smart Grid (99) (2011).
5. Eissa MM. Protection techniques with renewable resources and smart grids – a survey. Renew Sustain Energy Rev 2015; 52:1645–67.
6. Qinghua Li and Guohong Cao, "Multicast Authentication in the Smart Grid with One-Time Signature" IEEE Transaction on Smart Grid, vol. 2, no. 4, pp. 686–696, December 2011.
7. Hasen Nicanfar, Paria Jokar, Victor C.M. Leung, "Efficient Authentication and Key Management for the Home Area Network" IEEE Trans. pp. 878–882, 2012.
8. Hongwei Li, Rongxing Lu, Liang Zhou, Bo Yang and Xuemin (Sherman) Shen, "An Efficient Merkle-Tree- Based Authentication Scheme for Smart Grid", IEEE system Journal, pp. 1–9, 2013.
9. http://www.sciencedirect.com/ Chakib Bekara / Procedia Computer Science 34 (2014) 532–533.
10. Hrg, D., Budin, L., & Golub, M., "Quantum cryptography and security of information systems", IEEE Proceedings of the 15th Conference on Information and Intelligent System, 2004, pp. 63–70.
11. Papanikolaou, N., "An introduction to quantum cryptography", ACM Crossroads Magazine, Vol. 11 No. 3, 2005, pp. 1–16.
12. Wootters, W. K., & Zurek, W. H., "A single quantum cannot be cloned". Nature, 299, 1982, pp. 802.
13. http://www.ukessays.com/essays/computer-science/overview-of-quantum-key-distribution-protocol-computer-science-essay.

A Comparative Study on Face Detection Techniques for Security Surveillance

Dimple Chawla and Munesh Chandra Trivedi

Abstract Automatic Face Recognition and Surveillance helps providing a secure system for next generation. The critical technology used for this purpose is pretty poor, but it is slowly improving. In this paper, we are making a comparison on ADABOOST, PCA, LDA, Elastic Bunch Graph Matching algorithms for face recognition on the basis of origin, success rate, Eigen values, and vector score as output of an image. Further study from above said algorithms, it is clear to state that significant result is least expected to perform well on small sample data.

Keywords FERET · SMQT · SNOW · ADABOOST · PCA · LDA · EBGM

1 Introduction

A facial recognition system is a process to identify and verify a person from a given source image or from a video source. The best ways are to get competent selected facial features from the image and a facial reference database. Advancement of face recognition system is not just to be recognized by an individual face but also in computing various new features over the past few decades. After 1960, automated facial recognition changed the whole scenario. It is the first semi-automated system which helped the administrator to locate and compare physical features, i.e., ears, eyes, nose, etc., with the referenced database. In early era, the algorithm for facial recognition is the simple geometric models, but the process is bit advanced in mathematical representations and matching processes [1].

D. Chawla (✉)
Pacific Academy of Higher Education & Research University, Udaipur, India
e-mail: chawladel@gmail.com

M.C. Trivedi
ABES Engineering College, Ghaziabad, India
e-mail: munesh.trivedi@gmail.com

© Springer Nature Singapore Pte Ltd. 2018
S.K. Bhatia et al. (eds.), *Advances in Computer and Computational Sciences*,
Advances in Intelligent Systems and Computing 554,
https://doi.org/10.1007/978-981-10-3773-3_52

531

The recognition system uses spatial geometry for distinguishing between features of the particular face to identify or to authenticate the individual in source image. The system also gets different results where faces taken from some distance away like in real-time video camera. Therefore, system can be applied without knowing subject involved while considering it well.

1.1 Face Recognition Approaches

According to earlier face recognition techniques, it can be categories in two prominent approaches Geometric (feature based) in (V.V. Starovoitov et al. [2]) and Photometric (view based) mentioned in [2].

Feature-based approach (V.V. Starovoitov et al. [2]) uses geometric features, which is generated on the basis of segments, perimeters, and areas of figures generated by the various points as control vertices (Fig. 1). There are 37–39 points that are used for recognition. They include approximately 15 segments between the vertices and its mean values of symmetrical pairs [2]. In this, each image is of dimension 92 × 112 pixels with quantized 256 gray-scale levels. By following FERET (Face Recognition Sponsored Algorithm Development Research Technology), five similar face images were tested and if they were tested, then the result is considered as positive. Each image was tested as a query and compared with other input source and its recognition rate is nearly at 98.5%.

Photometric Stereo effect shown in (Fig. 2) is a computer vision technique; given in [3] uses high observing object under different lighting conditions. The fact about this technique is that, the more the reflection of light more the proportional to the surface orientation. To determine the light amount reflected into the camera and possible surface space orientation by providing appropriate light source from prominent angle of deviation, the orientation constrains to a single or even over-constrained due to multiple reasons.

Fig. 1 Feature-based technique

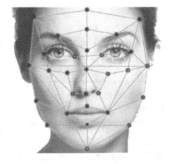

Fig. 2 Photometric stereo
effect

2   Literature Review

Majorly this study is selected for security under image processing for supporting both industry expert as well as to the academician who are in the similar research field. A review of the literature is based as follows:

As explained in (M. Singh et al. [4]) with the change on time, age, circumstances changes reflect on each individual faces, skeleton structure, muscle mass, and body fat. The mounting number of face recognition system, both images and videos are very eminent to work with. Different set of poses, expression, and illuminations are issues mostly studied under this system [4]. It has been observed in face recognition that the effects of compression through the images are directly stored and transported in a compressed form; whereas still pictures have been experimented so frequently, but only in uncompressed image formats. Even working on still-to-videos, it deals with issues like tracking and recognizing faces in a sense used as a gallery and compressed video segments on each blobs examining.

2.1   Face Detection Techniques

Face Detection Techniques in (Fig. 3) can be divided on the basis of feature-based technique where features given by (M. Antón Rodríguez et al. [5]) reflect to the personal identity of an individual and another as image-based technique [5].

Feature-based technique from (Divyarajsinh N. Parmar and Brijesh B. Mehta [6]) defines how local features are extracted and their locations with its statistics and structural classifier [6]. Similarly, image-based techniques in (Philippe Carré et al. [7]) have formulated color alterations with algebraic operations. The generalized linear filtering algorithms are defined with quaternions and define a new color edge detector [7]. A group of author in [2] have trained a system of small group of people using Multilayer Perceptron (MLP) in Neural Network (NN) [2]. Also, author in (Fahad Shahbaz Khan et al. [8]) has promoted an idea by introducing a model with

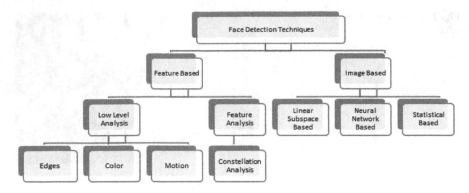

Fig. 3 Face detection technique

both bottom-up and top-down components of color attention employed to modulate the weights of local shape features [8].

2.2 Face Recognition Algorithms

The face recognition algorithm has been broadly categorized by researcher in (M. Singh et al. [4]) with two invariant approaches, i.e., discriminative and generative approaches [4]. Discriminative approaches basically utilize the basic information like age, weight, skeletal structure, and body mass which is given to be analyzed whereas generative approach defines the method to use the information to the model. The model built by designing efficient database is named as WhoIsIt database, the database contains age and weight information with the face images. The database has only focused on public figures in order to depict variation in age as well as in their weight over a period of time. The algorithm attempts to recognize images with the variation with respect to age and weight change. The effect of weight being is measured and then trained in neural networks. Learning-based algorithm is defined by training first, then further testing. The researcher has retrieved 28.53% Rank-I identification performance accuracy with 3.4% minimal as better to other approaches.

The system has worked over several decades to study given by authors (R. Singh, M. Vatsa et al. [9] on covariates of images by identifying individual's facial expression, pose, aging, real or artificial, disguise and plastic surgery [9].

2.3 Color Images

Face detection in color images is difficult when images contain complex background under various luminance with skin detection reduced and gives false positives result. A novel algorithm presented by authors (Li Zou and Seiichiro Kamata [10]) proposed parallel structure algorithm for skin color detection in order to improve the detection accuracy also to obtain a classifier from Gaussian-mixture model and Adaboost training algorithm followed in order to reduce the false positives [10]. Face detection algorithm for skin color model is experimented by Face Candidates algorithm and then its verification algorithm of classifier by Adaboost trained algorithm on various images as training samples.

Gabriela Csurka et al. in [11] and group of authors have described novel approach for building color palettes with varied applications like image retrieval, color palette, color transfer, etc. [11]. The combination of colors defined into abstract category with typical set of color conveys look and feel color combinations. The color plates defined in the dataset of color space with varied variety of color plates representation are called as swatches, which is 3D vector color space such as RGB, HSV, CMYK, etc., to form Pantone Matching System (PMS). The experimental comparison is explained between the generative and the discriminative based model for color plate ranking of BagofColors and Fisher Vectors based images.

2.4 Classification Methods

Many methods for face detection are classified by (Ming-Hsuan Yang et al. [12]) in four major categories, i.e., Knowledge-based method, Invariant feature method, Template matching, and Appearance-based methods. It detects faces to find related facial features, to find the invariant features while change in pose, change in luminance or any other changes in faces [12]. The pattern matching method describes the whole facial features to correlate between input and reference pattern being calculated for face detection.

The most challenging task for normal human being is to follow the facial recognition retrieval model for correct match in the least running time. Especially while dealing with moving or non-static environment like live video, webcam recording, or accessing real-time video in which facial features are not clear as to take as input image. In order to develop such model, the study given by (Petcharat Pattenasethanon and Charuay Savithi [13]) has designed a model for solution to both the stages, Facial Detection stage and Facial Recognition stage [13]. Facial detection stage defined by adopting pattern recognition in videos files implemented using single image matching algorithm. Second phase to consider input image through the camera, which started GUI for cropped square frame design to transfer the potential area to separate the facial features from complex background. Second stage recognizes the output image taken from the data source, followed to calculate mean through Successive Mean Quantization Transform (SMQT) and Eigen techniques applied to the images. Later splits up through Sparse Network of Windows (SNOW) classifier for facial detection without any effect on the background environments in high speed rate. The system has proven 100% recognition accuracy where they have tested 150 frontal faced images taken from webcam.

2.5 Face Recognition Algorithms

As per the literature reviewed, there were many organizations who have built various techniques and algorithms on the face recognition. From the treasure of various algorithms and technique, some of the famous algorithms are as follows:

- ADABOOST Algorithm (Adaptive boost Algorithm)
- LDA (Linear Discriminate Analysis)
- PCA (Principal Components Analysis)
- EBGM (Elastic Bunch Graph Matching)

i. **ADABOOST Algorithm**

Robert Schapire and Yoav Freund has announced Adaptive boost meta-machine learning algorithm [14]. This algorithm basically works on improving the performance of algorithm. In this, the final outcome of other algorithm weighted sum combined with this algorithm and then a boosted classifier results are calculated. ADABOOST is subtle on noisy data and the outliers, with its adaptiveness proven to subsequent weak learners in favor of those instances misclassified from base classifiers. In some problems, it is vulnerable to the overfitting problem than other learning algorithms. If the individual learner's performance is weak, it means the performance of each one is slightly better than random guessing error rate which is smaller than 0.5 for binary classification, with this final model has proven to converge as strong learner.

ADABOOST basically trained a boosted classifier in the form mentioned in (A1)

$$F^n = \sum_{t=1} f_{T(x)}, \tag{A1}$$

where function F^n defines the class of object as real valued result that takes an object x as input each to f_T is a weak learner.

The output of predicted class object gives an absolute classification value to the weak learner. Correspondingly, T-layer classifier shows the positive values if the sample is in the positive class otherwise the negative class. On each learner produces an hypothetical output as $h(x_i)$. At each iteration t, a weak learner is nominated and then assigned a coefficient αt, so that sum of training error Et of resulting t-stage boost classifier as minimum [3] stated in (A2).

$$Et = \sum_i E(Ft - 1(xi) + \alpha t \, h(x_i)) \tag{A2}$$

ii. LDA (Linear Discriminate Analysis)

Linear Discriminate Analysis (LDA) is linear discriminant method used in statistics for pattern recognition and machine learning. The analysis helps in characterizes between classes of objects to find a linear combination of features. The resultant combination is used as a linear classifier for dimensionality reduction and classification. Basically analysis relates to analysis of variance (ANOVA) and regression analysis, which works on dependent variable as a linear combination of other features whereas LDA works well on independent variables for each observation on continuous quantities.

As in [15], has shown working with two approaches by assuming the conditional density probability functions $\rho(\chi| \, y = 0)$ and $\rho(\chi| \, y = 1)$ both normally distributed with mean and its covariance as ($\mu 0$, $\Sigma 0$) and ($\mu 1$, $\Sigma 1$) respectively shown in (B1). Under this assumption, the Bayes optimal solution predicts that the points from the ratio is lesser to threshold T, so that

$$(\chi - \mu 0) \, T \sum 0 - 1(\chi - \mu 0) + In \geq |\sum 0 - (\chi - \mu 1) \, T \sum 1 - 1(\chi - \mu 1) - In|\Sigma 1| > T| \tag{B1}$$

LDA helps in find the 'boundaries' around clusters of classes. It projects your data points on a line so that your clusters 'are as separated as possible', with each cluster having a relative (close) distance to a centroid [16].

Multiclass LDA

If in the case when two or more than two classes are involved, the analysis appears on the class of variability used to find subspace in derivation of the Fisher discriminated extension. This generalization of class has a mean μi and the same covariance, here the class variability is defined by the sample covariance of class means as shown in (C1)

$$\sum b = 1/C \sum_{i=0} (\mu i - \mu)(\mu i - \mu)^T, \tag{C1}$$

where μ is the mean of the class mean [16].

iii. PCA (Principal Components Analysis)

Principal Component Analysis is a statistical tool for identifying patterns in such a way to highlight each and every similarities and differences between data and its interpretations. As the patterns of image data is hard to find in high dimension images. While finding patterns in the data, and compressing it by reducing the number of dimensions, without much loss of information is the importance of PCA [16]. This technique used in image compression.

It follows some steps of method which is as follows:

a. Get source data
b. Subtract the mean
c. Calculate covariance vector
d. Calculate Eigen values
e. Demo the new data set

iv. EBGM (Elastic Bunch Graph Matching)

It is the biological-inspired algorithm for object recognition which basically draws two major concepts, i.e.,

- Gabor wavelets which works on visual features like model of early visual processing in the brain.
- Dynamic Link Matching algorithm which works for a model of invariant object recognition in the brain.

In EBGM, graph represents as objects, whereas node represents the local textures based on Gabor wavelets and the edges represent distances between the node locations on an image. The purpose of extracting local texture arrangements from graph provides the rule composition of new objects. However, object of an image acts as the collection of local textures in spatial arrangement. Whenever a new object is recognized in image, the labeled graph of stored graph is called as *model graph* [17].

3   Comparative Result

The comparative analysis of above discussed algorithm, See (Table 1).

Table 1 The comparative analysis of above discussed algorithm

Basis	ADABOOST	PCA	LDA	EBGM
Definition	It is a learning algorithms to improve their performance [18]	It is powerful analyzing tool to convert set of observations of correlated variables into a set of values of linearly uncorrelated variables [19]	It is a generalization method used in statistics for pattern recognition [20]	It is a graph based matching algorithm for object recognition in the field of computer vision [21]
Originated	Originated in 1808, by the famous physician scientist Hughes, G.F. invented this algorithm	Karl Pearson invented an analogue of the principal axis theorem in 1901, and later independently renovated and named by Harold Hotelling in the 1930s	Originated in 1914 by Ronald Fisher in the form of ANOVA	Originated in 1920 by Prof. Laurenz Wiskott, Institute Neuro-informatics, Ruhr-University Bochum, Bochum in Germany
Input	Vector equation in form of Bayes theorem	Number in form of Eigen value	Number in form of possibility percentage	Number in form of Gabor wavelet
Output	In decision tree and graphs	In PCA vector in % form	It will be possibility density	In the form of Gabor wavelet transform
Framework based analysis	Viola-Jones object detection framework	Explanatory data analysis	Discriminate correspondence analysis	Jets, graphs
Benefits	It is suitable for smaller places i.e. between 1–10 km. Also deals in image with vector complexity and give 35% details for recognition	It is suitable for larger scale area to smaller database with complete information. Also reduced data complexity through images. It minutely identifies the features at time of verification	It is suitable for smaller places i.e. between 5–15 km. Also deal image with numerical complexity with 50% information for recognition	It is suitable for smaller places i.e. between 10–20 km. Also deal image with gabor wavelet complexity with 45% details for recognition [21]
Drawbacks	Maximum times have blur or shaded results	Not suitable for moving scenes	Sometime have blur image results	Minimum chances of recognition of correct image
Success rate (%)	50–65	85–95	70–80	55

4 Conclusion

This paper has attempted to review of significant algorithm which covers the development in the field of face recognition techniques. It is clear to state that comparison between ADABOOST, PCA, LDA, and EBGM algorithm that has shown significant but least expected to perform LDA method in face recognition due to linear projection from the image space to a low-dimensional space. But major drawback experienced while applying LDA is the small sample size. Whenever dealing with sample size data, the image data distribution in practice is highly complex because of illumination, facial expression, occluded image, and pose variation. This makes face recognition suitable for finding or tracking down fugitive using surveillance cameras.

Acknowledgements This work would not possible with correct guidance given by true expert in this research field area.

Statement on Consent: The research is no more than minimal risk of harm to subjects and does not involve any procedures for which written consent is normally/required outside the research setting other than the standard Lisa image. If still required submission of consent forms and documentation of the subject wishes.

References

1. https://www.fbi.gov/aboutus/cjis/fingerprints_biometrics/biometric-center-of-excellence/modalities/facial-recognition.
2. V. V. Starovoitov, D. I Samal, D. V. Briliuk: Three Approaches for Face Recognition: The 6-th International Conference on Pattern Recognition and Image Analysis October 21–26, 2002, Velikiy Novgorod, Russia, pp. 707–711.
3. https://en.wikipedia.org/wiki/Photometric_stereo.
4. M. Singh, S. Nagpal, R. Singh, and M. Vatsa: On Recognizing Face Images with Weight and Age Variations, IEEE Access, vol. 2, pp. 822–830, 2014.
5. M. Antón Rodríguez, D. González Ortega, F. J. Díaz Pernas, M. Martínez Zarzuela, and J. F. Díez Higuera: Color Texture Image Segmentation and Recognition through a Biologically Inspired Architecture: Pattern Recognition and Image Analysis, 2012, Vol. 22, No. 1, pp. 54–68., Pleiades Publishing, Ltd., 2012. ISSN 10546618.
6. Divyarajsinh N. Parmar, Brijesh B. Mehta: Face Recognition Methods & Applications: Int.J. Computer Technology & Applications, Vol 4 (1), 84–86 IJCTA: Jan-Feb 2013 Available online@ www.ijcta.com 84 ISSN:2229-6093.
7. Philippe Carré, Patrice Denis, Christine Fernandez-Maloigne: Spatial color image processing using Clifford algebras: application to color active contour: Springer-Verlag London Limited 2012, SIViP (2014) 8:1357–1372.
   DOI 10.1007/s11760-012-0366-5
8. Fahad Shahbaz Khan, Joost van de Weijer, Maria Vanrell: Modulating Shape Features by Color Attention for Object Recognition: Springer Science and Business Media, LLC 2011, Int J Comput Vis (2012) 98:49–64 DOI 10.1007/s11263-011-0495-2.
9. R. Singh, M. Vatsa, H. S. Bhatt, S. Bharadwaj, A. Noore, and S. S. Nooreyezdan: Plastic Surgery: A new dimension to face recognition: IEEE Trans. Inf. Forensics Security, Vol. 5, No. 3, pp. 441–448, Sep. 2010.

10. Li Zou and Sei-ichiro Kamata: Face Detection in color Images based on skin color models: TENCON 2010 - 2010 IEEE Region 10 Conference, ISSN: 2159-3442, Print ISBN:978-1-4244-6889-8, pp. 681–686, DOI:10.1109/TENCON.2010.5686631.
11. Gabriela Csurka, Sandra Skaff, Luca Marchesotti, Craig Saunders: Building look and feel concept models from color combinations with applications in image classification, retrieval and color transfer: The Visual Computer, December 2011, Volume 27, Issue 12, pp. 1039–1053.
12. Ming-Hsuan Yang, David J. Kriegman, Narendra Ahuja: Detecting Faces in Images: A survey: IEEE Transaction on Pattern Analysis and Machine Intelligence. Vol. 24, No. 1, January 2002, pp. 34–58.
13. Petcharat Pattenasethanon and Charuay Savithi: Human Face Detection and Recognition using Web-Cam: Journal of Computer Science 8 (9), 2012, ISSN 1549-3636, pp. 1585–1593.
14. Robert E. Schapire, Yoav Freund: Boosting: Foundations and Algorithms: The MIT Press 2012, Pages 496, ISBN0262017180 9780262017183.
15. K. Susheel Kumar, Shitala Prasad, Vijay Bhaskar Semwal, R C Tripathi: Real Time Face Recognition Using ADABOOST Improved Fast PCA Algorithm: International Journal of Artificial Intelligence & Applications (IJAIA), Vol. 2, No. 3, July 2011, pp. 45–58, DOI:10. 5121/ijaia.2011.2305.
16. Aleix M. MartõÂnez, Avinash C. Kak: PCA versus LDA: IEEE Transactions On Pattern Analysis and Machine Intelligence, VOL. 23, NO. 2, FEBRUARY 2001, page 228–233, http://www2.ece.ohio-state.edu/~aleix/pami01.pdf.
17. http://www.scholarpedia.org/article/Elastic_Bunch_Graph_Matching.
18. Guo-Dong Guo, Hong-Jiang Zhang: Boosting for Fast Face Recognition: Microsoft Research China, http://www.face-rec.org/algorithms/boosting-ensemble/ratfg-rts01guo.pdf.
19. http://users.utcluj.ro/~tmarita/IOC/C7/C7.pdf.
20. Suman Kumar Bhattacharyya, Rahul Kumar: Face Recognition by Linear Discriminant Analysis: International Journal of Communication Network Security, ISSN: 2231 – 1882, Volume-2, Issue-2, 2013 http://www.interscience.in/IJCNS_Vol2Iss2/31-35.pdf.
21. http://www.it.iitb.ac.in/frg/wiki/images/0/01/Report_04_113050033_Anshita.pdf.

Proposed Approach for Book Recommendation Based on User k-NN

Rohit, Sai Sabitha and Tanupriya Choudhury

Abstract Large data repositories helped us in support systems but created a huge problem for meaningful information retrieval. Filtering of data based on user requirements solved this problem. This process of data filtering when combined with prediction developed recommendation systems. Initial work in recommendation systems can be listed in the areas of cognitive science, approximation theory, marketing models, and automatic text processing. This paper focuses on recommendation system for books. In this paper, training and testing models are designed to predict user ratings for new users. The predicted user ratings are used to propose three types of recommendations based on three different user attributes.

Keywords Recommendation system · Collaborative filtering · Pearson similarity · Cosine similarity · User k-nn

1 Introduction

In present world each person wants quick supplies for his requirements in every field of life including shopping or renting of books. Recommendation systems provide best possible solution to this problem. These are kind of expert systems which help in gathering the related information [1]. Most of recommendation systems work for almost similar purpose that is to recommend items which are most relevant to the users. To fulfill this purpose recommendation systems use different approaches including collaborative, item-based, and hybrid filtering.

Rohit (✉)
Department of CS&E, Amity University, Noida, India
e-mail: rohit.ahlawat@live.in

S. Sabitha · T. Choudhury
Faculty, Department of CS&E, Amity University, Noida, India
e-mail: assabitha@amity.edu

T. Choudhury
e-mail: tchoudhury@amity.edu

© Springer Nature Singapore Pte Ltd. 2018
S.K. Bhatia et al. (eds.), *Advances in Computer and Computational Sciences*,
Advances in Intelligent Systems and Computing 554,
https://doi.org/10.1007/978-981-10-3773-3_53

Table 1 Output for recommendation

User Id Prediction	Age	Author	Book Title		
4017	48	A. Manette Ansey	Midnight Champagne: A Novel (Mysteries & Horror)	New Orleans, Louisiana, USA	4.102143
4017	48	A. Manette Ansey	Sister (Mysteries & Horror)	New Orleans, Louisiana, USA	3.450125
4017	48	A. Manette Ansey	Vinegar Hill (Oprah's Book Club (Paperback))	New Orleans, Louisiana, USA	3.355193
4228	41	A. Manette Ansey	Unwanted Company	Austin, Texas, USA	3.151324

In this paper we are using collaborative filtering approach to provide recommendations to the users. We are training a book rating data with our training model. This trained data will be sent to testing model. The testing model will predict user ratings for new users. On the basis of these predicted values, a system is proposed to recommend books to new users on their personal attributes which are age, location, and interest. Using these three attributes we are proposing three different models. All models include dataset provided by our training and testing models. To create this training model we used a real-time dataset of books as described in Fig. 5. It has large number of entries which are feasible for our analysis. Main objective of this proposal is to assist new users of any book repository in finding their desired books. Research works have been accomplished by many researchers with similar objective as shown in Table 1. Main purpose of this research work is to design a different approach in the creation of recommendation systems. Our work will provide a base in creation of recommendation systems using User k-NN prediction model.

2 Theoretical Background

2.1 Recommendation System Overview

Lot of work has been done in recommendation systems but interest remains same as it is a problem-rich field and having limitless possibilities both in research and industry. It has large number of practical implementations to solve the problem of information overloading and providing personalized information [2]. Following list of different research works in the field of recommendation systems will support the fact that the recommendation system using user k-NN prediction is least touched and thus have large opportunities for research work (Fig. 1).

Initial work in recommendation system can be listed in the areas of cognitive science [3], approximation theory [4], marketing models [5], and automatic text processing [6]. This work later became the rating estimation for new entries on the basis of different attributes and likes of already present entries similar to them.

S.No.	Author	Year	Area	Based on
1	ZHEN ZHU	2007	Book Recommendation	Apriori Algorithm
2	Binge Cui	2009	Online Book Recommendation	Web Services
3	Yongcheng Luo	2009	Privacy-Preserving Book Recommendation	Multi Agent
4	Maria Soledad Pera	2011	Personalized Book Recommendations	Word Similarity
5	CHENG Qiao	2013	Simulation Resource Recommendation	Collaborative Filtering
6	Pijitra Jomsri	2010	Recommendation system for Digital liabrary	Association Rule
7	Salil Kanetkar	2014	Web based recommendation system	Hybrid
8	Anand Shanker Tewari	2014	Opinion based book recommendation	Naïve Bayes Classifier
9	Kumari Priyanka	2015	Personalised book recommendation	Opinion Mining

Fig. 1 Literature survey

Recommendation systems can be categorized based on how recommendations are made [7]:

- Content-based recommendations: Items are recommended on the basis of past preferences of the user.
- Collaborative recommendations: Items are recommended on the basis of past preferences of users with similar taste.
- Hybrid recommendations: These are the combinations of both content-based and collaborative recommendations.

We are using collaborative recommendations and user k-NN method for our system which is explained in Sects. 2.3 and 2.4 respectively.

2.2 Performance Measures

RMSE: Root-mean-squared error is a very good general-purpose error matric for numerical predictions [8]. Its value lies between 0 and ∞, 0 is the best value for any prediction and ∞ is the worst. Hence, this value should be minimized to prove performance of our model better.

MAE: Mean absolute error measures the average of magnitude of errors in a specific prediction [9]. Value of MAE also lies between 0 and ∞, 0 is the best values for any prediction and ∞ is the worst. So, our motive is to minimize this value for the better performance.

2.3 Similarity Measures

There are two main similarity measures which are present in Rapid Miner:

- Cosine-based similarity: This treats the two items as different vectors and the similarity is calculated on the basis of angle between these two vectors. It is also known as vector-based similarity.
- Pearson-based similarity: It checks how much the rating provided by a common user is different from the average rating of that item.

We used Pearson correlation mode because it provided more accurate results than Cosine for our dataset. Value of RMSE in case of Pearson is less than Cosine by a percentage of 10.66 as shown in Figs. 2 and 3.

2.4 Collaborative Recommendation

Collaborative recommendations are provided on the basis preferences of users which are similar in taste to new users [10]. We chose this over content-based because content-based cannot find out the quality of the item [11]. Collaborative recommendations work on collaborative filtering (CF) algorithm which works as follows [12]:

Fig. 2 RMSE of Pearson

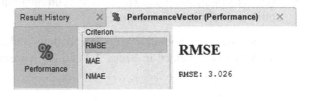

Fig. 3 RMSE of Cosine

- Similarity values are calculated between two or more items in a dataset using one of the similarity measures. These measures are explained in Sect. 2.4.
- These similarity values are used to predict ratings for the entries not present in dataset.

In this paper, collaborative filtering is used along with the user k-NN to provide an approach for recommendation system. Collaborative filtering solves most of the shortcomings present in the content-based filtering [13]. Since feedback of other users creates difference between recommendations, there is a possibility of maintaining the effective performance. The approach of this research is as follows.

2.5 k-NN Algorithm

K-nearest neighbors is the method used for both regression and classification [14]. It is a type of instance-based learning and also called lazy learning. Following is the algorithm for k-NN approach.

It is a technique which uses K-instances as represented points in a Euclidean space.

- In K-NN classification, an object is classified by a majority vote of its neighbors, and the object is assigned to the class most common among its K nearest neighbors for discrete value.
- For real value, it returns the mean values of the K nearest neighbors (K is a positive integer, typically small). If K = 1, then the object is simply assigned to the class of that single nearest neighbor.

3 Methodology

The methodology to adopt for the research is depicted in Fig. 4:

Datasets from three excel sheets of BX-Book-Ratings, BX-User, and BX-Books details are integrated using data integration techniques.

1. The integrated data is pre-processed.
2. User k-NN algorithm is used for predictive analysis of training samples book ratings.
3. The predictive model is designed using rapid miner.
4. The model is tested using testing samples.
5. Performance of the model will be measured using performance measures named RMSE, MAE, and NMAE.

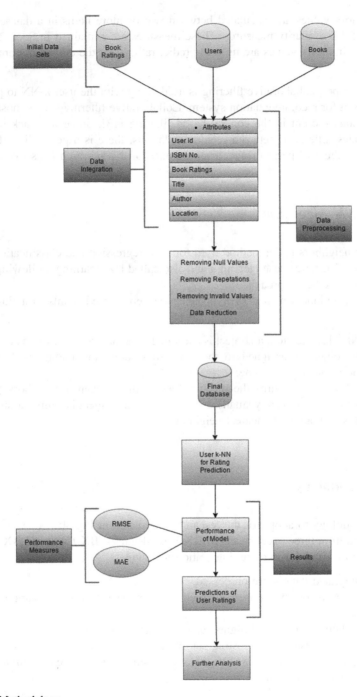

Fig. 4 Methodology

3.1 Data Integration

There were three files in the initial dataset with different attributes in them. Description of those files is provided in Fig. 5. To select most suitable attributes Pearson R Test is performed to calculate the similarity between attributes.

Attributes with high similarity were reflected as single attributes. Formula for Pearson R Test is given below:

$$r = \frac{\sum (x - \bar{x})(y - \bar{y})}{\sqrt{\sum (x - \bar{x})^2 (y - \bar{y})^2}}.$$

Manual integration is also performed to get most suitable attributes. For example, there were image URL in BX-Books excel files which are not usable to this research. Other attributes such as publisher details and year of publication were not relevant to this approach, and hence removed from the attribute list.

3.2 Data Pre-processing

- The dataset of book rating, user details, and book details had 1,149,780 ratings for 271,379 books.

Original Data	
Source	http://www2.informatik.uni-freiburg.de/~cziegler/BX/
No. of Excel Files	3
Names of Files	BX-Users, BX-Books, BX-Book-Ratings
No. of Attributes in BX-Users	3
Entries in BX-User	168097
No. of Attributes in BX-Books	8
Entries in BX-Books	271380
No. of Attributes in BX-Book-Ratings	3
Entries in BX-Book-Rating	1048576
Processed Data	
No. of Excel Sheet -	1
No. of Attributes -	6
Names of Attributes -	User-Id, ISBN, Book-Rating, Book-Title, Author, User-Location
Reduction Range -	Up to User-Id 5000
Total Entries -	8660

Fig. 5 Metadata of dataset

- The user ids are made anonymous and mapped to integers.
- Six attributes User Id, ISBN No, Book Ratings, Title, Author, and Location were selected from set of different attributes.
- Data cleaning was performed and repeated; invalid and null values were removed.
- The dataset is reduced till 5000 user ids for better understanding of results.

4 Experimental Setup

4.1 Dataset Used

The dataset was collected in 4-week crawl from the Book-Crossing community. It was downloaded from official website of IIF [15]. The metadata of the original dataset is given and the pre-processed dataset is shown in Fig. 5.

4.2 Tool Used

The Rapid Miner data mining tools are used for the purpose of research and analysis in data mining. It is a tool with integrated environments for data mining, machine learning, predictive analysis, and text mining. It is used for information mining process including results, presentations, validation, and optimization. It provides a large pool of data loading, data transformation, data modeling, and data visualization methods [16].

4.3 Model Construction for Training

Model constructed in Rapid Miner for training of data which will be used to predict user ratings is shown in Fig. 6. Following steps describe the working and flow of the model:

1. "Read Excel" is used to import an excel file in the Rapid Miner process.
2. Set Role method specifies the role of each attribute present in the excel file [17]. In this model Book Ratings are specified as "label", ISBN as "item identification", User Id as "user identification" and all other attributes as "regular".
3. User k-NN is a model for rating prediction and can be used after installing an extension called "Recommender" in your Rapid Miner tool.
4. Apply Model implements the model selected and provides the final result of that model. Here User k-NN model is User k-NN and result is prediction.
5. "Performance" shows the accuracy and validity of your model.

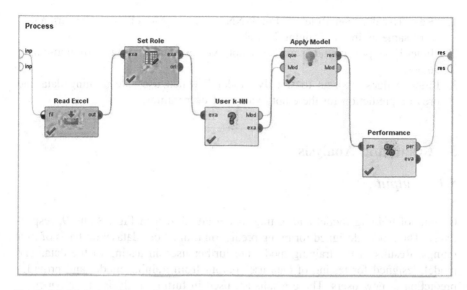

Fig. 6 Model designed for training of data

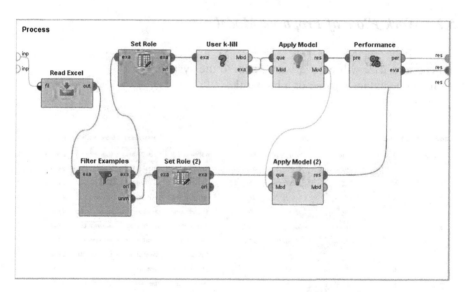

Fig. 7 Model designed for testing of data

4.4 Model Construction for Testing

Model constructed in Rapid Miner for testing of data is shown in Fig. 7. This model tests the prediction of ratings for the new users. Following steps describe the working and flow of the model:

1. "Read Excel", "Set Role", "User k-NN", "Apply Model" and "Performance" work same as in the Training Model.
2. "Filter Example" method separates empty values of user ratings from non-empty values.
3. Empty values are sent to "Apply Model2" which uses the training data and provide prediction for the empty values of user ratings.

5 Result and Analysis

5.1 Output

Outputs of training model and testing model are shown in Figs. 8 and 9, respectively. The model designed for rating prediction trained our dataset on basis of user ratings. Results of the training model are further used in testing of the data. The model designed for testing of data uses output from training model and provides prediction to new users. These results are used in further analysis in the paper.

5.2 Work Flow of Proposed Model

- The new user will enter a search item to the system.
- It can be author's name or a book title.

ExampleSet (8659 examples, 4 special attributes, 4 regular attributes)   Filter (8,659 / 8,659 examples): all

Row No.	Book-Rating	Age	ISBN	User-ID	Author	Book_Title	User_location	prediction
1	0	18	195153448	2	Mark P. O. Mo...	Classical Myt...	stockton, calif...	2.649
2	0	26	1841721522	10	Celia Brooks ...	New Vegetari...	albacete, wis...	2.269
3	7	14	375759778	19	ARTHUR PHI...	Prague : A No...	weston, ,	4.506
4	0	19	425163091	20	Stephan Jara...	Chocolate Je...	langhorne, p...	2.649
5	0	24	067176537X	36	Dolores Krie...	The Therape...	montreal, qu...	2.649
6	7	17	553582747	42	Dean Koontz	From the Cor...	appleton, wis...	3.513
7	0	51	425182908	44	Patricia Corn...	Isle of Dogs	black mounta...	2.898
8	0	51	042518630X	44	J.D. Robb	Purity in Death	black mounta...	2.392
9	8	51	440223571	44	Maeve Binchy	This Year It W...	black mounta...	3.992
10	0	51	812523873	44	Laura J. Mixon	Proxies	black mounta...	2.392
11	0	51	842342702	44	Tim Lahaye	Left Behind: A...	black mounta...	1.868
12	9	34	440225701	51	JOHN GRISH...	The Street La...	renton, washi...	4.256
13	7	24	671623249	56	Larry McMurtry	LONESOME ...	cheyenne, wy...	4.576
14	0	24	679810307	56	SUZANNE FI...	Shabanu: Da...	cheyenne, wy...	3.176
15	9	24	679865691	56	SUZANNE FI...	Haveli (Laure...	cheyenne, wy...	4.976
16	7	32	2070423204	64	Michel Tournier	Lieux dits	lyon, rhone, fr...	4.506

Fig. 8 Output of training model

ExampleSet (2660 examples, 3 special attributes, 5 regular attributes)   Filter (2,660 / 2,660 examples

Row No.	ISBN	User-ID	Age	Author	Book_Title	User_location	prediction
382	1551665077	3371	25	Nora Roberts	Last Honest ...	groveland, m...	3.454
383	1551665638	3371	25	Jayne Ann Kr...	Call It Destiny...	groveland, m...	3.454
384	1551665794	3371	25	Barbara Deli...	Twelve Across	groveland, m...	4.006
385	1558747109	3371	25	Jack Canfield	Chicken Sou...	groveland, m...	3.454
386	1583486259	3371	25	Tom Maremaa	Imagined	groveland, m...	3.454
387	1854879820	3371	25	Anne Styles	That Cinderel...	groveland, m...	3.454
388	1885983212	3371	25	Jack Kersh	Hotel Sarajevo	groveland, m...	3.454
389	006028871X	3373	30	Louise Renni...	Angus, Thon...	elk grove, cali...	3.454
390	60958022	3373	30	Joanne Harris	Five Quarters...	elk grove, cali...	4.155
391	61013420	3373	30	Stuart Woods	Worst Fears ...	elk grove, cali...	3.454
392	006109157X	3373	30	Stuart Woods	Dead Eyes	elk grove, cali...	4.214
393	61099368	3373	30	Stuart Woods	Palindrome	elk grove, cali...	3.454
394	61099805	3373	30	Stuart Woods	Swimming to ...	elk grove, cali...	3.454
395	312291639	3373	30	Emma McLa...	The Nanny Di...	elk grove, cali...	4.283
396	312957955	3373	30	William Hjort...	Falling Angel ...	elk grove, cali...	3.454
397	312995423	3373	30	Dan Brown	Digital Fortre...	elk grove, cali...	3.454
398	316789089	3373	30	Anita Shreve	The Pilot's Wi...	elk grove, cali...	4.168

Fig. 9 Output of testing model

- Then the user is asked for the required attributes which are age, location, and area of interest.
- Then the dataset which was created by the models will come in picture and will be used for the recommendation.
- Highest rated books of that author will be recommended to the user if he searched by the author.
- If he searched by title, then the books which are categorized in that group are recommended to the user.

Example: New user XYZ asks for following author:
"Manette Ansay"

Then all the books written by A. Manette Ansay will be searched from the dataset created by testing model and following is the sample of that data:

Here we have four books by requested author but the three books with highest rating will be sent as recommendation. The recommendations will be

1. Midnight Champagne by A. Manette Ansay
2. Sister by A. Manette Ansay
3. Vinegar Hill by A. Manette Ansay

Models	RMSE	MAE	NMAE
Training model	3.025	2.652	0.295
Testing model	2.990	2.631	0.292

Table 2 Values of performance measures

5.3 Performance Measures

Performance of prediction model is measured on factors defined in Sect. 2(B). Following table mention performance measures for both models (Table 2):

5.4 Analysis

We are following below-defined procedures for our further analysis and research work. On first access user is asked for following attributes:

- Age
- Location
- Area of Interest

These three possibilities are proposed using above-defined attributes and data created by our training and testing models.

Case study 1: Recommendation using age. When recommendations are provided to new user it cannot use ratings as a total base. Suppose new user is 25 years old and recommended item is rated high by persons of more than 60 years old. Then it will not be a fair recommendation for that user. So using output of testing model, new proposal is made which uses age of new user as a main attribute.

In Fig. 10, predictions provided by testing model are put together with users with different age to show the distribution between them.

The model shown in Fig. 11 uses age as an attribute of test data and finds similar objects in data trained by our model.

1. Age groups are created of range 10 using data of Fig. 10.
2. Suppose user lies in Group 1 which is of 0–10, then three books with highest ratings in that age group are fetched from training dataset.
3. These results are provided to the recommender system and will be produced as recommendations to the new user.
4. Next top three books are recommended in case user does not like provided recommendations.

Case study 2: Recommendation using location: As stated in case study 1, it is necessary to have an attribute which helps in providing more relevant recommendations. In this case, it is location of new user. On the basis of this, a proposal is made for better recommendations.

In Fig. 12, predictions provided by testing model are put together with users with different locations to show the distribution between them.

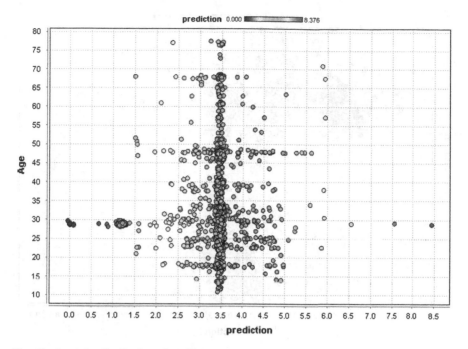

Fig. 10 Age-wise distribution of prediction

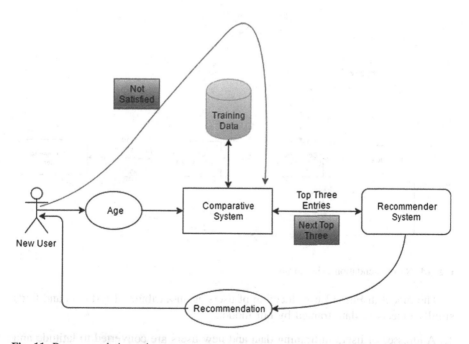

Fig. 11 Recommendation using age

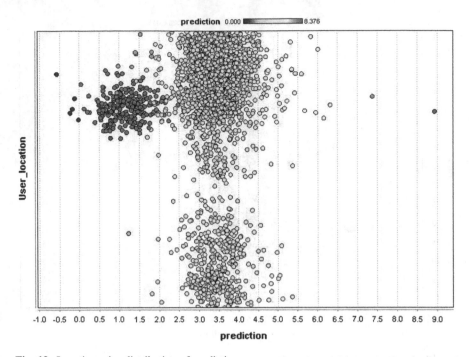

Fig. 12 Location-wise distribution of prediction

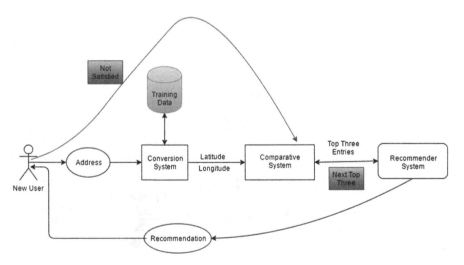

Fig. 13 Recommendation using location

The model in Fig. 13 uses location of users as an attribute of test data and finds similar objects in data trained by our model.

1. Addresses of users in training data and new users are converted to latitude and longitude values using data provided by Fig. 12.

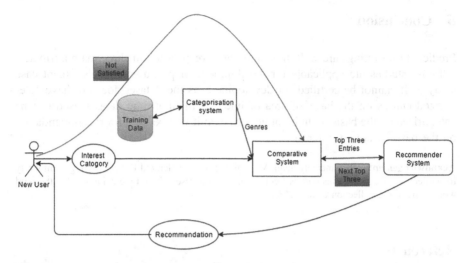

Fig. 14 Recommendation using interest

2. 10 values which are closest to the values of new user are selected.
3. Three books with highest ratings in those 10 entries are selected and sent to recommender system.
4. These results will be produced as recommendations to new user.
5. Next top three books are recommended in case user does not like provided recommendations.

Case study 3: Recommendation using interest:
This model uses Area of Interest as an attribute of test data and finds similar objects in data trained by our model (Fig. 14).

1. All books present in training data are categorized in different genres.
2. System provides list of genres and new user selects one of them according to related interest.
3. Three books with highest rating in that genre are selected and sent to recommender system.
4. These results will be produced as recommendations to new user.
5. Next top three books are recommended in case user does not like provided recommendations.

6 Conclusion

Predicted user ratings are well distributed with respect to our three main attributes. All case studies are applicable for development of proposed models except case study 3. It cannot be certified for development as the dataset does not have categorized entries on the basis of area of interest. In future the dataset used can be categorized on the basis of different genres, then it will be used for recommendation on the basis of area of interest.

Acknowledgements We sincerely thank Mr. Cai-Nicolas Ziegler and Book-Crossing community for collection of dataset. This data is freely available for research and we acknowledge the hard work done in the collection of data [18].

References

1. Zhang Haiyan, "Research on the Recommendation System Based on Social Tag (in Chinese)", Information Studies: Theory &Application, vol. 35, no. 5, pp. 103–106, 2012.
2. Adomavicius, G., & Tuzhilin, A. (2005). Toward the next generation of recommender systems: A survey of the state-of-the-art and possible extensions. *Knowledge and Data Engineering, IEEE Transactions on, 17*(6), 734–749.
3. Rich, E. (1979). User modeling via stereotypes*. *Cognitive science, 3*(4), 329–354.
4. Powell, M. J. D. (1981). *Approximation theory and methods*. Cambridge university press.
5. Lilien, G. L., Kotler, P., & Moorthy, K. S. (1992). *Marketing models*. Prentice Hall.
6. Salton, G. (1989). Automatic Text Processing. Addison Welsley. *Reading, Massachusetts, 4.*
7. Balabanović, M., & Shoham, Y. (1997). Fab: content-based, collaborative recommendation. *Communications of the ACM, 40*(3), 66–72.
8. https://www.kaggle.com/wiki/RootMeanSquaredError.
9. http://www.eumetcal.org/resources/ukmeteocal/verification/www/english/msg/ver_cont_var/uos3/uos3_ko1.htm.
10. Tewari, A. S., Kumar, A., & Barman, A. G. (2014, February). Book recommendation system based on combine features of content based filtering, collaborative filtering and association rule mining. In *Advance Computing Conference (IACC), 2014 IEEE International* (pp. 500–503). IEEE.
11. Sarwar, B., Karypis, G., Konstan, J., & Riedl, J. (2001, April). Item-based collaborative filtering recommendation algorithms. In *Proceedings of the 10th international conference on World Wide Web* (pp. 285–295). ACM.
12. Xin, L., Haihong, E., Junde, S., Meina, S., & Junjie, T. (2013, December). Collaborative Book Recommendation Based on Readers' Borrowing Records. In *Advanced Cloud and Big Data (CBD), 2013 International Conference on* (pp. 159–163). IEEE.
13. Su, X., & Khoshgoftaar, T. M. (2009). A survey of collaborative filtering techniques. *Advances in artificial intelligence, 2009, 4.*
14. Keller, J. M., Gray, M. R., & Givens, J. A. (1985). A fuzzy k-nearest neighbor algorithm. Systems, Man and Cybernetics, IEEE Transactions on, (4), 580–585.
15. http://www2.informatik.uni-freiburg.de/~cziegler/BX/.
16. https://RapidMiner.com/products/studio/.
17. http://docs.rapidminer.com/studio/operators/.
18. Cai-Nicolas Ziegler, Sean M. McNee, Joseph A. Konstan, Georg Lausen; *Proceedings of the 14th International World Wide Web Conference (WWW '05),* May 10–14, 2005, Chiba, Japan.

Improved FP-Linked List Algorithm for Association Rule Mining

Aditya Gupta, Kunal Gusain and Lalit Mohan Goyal

Abstract One of the more important techniques used in Data Mining is, Association Rule Mining and it involves the finding of frequent item sets from the database. A linked list version of one of the most widely used association mining algorithms that are the FP-Growth algorithm was proposed, called the FPBitLink Algorithm. It uses a bit matrix along with linked lists to find the desired item sets. In this paper, we propose two things, first is a variant of the FPBitLink Algorithm, in which instead of treating the items in the datasets as individual nodes, we take a single transaction to be the node in the linked list, and finally use UNION set operation to obtain the frequent pattern set. Since this transactional version is a highly efficient alternative when the number of transactions are greater than the number of items, as it saves valuable space and time, whereas the original FPBitLink algorithm is more efficient when the number of items is greater than the transactions, we further propose the installation of a checkpoint in the beginning, such that depending upon the data either of the two algorithms can be chosen. This way we arrive at a frequent pattern set finding procedure, which is both, highly efficient and extremely efficacious.

Keywords Data mining · Association rules · Frequent itemset · FP-Growth algorithm · Linked list · Algorithm

A. Gupta (✉) · K. Gusain · L.M. Goyal
Department of Computer Science Engineering, Bharati Vidyapeeth's
College of Engineering, Delhi, India
e-mail: adityag95@gmail.com

K. Gusain
e-mail: kunalgusain1995@gmail.com

L.M. Goyal
e-mail: lalitgoyal78@rediffmail.com

1 Introduction

Data Mining is one of the most widely talked about and researched fields these days, and it is so because of its huge potential in real-world applications and the limitless possibility of discovering knowledge in data. Over time, a number of methods and algorithms have been proposed to go about finding useful information in and from, the data, and thus it has always been the aim of researchers to find the most efficient way to do so. Decision support problems, another key area of interest with regard to data mining in real-world applications, makes this field a particularly fascinating one. However, since most of the times, the size of the database under consideration is huge, many of the algorithms proposed entail huge computation overheads and are thus not practically very feasible.

Association rule mining, first introduced for evaluating associative properties between items of basket data, which was nothing but a list of items bought by a particular customer, using this we find a list of items that were most often purchased together. First proposed by Agrawal [1] in 1993, the idea was to identify frequently occurring patterns in the database. For a set of transactions, its association rule is expressed as $X \Rightarrow Y$, where X and Y are the sets of items under consideration. The implication being that the items present in the transaction X, also have the items present in the Y transaction. Finally, we get

$$X \Rightarrow Y,$$

where $X, Y \subseteq I$ and $X \cap Y = \theta$.

The frequent item set finding problem can be broadly classified to contain two fundamental approaches. The first is where we generate a set of candidates, by joining a single item in each successive candidate step generation procedure and then checking it against the database for verification. Algorithms like AIS [2] and DIC [3] use this approach. However, since a number of candidate itemsets have to be generated, it consumes a lot of time, and the need for multiple scanning of the database does not help its case either. The second approach involves a frequent pattern growth method in which we generate a conditional database for each frequent set to be used as a prefix for further comparisons, in the process circumventing the process of having to generate a huge number of candidate itemsets. FP-Growth is on such algorithm that works on the second approach [4].

To find these frequent item sets, a number of algorithms were proposed. Apriori was the first one and used breadth-first search strategy, extending frequent item sets one at a time, resembling a Hash tree, to test these candidate solutions against the dataset [2]. Thus, Apriori's bottom-up approach made way for other algorithms to come in and further improve the mining process. However, it involved repeated scanning of the database to produce candidate sets, which turned out to be a cumbersome task and thus adversely affected the running speed for mining the datasets.

Eclat or the Equivalence Class Transformation algorithm was the next one to arrive and instead of doing a breadth-first search like Apriori, it started with a depth-first search using set intersection [5]. It was an effective strategy and was modeled on a vertical structure fashion, efficient in both sequential and parallel execution.

The FP-Growth (Frequent pattern) algorithm [4] compresses the data sets into a tree, called FP-tree in the first pass and extracts frequent item sets in the second. Since it forms the base of our proposed work, we take a closer look.

2 FP-Growth Algorithm

This algorithm works on a divide-and-conquer strategy [6] and is a deep-search-based method. The idea behind it was to eliminate the need of finding frequent item sets using the candidate generation steps and in the process reducing the computational overhead. It is a two-step approach

Step 1: Construction of FP-Tree

(1) In the first pass over the database, our aim is to find the support for each item set in the data. The original database can be seen in Table 1. Support here refers to the minimum support, in this example, we take it as 3. Then after getting a list of all the item sets, the infrequent ones are discarded since they are not going to be used. The remaining frequent item sets are then ordered in a decreasing manner based on their support. The priority is assigned based on their frequency of occurrence, as shown in Table 2. This table has to be used while building the FP-Tree so that we can share the common prefixes between the items.

(2) In the second pass, the algorithm starts by reading a single transaction at a time and starts mapping it to a path. The first node is made null. A definitive and fixed order in allocating nodes has to be employed, such that if a condition arises where the paths could overlap as transactions share some items, they

	TID	List of item ID's	Ordered items (based on priority)
Table 1 List of transactions	T1	I1, I2, I5	I2, I1, I5
	T2	I2, I4	I2, I4
	T3	I2, I3	I2, I3
	T4	I1, I2, I4	I2, I1, I4
	T5	I1, I3	I1, I3
	T6	I2, I3	I2, I3
	T7	I1, I3	I1, I3
	T8	I1, I2, I3, I5	I2, I1, I3, I5
	T9	I1, I2, I3	I2, I1, I3

Table 2 Frequency of occurrence of items

Item	Frequency	Priority
I1	6	2
I2	7	1
I3	6	3
I4	2	4
I5	2	5

must be allowed to do so and in the process the counters associated with the prefix must be incremented. We allocate and maintain pointers between the nodes that have the same items so that in the end a singly linked list can be created depicted by sets of dotted lines. Finally, all the frequent item sets are evaluated and extracted from this FP-Tree. The final FP-Tree looks the one in Fig. 1.

Step 2: Generation of frequent item sets from the tree

The second part of the FP-Growth algorithm starts the process of extraction of the frequent item sets from this FP-Tree. We employ a bottom-up approach that involves moving up from the leaves of the tree towards its root. This is the step where the divide-and-conquer approach comes in, we start by looking for frequent item sets ending in a particular item, then we look at the item set where this item is the suffix and so on for each item. We now extract the prefix path portion of the entire tree in the form of a sub-tree that ends in the required item set. The solution of frequent patterns generated is given in Table 3.

Fig. 1 Finally constructed FP-Tree

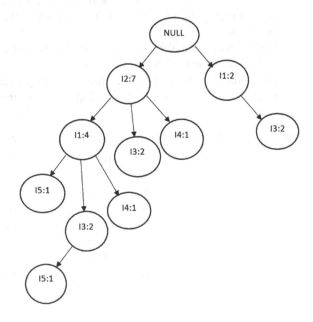

Table 3 Frequent item sets generated

Item ID's	Frequent item sets generated
I5	{I2, I5: 2}, {I1, I5: 2}, {I2, I1, I5: 2}
I4	{I2, I4: 2}
I3	{I2, I3: 4}, {I1, I3: 4}, {I2, I1, I3: 2}
I1	{I2, I1: 4}

The FP-Growth algorithm is much faster than Apriori and one the reasons for this is, that only passes over the data twice, whereas Apriori had to go over it again and again. Also, since there is no need for generation of candidate sets we save both valuable time and resources. However, like all algorithms, this too has its fair share of disadvantages. Since it needs to build a tree as well as a table, FP-Tree is heavy on the memory, it also building the tree is an expensive process.

Since FP-Growth was quite a fast and efficient algorithm, many subsidiaries of it were also proposed. A dynamic version of this algorithm was proposed by Gyorodi C. et al. where the principal idea was to ensure that the FP-tree was not same for different logically equivalent databases [7]. One of the major improvements was in the AFOPT algorithm [8] which made improvements in the item search order, representation of the conditional databases, their construction, and minimizing the tree traversal time.

3   FPBitLink Algorithm

The use of linked lists for FP-Growth was first done in the New FP-Linked List Algorithm given by Sohrabi and Marzooni [9]. The New FP-Linked List Algorithm uses a bit matrix and a linked list structure to extract the frequent patterns. The bit matrix transforms the data set and using this bit matrix we then construct a linked list. For understating this procedure, let us start by considering an example, as in [9], given below in Table 4.

The first step in making a Bit-TDB (Transaction Database) matrix is to remove all the items from the TDB whose number is less than the Minsup (Minimum Support). In the next step, we make an $N * M$ matrix where N is the number of transactions and M is the number of items in TDB. Now we can start filling out the matrix in such a way that, if 'i' and 'j' are the item and transaction under consideration then the value

Table 4 Transactions along with their respective items

TID	Items
T1	a, b, f, e, c
T2	a, b, d
T3	b, c, d, e
T4	b, c, e, f
T5	a, c, e, f
Minsup = 3	

Table 5 Unsorted BIT-TDB matrix

TID	Items				
	a	b	f	e	c
T1	1	1	1	1	1
T2	1	1	0	0	0
T3	0	1	0	1	1
T4	0	1	1	1	1
T5	1	0	1	1	1
Summation	3	4	3	4	4

Table 6 Sorted BIT-TDB matrix

TID	Items				
	b	c	e	a	f
T1	1	1	1	1	1
T2	1	0	0	1	0
T3	1	1	1	0	0
T4	1	1	1	0	1
T5	0	1	1	1	1
Summation	4	4	4	3	3

Table 7 Unsorted BIT-TDB Matrix

Items	TID				
	T1	T2	T3	T4	T5
a	1	1	0	0	1
b	1	1	1	1	0
c	1	0	0	1	1
d	1	0	1	1	1
e	1	0	1	1	1
Summation	5	2	3	4	4

of matrix at position (i, j) will be equal to one if that item appears in the transaction, and zero, if that item does not appear in the current transaction. After the completion of this step, our matrix would look something like in Table 5. (The database used in Tables 5, 6 and 7 are derived from data in Table 4.)

The second step is to create a summation array which shows the numbers of items in the TDB. Its value is the sum of elements of each column. Based on the summation value the order of the columns is changed in such a way that the item with highest summation value comes in the front and if two columns have same summation value then the alphabet which comes first in the Latin Alphabet will come first. After completion of this step, we obtain Table 6.

The third step is to make a linked list of frequent patterns using the above sorted Bit-TDB matrix. In the linked list for each column (items), we have a linked list. Now we can take Boolean AND between each column, in the matrix and a column matrix can be made. For this example, we can take the column of item b and

Fig. 2 **a** Checking pattern with minsup. **b** FP-Linked List

(a)

b		c		bc
1		1		1
1	And	0	=	0
1		1		1
1		1		1
0		1		0

(b)

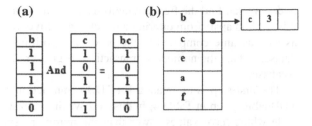

Fig. 3 Final FP-Linked List

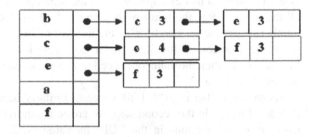

calculate its Boolean AND operation will all other items, i.e., c, e, a, f, then for each of the 'AND matrix', the sum of elements is computer, if its value is greater than or equal to the *minsup* value, the second item will be added to a list which the first item is the beginning member. Let us see it using diagrams.

According to Fig. 2a which is given in the paper given by Sohrabi and Marzooni [9], since $\Sigma_{k=0}^{n}(bc) = 3 > = $ Minsup, where 'n' denotes the total of transactions in the database, then item c will be put in the list with start as b. The final representation of '*c*' in front of '*b*' will look like as shown in Fig. 2b.

The same procedure will be followed for the rest of the columns as well. After the process is completed, logically the data structure will look like in Fig. 3.

4 Transactions-Based FPBitLink Algorithm

The usage of linked lists for finding frequent data sets was indeed a marvelous new approach, and it turned out to be both effective and efficient as well, compared to its counterparts [9]. The authors of this paper [9] used nodes of a linked list to denote different item sets and using this eventually found frequent patterns. In our proposed work, we took a different approach, both towards visualizing the list as well as the way to find the required data sets. We consider that any transaction of items is with its associated transaction id or TID, can be visualized to be a node in itself and thus would have its own matrix based on the minimum support or minsup, and the summation of their occurrences. Next, we use the Boolean AND operation between transactions to find the linear chain in which the nodes are being arranged. Finally, we take the UNION set operation between the lists, so that from the transactions as nodes we can find the frequent items inside them. One of the distinct advantages of this method apart from offering a novel perspective to FP-Growth and saving space as well as time compared to the previous algorithms that were used is for those datasets where the number of transactions is greater than the number of items in the database.

The first step is to create a Bit-TDB (Transactional Database) matrix using the TDB table given in Table 4. In this step, we first remove all the items from the TDB table which have values lower than the minimum support value or minsup, calculated earlier. In this example, we take the minsup to be three (3). After removing all the items below the minsup value, an $N * M$ matrix is made where 'N' is the number of transactions and 'M' is the number of items. The matrix is filled in the following manner, if the item under consideration, 'j' appears in the current transaction 'i', then the value of the cell (i, j) is set to one, else its value is set to zero as shown in Table 7.

As we did earlier in FPBitLink version, similarly here it is done for transactions instead of items. In the second step, we create a summation array which shows the numbers of transactions in the TDB. Its value is the sum of elements of each column. Based on the summation value the order of the columns is changed in such a way that the item with highest summation value comes in the front and if two columns have same summation value then the transaction that appeared first in the order comes first. After completion of this step, we obtain Table 8.

The third step in the algorithm is to create a linked list of transactions using this sorted Bit-TDB matrix. In order to find those values of transitions that are above the minsup value, we employ the Boolean AND operation between each column in the matrix, thus, we finally get the required column matrix with values above the minsup value. For this example, we can take the column of transactions T1 and calculate its Boolean AND with all the other transactions on its right, i.e., T2, T3, T4, T5. However, it must be kept in mind that in each evaluation of the Boolean AND, only the transactions to the right of the transaction under consideration are considered, and not the ones before it, for example, let us say for the transaction T3, we only consider transactions T4, T5 and so on, and not T2 or T1. Now for each of the 'AND matrix', the sum of elements has to be computed, if the values are greater than or equal to the minsup value, then the second transaction will be added to a list which the first item is the beginning member. Let us see it using diagrams.

Table 8 Sorted BIT-TDB matrix	Items	TID				
		T1	T4	T5	T3	T2
	a	1	0	1	0	1
	b	1	1	0	1	1
	c	1	1	1	0	0
	d	1	1	1	1	0
	e	1	1	1	1	0
	Summation	5	4	4	3	2

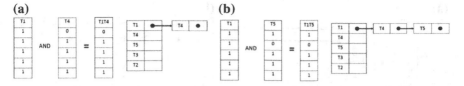

Fig. 4 In (**a**) Checking T1T4 pattern with minsup and the FP-Linked List after checking. In (**b**) Checking T1T5 pattern with minsup and the FP-Linked List after checking

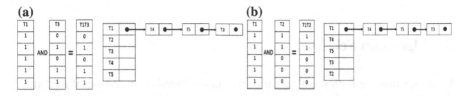

Fig. 5 In (**a**) Checking T1T3 pattern with minsup and the FP-Linked List after checking. In (**b**) Checking T1T2 pattern with minsup and the FP-Linked List after checking

According to Fig. 4, since $\sum_{k=0}^{n}$ (T1T4) = 3 >= Minsup, then transaction T4 will be put in the list with the start as T1, and as $\sum_{k=0}^{n}$ (T1T5) = 3 >= Minsup, then transaction T5 will be put in the list with the start as T1.

According to Fig. 5, since $\sum_{k=0}^{n}$ (T1T3) = 3 >= Minsup, then transaction T3 will be put in the list with start as T1, and since $\sum_{k=0}^{n}$ (T1T2) = 2 >≠ Minsup, then transaction T4 will not be listed at the end of any list.

According to Fig. 6, since $\sum_{k=0}^{n}$ (T4T5) = 3 >= Minsup, then transaction T5 will be put in the list with start as T4, and since $\sum_{k=0}^{n}$ (T4T3) = 3 >= Minsup, then transaction T3 will be put in the list with start as T4.

According to Fig. 7, since $\sum_{k=0}^{n}$ (T5T3) = 2 >≠ Minsup, then transaction T3 will not be listed at the end of any list. After this step we get the final linked list, using which we can obtain the desired frequent item sets.

The final step in our Transactional FPBitLink Algorithm is to merge all the transactions present in the linked list. We separately maintain a counter which is incremented each time a transaction appears in the final merger, and the value of that counter is saved in a separate dictionary denoting the frequent item sets which will then be used for mining of association rules.

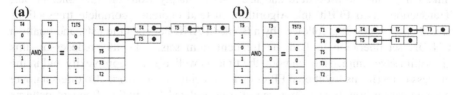

Fig. 6 In (**a**) Checking T4T5 pattern with minsup and the FP-Linked List after checking. In (**b**) Checking T5T3 pattern with minsup and the FP-Linked List after checking

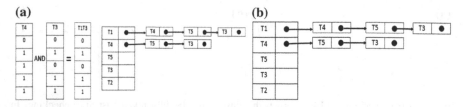

Fig. 7 In (**a**) Checking T4T3 pattern with minsup and the FP-Linked List after checking. In (**b**) we have the final linked list that will be used to find frequent patterns

5 Algorithm Selector

When we observe both the algorithms, that is the Item-based version of FPBitLink Algorithm and the other transaction-based one, one thing is apparent, that for a particular scenario in which the number of transactions is greater than the number of items in a database, it makes sense to choose that algorithm which works primarily in terms of transactions to improve on efficiency, and when items are greater, choose the one which works primarily with items. So instead of trying to make one shoe fit all, it would be efficient and logical to decide which algorithm runs in which case, this decision can simply be called the algorithm selection step. Since Item-based FPBitLink Algorithm works better than its counterpart when the number of items are lesser as compared to the transactions, select this algorithm in those case, and for where transactions are lesser chose the Transaction FPBitLink version. As FPBitLink had already proven experimentally [9] to be faster and more efficient than the traditional FP-Growth algorithms, so an optimized result using the selector was not surprising.

6 Conclusion

In this paper, we have devised a procedure which can use either of two algorithms: One is the FPBitLink Algorithm which, using items as nodes of linked lists and then taking the Boolean AND between the lists, finds out the frequent item sets and the second is the algorithm we have proposed ourselves which although still uses linked lists, but the procedure and use differs sharply from the first one. In our, Transaction-based FPBitLink Algorithm, instead of items, complete transactions are taken as nodes. It is followed by a Boolean AND between the lists, and finally a UNION set operation to find the frequent item sets. Like its predecessor, this algorithm saves immensely on both, the time as well as space complexity and in the process outperforms traditional FP-Growth algorithms. Our work also offers a novel perspective on how transactions can also be utilized for finding frequent patterns and opens the door to a fascinating research area as well.

To further find a real-world ready procedure for employing association rule mining in our day-to-day lives as well as in Data Mining applications, we observed how both the algorithms performed on different databases. In those data sets where the number of transactions were significantly larger than the number of items, FPBitLink ran extremely efficiently, and as something that could be inferred from the way Transactional FPBitLink algorithm works, in those cases where items are greater than the number of transactions, then choosing transactions, which are in lesser number, to be the nodes, saves on the space taken by the algorithm. Thus, an algorithm selector was proposed, which depending upon the data made the choice of which algorithm is to be run, and in this way gave us an efficacious real-world applicable way to mine for frequent item set in our databases.

References

1. R. Agrawal, T. Imielinski, and A. Swami. Mining association rules between sets of items in large databases. In Proceedings of the 1993 ACM SIGMOD International Conference on Management of Data, Washington, D.C., 1993.
2. Agrawal, Rakesh; and Srikant, Ramakrishnan; *Fast algorithms for mining association rules in large databases*, in Bocca, Jorge B.; Jarke, Matthias; and Zaniolo, Carlo; editors, *Proceedings of the 20th International Conference on Very Large Data Bases (VLDB), Santiago, Chile, September 1994*, pages 487–499.
3. S. Brin, R. Motwani, J. D. Ullman, and S. Tsur. Dynamic itemset counting and implication rules for market basket data. In *SIGMOD*, 1997.
4. J. Han, J. Pei, and Y. Yin. Mining frequent patterns without candidate generation. In *SIGMOD*, 2000.
5. Zaki MJ, Parthasarathy S, Ogihara M, Li W (1997) Parallel algorithm for discovery of association rules. Data mining knowl discov, 1:343–374.
6. Jiawei Han and Micheline Kamber, Data Mining: Concepts and Techniques. 2nd edition, Morgan Kaufmann, 2006.
7. Cornelia Gyorödi, Robert Gyorödi, T. Cofeey & S. Holban. Mining association rules using Dynamic FP-trees. in Proceedings of The Irish Signal and Systems Conference, University of Limerick, Limerick, Ireland, 30th June 2nd July 2003, ISBN 0-9542973-1-8.
8. Guimei Liu, Hongjun Lu, Jeffrey Xu Yu, Wei Wang, Xiangye Xiao. AFOPT: An Efficient Implementation of Pattern Growth Approach. Workshop on Frequent Itemset Mining Implementations - FIMI, 2003.
9. Sohrabi, Mohammad Karim, and Soodeh Akbari. "A comprehensive study on the effects of using data mining techniques to predict tie strength." *Computers in Human Behavior* 60 (2016): 534–541.

On Hierarchical Visualization of Event Detection in Twitter

Nadeem Akhtar and Bushra Siddique

Abstract The data generated from social networking services like Twitter, contains rich information of all kinds of events. Studies have been made for event detection events in Twitter, the focus being primarily to detect events and visually align them along a timeline. Since the events can be relatively large in number carrying unequal importance, it might be overwhelming for the user to go through all the events along the timeline. A better approach could be, if the user can get an overview of the timeline at different levels of detail and traverse to those segments in which he is more interested. In this paper, we propose a novel unified workflow in which events are detected and a hierarchy of the detected events is generated through recursive hierarchical clustering. The levels of hierarchy represent the timeline at different granularities of time. Comprehensive experiment on Twitter dataset demonstrates the effectiveness of our framework.

Keywords Social networking services · Event detection · Hierarchical clustering

1 Introduction

Microblogging websites such as Twitter, Tumblr, and Weibo are repositories of real-world events, responsible for generating rich and timely information. There have been a number of incidents in the past which were known to be reported almost in real time by the users of social networking platform. In order to obtain information about ongoing events, web search engines have been used since a long

N. Akhtar (✉) · B. Siddique
Aligarh Muslim University, Aligarh, India
e-mail: nadeemalakhtar@gmail.com

B. Siddique
e-mail: bushrasiddique006@gmail.com

© Springer Nature Singapore Pte Ltd. 2018
S.K. Bhatia et al. (eds.), *Advances in Computer and Computational Sciences*,
Advances in Intelligent Systems and Computing 554,
https://doi.org/10.1007/978-981-10-3773-3_55

time and work addressing webpage summarization [1] has been done in this regard. However, in the meantime, instead of using web search engines, users are more willing to obtain information from Twitter [2], calling for sophisticated event detection techniques with easy and better visualizations. Kerman et al. in [3] define an event as "a significant occurrence or large scale activity that is unusual relative to normal patterns of behavior". Based on the various definitions from the literature, we conclude that an event always shows a change in a particular aspect of the state of the world. This can be a social occurrence like a football match, a natural phenomenon like an earthquake disaster or something intangible like a change in opinion.

1.1 Contribution

There have been a number of recent studies on event detection on Twitter, the focus being primarily to detect events and visually align them along a timeline. As per our knowledge, our work is the first one that is concerned with the hierarchical visualization of the detected events. This makes our work remarkably different from previous works. The proposed approach offers following advantages:

1. A large data corpus may contain numerous events. Additionally, each event may not be equally important as far as user's interest is concerned. As a result, the events when placed along a timeline may not offer the best visualization. The hierarchical output results in progressive disclosure of information and therefore, does not overwhelm the user.
2. As far as a timeline of events is considered, the user does not have the freedom to navigate segments of timeline with varying details. On the other hand, the proposed framework takes care of this and the user could drill down the hierarchy for only those segments of timeline in which he is interested.

The main contributions of this work are as follows:

- We propose a framework which, given a microblog posts corpus, detects events, and produces a hierarchical visualization for the same. The levels of the hierarchy represent the event timeline at different granularities of time.
- We propose an event detection algorithm which detects events by monitoring two kinds of variations.
- Comprehensive experiment on Twitter dataset shows the effectiveness of the framework.

The remaining of the paper contents is organized as follows. Section 2 introduces related work. Overview of the framework is presented in Sect. 3. In Sect. 4, experimental results are reported and discussed. Finally, we conclude our work in Sect. 5.

2 Related Work

The easiest way to get insight into the massive microblog data calls for the development of sophisticated visualization techniques. One such popular technique is to detect events and generate a timeline for the same, making the analysis task quite easier as well as faster. Diakopoulos and Shamma in [4] have made early efforts in this area, using timelines to explore the 2008 Presidential Debates by Twitter sentiment.

In [5], Lin et al. have proposed a two-level solution to generate real-time storylines from the data of microblogs. However, their framework is query-driven contrast to ours which does not take any user input. A number of studies like in [6, 7] are centered on sports events which are known to be more structured as compared to other types of events. In [8], authors have proposed a system to generate timelines for complex events. In their work, they have used the user-generated context responsible for revealing interests and opinions, along with information from traditional media. Few studies [9, 10] have considered the use of rapid increases, so called "spikes", in the volume of status updates to detect important instance of time.

In [11], J. Christensen et al. have introduced hierarchical summarization, a paradigm for large-scale summarization of news documents. The paradigm allows the user to navigate a hierarchy of relatively short length summaries. Our work is related to theirs in relation to the hierarchical clustering they have performed as the first step of summarization.

3 Framework

As shown in Fig. 1, our framework consists of three main modules: the preprocessing module, the event detection module, and the visualization module. In this section, we shall present each of them in detail.

3.1 Preprocessing

Tweeting is but a social activity. As a result, the data generated can be unpredictably noisy. In a raw tweet, words can be combined with any sort of hyphenation and punctuation. Furthermore, abbreviations, conventional word variations, and typos are inevitably present. To extract a bag of cleaner words, we have made use of Twokenizer tool [12] for removing stop words and punctuations, compressing redundant character repetitions, and removing mentions.

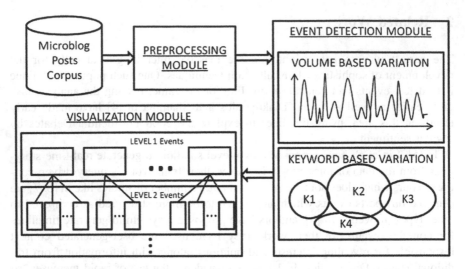

Fig. 1 The unified workflow of event detection and visualization

3.2 Event Detection

The event detection module implements a recursive hierarchical clustering algorithm. Since divisive as well as agglomerative clustering assumes a binary split at every node [13], they are not suitable in our case. The number of splits at each node should be according to the type of input data. For the purpose, we employ a recursive clustering algorithm that chooses the best possible number of split points for generating clusters at each level. Each cluster in turn corresponds to an event.

The idea is that we bin the tweets in interval size of varying lengths for different levels of the hierarchy. The interval length at level i is greater than that at level j, for $i > j$. For instance, the interval length for the first level may be 1 day, i.e., 24 h, for the second level, the interval length may be 1 h and so on. This allows the events to be viewed at finer granularity of time at subsequent levels of the hierarchy. To obtain the splitting points recursively at each level, two types of variations are monitored as discussed in subsequent section.

Variations Monitored for Clustering.

Volume-Based Variation. We employ the concept of rapid increases (or "spikes") in the volume of status updates, which is a common but useful technique in many of the existing event detection systems. A spike in the tweet volume over time indicates that something important has just happened prompting a large number of people to comment on it. For a given cluster c, the volume-based variation $vol_{var}(c)$ is calculated as the difference in the number of tweets generated in the interval preceding the first interval in c and the number of tweets generated in the first interval of c, as in (1)

$$vol_{var}(c) = gen(I_i) - gen(I_{i-1}), \tag{1}$$

where I is an interval indexed over time, such that I_j is an interval before I_{j+1}, and I_i is the first interval in c. $gen(I_i)$ is the number of tweets generated in the interval I_i.

Topical Keywords Based Variation. Given a text corpus, there exists a number of ways to extract meaningful topical themes. To this end, we first employ the most widely used topic model LDA [14], to extract the most relevant topic for the tweet volume in each interval. Furthermore, to quantify the variation in topical keywords in subsequent clusters, we make use of Hellinger distance which measures the distance between two probability distributions of topics as in Eq. (2).

$$topic\_distance(i,j) = \sum_{v=1}^{N} \left(\sqrt{\beta_{i,v}} - \sqrt{\beta_{j,v}} \right)^2, \tag{2}$$

where β is the probability of the ith topic over term v, and N denotes the collective set of terms corresponding to topics i and j. We argue that if the distance calculated is above a certain threshold, the cluster represents an appreciable change in topic. For a given cluster c, the topical keywords based variation $key_{var}(c)$ is calculated as in (3) making use of Eq. (2)

$$key_{var}(c) = topic\_distance(I_{i,p}, I_{i-1,q}) \tag{3}$$

where I is an interval indexed over time, such that I_j is an interval before I_{j+1}, and I_i is the first interval in c. $I_{i,p}$ corresponds to the most relevant topic p in interval I_i. We thus choose clusters $C = \{c_1, c_2, \ldots, c_k\}$ as follows:

$$maximize\ W_k = \alpha * vol_{var}(C) + (1 - \alpha) * \beta * key_{var}(C), \tag{4}$$

where, $vol_{var}(C) = \sum_{c \in C} vol_{var}(c)$ using (1), $key_{var}(C) = \sum_{c \in C} key_{var}(c)$ using (3), $\alpha \in [0, 1]$ is a constant, and $\beta > 1$ is a constant depending upon the dataset. The constant β is used to make the weight of key_{var} comparable to that of vol_{var}, since the two factors in the second part of Eq. (4) are both less than 1.

To determine the appropriate number, k which denotes the number of clusters for each level, we make use of the gap statistic in [15]. Specifically, the algorithm will cluster with varying values of k and return that value that maximizes the gap statistic

$$Gap_n = E_n^* \{\log(W_k)\} - \log(W_k), \tag{5}$$

where W_k is the score of the cluster set computed with Eq. (4), and E_n^* is the expectation value under a sample of size n from a reference distribution.

Ideally, the number of tweets in each cluster decides the maximum depth of the hierarchy. For the computational implementation of the reference distribution, we generate each reference feature uniformly over the range if the observed values for that feature.

3.3 Visualization

The levels of hierarchy represent the events at different granularities of time. For instance, for a data corpus spanning across 1 year, the top level of the events may represent events on a monthly basis, the second level representing events on a weekly basis, the subsequent level on daily basis and so on. As per the interest of the user, he can navigate to the particular segment of the timeline with finer detail by drilling down the hierarchy.

At each level, every event contains following set of fields associated with it: The starting timestamp of the event, the ending timestamp of the event, the duration of the event, and top-k tweets describing the event. We employ a tweet ranking method based on matching topical keywords and the tweets to select the best representative tweets.

4 Experimental Results

To test the performance of the framework proposed in Sect. 3, we tested it on Twitter dataset focused on a popular real-world event. We first present the dataset and describe the process of creating the ground truth. Then, we present the performance of methods, comparing between different implementations.

4.1 Evaluation Methodology

Data collection and extraction of ground truth. We extracted Twitter data making use of the public streaming API of Twitter for the major event that occurred on March 29, 2016: Egyptair MS181 flight incident. The dataset contained 514,902 English language tweets. Clearly, there is no standard way to detect the nature and number of events in the corpus. As a result, we relied on mainstream media reports to identify significant events that unfolded during the course. We identify a ground truth topic based on the time in which that topic emerged in mainstream news.

Comparison of output with ground truth. For comparison, we consider the events detected at the last level of the three-level hierarchy. The interval length at the first, second and third level were 5 h, 1 h, and 5 min, respectively. The automatically detected events are compared to the ground truth using two metrics:

Ground Truth Topic Recall (GTT Recall). Percentage of ground truth topics successfully detected.

Ground Truth Topic Precision (GTT Precision). Percentage of correctly detected ground truth topics out of the total number of ground truth topics in the time slot under consideration.

4.2 Parameter Tuning

The parameter under consideration is α in Eq. (4). We report three results based on the value of α.

- Value of α when set to 0, considers only Topical Keyword based Variation (TKV).
- Value of α when set to 1, considers only Volume-based Variation (VV).
- Value of α when set to 0.5 gives equal weight to both the variations (TKV + VV).

4.3 Results and Discussions

Table 1 shows the precision and recall values for the events at the last level of the three-level hierarchy.

As shown in Table 1, the precision and recall values for both TKV and VV are comparable. This shows that variation of topical keywords and variation in volume when monitored individually detect almost equal number of events correctly. Furthermore, both the variations when monitored collectively and given equal weight give best result. The volume-based variation correctly captures the social nature of tweeting in response to an event and the topical keywords based variation correctly captures the drift from one event to another. Since the events at higher level are connected to one or more events at the lower level in the hierarchy, we argue that the precision and recall values of the events at higher levels (level 1 and 2) is greater than or at least equal to that reported for the last level.

Table 1 Results

Method	GTT recall (%)	GTT precision (%)
TKV	66.66	40
VV	70.59	41.67
TKV + VV	100	60

5 Conclusion

Owing to the immense popularity of Twitter in recent years, the need for sophisticated techniques for making use of the rich information generated is but necessary. In the paper, we have proposed a unified workflow which detects events in a given Twitter posts corpus and produces a hierarchical visualization for the detected events. This makes our work considerably different from existing work which primarily focuses on producing a flat timeline. The hierarchical output represents the event timeline at varying granularities of time and thus results in progressive disclosure of information. Furthermore, the framework was tested on Twitter dataset corresponding to a popular event and the result is reported for three different methods. The framework proposed is however limited to static datasets. Future work includes the adaptation of the framework to twitter stream and comprehensive experimentation on larger datasets.

References

1. Akhtar, N., Siddique, B., Afroz, R.: Visual and textual summarization of webpages. In Data Mining and Intelligent Computing (ICDMIC), 2014 International Conference on, pp. 1–5. IEEE (2014).
2. Teevan, J., Ramage, D., Morris, M.R: #TwitterSearch: A comparison of microblog search and web search. Proc. fourth ACM Int. Conf. Web search data Min. - WSDM '11, p. 35 (2011).
3. Kerman, M.C., Jiang, W., Ph.D., Blumberg, A.F., Buttrey, S.E.: Event Detection Challenges, Methods, and Applications. Natural and Artificial Systems Topic Categorization : System of Systems Engineering Lockheed Martin MS2 Stevens Institute of Technology Castle Point on Hudson Director, Center for Maritime System," no. March (2009).
4. Diakopoulos, N., Shamma, D.A.: Characterizing debate performance via aggregated twitter sentiment. CHI '10 Proc. SIGCHI Conf. Hum. Factors Comput. Syst., p. 1195 (2010).
5. Lin, C.C., Lin, C.C., Li, J., Wang, D., Chen, Y., Li, T.: Generating event storylines from microblogs. Proc. 21st ACM Int. Conf. Inf. Knowl. Manag., p. 175 (2012).
6. Alonso O., Shiells K.: Timelines as summaries of popular scheduled events. Proc. 22nd Int. Conf. World Wide Web Companion, pp. 1037–1044 (2013).
7. Chakrabarti D., Punera, K.: Event Summarization using Tweets. Diversity, pp. 66–73 (2011).
8. Wang, L., Cardie, C., Marchetti, G.: Socially-Informed Timeline Generation for Complex Events. Hum. Lang. Technol. 2015 Annu. Conf. North Am. Chapter ACL, pp. 1055–1065 (2015).
9. Marcus, A., Bernstein, M.S., Badar, O., Karger, D.R., Madden, S., Miller, R.C.: TwitInfo: Aggregating and Visualizing Microblogs for Event Exploration. Proc. 2011 Annu. Conf. Hum. Factors Comput. Syst., pp. 227–236 (2011).
10. Nichols, J., Mahmud, J., Drews, C.: Summarizing Sporting Events Using Twitter. Proc. 2012 ACM Int. Conf. Intell. User Interfaces, pp. 189–198 (2012).
11. Christensen, J., Soderland, S., Bansal, G., Mausam: Hierarchical Summarization : Scaling Up Multi-Document Scaling Up Multi-Document Summarization. In Proceeedings of the 52nd Annual Meeting of the Association for Computational Linguistics, pp. 902–912 (2014).
12. Connor, B., Krieger, M., Ahn, D.: TweetMotif : Exploratory search and topic summarization for Twitter. 4th Int. AAAI Conf. Weblogs Soc. Media, no. May, pp. 2–3 (2010).

13. Berkhin, P.: Survey of clustering data mining techniques. Group. Multidimens. Data Recent Adv. Clust., vol. 10, pp. 25–71 (2006).
14. Blei, D.M. Ng, A. Y. Jordan M.I.: Latent dirichlet allocation. J. Mach. Learn. Res., vol. 3, pp. 993–1022 (2003).
15. Tibshirani, R., Walther, G., Hastie, T.: Estimating the number of clusters in a data set via the gap statistic. Journal of the Royal Statistical Society: Series B (Statistical Methodology), vol. 63. pp. 411–423 (2001).

Audio Steganography Techniques: A Survey

Shilpi Mishra, Virendra Kumar Yadav, Munesh Chandra Trivedi
and Tarun Shrimali

Abstract Today's web group the protected information exchange is limited
because of its assault on information correspondence. Security of information can
be accomplished by executing steganography strategies. The greater part of the
current steganographic strategies utilize the computerized sight and sound documents as a spread mediums to hide mystery information. There are more potential
outcomes to conceal huge measure of information inside advanced sound document. Signals and computerized sound documents make suitable mediums for
steganography due to its abnormal state of excess and high information transmission rate. In this paper a study have done on overall standards of concealing
mystery facts in sound document utilizing sound information concealing systems,
and convey an outline of present strategies and capacities furthermore talk about the
favorable circumstances and disservices of diverse sorts of audio steganography
method.

Keywords Steganography · Human auditory system · LSB · Parity coding ·
Echo hiding

S. Mishra (✉) · V.K. Yadav · M.C. Trivedi
ABES Engineering College, Ghaziabad, Uttar Pradesh, India
e-mail: shilpi.mishra27@gmail.com

V.K. Yadav
e-mail: virendra.yadav@abes.ac.in

M.C. Trivedi
e-mail: munesh.trivedi@gmail.com

T. Shrimali
Sunrise Group of Institutions, Udaipur, Rajasthan, India

© Springer Nature Singapore Pte Ltd. 2018   581
S.K. Bhatia et al. (eds.), *Advances in Computer and Computational Sciences*,
Advances in Intelligent Systems and Computing 554,
https://doi.org/10.1007/978-981-10-3773-3_56

1 Introduction

Steganography is basically made from two Latin words steganos and grapter. Steganos means covered and grapter means writing. The objective of steganography is to conceal the content by introduced messages in articles, for example, advanced pictures, audio, video, or content files [1, 2]. The other range of steganography is a copyright checking where the message is embedded to copyright over a record. Steganography and watermarking portray routines to insert data straightforwardly into a bearer signal. Data stowing away in sound signs is increasing boundless significance for secure correspondence of data, for example, clandestine front line and managing an account exchanges by means of open sound approach. Steganography and cryptography are nearly related only the difference in their objectives. Both are used for security purpose but with different approach or implementation. Steganography shifts from cryptography as in wherein cryptography specializes on preserving the substance of an information secret.

2 Audio Steganography

Audio steganography is a method for hiding facts interior an audio signal. Present audio steganography software program can embed message in Wav, Au, and maybe Mp3 sound data. Embedding hidden message in analog sound is generally a more challenging manner then embedding message in different data together with virtual photographs. It is important to get routines that limit access to those sound documents furthermore for its security. Generally information is inserted in audio records with the end goal of copyright insurance or for confirmation of computerized media. In a PC-based audio Steganography framework, hidden messages are set up in automatic sound. In Audio Steganography, the shortcoming of the HAS is utilized to conceal data inside the sound.

2.1 Characteristics

A viable audio steganographic plan ought to going with the taking after three attributes.

Inaudibility of distortion (Perceptual Transparency): It assesses the perceptible contortion because of sign changes like message implanting or assaulting. The information concealing plan needs to embed extra information without influencing the intuitive nature of the host sound sign.

Robustness: It gauges the capacity of the inserted information opposed to purposeful and inadvertent assaults. Inadvertent assaults for the most part incorporate normal information controls, for example, re-inspecting, re-quantization and so forth. Purposeful assaults incorporate expansion of commotion, resizing, rescaling, and so on.

Data Rate (**Capacity**): It alludes to the measure of data that an information concealing plan can effectively implant without presenting perceptual mutilation. As it were the sample rate of the information is the quantity of the installed sample inside a unit of time and is generally given in sample every second (bps).

3 Techniques of Data Hiding in Audio

There are a few strategies that are accessible for audio steganography. Further a concise presentation on a percentage of the techniques is explained.

3.1 Spatial Domain Technique

Spatial domain technique is additionally called temporal domain technique and substitution technique. In spatial area strategy, the concealed information is shrouded specifically into host record in which the steganography method is basic and simple to implement.

LSB coding: LSB hiding is an easy and quick procedure for inserting data in a sound sign [3]. In LSB approach LSB of binary series of every illustration of digital audio data is interchanged with binary equivalent of secret data [4]. The limit is stand out bit every sample of the spread sound which could be less for some applications. In LSB code, the perfect information transmission rate is 1 kbps in line with 1 kHz. In a few executions of LSB code, on the other hand, the two least significant bits of a series are supplanted with two data bits. This will increase the quantity of information that may be encoded however conjointly will increase the quantity of ensuing noise within the audio file additionally. A more modern methodology is to utilize a pseudorandom amount generator to extend the information over the sound document in an arbitrary way. One prevalent methodology is to utilize the random temporary strategy, in which a secret key controlled by the sender is utilized as a concept as a part of a pseudorandom amount generator to make an irregular succession of test lists. This approach is two deprivation connected with the utilization of systems like LSB coding. The human ear is exceptionally delicate and can frequently distinguish even the scarcest bit of commotion brought into a sound document, second deprivation on the other hand is that this is not strong. In the event that a sound record installed with a mystery message utilizing either LSB coding was resample, the implanted data would be lost.

Parity coding: The signal is separated into gathering of specific pattern or test known as pattern location and this sample region every bit encoded from the concealed message in a pattern location parity bit. In the event that the equality bit of a choose area does not coordinate, the mystery bit to be encoded procedure flips the LSB of one of the samples inside the location. So the sender has to a greater extent a decision in encoding the secret bit.

Advantage: The sender has even more a decision in encoding the mystery bit and the sign can be modified in a more subtle way.

Limitation: This technique like LSB code is not strong in quality. The limit remains the same as that of LSB strategy.

Echo Hiding: This technique acquaints a shorten echo with the host signal and afterward inserts information in it. Three operation of echo signal are controlled for concealing information: Initial Amplitude, the offset (delay), the decay rate. Echo hiding strategy inserts information inside mask audio sign by presenting an echo (Figs. 1 and 2).

The offset is varied to symbolize the binary message to be encoded. One setoff value represents a binary one, and a second setoff value represents a binary zero [4]. If only one echo changed into made from the original sign, simplest one little bit of

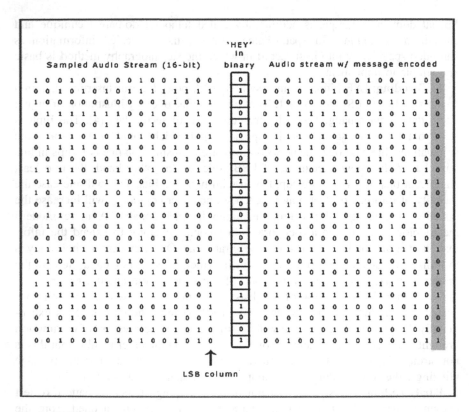

Fig. 1 LSB coding

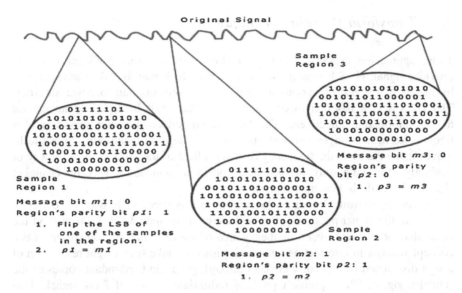

Fig. 2 Parity coding

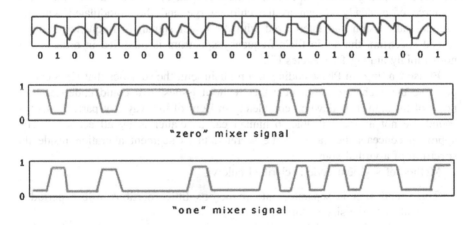

Fig. 3 Echo hiding

data can be encoded. Consequently, the authentic sign is damaged down into blocks before the encoding approach begins and as soon as the encoding way is finished, the blocks are connected once more collectively to create the final sign [5]. Amplitude and decay rate can be set to esteem which are under capable of being heard threshold of human ear (Fig. 3).

Strength: It takes into account a high information transmission rate and gives unrivaled strength.

Limitations: It gives low security and embedding rate.

3.2 Transform Domain Technique

Those approaches conceal records along inside the recurrence allocation of the provider signal. HAS has numerous tendencies which may be taken advantage of by means of the use of various method of transform domain to cover statistics. The HAS device has convinced peculiarities that must be taken advantage for hiding information effectively. The "overlaying effect" peculiar masks weaker frequencies near stronger resonant ones [6]. To reap the inaudibility, those techniques make the most the frequency covering effect of the HAS at once by way of explicitly modifying only masked areas or in a roundabout way by changing slightly the audio indicators samples.

Spread Spectrum: Spread spectrum approach is specifically used for correctly improving the signal pass away the loud channel. In spread spectrum method the concealed facts is sent over a frequency area of audio signal. Spread spectrum is a concept superior in information communications to make sure a right restoration of a sign dispatched over a noisy channel through producing redundant copies of the statistics signal. This approach produce redundant copies of facts signal. Two models of spread spectrum may be used in audio sign—the direct collection and frequency hopping collection. In DSSS, the secret data is unfold out with the useful resource of a steady known as the chip charge and then modulated with the pseudorandom sign and interleaved with the cover signal. In frequency hopping spread spectrum, the audio file's frequency spectrum is modified in order that it hops rapidly among frequencies [7].

Phase Coding: In Phase coding accomplishment, the statistics that HAS cannot perceive the exchange in segment that without issues so it could understand the noise of the sign. It is develop completely in light of the way that part segment of sound are not as recognizable to human ear as clatter is by all accounts. This approach conceals the name of the secret data as segment alteration inside the spectrum of a virtual sign.

Method of segment code is clarified below:

1. The sound sign is separated into numerous littler portions whose period is equivalent to the size of thriller message.
2. Discrete Fourier remodel is carried out to every section. As a stop end result, a matrix of phase and Fourier transform significance are framed.
3. Section variations among adjoining portions are figured.
4. Relative section variations among adjoining fragment cannot be changed. However we will exchange absolutely the stages of the section. Subsequently, mystery data can be embedded within the first segment of the sign section.
5. New fragment matrix is made having segment of first section and particular phase variation (Fig. 4).

Discrete Wavelet Transform: Latest, the audio steganography is construct absolutely in light of discrete wavelet transform (DWT). Information embedded is carried out inside the LSB of the wavelet coordinate produce excessive potential of

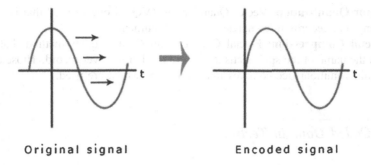

Original signal   Encoded signal

Fig. 4 Phase coding

200 kbps in 44.1 kHz audio sign. Wavelet generally provide to small-scales stream. This approach is used to cover statistics in remodel belonging of the audio sign. Fourier remodel is used for evaluate additives of table certain characteristic. A static sign may be portrayed at the time that sign with no alteration in the view of the sign. We are able to express, Fourier remodel [8] is the effective device for coping with sign which might be composed of some sine or cosine indicators or mixture of the two signals.

Tone Insertion: Tone insertion systems based upon the incoherence of lower energy tones inside the proximity of considerably higher ones. Tone insertion technique can oppose assaults which includes low-bypass filtering and bit truncation. Tone insertion approach has low embedding potential. Additionally, the installed information can be effortlessly extracted following embedded tones are anything but difficult to distinguish.

Amplitude Coding: The HAS features based upon the recurrence values as it's far more perceptive to mass segment. Subsequent this assumption, authors in [8] advise a steganographic set of rules that embeds large-potential data within the value speech spectrum although making sure the hidden-statistics protection and controlling the bias of the covering area. The concealed data (payload) may have being of any kind together with: encoded facts, compressed records, corporations of facts (LPC, MP3, AMR, CELP, specification of speech reputation, and lot of others). The proposed set of rules is depends on finding at ease spectral embed-range in a wideband significance speech spectrum the use of a recurrence masks represent at 13 dB beneath the specific sign spectrum.

3.3 Compressed Domain Technique

In compressed area, forms of techniques are superior. In this domain, cover records or mystery statistics is compressed the usage of one of a kind compression strategies to extend steganography techniques and bring high ability and compression ratio.

Vector Quantization: Vector Quantization (VQ) is one method that is used to cover mystery records in compressed cover document.

Fractal Compression: Fractal Compression (FC) is other method that compresses the name of the sport statistics before hidden in cover record. Those are the maximum commonplace compression techniques which are used.

3.4 Coded Domain Technique

At the same time as thinking about records hiding for real period conversation, voice encoders which encompass: AMR, ACELP, and SILK at their specific encoding charge are employed. Even as passing through one of the encoders, the pass on audio sign is coded in step with the encoder rate then decomposed on the decoder end. The two codec area method may be dignified as in-encoder and post-encoder.

In-encoder: This method uses sub-band amplitude modulation on the manner to conceal statistics in various speech and audio. Facts embedding is executed the use of a LPC vocoder. Pitch detection is accomplished the use of an autocorrelation strategy which is utilized to area discourse into voiced/voiceless. The sign is viewed as the utilization of the unmodified LPC filter through coordinate. The procedure gives a solid concealing rate of 2 kbps [9]. This methodology installs records inside the LSB of the Fourier redesign in the forecast lingering of the host sound sign.

Post-encoder: This method utilizes ACELP codec to be able to insert records inside the bitstream of spread facts. Information inside the bitstream of an ACELP code which helps the evaluation-via-compound codebook search. Experiences for this circumstance is addressed through given way combined code which codes each sample with a charge between −127 and 127 which merges −0 and +0.

4 Conclusion

Audio steganography techniques deal with problems associated with the want to comfort and keep the integrity of statistics hidden in voice technology especially. From our segment of perspective, the variety and full-size form of present audio steganography strategies make bigger software opportunities. The gain on the use of one technique over each other one is predicated upon at the software limitations being used and its necessity for masking ability, embedded facts protection diploma, and encountered assaults resistance. Future work can be developing some robust and reliable algorithm which can withstand with steganalysis.

References

1. Pooja P. Balgurgi, Prof. Sonal K. Jagtap. "Intelligent Processing: An Approach of Audio Steganography" International Conference on Communication, Information & Computing Technology (ICCICT), 2012, IEEE.
2. Anupam Kumar Bairagi, Saikat Mondal "A Dynamic Approach In Substitution Based Audio Steganography" IEEE/OSA/IAPR International Conference on Informatics, Electronics & Vision, 2012.
3. Shailendra Gupta, Ankur Goyal, Bharat Bhushan, 2012, Information Hiding Using Least Significant Bit Steganography And Cryptography, 2012 I. J MECS.
4. H.B. Kekre, A. Athawale, S. Rao, U. Athawale, "Information Hiding in Audio signals" International Journal of Computer Applications, (0975–8887) Volume 7– No. 9, October 2010.
5. P.K. Singh, R.K. Aggrawal, "Enhancement of LSB based Steganography for Hiding Image in Audio", International Journal on Computer Science and Engineering, Vol. 02, No. 05, 2010.
6. Ifra Bilal, Mahendra Singh Roj, Rajiv Kumar, P K Mishra "Recent Advancement in Audio Steganography" International Conference on Parallel, Distributed and Grid computing 2014 IEEE.
7. Ahmed Hussain Ali, Mohd Rosmadi Mokhtar, LoayEdwar George, "A Review on Audio Steganography Techniques" Research Journal of Applied Sciences, Engineering and Technology 12(2): 154–162, 2016.
8. Fatiha Djebbar, Beghdad Ayad, Karim Abed Meraim, Habib Hamam, "Comparative study of digital audio steganography techniques" Djebbar et al. EURASIP Journal on Audio, Speech, and Music Processing 2012.
9. Nikita Atul Malhotra, Nikunj Tahilramani, "Survey on Speech and Audio Steganography Techniques in Temporal, Transform and Coded Domains" International Journal of Advanced Research in Computer science and Software engineering, Volume 4, Issue 3, March 2014.

Role of Clustering in Crime Detection: Application of Fuzzy K-means

Nidhi Tomar and Amit Kumar Manjhvar

Abstract The rising rate of crime has devastated everything seriously. The reason of working on crime dataset is to make a better system, which can make people more aware about the increasing type of crime and crime rate in various fields. The proposed paper works on the detection of crime count and the factors that decide the increasing nature of crime in a better way. Utilization of fuzzy k-means has lead to a better technology that detects the crime rate in a better and effective way. The termination measure is an important factor to define the clusters that are formed over the years. They help in easy detection whether the crime is increasing over the years or not. The dataset from the Indian government's website is taken and been processed so that the results that are calculated can be as correct and near to reality as possible.

Keywords Data mining · Crime detection · Fuzzy · K-means · Clustering · Termination measure

1 Introduction

Crime is a very serious problem in the world. Malefaction is a crime against the society that is often prosecuted and realizable by the law [1]. Criminals commit crime at the place anywhere in type. Traditionally solving crime has been the privilege of the crime equity and law enforcement specialists. With the incrementing utilization of the computerized system to track crime, computer data analysis has commented availing the law enforcement [2]. Many challenges are increasing encountered by decision-makers in the law enforcement department in detecting, identifying the public crime or and tracing or tracking the social crime or

N. Tomar (✉) · A.K. Manjhvar
Department of Computer Science and Engineering, MITS College, Gwalior, India
e-mail: tomarnidhi4@gmail.com

A.K. Manjhvar
e-mail: mitkumar@mitsgwalior.in

© Springer Nature Singapore Pte Ltd. 2018
S.K. Bhatia et al. (eds.), *Advances in Computer and Computational Sciences*,
Advances in Intelligent Systems and Computing 554,
https://doi.org/10.1007/978-981-10-3773-3_57

actions according to their timeline is becoming a tedious task [3]. The Process of information divided into similar object groups known as clusters. Object consist that are similar to one another and exceptional to objects of other collection is called clustering in this implementing work we differentiate a cluster through a c-means clustering methods using the concept of fuzzy [4].

1.1 Fuzzy K-means Approach

It is the type of clustering algorithms, it is the procedure of partitioning the points of data into the clusters like k and S1 (l = 1, 2,..., k) and clusters Sl are related to representatives (cluster center) Cl. The relationship between a cluster and data points belong to a fuzzy [5]. That is, a membership is $u_{i,j} \in [0, 1]$ is used to show the degree is belong to data point Xi and cluster center C_j. Denote data points set as S = {Xi}. Fuzzy K-means method is based on minimizing following distortion:

Clustering methods Fuzzy k-means is used for division of points of records into the k clusters S_1 ($l = 1, 2,....,k$) and clusters S_1 are related with a representative (cluster center) C_1. The correlation between a data point and cluster representative is fuzzy [6]. That is a membership $u_{ij} \in [0,1]$ is used to show the degree which is belongings of data point X_i and cluster center C_j—Denote the set of data points as S = {X_i}. This algorithm of Fuzzy k-means is based on the minimizing the following distortion:

$$J = \sum_{j=1}^{k} \sum_{i=1}^{N} u_{i,j}^m d_{ij} \tag{1}$$

With respect to the memberships u_{ij} and cluster representatives C_j, where N is various data points; m is the fuzzifier parameter; k is numerous clusters; and d_{ij} is squared Euclidean distance between data points X_i and also representative of cluster C_j. It is noted that u_{ij} should satisfy the following constraint:

$$\sum_{j=1}^{k} u_{i,j} = 1, \, for \, i = 1 \, to \, N. \tag{2}$$

1.2 Working Steps of Fuzzy K-Means

(1) Input is a set of initial cluster centers $SC_0 = \{C_j(0)\}$ and the value of is set P = 1.
(2) Set of cluster centers SC_p, are given compute d_{ij} for i = 1 to N and j = 1 to K. Update memberships value $u_{i,j}$ by the following equation:

$$u_{i,j} = \left((d_{ij})^{\frac{1}{m}-1} \sum_{l=1}^{k} \left(\frac{1}{d_{il}} \right)^{\frac{1}{m}-1} \right)^{-1} \tag{3}$$

If $d_{ij} > \eta$, set $u_{ij} = 1$, where n is a small positive number.

(3) Calculate center of all clusters applying Eq. (4). To find a new cluster set representatives SC_{p+1}.

$$Cj(P) = \frac{\sum\limits_{i=1}^{N} u_{ij}^m X_i}{\sum\limits_{i=1}^{N} u_{ij}^m} \tag{4}$$

(4) If $\| C_j (p) - C_j(p-1) \| >$ and for $j = 1$ to K, stop, where $\varepsilon > 0$ is a very small positive number. Otherwise, set $P + 1 \rightarrow P$ and go to step 2.

The computational complexity of FKM in the form of phase 2 and 3. Though, the computational complexity of phase 3 is much less than that of phase 2. For that cause, this complexity, in various distance calculations terms, of FKM is O (Nkt), where t is the amount of iterations [7].

2 Literature Survey

Wang Shunye et al. [8] Stimulated with the random determination problem of initial centroid and similarity measures, the researcher presented a make novel k-means methods of clustering dissimilarity based. The algorithm which proposed gives enhanced accuracy rates and results. Pallavi Purohit and Ritesh Joshi et al. [9] proposed an enhanced approach designed for original K-means algorithm due to its certain limitations. The main reason of this method is poor performance in the initial centroid random selection. The proposed algorithm deals with this problem and improves the performance and cluster quality of original k-means methods. It first finds out the closest data point by calculating Euclidian distance between each data point and then these points are deleted from population. Juntao Wang and Xiao log [10] discussed about an improved version of k-means clustering algorithm to deal with the problem of outlier detection of existing k-means algorithm. The proposed algorithm usages noise information filter to the deal with this issue [11]. Density based outlier detection method is applied to the data to be clustered so as to remove the outliers [12]. The motive of this method is that the outliers may not be engaged in computation of initial cluster centers.

3 Proposed Work

The proposed work show that the number of clusters formed in the graph displays
the crime count yearwise in all the cities. The proposed algorithm is applied fuzzy
k-means of the crime data which leads to cluster formation in a better way. The crime
detection being an important aspect needs to be taken seriously. The main motive of
providing this proposed is to highlight the use of clustering technique on real-world
based dataset which is based on crime. The database contains crime type, location,
city, crime id and few more columns. The database seems like this: (Fig. 1).

A flowchart representation has been shown below here for the proposed work (Fig. 2).

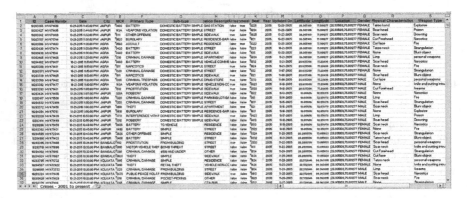

Fig. 1 Crime dataset

Fig. 2 Flowchart of
proposed algorithm

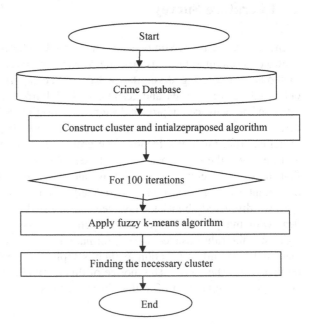

Proposed Algorithm Pseudocode:
Pseudo_ Yf_FKMC1 (X, c, options, init_V)

1. Define maximum iterations max_it = 100;
    termination threshold = 1.0e+03*;
    initial_velocity = 0.
2. Define n: Number of feature vectors
    p: Length of each feature vector
3. if use_init_V,
      V = init_V;
      else
    V = Yf_FKMC1_InitV (c, p); % Initial cluster centers
    end
4. for i = 1:max_iter,
    [V, U, E(i)] = Yf_FKMC1_Step (X, V, c, m);
    Show iteration count & termination measure value.   End.

Pseudo_ Yf_FKMC1_InitV(c, p)
Generate initial cluster centers for FKM clustering using formula:
V = rand(c, p)

Pseudo_Yf_FKMC1_Step(X, V, c, m)

1. Initialize n = size (X, 1)
   p = size (X, 2), where, X is the input data of the crime
2. Distance calculation using Euclidean distance formula
   dist = Yf_EuDistArrayOfVectors1 (V, X)
3. Now calculating the new membership degrees using a variable temp.
   tmp = dist.^(−2/(m−1))
   U = tmp./(ones(c, 1)*sum(tmp))
4. Check constraint by checking
   tmp = ((sum (U)−ones (1, n)) > 0.0001)
5. Update V, mf, and E
   mf = U.^m % MF matrix after exponential modification
   V = mf * X./((ones(p, 1) * sum(mf'))') % new center
   E = norm (V−V_old, 1)
6. End.

4   Result Analysis

The results show various factors and results:

1. Cluster formation in the base: these are the crime counts on an overall basis. The
   x-axis shows the year and y-axis shows the total number of crimes (Fig. 3).
2. Termination measure: this can be defined as the value of the termination measure that goes along with the number of iterations. The value denotes the

Fig. 3 Crime count yearwise

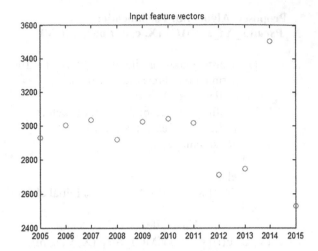

Fig. 4 Termination measure

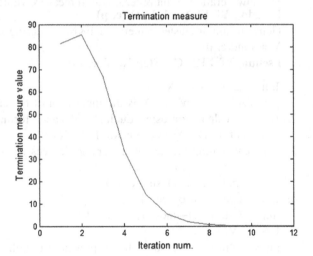

termination of the proposed algorithm. The termination value is set initially and the algorithm is terminated when its value reaches a very less point. The termination value taken in proposed algorithm is 1.0e + 03* (Figs. 3 and 4).

3. Membership function: membership function is used in fuzzy as the degree of truth and evaluation of extension of the outcomes. The figure shows the membership values for both base and proposed respectively. From the figure it can be derived as the fact that membership value for proposed is better than previous values (Fig. 5).

4. Cluster formation: Number of clusters defined = 2. The clusters formed are shown in the Fig. 6.

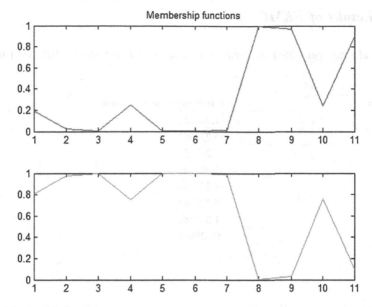

Fig. 5 Membership function

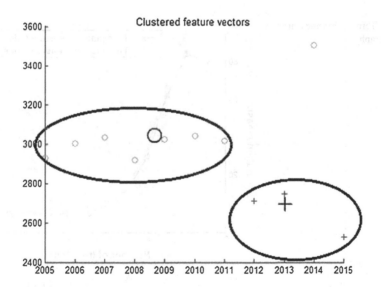

Fig. 6 Clustered feature vectors

4.1   *Results of FKMC*

The results are generated in iterative manner which are shown below in tabular form:

Iteration	Termination measure value
1	81.444251
2	85.274352
3	67.243550
4	33.692983
5	14.442164
6	5.292763
7	1.842184
8	0.629867

4.2   *Comparison Results*

Fig. 7 Termination measure value graph

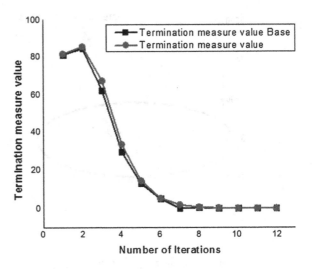

5 Conclusion

Crime detection plays a vital role in our lives for the reason that of the increasing rate of the crime happening in every area of the country. The crime is also of various types which has made it complex further. The government is not able to find a proper way out towards the removal of the crime and have better control over it. The proposed works above is a step ahead in this field. A further work needs also to be done so as to detect the category of crime, crime count, citywise distribution, increasing rate, etc. The various factors which affect crime rate are discussed.

References

1. R. G. Uthra "Data Mining Techniques to Analyze Crime Data.", International Journal For Technological Research In Engineering Volume 1, Issue 9, (2014).
2. Gupta, Manish, B. Chandra, and M. P. Gupta. "Crime mining for the Indian police information system." *Proceeding of the 2008 CSI* (2008).
3. Ismail,M,A, and Selim, S,Z. Fuzzy c-means optimality of solutions and effecttive termination of the algorithm. Pattern recognition, 19(6), pp (1986). 481–485.
4. Sharma, Anshu, and Shilpa Sharma. "An Intelligent Analysis of Web Crime Using Data Mining." International Journal of Engineering and Innovative Technology (2012).
5. Jain, Anil K., and Richard C. Dubes "Algorithm for clustering data." Prentice-Hall, Inc., (1998).
6. Pal, Sankar K., and Dwijesh K Dutta-Manjumder" Fuzzy mathematical approach to Pattern recognition Halsted Press, (1986).
7. Hall, L.O, Bensaid, A.M., Clarke, L.P. Velthuizen, R.P., Silbiger, M.S. and Bezdek, J.C, 1992, A comparison of neural network and fuzzy clustering techniques in segmenting magnetic resonance images of the brain., IEEE Transactions on 3(5), (1992) pp. 672–682.
8. Wang, Juntao, and Xiaolong Su. "An improved K-means clustering algorithm." In communication software and networks (ICCSN), IEEE, (2011).
9. Purohit, P. And Joshi, R., A new efficient approach towards K-means clustering algorithm. International Journal of computer applications, 65 (11) (2013).
10. Wang, J. And Su, X., 2011, May. An improved K-Means clustering algorithm. In Communication Software and Networks (ICCSN), IEEE 3rd International Conference on (pp. 44–46). IEEE. (2011).
11. Nath, Shyam Varan. "Crime pattern detection using data mining." In Web Intelligence and Intelligent Agent Technology Workshops, 2006. WI-IAT Workshops. IEEE/WIC/ACM International Conference on, pp. 41–44. IEEE, (2006).
12. Fahim, A. M., A. M. Salem, F. A. Torkey, and M. A. Ramadan. "An efficient enhanced k-means clustering algorithm." Journal of Zhejiang University *SCIENCE A* 7, no. 10 (2006): 1626–1633 (2006).

Implementation of Modified K-means Approach for Privacy Preserving in Data Mining

Shifa Khan and Deepak Dembla

Abstract Recent concerns regarding privacy breach issues have motivated the development of data mining methods, which preserve the privacy of individual data item. A cluster is gathering of information in such a way that the objects with similar properties are grouped into similar clusters and objects with dissimilar properties are placed into different clusters. The K-Means clustering algorithm is a broadly utilized plan to solve the clustering problem. In this paper, a comparative study of three clustering algorithms—K-means, Hierarchical and Cobweb across two different datasets is being performed. To form Clusters WEKA API has been used. The comparison is made with the variant of standard K-means technique that is Modified K-means technique. The Modified K-means technique has been developed to give better results as compared to existing K-means, Hierarchical and Cobweb techniques. This work also includes encryption and decryption of the formed clusters using AES algorithm to provide privacy to the data while transferring over networks. Experimental result proves that the performance of Modified K-means algorithm is better as compared to the existing K-Means and better than the hierarchical and Cobweb when tested on two datasets. K-Means and Hierarchical clustering is forming less number of clusters. In contrast, Cobweb is forming many clusters, which create memory issues. Therefore, Modified K-means forms an appropriate number of clusters in an organized manner and also takes minimum amount of time.

Keywords K-means · Clustering · AES · File splitting · File joining

S. Khan (✉) · D. Dembla
JECRC University, Jaipur, India
e-mail: Shifakhan1390@gmail.com

D. Dembla
e-mail: Hod.it@jecrcu.edu.in

© Springer Nature Singapore Pte Ltd. 2018   601
S.K. Bhatia et al. (eds.), *Advances in Computer and Computational Sciences*,
Advances in Intelligent Systems and Computing 554,
https://doi.org/10.1007/978-981-10-3773-3_58

1 Introduction

Privacy Preserving Clustering [1, 2] is a technique used to apply privacy to the formed clusters in order to provide surety to the data owners that their data is being transferred securely to the other end. The main aim of privacy preserving is to protect object values that are used for clustering analysis and to achieve this, each data object needs to be protected.

The goal is to transform D into D', i.e., transferring D dataset into D' dataset by applying some pattern P to the dataset to achieve privacy. Clustering [3] is a procedure of collecting information objects into incoherent clusters so that the information in the same cluster is comparative, yet having a place with various cluster contrast. A cluster is a gathering of information in a way that the objects with similar properties are grouped into similar clusters and objects with dissimilar properties are placed into different clusters.

K-means [4] is the most famous dividing technique for clustering. It is an unsupervised, non-deterministic, numerical, iterative technique for clustering. In K-means every cluster is identified by the mean estimation of attributes in the cluster. K-means is widely used because of its simplicity and the ability to give quick result [5, 6].

Algorithm for K-means:

(1) The first step demands to choose k data item from D database.
(2) In the second step it arranges the k data items into more groups on the basis of centroid value.
(3) Now, place these data items into clusters where similar values are placed in same cluster and different values are placed in different clusters.
(4) In the next step, calculate the mean value of the each data items for each cluster and update the mean value.
(5) Continue this process until there is no change in the cluster.
(6) Gives output in the form of clusters [7].

In Hierarchical Clustering [8] data objects are split into clusters and thus forming a hierarchy of clusters. Hierarchy of clusters is known as Dendrogram because it forms a tree of clusters. The Cobweb forms a tree structure in which each node is representing its concept, which in return represents a set of objects.

2 Related Work

Lavi Tyagi and Munesh Chandra Trivedi [9] have compared two algorithms—modified K-means algorithm and improved K-means algorithm. They proposed a new algorithm formed out of these two and named as Hybrid K-means algorithm. The modified K-means is used to solve the problem of empty cluster formation. The improved K-means is used to reduce the Euclidean distance between the data objects. Therefore, using these two techniques a new technique has been introduced to have a color image clustering and to get the refined clusters.

Jain et al. [3] have presented a data clustering techniques in broad manner. They have focused on important clustering applications such as image segmentation, object recognition, and information retrieval. An overview of pattern clustering methods has been shown in the paper with the goal of providing useful information or references to the broad community of clustering users.

Yu et al. [10] showed that the K-means clustering algorithm is generally utilized plan to take care of the grouping issue which characterizes a given arrangement of n information focuses in m-dimensional space into k bunches. In this paper, they have applied the idea of parallel computing to comprehend the protection safeguarding multi-party K-means grouping issue.

Turaga et al. [11], analysed under what conditions compression methodologies can held the result of cluster operations. They concentrated on the well-known K-means grouping algorithm and exhibit how a legitimately built compression technique is used for maintaining the global cluster structure. Their analytical derivations showed that a 1-bit moment preserving quantizer per group is adequate to hold the original data cluster.

Li et al. [12] presented a technique that permits the information proprietor to scramble its information with a homomorphic encryption plan and the adminis-tration supplier to perform K-means bunching specifically over the encoded information. They proposed a methodology that empowers the administration supplier to contrast encoded separations and the trapdoor data ace vided by the information proprietor.

Erkin et al. [13] introduced a variant of K-means clustering which comprises of two operators one is server and the other is multiple user where users are grouped into K-clusters. The server is not allowed to learn the individual data and the users are unaware of cluster centers. Implementation of this clustering method with movie lens data set gives promising running time results even on the modest hardware programs and has a reliable accuracy of the system.

3　Proposed Solution

In the proposed concept, clustering-based security framework has been imple-mented. The activity of clustering task is carried out in the following steps:

- **Pattern Selection**: Pattern Selection refers to the selection of appropriate pattern from the available number of patterns.
- **Pattern Definition**: It defines the properties of individual pattern. For example, In K-means, Euclidean distance is used to find dissimilarity between two patterns.

- **Grouping**: It means clustering, making clusters in a way that similar data objects are placed in same cluster and dissimilar data objects are placed in different cluster.
- **Information Abstraction**: In this section extract the useful information according to the requirement using the above formed clusters.

Therefore, to carry out the above process a new approach has been introduced named as Modified K-means approach. It is based on the alphanumeric data and number of clusters. In this the performance of the algorithm is evaluated on the basis of the number of clusters and time parameters to compare the proposed work with the existing work. On the basis of number of clusters, two tasks were performed.

- **Splitting the file**: It allows the sender to split the information into clusters in such a way that it simultaneously encrypts the file using AES encryption technique.
- **Joining the file**: It allows the receiver to join the files to get the original data using the same technique.

K-means is used as the base algorithm to make the comparison with the modified algorithm. The proposed work also makes the comparison of two more algorithms, i.e., Hierarchical and Cobweb. For each algorithm comparison is made on the basis of same number of clusters and time parameters.

3.1 Algorithm for Modified K-means

Step 1:   Read the Excel .csv file containing the Sample data.
Step 2:   Select the base file which becomes the basis for clustering.
Step 3:   Select the alphanumeric field from the drop down and save it as .csv form.
Step 4:   Obtain the clusters for each algorithm.

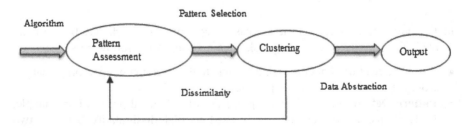

Fig. 1 Clustering process

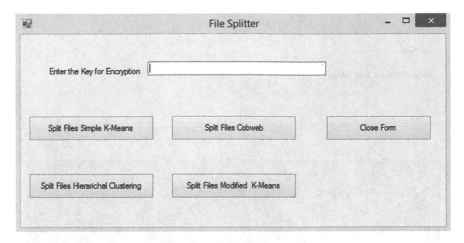

Fig. 2 Form for splitting the files on the basis of clusters

Step 5: Split the base file on the basis of the clusters and encrypt the file using AES algorithm and the private key concept.
Step 6: Resultant encrypted files are then transferred to receiver.
Step 7: Receiver decrypts the file using the same private key and obtain the original file.

4 Experimental Setup

In order to verify and specify the utility of the proposed concept, two dataset files have been taken. The first file contains around 1100 records and the second file consists of 800 records. The Whole framework has been designed in Visual Studio 2010 and tested the data using the GUI created using the VS2010. WEKA API has been used for calling other clustering algorithms.

In Fig. 1, base file is browsed to form clusters then choose the alphanumeric column from the drop down and click on the Save CSV button to form the CSV file. After this, click on the K-means, Hierarchical, Cob Web and Modified K-means button one by one to get the desired clusters. In Fig. 2, the first task to be performed is to enter the private key in order to secure the data. Now click on the Split Files Simple K-means button to split the files into clusters and in encrypted form. In Fig. 3, the first task is to enter the key which was previously entered at the sender side to access the records. Next, click on Join Files Simple K-means button to get the original file from the encrypted clusters.

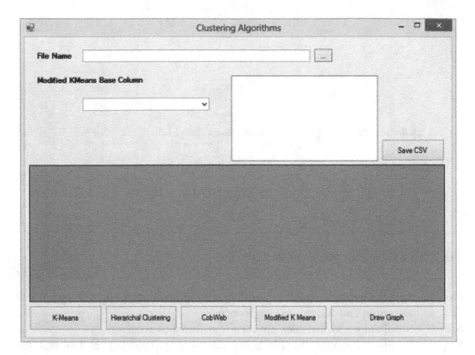

Fig. 3 Form for creating the clusters using various clustering algorithms

Table 1 Comparison table for dataset 1

Parameters	K-Means	Hierarchical	Cobweb	Mod. K-Means
Number of clusters	2	2	334	21
Time taken for encryption + splitting in milliseconds	61	59	12321	570
Time taken for decryption + joining in milliseconds	59	57	15167	567

5 Results and Analysis

According to the results, it is clear that the proposed algorithm results in the decent number of clusters, which are well enough for the security and handling purpose. The comparison of results is shown in the following two tables for the two datasets: (Tables 1 and 2).

The comparison of four algorithms is shown in the tables. Modified K-means is the proposed algorithm. The concept behind is to provide better organized clusters in order to use data in an efficient manner. In modified K-means, 21 clusters (Dataset 1) are formed but in existing K-means only 2 cluster (Dataset 1) are

Table 2 Comparison table for dataset 2

Parameters	K-Means	Hierarchical	Cobweb	Mod. K-Means
Number of clusters	2	2	397	20
Time taken for encryption + splitting in milliseconds	68	58	17077	546
Time taken for decryption + joining in milliseconds	71	63	18296	547

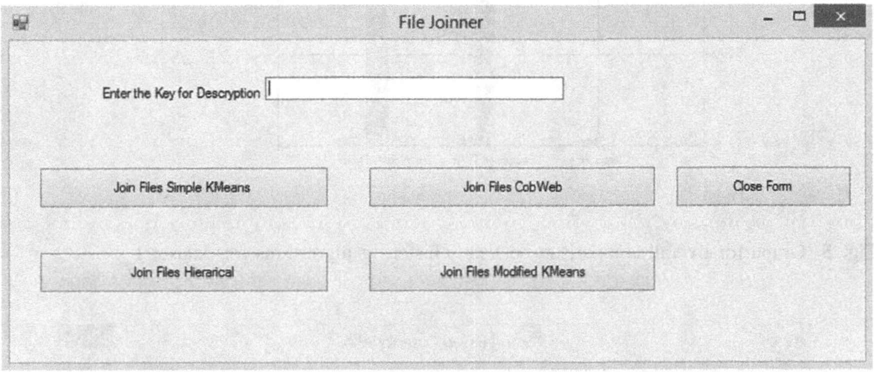

Fig. 4 Form for joining the files on the basis of clusters

formed. In modified K-means clusters are formed on the basis of alphabetically order thus giving much better result as compared to existing K-means which is forming 2 clusters in some random order. Hierarchical clustering is giving hierarchy of clusters but not in an organized manner. Cobweb is forming too many clusters which create memory issues. Therefore, Modified K-means is better as compared to the K-means, Hierarchical, Cobweb and Modified K-means algorithm because it is giving appropriate number of clusters in a well-organized form.

5.1 Graphs

The above graphs show the graphical representation of all compared parameters, i.e.: Number of clusters, Encryption time (time taken for splitting), Decryption time (time taken for joining). In Fig. 4, graphs according to the number of clusters are formed. There are two clusters formed for the K-Means and Hierarchical clustering. For Cobweb 334 clusters are formed which consumes a lot of memory. Lastly in case of modified K-Means only 21 clusters are formed in an organized manner.

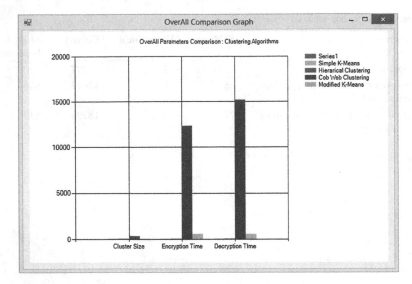

Fig. 5 Graph for overall comparison between clustering algorithms for dataset 1

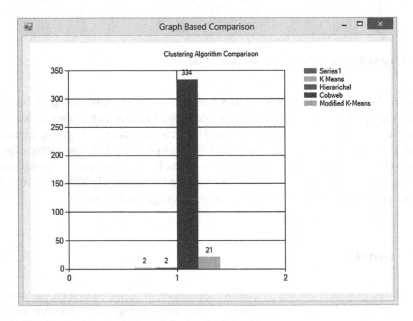

Fig. 6 Graph according to number of clusters for dataset 1

In Fig. 5 same comparison is made on the basis of dataset 2. In Fig. 6, comparison is made on the basis of number of clusters, encryption time and decryption time (Figs. 7 and 8).

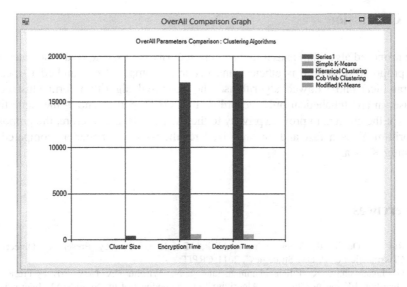

Fig. 7 Graph for overall comparison between clustering algorithms for dataset 2

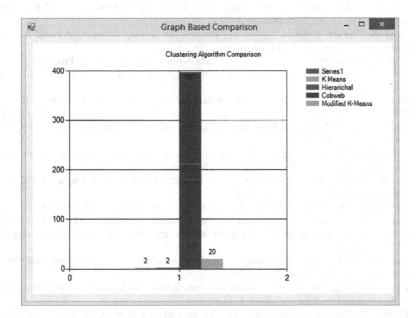

Fig. 8 Graph according to number of clusters for dataset 2

The Modified K-means algorithm gives appropriate result as compared to the other three algorithms. The Proposed algorithm forms desired number of clusters to give good efficiency to the users. This algorithm is also performing well when measured on the basis of time parameters as compared to the other algorithms.

6 Conclusion

The proposed algorithm is useful in solving security issues. The result indicates that the proposed algorithm is efficient and better as compared to standard K-means, Hierarchical and Cobweb algorithms. The proposed algorithm form clusters of dataset in an alphabetical order according to their properties and at the same time encrypt the clusters to provide privacy to the owner's data. Therefore, the proposed algorithm gives a fast and an organized result, which is better as compared to existing K-means.

References

1. Rui Li, Denise de Vries, John Roddick," Bands Of Privacy Preserving Objectives: Classification of PPDM Strategies", 2011 CRPIT.
2. G. Jagannathan, K. Pillaipakkamnatt, and R.N. Wright, "A New Privacy-Preserving Distributed K-means Clustering Algorithm," in Proceedings of the Sixth SIAM International Conference on Data Mining, 2006.
3. A.K. Jain, M.N. Murty and P.J. Flynn, "Data Clustering: A Review", ACM Computing Surveys, Vol. 31, No. 3, September 1999.
4. Neha B. Jinwala, Gordhan B. Jethava, Privacy "Preserving Using Distributed K-means Clustering for Arbitrarily Partitioned Data", 2014 IJEDR.
5. P. Bunn and R. Ostrovsky, "Secure Two-Party K-means Clustering," in Proceedings of the 14th ACM Conference on Computer and Communications Security. ACM New York, NY, USA, 2007.
6. Geetha Jagannathan and Rebecca N. Wright, "Privacy-Preserving Distributed K-means Clustering Over Arbitrarily Partitioned Data," in Proceedings of the Eleventh ACM SIGKDD International Conference on Knowledge Discovery in Data Mining, New York, NY, USA, 2005, pp. 593–599, ACM.
7. JyotiYadav, Monika Sharma, "A Review of K-mean Algorithm", International Journal of Engineering Trends and Technology (IJETT)–Volume 4 Issue 7-July 2013.
8. George Karypis, eui-hong(sam)Han, Vipin Kumar, "Chameleon: Hierarchical Clustering Using Dynamic Modeling" 2009 IEEE.
9. Lavi Tyagi, Munesh Chandra Trivedi, "Hybrid K-mean and Refinement Based on Ant for Color Image Clustering" published in the springer proceedings(AISC), International Conference on Information and Communication Technology for Sustainable development (ICT4SD 2015).
10. Teng-Kai Yu, D.T. Lee, Shih-Ming Chang, "Multi-Party k-Means Clustering with Privacy Consideration", IEEE DOI 10.1109/ISPA.2010.8.
11. Deepak S. Turaga, Michail Vlachos, Olivier Verscheure, "On K-Means Cluster Preservation using Quantization Schemes", 2009 IEEE.
12. Dongxi Li, Elisa Bertin, Xun Yi, "Privacy of Outsourced K-Means Clustering", ASIA CCS'14.
13. ZekeriyaErkin, Thijs Veugen1, Tomas Toft, Reginald L. Lagendijk1, "Privacy-Preserving User Clustering In a Social Network", 2009 IEEE.

Cross-Lingual Information Retrieval: A Dictionary-Based Query Translation Approach

Vijay Kumar Sharma and Namita Mittal

Abstract The rapidly increasing demographics of the Internet population and the abundance of multilingual content on the web increased the communication in multiple languages. Most of the people use their regional languages to express their need and the language diversity becomes a great barrier. Cross-Lingual Information Retrieval (CLIR) provides a solution for that language barrier which allows a user to ask a query in the native language and get the relevant documents in the different language. In this paper we build a dictionary-based query translation system. Queries are tokenized and multi-words query terms are created using N-gram technique. Out of vocabulary (OOV) terms are transliterated using the proposed OOVTTM technique. Target documents are retrieved using vector space retrieval model. Experiment results represent that the proposed approach achieves better results.

Keywords Dictionary-based query translation · OOVTTM transliteration mining · CLIR

1 Introduction

Information Retrieval (IR) is a reasoning process that is used for storing, searching and retrieving the relevant information from a set of documents [1]. Global Internet Usage statistics shows that number of web access by the non-English users is tremendously increased. But, all of them are not able to express their query in English.[1] So information retrieval tasks are not restricted to only monolingual but

---

[1]Internet World Stats: http://www.internetworldstats.com.

V.K. Sharma (✉) · N. Mittal
Department of Computer Science and Engineering, MNIT Jaipur, Jaipur, India
e-mail: sharmavijaykumar55@gmail.com

N. Mittal
e-mail: mittalnamita@gmail.com

© Springer Nature Singapore Pte Ltd. 2018 611
S.K. Bhatia et al. (eds.), *Advances in Computer and Computational Sciences*,
Advances in Intelligent Systems and Computing 554,
https://doi.org/10.1007/978-981-10-3773-3_59

also multilingual. The classical IR normally regards the documents and sentences in other languages as unwanted "noise" [2]. The needs for handling multiple languages introduce a new area of IR that is CLIR. CLIR deals with user queries in one language and target documents in different languages. A CLIR approach includes a translation approach followed by monolingual information retrieval. There are two types of translation approaches, namely query translation and documents translation. A lot of computation time and space is elapsed in document translation approach so query translation approach is preferred [3]. Three query translation approaches are discussed in the State-of-art CLIR, i.e. *Dictionary-Based Translation* (DT), *Corpus-Based Translation* (CT) and *Machine Translation* (MT). MT and CT approaches need a parallel corpus, which is not available for resource-poor languages like Hindi. It is very cumbersome to create such kind of parallel corpus so DT approach is preferred for fast computation [3, 1, 4, 5].

Source language queries are tokenized and multi-word terms are created. These query terms are searched in a bilingual dictionary and replaced by corresponding translated terms. The terms, which are not found in dictionary, are called OOV terms. OOV terms are transliterated by the proposed Sub OOVTTM Technique. Vector space retrieval model is used for target document retrieval. Related work is discussed in Sect. 2. Proposed approach is discussed in Sect. 3. Experiment results and discussion are presented in Sect. 4.

2 Related Work

Pingali et al. [6, 7] were experimented with Hindi and Tamil to English language. They used Bilingual dictionary for query translation. OOV terms were transliterated using probabilistic algorithm. Target documents were retrieved using extended Boolean model and Vector based ranking model. Makin et al. [8] were experimented with Hindi document collection. They were concluded that bilingual dictionary with cognate matching and transliteration achieved better performance than the bilingual dictionary alone. Sethuramalingam et al. [9] were experimented with FIRE 2008 data. Combinations of dictionaries were used for query translation. Named entities and OOV words were translated using CRF-based named entity recognition tool. Documents were retrieved using Lucene's OKAPI BM25. Jagarthanam et al. [10] were exploited Compressed Word Format (CWF) algorithm for named entity transliteration. Jagarlamudi et al. [11] were prepared a Statistical Machine Translation (SMT) system which trained on aligned parallel sentences and a word alignment table was created. Queries were translated in target language with the use of SMT and transliteration technique. Relevant documents were retrieved using a language modelling based retrieval algorithm. Pattabhi et al. [12] were experimented with FIRE 2010 Tamil–English language pair. Named entity terms were extracted from Tamil queries and translate them individually. Bajpai et al. [13] were analysed the CLIR system for various Indian language and a prototype model was suggested. Queries were translated using any one technique including MT, dictionary-based and corpora-based. A common problem of word ambiguation was resolved using Word Sense Disambiguation (WSD) technique further Boolean, Vector space and Probabilistic model was used for IR.

3 Proposed Approach

The proposed approach for dictionary-based CLIR is expressed in four steps. (1) *Pre-processing*, where a query string is tokenized. Multi-word terms are created using N-gram technique. Stop words are eliminated. (2) *Query Translation*, where a query term is searched in bilingual dictionary and replaced by the corresponding target language word. (3) *OOV Term Transliteration Mining* and (4) *Indexing, Retrieval and Evaluation*. The proposed approach is also depicted in Fig. 1.

3.1 Pre-processing

Source language query string is tokenised and multi-word terms are created using n-gram technique. Multi-word terms are created before stop word removal for appropriate translation. For example term "के विरुद्ध" is translated as "against" but if stop words are removed first, then term becomes "विरुद्ध" and transalation is

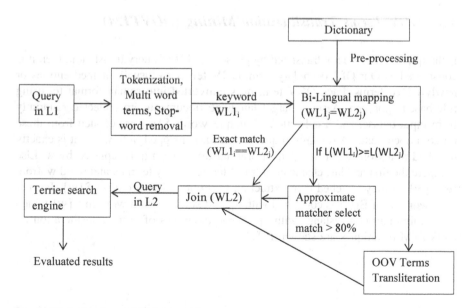

Fig. 1 Dictionary-based CLIR approach

"opposite" and "repugnant" which is less appropriate then "against". Stop words are eliminated in the case of unigram. A set of keywords WL1 is prepared after pre-processing of query string.

3.2 Query Translation

The only goal of query translation is to obtain such a target language translation which enable proper retrieval. Source language query terms are mapped to a bilingual dictionary. Every query term is routed to two phases.

In *phase I*, if a source language query terms (WL1) is exactly matched in bilingual dictionary then it is replaced by the corresponding translation (WL2). For example query word "समुदाय" is exactly matched two times in the bilingual mapping and corresponding translations are "fold" and "tribe". If a term is not translated by phase I then it is routed to phase II.

In *phase II*, if length of source language query term $WL1_i \geq$ length of bilingual dictionary word $WL1_j$ then a Longest Common Subsequence (LCS) $WL1_{LCS}$ is extracted and calculate the percentage match of $WL1_{LCS}$ in $WL1_i$. A threshold is defined empirically to select source language word from bilingual dictionary, i.e. 80%. For example query term "प्रतिनिधी" matched with a dictionary entry "प्रतिनिधि".

3.3 OOV Terms Transliteration Mining (OOVTTM)

If the query terms are not translated by phase I and II of query translation, then it is considered as Our Of Vocabulary term. OOV terms are usually named entities or newly added terms. Each OOV term q_i is converted into roman format tq_i using rule-based approach. A set of target language documents are collected randomly from input dataset and a collection of unique words $S\{w\}$ are created from these random documents. A romanised query term tq_i is mapped in $S\{w\}$. If it is exactly matched to a word w_i in $S\{w\}$, then original query term q_i is replaced by w_i Else compare the first and last character of transliterated query term tq_i and word w_i from set $S\{w\}$. If they matched then extract all the word $E\{w\}$ from set $S\{w\}$. Further each word w_i of $E\{w\}$ is compared with tq_i and select a word w_i from $E\{w\}$ which has a minimum edit distance with tq_i. Some examples of terms transliteration by OOVTTM techniques are shown in Table 1.

Table 1 Transliteration of terms by OOVTTM technique

Query word (q_i)	Transliterated word (tq_i)	Word (w_i from $S\{w\}$)
मीणा	Meena	meena
शंकर	shnkr	Shankar
भारत	bhart	Bharat

3.4 Indexing, Retrieval and Evaluation

Terrier[2] search engine is used for indexing, retrieval and evaluation. Vector space model is used for retrieving target language documents.

4 Experiment Results and Discussions

FIRE[3] 2010 dataset is used for evaluation the proposed approach, which contain a topic set of 50 Hindi language queries and a set of 1,25,638 English language documents. Topic set includes <title>,<desc> and <narr> tag field in each query but only <title> tag field is considered in the experiment. Shabdanjali[4] and English–Hindi mapping[5] are used as a translation resources. N-gram technique is applied on each query and tri-grams, bi-grams and unigrams are constructed. Bi-lingual dictionary is searched for tri-grams and if tri-grams are not found then search for bi-grams and then for unigrams. OOV terms is transliterated by the LCS or by the proposed OOVTTM technique. A random set of 139 english documents are selected from 1,25,638 English documents for preparing of unique set of words $S\{w\}$ in the proposed OOV word transliteration mining technique. Terrier search engine is used for indexing, retrieval and evaluation. CLIR system is evaluated by using Mean Average Precision (MAP). MAP for a set of queries is the mean of the average precision score of each query. Precision is the fraction of retrieved documents that are relevant to the query. CLIR system for Hindi–English language achieves better MAP with the proposed approach. The experiment results are shown in Table 2. A maximum 0.1172 MAP is achieved with only query <title> tag while as discussed in [9] 0.0907, 0.1204 and 0.1112 MAP was achieved with <title,desc>,<title,narr> and <title,desc,narr> tag.

---

[2]http://terrier.org/.

[3]http://fire.irsi.res.in/fire/home.

[4]http://ltrc.iiit.ac.in/onlineServices/Dictionaries/Dict_Frame.html.

[5]http://www.cfilt.iitb.ac.in/Downloads.html.

Table 2 Hindi–English CLIR experiment results with FIRE 2010 dataset

S.no.	Experiments (Query Translation + N-gram + Transliteration)	Result (MAP)
1	Monolingual (English–English)	0.3705
2	Shandanjali + Unigram + LCS	0.0851
3	Shabdanjali + Unigram + OOVTTM	0.0971
4	Shabdanjali + N-gram + OOVTTM	0.1172
5	Shabdanjali and English–Hindi mapping + N-gram + OOOVTTM	0.0535
6	Shabdanjali and English–Hindi mapping + N-gram + OOVTTM and rule based transliteration	0.0579

Table 3 Word translations by dictionary

Query word	अनारा	गुप्ता	मामले	राज़्यों
Dictionary word	अनार	गुप्त	मामला	राज़्य
Translated word	Pomegranate	Secret, covert, undercover	Case	State

The terms, which are exactly not matched in the dictionary may be wrongly translated by phase II of query translation step, as shown in the Table 3. But these wrongly translated terms are always non-dictionary words or named entities. Dictionary words are correctly translated.

As presented in experiment second and third, OOVTTM technique for transliteration achieves better result than the LCS. LCS technique achieves good results for good length terms. For example transliteration of term "नागालैंड" by rule-based is "Nagalaind" and we get a transliterated word "Nagaland" from set S{w}. OOVTTM technique increases the probability of correct translation and reduces the transliteration mining time due to selection of set E{w} from set S{w}.

As presented in experiment 4th, Query translation with N-grams achieves better result because word like "हवाई अड्डा" translated as "airbase", which can not be translated appropriately by unigram technique. Multiple dictionaries are included for enriching dictionary words pairs but MAP is got decreased due to multiple translation for a single query word as reported in Table 4. Rule based transliteration could not significantly improve the result, since most of the words are wrongly transliterated by rule based technique.

Table 4 Multiple translation of a single word

S.no.	Word	Shandanjali translation	(Shabdanjali + English–Hindi) mapping translation
1	समुदाय	Fold, tribe	Fold, tribe, category, plurality, battalion, pack, community
2	विध्वंस	Annihilation, collapse	Annihilation, collapse, destruction, wipeout, demolition

5 Conclusion

Experiment results for dictionary based CLIR system represents that including multi-words term translation and OOVTTM technique for transliteration improves the CLIR system. A maximum of 0.1172 MAP is achieved with Shabdanjali dictionary, N-gram, OOVTTM technique. A large set of S{w} is needed for better performance of OOVTTM. Named entity terms are wrongly translated by dictionary so there is a need of distinguish between named entities and dictionary term. Including more resources for enriching dictionary lead to query diversification issue. Based on the current experience, the proposed CLIR system will be improved in future by distinguishing named entities and by selection of N-number of appropriate translation.

References

1. Nagarathinam A. and Saraswathi S., State of art: Cross Lingual Information Retrieval System for Indian Languages, International Journal of computer application, Vol. 35, No. 13 (2011), pp. 15–21.
2. Abusalah, M., J. Tait, and M. Oakes, Literature Review of Cross Language Information Retrieval. World Academy of Science, Engineering and Technology 4 (2005), 175– 177.
3. Nasharuddin N. Amelina and Abdullah M. Taufik., Cross-lingual Information Retrieval State-of-the-Art, electronic Journal of Computer Science and Information Technology (EJCSIT), Vol. 2, No. 1 (2010), pp. 1–5.
4. Sujatha P. and Dhavachelvan P., A review on the Cross and Multilingual Information Retrieval, International Journal of Web & Semantic Technology (IJWesT) Vol.2, No.4 (2011), pp. 155–124.
5. Sharma VK, Mittal N., Cross Lingual Information Retrieval (CLIR): Review of Tools, Challenges and Translation Approaches. In Information System Design and Intelligent Application (2016), pp. 699–708.
6. Pingali P. and Varma V., Hindi and Telugu to English cross language information retrieval at CLEF 2006, In working notes of CLEF, LTRC, IIIT Hyderabad (2006).
7. Pingali P. and Varma V., IIIT Hyderabad at CLEF 2007- Adhoc Indian Language CLIR Task, In working notes of CLEF, LTRC, IIIT Hyderabad (2007).
8. Makin R., Pandey N., Pingali P., and Varma V., Approximate String Matching Techniques for Effective CLIR, International Workshop on Fuzzy Logic and Applications, Italy, Springer-Verlag (2007), pp. 430–437.
9. Sethuramalingam S. and Varma V., IIIT Hyderabad's CLIR Experiments for FIRE-2008, FIRE 2008.
10. Janarthanam S.C., Sethuramalingam S. and Nallasamy U., Named Entity Transliteration for Cross-Language Information Retrieval using Compressed Word Format Mapping Algorithm, in proceedings of the 2nd ACM workshop on Improving non English web searching, ACM (2008), pp. 33–38.
11. Jagarlamudi J. and Kumaran A., Cross-Lingual Information Retrieval System for Indian Languages, Advances in multilingual and multi modal information retrieval (2008), pp. 80–87.

12. Pattabhi R. K. Rao and Shobha L., AU-KBC FIRE2010 submission – Cross Lingual Information Retrieval Track: Tamil-English, Fire 2010.
13. Bajpai P. and Verma P., Cross Language Information Retrieval: In Indian Language Perspective, International Journal of Research in Engineering and Technology, Vol. 3 (2014), pp. 46–52.

Predictive Classification of ECG Parameters Using Association Rule Mining

Kratika Tyagi and Sanjeev Thakur

Abstract Data mining is the procedure of extricating valuable information from the tremendous information stored in the database. Association rule mining is one of the most important and powerful data mining techniques. Association rule mining is normally carried out in two stages: first is to find frequent item set and second is to utilize those item sets to recognize the association rules. In recent medical history of cardiac arrests it has been observed that a huge gap exists in interpreting ECG data among differently skilled doctors. In this paper we will use the principle of meta-analysis and will reduce the gap between the interpretations of different doctors by employing statistical techniques like correlation and multiple linear regression. We would also generate rules using predictive apriori association rule mining among the various attributes of ECG to classify whether a patient requires an ECG before a cardiac arrest or not. The purpose of carrying out this work is to reduce the fatality rate and be able to predict that whether a patient requires ECG before actually facing a cardiac arrest and to minimize the cases of wrong interpretation.

Keywords Data mining · Association rule mining · Predictive apriori · ECG data

1 Introduction

Data mining is the procedure of fascinating knowledge or patterns from extensive databases. There are various techniques that have been utilized to find such sort of information, majority of them results from the machine learning and statistical area [1].

K. Tyagi (✉) · S. Thakur
Department of Computer Science & Engineering, ASET Amity University,
Noida, Uttar Pradesh, India
e-mail: kratikatyagi4192@gmail.com

S. Thakur
e-mail: sthakur3@amity.edu

© Springer Nature Singapore Pte Ltd. 2018   619
S.K. Bhatia et al. (eds.), *Advances in Computer and Computational Sciences*,
Advances in Intelligent Systems and Computing 554,
https://doi.org/10.1007/978-981-10-3773-3_60

Association rule mining is the powerful data mining techniques which are normally carried out in two steps: first is to find frequent item set and second is to utilize those item sets to recognize the association rules. Keeping in mind the end goal to produce relationship among different work factors, association rule mining can help us to create association rules utilizing the idea of CAR. The extent of this paper is to generate the association rules using predictive apriori algorithm that will produce best rules based on increasing support–confidence threshold over a dataset generated by doctor responses. The paper is divided into five sections. The second section following this introduction is a detailed literature discussing the factors and the predictive apriori algorithm. The third section is a simple experimental setup followed by results in the fourth section. The final section will be a detailed discussion on the results along with plausible conclusion.

2 Literature Survey

In this section we will discuss about the essentials of association rule mining and basic terminologies related to it. Further we will try to investigate the apriori and predictive apriori association rule mining. We will continue with the literature survey and have followed the work done by Banerjee et al. [2].

2.1 Association Rule Mining

Let us interpret what we precisely mean by association rule mining. The most straight forward answer would be that it is a way to deal which decides the relationship among the items in a set of transaction of a database taking into account the interestingness [2]. Association rule mining algorithms mine the different association rules based on support and confidence parameters by generating item sets. At every emphasis the support of the selected item is compared with threshold support. This procedure is iterated and every association rule that can accomplish the minimum confidence threshold is recorded. Higher the confidence, the better chance of generating a good association rule [2].

Consider a database containing items {A, D, E, L, M, X, Y, W, W, J}. Then

1. Each tuple is a transaction.
2. Each transaction contains an attribute–value pair known as an item.
3. An association rule will be represented in the form of BODY=>HEAD[support, confidence]. For example, eat(y, "icecream") =>sugar(y, "high") [0.2, 0.8].
4. Support: Consider two items A and D such that A=>D: P(Q.W). Then, support is equal to number of transactions containing (AUD) divided by total number of transactions.

5. Confidence: Consider two items A and D such that A=>D: P(W|Q). Then confidence is equal to number of transactions containing (AUD) divided by number of transactions containing A.
6. So the goal is to determine the association rules that fulfill a pre-defined support and have high confidence.

To understand the working of association rule mining, one may refer to the following example:

1. Download and install Weka 3.6 which is available online.
2. Run Weka 3.6 and select "Explorer". Click on preprocess.
3. Now click on Open file and load the data.csv data file.
4. Click on Associate and select the apriori algorithm and click start.
5. Four association rules will be displayed on the Associator output window as shown in Fig. 1.

With reference to the rules generated in Fig. 1, we can easily observe that item sets with minimum support have been associated. This basically implies that higher the support more will be the coverage of items and confidence acts as a parameter for exactness. For n item sets, the numbers of possible rules that can be generated are 2n−1 [3]. It is clear that why association rule mining is important for real-world application since it can mine relationships among attributes for any data. However, the disadvantage with rule mining is that it generates an excessive number of classes for attribute–value pairs for which subsequently an excessive number of rules are generated. It implies that these algorithms are computationally unmanageable [4].

2.2 Apriori Association Rule Mining

Apriori rule is based on the fact that "any subset of a frequent item set is frequent." To understand its working, a portion of the related terms is discussed below:

1. Frequent item set is the set of items that have fulfilled the minimum support and is represented by L_k for the kth item set.

```
Best rules found:

1. Blood Pressure=No Thyroid=No 116 ==> Class=ECG 73      conf:(0.63)
2. Heart Rate=High Blood Pressure=No 128 ==> Class=ECG 79   conf:(0.62)
3. Triglyceride=No Thyroid=Yes 135 ==> Class=No ECG 83      conf:(0.61)
4. Height=Normal Blood Pressure=No 125 ==> Class=ECG 76     conf:(0.61)
```

Fig. 1 Association rules generated using apriori algorithm

2. Join operation focuses on performing a join item L_k-1 with itself to generate candidate item C_k.
3. Prune operation basically confirms that if any k−1 item set is frequent, then it should belong to a subset of a successive k item set.

The algorithm for implementing the apriori association rule with reference to Fig. 2 is as follows:

C_j = Item sets of unary size in I;
Identify every large item sets with unary size, L_j;
j = 1;
Repeat
j = j + 1;
C_j = Apriori (L_j−1);
Apriori (L_j−1)

Generate candidates of size j + 1 from large item sets of size j.
Perform join operation on large item sets of size j if they agree on j−1.
Perform pruning of candidates who have subsets that are not large.

Count C_j to determine L_j unless no large item sets are determined;

Assume a scenario where for a database some association rules with sup = 0.5 and conf = 0.8 [5]. To generate the association rule from frequent item set, first for all frequent item set L the non-empty subset of L is determined as shown in Fig. 3 such that a rule in the form of M => (L−M) is generated if and only if (given in Eq. 1)

$$Supp(L)/Supp(M) > = Min\ Conf.$$ 　　　　(1)

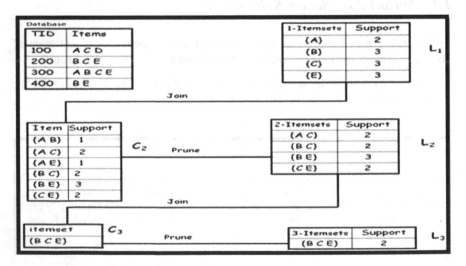

Fig. 2 Implementation of apriori algorithm [5]

Rules	Support (X Y)	Support(X)	Confidence
{A} -> {C}	2	2	100
{B} -> {C}	2	3	66.66666667
{B} -> {E}	3	3	100
{C} -> {E}	2	3	66.66666667
{B} -> {C E}	2	3	66.66666667
{C} -> {B E}	2	3	66.66666667
{E} -> {B C}	2	3	66.66666667
{C} -> {A}	2	3	66.66666667
{C} -> {B}	2	3	66.66666667
{E} -> {B}	3	3	100
{E} -> {C}	2	3	66.66666667
{C E} -> {B}	2	2	100
{B E} -> {C}	2	3	66.66666667
{B C} -> {E}	2	2	100

Fig. 3 Generation of association rules [5]

2.3 Predictive Apriori Association Rule Mining

Predictive apriori rule mining is another confidence-based association rule mining algorithm that generates frequent item sets in the same fashion as apriori algorithm does with the only difference that predictive apriori estimates the confidence of an association rule differently [3, 6]. The objective of predictive apriori is to increase the correctness of the association rule rather than the correctness of data that is used as an input [3]. Let I be a database which contains some records t generated by process C, let A=>B be an associated rule. The predictive accuracy pra(A=>B) is equivalent to the probability that t satisfies A such that it also satisfies B. This conditional probability of B being a subset of t given that A is also a subset of t is governed when process C distributes t [6]. Where apriori prefers generating more general rules, predictive apriori generates the best n rules based on the following criteria as described by Mutter et al. [3].

1. Among the n best the predictive accuracy is calculated and
2. In case of same predictive accuracy, rules generated by predictive apriori are not subsumed.

When compared with apriori we observe that predictive apriori generates the association rules and puts them in the set of n best rules on increasing minimum support threshold. If a particular rule is to be discarded from the list of n best rules, then the algorithm has to re-run recursively again for generation of rules. Also the classification of association rules becomes easier [3] when using predictive apriori. Also the quality of rules generated with predictive apriori is better when using small data [7] sets which fits perfectly for our work. However, the drawback is that predictive apriori has higher time complexity as compared to apriori algorithm [3]. Now we will move forward with predictive apriori algorithm and apply over the ECG data set we have generated by collecting doctors feedback in the next section.

3   Experimental Setup

In this part, we will discuss about our ECG data set and the factors that have been considered. We prepared a questionnaire for 600 participants out of which, 500 responses were selected due to their completeness. 49.77% males and 40.23% of the females had participated between the ages of 30–55 years. Each participant is working in some hospital with different work environment variables. The data set covers the following factors in the form of questions which can be answered by a participant in yes/no, or low, medium, high:

1. Age between 30–55: This describes the age of the patient.
2. Gender: It will depict the gender of the patient whether he is male or female. For male it is represented by 0 and for female it is represented by 1.
3. Heart Rate: It indicates the patient's heart rate whether it is high or low. For low it is represented by 1 and for high it is represented by 2.
4. Blood Pressure: It indicates the blood pressure of the patient whether it is low, medium, or high. For low it is represented by 0, for medium it is represented by 1, and for high it is represented by 2.
5. Diabetes: It indicates whether a patient is having diabetes or not. If the patient is a diabetic patient, it is represented by 0, else it is represented by 1.
6. Triglyceride: It indicates whether a patient is having triglyceride or not. It is represented by 1 for low, 2 for medium, 3 for high, and 4 for very high.
7. Thyroid: It indicates the thyroid status of the patient whether the patient is having low, medium, or high. For low it is represented by 1, for medium it is represented by 2, and for high it is represented by 3.
8. Cholesterol: It indicates whether a patient is having low cholesterol, or medium or high. For low it is represented by 1, for medium it is represented by 2, and for high it is represented by 3.
9. ECG Needed: It represents the class whether a patient requires ECG or not. If the patient requires ECG then it is represented by 1, if the patient does not require ECG then it is represented by 2, and 3 will represent the doubtful case in which he/she might requires ECG or not.

4   Results

In this section we will operate the predictive apriori algorithm using Weka 3.6 which is an open-source data mining tool. We will perform the following steps:

1. Convert the excel data into csv file format.
2. Load the dataset in csv file format into Weka under preprocess tab.
3. Set CAR = True and set numrules = 10 for predictive apriori for classifying the association rules under associate tab.
4. Click start to build the model for the dataset and generate the association rules in the output window.

100 association rules are generated using predictive apriori. Out of them top 20 association rules generated are as follows:

1. Sex = Female Height = Abnormal Heart Rate = Low Blood Pressure = Yes Diabetes = Yes 16 ==> Thyroid = Yes 16 acc:(0.99295)
2. Height = Abnormal Weight = Normal Heart Rate = High Blood Pressure = No Cholesterol = No Class = ECG 14 ==> Age = Young 14 acc:(0.99167)
3. Height = Abnormal Heart Rate = High Blood Pressure = No Triglyceride = No Cholesterol = No Class = ECG 14 ==> Age = Young 14 acc:(0.99167)
4. Age = Old Height = Abnormal Weight = Obese Heart Rate = High Blood Pressure = No 13 ==> Diabetes = No 13 acc:(0.99067)
5. Age = Old Sex = Male Height = Abnormal Blood Pressure = No Triglyceride = Yes 12 ==> Diabetes = No 12 acc:(0.98929)
6. Heart Rate = High Diabetes = Yes Triglyceride = Yes Thyroid = No Cholesterol = Yes 12 ==> Weight = Normal 12 acc:(0.98929)
7. Age = Old Sex = Male Height = Normal Weight = Normal Heart Rate = High Diabetes = Yes 12 ==> Cholesterol = Yes 12 acc:(0.98929)
8. Sex = Female Height = Abnormal Heart Rate = Low Diabetes = Yes Thyroid = No 11 ==> Blood Pressure = No 11 acc:(0.98736)
9. Age = Old Sex = Male Weight = Obese Blood Pressure = No Triglyceride = Yes Thyroid = Yes 11 ==> Diabetes = No 11 acc:(0.98736)
10. Age = Young Sex = Female Height = Normal Diabetes = No Triglyceride = Yes Cholesterol = No 11 ==> Weight = Obese 11 acc:(0.98736)
11. Age = Young Sex = Female Height = Normal Triglyceride = Yes Thyroid = Yes Class = ECG 11 ==> Weight = Obese 11 acc:(0.98736)
12. Age = Young Sex = Female Heart Rate = Low Blood Pressure = No Diabetes = No Triglyceride = Yes 11 ==> Height = Normal 11 acc:(0.98736)
13. Age = Young Sex = Male Weight = Normal Blood Pressure = No Cholesterol = No Class = ECG 11 ==> Heart Rate = High 11 acc:(0.98736)
14. Age = Young Height = Normal Blood Pressure = Yes Triglyceride = No Cholesterol = Yes Class = ECG 11 ==> Sex = Male 11 acc:(0.98736)
15. Age = Young Weight = Obese Blood Pressure = No Diabetes = No Triglyceride = Yes Cholesterol = No 11 ==> Sex = Female 11 acc:(0.98736)
16. Sex = Male Blood Pressure = Yes Diabetes = Yes Triglyceride = No Thyroid = Yes Class = No ECG 11 ==> Heart Rate = High 11 acc:(0.98736)
17. Age = Old Sex = Female Blood Pressure = No Diabetes = Yes Class = No ECG 10 ==> Heart Rate = Low 10 acc:(0.98462)

18. Age = Old Height = Normal Blood Pressure = No Diabetes = Yes Class = No ECG 10 ==> Cholesterol = Yes 10 acc:(0.98462)
19. Age = Old Sex = Male Heart Rate = High Thyroid = Yes Cholesterol = No Class = No ECG 10 ==> Weight = Obese 10 acc:(0.98462)
20. Age = Old Heart Rate = Low Triglyceride = No Thyroid = No Cholesterol = No Class = ECG 10 ==> Sex = Female 10 acc:(0.98462)

5 Observations

From the above execution of the predictive apriori algorithm as depicted in Fig. 4, we have gathered the following observations:

1. Each association rule has been ranked on the basis of predictive accuracy.
2. The algorithm does not take support as input parameters but starts increasing the minimum support threshold.
3. Each attribute–value pair shares a relationship with another attribute–value pair.
4. Predictive apriori generates more rules than apriori algorithm.
5. Predictive accuracy drops when applying CAR.
6. The predictive apriori algorithm is worse than apriori algorithm in terms of time complexity.
7. Predictive apriori can generate more rules than apriori Algorithm.

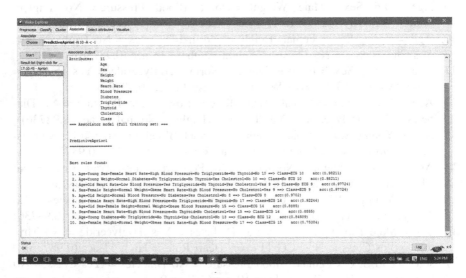

Fig. 4 Association rules generated by predictive apriori

6 Conclusion

In this paper, we have discussed the use of association rule mining to identify the relationships among factors that suggests whether a patient requires ECG or not. A detailed discussion regarding predictive apriori and standard apriori algorithm is discussed. The outcome from the experimental setup provided some observations that will help researchers to develop a forecasting system that can help doctors to identify whether a patient requires an ECG or not.

References

1. Jain, J., Tiwari, N., Ramaiya, M.:A Survey on Association Rule Mining, Published in International Journal of Engineering Research and Applications (IJERA), pp. 2065–2069, 2013.
2. Banerjee, S., Thakur, S,.: A Critical Study of Factors Promoting Cyberloafing in Organizations. International Conference on Cyber Security and Digital Forensic, ACM Proceedings (2016) (In press).
3. Mutter, S., Hall, M., Frank, E.: Using Classification to Evaluate the Output of Confidence-Based Association Rule Mining. Lecture Notes in Computer Science. 538–549 (2004).
4. Ignatko, I.: Applying Data Mining Techniques to discover Patterns in Context Activity data, Master Thesis, Aachen University, 2013.
5. Salem, O.:Apriori Algorithm-CodeProject, http://www.codeproject.com/Articles/70371/ Apriori Algorithm.
6. Scheffer, T.: Finding Association Rules That Trade Support Optimally against Confidence. Published in Principles of Data Mining and Knowledge Discovery. 424–435 (2001).
7. Gyenesei, A.Teuhola, J.: Interestingness Measures for Fuzzy Association Rules. Published in Principles of Data Mining and Knowledge Discovery. 152–164 (2001).

Two-Level Diversified Classifier Ensemble for Classification of Credit Entries

Pramod Patil, J.V. Aghav and Vikram Sareen

Abstract Classification of creditable customers from the customers which have applied for the loan is the first step for assessing the potential losses and credit exposure faced by financial institutions. So in the present scenario, it is very important for the banks and the financial institutes to minimize the loan defaults. One of the important strategies is to predict the likely defaulters so that such loans are either not issued or monitored closely after the issuance. In this paper, we have surveyed various classification algorithms used in financial domain and various ensemble techniques like bagging, boosting and stacking. The experimental results and statistical tests show that this new proposed classifier ensemble constitutes a proper solution for classification problem of credit entries, performing better than the traditional single ensembles like bagging, boosting and more significant than individual classifiers.

1 Introduction

The Indian banking system consists of 26 public sector banks, 20 private sector banks, 43 foreign banks, 56 regional rural banks, 1,589 urban cooperative banks and 93,550 rural cooperative banks [1].

Total lending and deposits increased at a compound annual growth rate (CAGR) of 20.7% and 19.7%, respectively, during 2007–2014 and are further poised for growth, backed by demand for housing and personal finance. Total asset size of banking sector assets is expected to increase to US$ 28.5 trillion by 2025. Along with the growth of the market, the default in the finance market is also becoming an issue of concern. Many players in the market depend on strong monitoring and control processes to minimize the non-performing assets [2]. However, it is

P. Patil (✉) · J.V. Aghav
Department of Computer Engineering, College of Engineering, Pune, India
e-mail: patilpramod2157@gmail.com

V. Sareen
Bluebricks Technologies Pvt. Ltd., Pune, India

© Springer Nature Singapore Pte Ltd. 2018
S.K. Bhatia et al. (eds.), *Advances in Computer and Computational Sciences*,
Advances in Intelligent Systems and Computing 554,
https://doi.org/10.1007/978-981-10-3773-3_61

important to evolve effective strategies to minimize the non-performing assets in the industry. Predictive Analytic centered on data mining techniques have been used effectively to predict the default.

So our motivation behind writing this paper is to give an overview of credit scoring system in India and machine learning algorithms used in credit scoring. We have also covered ensemble methods in machine learning to ensemble multiple base classifiers. For every machine learning algorithm, there is some limit beyond that it cannot fit that data. So accuracy stops at that point. If still we try to fit it more, it leads to an over fit problem. Ensemble tries to overcome this problem with the use of multiple models with different ensemble techniques like bagging, boosting or stacking. As each ensemble method is having some advantages and disadvantages over each other. So our focus is to make an enhancement of earlier existing ensemble methods, that will result in better results. An ensemble is a supervised learning algorithm, because it can be trained and then used to make predictions. The trained ensemble, therefore, represents a single hypothesis. This hypothesis, however, is not necessarily contained within the hypothesis space of the models from which it is built. So ensembles have shown that they have more edibility in the functions that they can represent. Bagging, boosting, stacking, blending and random forest are the well-known ensemble methods which have given better accuracy than the individual algorithms. In this paper, we surveyed classification algorithms linear regression with threshold classification, discriminant analysis, logistic regression, decision tree, neural network and support vector machine. Also we have surveyed ensemble techniques bagging, boosting, stacking and random forest. Empirically, ensembles tend to yield better results when there is a significant diversity among the models [3]. So we have proposed a simple solution of an ensemble of three diverse classifiers for two class classification problem and which has shown significant performance improvement than traditional ensemble and individual classifiers.

2 Methodology

Three real-world credit data sets have been taken to compare the performance of two-level hybrid ensemble with other classifier ensembles and base classifiers. The widely used Australian, German and Japanese data sets are from the UCI Machine Learning Database Repository (http://archive.ics.uci.edu/ml/) Table 1.

Table 1 Some characteristics of data sets used in experiments

Data set	#attributes	#good	#bad	%good%bad
German	24	700	300	70.0 30.0
Australian	14	307	383	44.5 55.5
Japanese	15	296	357	45.3 54.7

2.1 Experimental Protocol

The standard way to assess credit scoring systems is to employ a holdout sample since large sets of past applicants are usually available [1]. But here data sets are too limited to build an accurate scorecard. So, strategies like cross-validation are applied in order to get a good performance measure of the classification.

2.2 Model Evaluation

The main objective of the data mining models in this study is to be able to predict the defaulters and non-defaulters with as much high accuracy as possible. Since there are two categories namely defaulters and non-defaulters, four possibilities can be defined to measure the effectiveness of the models. The first category is called the True Positives (TP) consisting of the defaulters that were correctly predicted as defaulters by the model. The second category is the True Negatives (TN) which consist of non-defaulters who were correctly predicted as non-defaulters [4]. The remaining two categories are False Positives (FP) and False Negatives (FN). FP are non-defaulters who were wrongly classified as defaulters by the model. Similarly, FN are defaulters who were misclassified by the model as non-defaulters. High percentage of FP implies large number of non-defaulters misclassified as defaulters resulting in more diligent follow-up which, in turn, will lead to unnecessary expenditure to the company [5]. On the other hand, high percentage of FN will result in paying less attention to the potential defaulters and consequently lead to a higher default rate. By the same token, it is important to maximize the TP and TN. Needless to say, maximizing TP and TN will automatically lead to minimizing of FP and FN.

2.3 Ensemble Models

A classifier ensemble (also referred to as committee of learners, mixture of experts, multiple classifier system) consists of a set of individually trained classifiers (base classifiers) whose decisions are combined in some way, typically by weighted or non-weighted voting, when classifying new examples (Kittler 1998; Kuncheva 2004). It has been found that in most cases the ensembles produce more accurate predictions than the base classifiers that make them up (Dietterich 1997). Nonetheless, for an ensemble to achieve much better generalization capability than its members, it is critical that the ensemble consists of highly accurate base classifiers whose decisions are as diverse as possible (Bian and Wang 2007; Kuncheva and Whitaker 2003) [6]. In statistical pattern recognition, a large number of methods have been developed for the construction of ensembles that can be applied to any base classifier. In the following sections, the ensemble approaches relevant for this study are briefly described.

Bagging: In its standard form, the bagging (Bootstrap Aggregating) algorithm (Breiman 1996) creates M bootstrap samples T_1, T_2, ..., T_M randomly drawn (with replacement) from the original training set T of size n [7]. Each bootstrap sample T_i of size n is then used to train a base classifier C_i. Predictions on new observations are made by taking the majority vote of the ensemble C built from C_1, C_2, ..., C_M. As bagging re-samples, the training set with replacement, some instances may be represented multiple times while others may be left out [8]. Since each ensemble member is not exposed to the same set of instances, they are different from each other. By voting the predictions of each of these classifiers, bagging seeks to reduce the error due to variance of the base classifier.

Boosting: Similar to bagging, boosting also creates an ensemble of classifiers by re-sampling the original data set, which are then combined by majority voting. However, in boosting, re-sampling is directed to provide the most informative training data for each consecutive classifier [9]. The Ada-boost (Adaptive Boosting) algorithm proposed by Freund and Schapire (1996) constitutes the best known member in boosting family. It generates a sequence of base classifiers C_1, C_2, ..., C_M by using successive bootstrap samples T_1, T_2, ..., T_M that are obtained by weighting the training instances in M iterations. Ada-boost initially assigns equal weights to all training instances and in each iteration; it adjusts these weights based on the misclassification made by the resulting base classifier [10]. Thus, instances misclassified by model C_i are more likely to appear in the next bootstrap sample T_i. The final decision is then obtained through a weighted vote of the base classifiers (the weight w_i of each classifier C_i is computed according to its performance on the weighted sample T_i it was trained on).

Stacking: In Wolpert's stacked generalization (or stacking), an ensemble of classifiers is first trained using bootstrapped samples of the training data, creating Tier 1 classifiers, whose outputs are then used to train a Tier 2 classifier (meta classifier) (Wolpert 1992) [11]. The underlying idea is to learn whether training data have been properly learned. For example, if a particular classifier incorrectly learned a certain region of the feature space, and hence consistently misclassifies instances coming from that region, then the Tier 2 classifier may be able to learn this behavior, and along with the learned behaviors of other classifiers, it can correct such improper training. Cross-validation type selection is typically used for training the Tier 1 classifiers: the entire training data set is divided into T blocks, and each Tier-1 classifier is first trained on (a different set of) T-1 blocks of the training data. Each classifier is then evaluated on the Tth (pseudo-test) block, not seen during training [12]. The outputs of these classifiers on their pseudo-training blocks, along with the actual correct labels for those blocks constitute the training data set for the Tier 2 classifier.

3 Related Work

Randomized weighted majority algorithm is (RWMA) is a meta-learning algorithm which "predicts from expert advice" [13]. Our idea is motivated from randomized weighted majority. [14] discusses that various versions of the Weighted Majority Algorithm and prove mistake bounds for them that are closely related to the mistake bounds of the best classifiers of the pool. Randomized majority voting picks the expert from the pool of classifiers. We have taken same approach but rather than picking up an overall expert, we picked up class-wise expert. In RWMA, for each wrongly predicted entry some penalty is applied to weight of each individual classifier which is an iterative process. But with our approach, we are using caret Ensemble and confusion matrix. So with prediction of validation set, we can directly derive the weights with obtained confusion matrix. So the proposed approach also has better efficiency.

4 Constructing Two-Level Hybrid Ensemble

In their most classical form, the base classifiers that comprise an ensemble correspond to simple prediction models such as linear discriminant analysis, logistic regression, support vector machines, decision tree, random forest, gradient boost, and XGBoost. However, the two-level ensemble approach to classify loan approval decision proposed extends the traditional approach of multiple classifier systems by using an ensemble as base classifier of a higher-level ensemble.

In order to exploit the advantages of the two different induction strategies previously mentioned (i.e., using different training sets and using different attribute subsets), here we propose to construct a prediction model that integrates the re-sampling based and the attribute-based methods into a unified two-level classifier ensemble [15]. In summary, a two-level hybrid ensemble will consist of positive and negative expert which are again ensemble of individual classification algorithm. Positive expert is an ensemble of different individual classifier ensemble with weighted averaging of sensitivity and same way negative expert will be build with ensemble by weighted average of specificity. General model is ensemble by weighted average of accuracy of model.

Figure 1 shows an architecture of two-level hybrid ensemble that combines positive expert and negative expert with a general model. We will achieve the whole ensemble model with three algorithms and will be trained once. We will train three different and diverse algorithms. Then ours training part is complete. For construction of positive, negative experts and general model, we will use same three trained models. Positive expert's output will be calculated by weighted average of outputs of individual base classifiers with sensitivity of each model as a weight. Same way negative expert's output will be calculated with specificity of each model as a weight. And general model will be with accuracy as a weight.

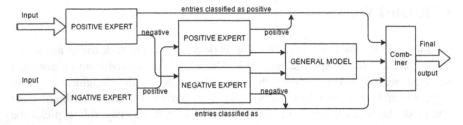

Fig. 1 Two level hybrid ensemble

We will start the process of prediction by passing the whole test set to both positive and negative expert. If they classify an entry into their expert class (e.g., positive expert classify entry as positive) then those entries will be directly considered as the desired outputted class. For entries which they make conflict will be sent to respective expert (e.g., positive expert classify entry as negative then that entry will be again send to negative expert) for their opinion. If respective expert agrees then labeled with that class and if conflicts, then general model's suggestion will be taken for those entries.

5   Experiments and Results

In order to test the validity and performance of the method just proposed, several experiments have been carried out. The objective of this paper is to compare the performance of method that is just proposed and the existing ensemble techniques bagging, boosting and stacking. We have used linear discriminant analysis, logistic regression, neural network, support vector machines, decision tree, random forest, gradient boost, and XGBoost, etc., as base classifiers for various ensembles.

The whole experimental analysis in performed in R statistical programming language. Table 2 shows AUC values of different prediction models. As it can be seen ensembles AdaBag, Random Forest, GBM, XGBoost, Blending (stacking) have performed better than individual algorithms linear discriminant analysis, logistic regression, neural network and support vector machine. That is as expected individual classifiers achieve lower AUC values than ensemble. Although differences in AUC may appear to be relatively low, it should be noted that even a small increase in prediction performance can yield substantial cost savings for financial institutions [15].

It seems therefore to be of sufficient interest the use of the two-level classifier ensembles in credit scoring applications. The highest differences are observed for the German and Australian credit data set, which corresponds to unbalanced data sets with 70% positive entries where as 30% negative entries and 44–56% respectively. For example, the two-level hybrid ensemble has performed better than bagging and boosting algorithms for majority of data sets, i.e., for german data set

Table 2 AUC values for classifiers

Classifiers/datasets	German	Australian	Japanese
LDA	0.6518	0.6180	0.6450
Logistic	0.6227	0.5871	0.6136
1NN	0.6040	0.6486	0.6151
NaiveBayes	0.5727	0.6112	0.7217
SVM	0.6312	0.6172	0.7411
Rpart	0.5557	0.5949	0.6675
C5.0	0.5789	0.5852	0.6881
AdaBag	0.5904	0.6087	0.6677
RandomForest	0.5918	0.6384	0.6974
GBM	.6305	0.6507	0.7537
XGBoost	0.6381	0.6513	0.7581
Blending	0.6418	0.6412	0.7718
2LevelEnsemble	0.6503	0.6418	0.7611

proposed model has achieved 0.0599 than bagging, 0.0198 than boosting and 0.0085 than Blending.

Blending model: Blending model is ensemble made up of 3 algorithms GBM, rpart and treebag as base classifiers. Another GBM is used as meta learner.

2LevelEnsemble: Two-level hybrid ensemble model is built of three algorithms namely GBM, rpart and treebag. Classifier is built as mentioned in Sect. 3

Same algorithms are used for implementation of construction of blending model and our proposed model, so that comparison of new ensemble idea will be made based on idea of ensemble rather than on base classifiers.

6 Conclusions and Further Extensions

In this work, a new ensemble idea for classification of credit entries is developed with combination of different ensemble methods like boosting, random forest and bagging, by keeping target of obtaining enhanced performance than single ensembles. This strategy can also be viewed as ensemble of ensemble or a two-level ensemble approach which combines already existing ensemble techniques. In general, the proposed ensemble has produced the better results in terms of area under curve, which leads to better cost savings in credit scoring applications. Since its simplicity of building it makes it easy to build, easy to interpret. Further research can be emerged from this study: (i) To compare the ensembles studied in the present work with different other methods as base classifiers (for example, blending or stacked generalization as a base classifier); and (iii) To explore the chances of using multiple levels instead of two levels.

References

1. Vishnuprasad Nagadevara, Application Of Hybrid Methodology To Predict Housing Loan Defaults In India, Journal Of International Management Studies, Volume 15, Issue 3, pp. 43–50, December 2015.
2. G.V. Bhavani Prasad, D. Veena, NPAs Reduction Strategies for Commercial Banks in India, IJMBS Vol. 1, Issue 3, September 2011.
3. Nan-Chen Hsieh a, Lun-Ping Hung, A data driven ensemble classifier for credit scoring analysis, Journal Expert Systems with Applications: An International Journal Volume 37 Issue 1, January, 2010.
4. Akhil Bandhu Hens, Manoj Kumar Tiwari, Computational time reduction for credit scoring: An integrated approach based on support vector machine and stratified sampling method, Expert Systems with Applications 39 (2012) 67746781.
5. AI Marques, V Garca and JS Sanchez, A literature review on the application of evolutionary computing to credit scoring, Journal of the Operational Research Society (2013) 64, 13841399.
6. Jue Wanga, Abdel-Rahman Hedar, Shouyang Wang, Jian Ma, Rough set and scatter search metaheuristic based feature selection for credit scoring, Expert Systems with Applications 39 (2012) 61236128.
7. A.I. Marqus, V. Garca, J.S. Snchez, Exploring the behaviour of base classifiers in credit scoring ensembles, Expert Systems with Applications 39 (2012) 1024410250.
8. A.I. Marqus, V. Garca, J.S. Snchez, Two-level classifier ensembles for credit risk assessment, Expert Systems with Applications 39 (2012) 1091610922.
9. David Pardoe, Peter Stone, Boosting for Regression Transfer, Proceedings of the Twenty-Seventh International Conference on Machine Learning (ICML 10), Haifa, Israel, June 2010.
10. A.I. Marqus, V. Garca, J.S. Snchez, Exploring the behaviour of base classifiers in credit scoring ensembles, Expert Systems with Applications 39 (2012) 1024410250.
11. Amjath Fareeth Basha, Gul Shaira Banu Jahangeer, Face Gender Image Classification Using Various Wavelet Transform and Support Vector Machine with various Kernels, IJCSI, Vol. 9, Issue 6, No 2, November 2012.
12. Asif Ekbal, Sriparna Saha, Stacked ensemble coupled with feature selection for biomedical entity extraction, Knowledge-Based Systems Volume 46, July 2013, Pages 2232.
13. https://en.wikipedia.org/wiki/Randomizedweightedmajorityalgorithm.
14. Nick Littlestone, Manfred K. Warmuth, The Weighted Majority Algorithm, IEEE 1989.
15. A.I. Marqus, V. Garca, J.S. Sanchez, Two-level classifier ensembles for credit risk assessment, Expert Systems with Applications 39 (2012) 1091610922.

P-RED: Probability Based Random Early Detection Algorithm for Queue Management in MANET

Neelam Sharma, Shyam Singh Rajput, Amit Kumar Dwivedi and Manish Shrimali

Abstract Thousands of flows could traverse an Internet gateway at each given time. Ideally, every single of these flows should dispatch at precisely its fair allocate, and the gateway should not demand to make each decisions. In exercise, the burden inclines to fluctuate and the traffic origins incline to be greedy. Therefore, a gateway queuing strategy has to permit buffering of provisional excess burden but furnish negative feedback if the excess burden persists. Such a strategy has to stop elevated stay by manipulating the queue size, it has to circumvent compelling queues to be too short, that can cause low utilization; and it have to furnish negative feedback fairly. As RED queues precisely present larger than drop_tail queues, it is tough to parameterize RED queues to give good presentation below disparate Congn scenarios. One of the aims of RED is to stabilize the queue lengths in routers, but that does not prosper because the equilibrium queue length powerfully depends on the number of TCP connections. We present Probability based Random Early Detection (P-RED), a modification to RED that enhances fairness after disparate traffic kinds allocate a gateway. P-RED is extra competent in isolating ill-behaved flows that turns out to protect the bursty and low speed flows in a larger amount.

Keywords Congestion Control · Random early detection (RED) · MANET · Queue management

N. Sharma (✉) · S.S. Rajput · A.K. Dwivedi
ABES Engineering College, Ghaziabad, India
e-mail: sharma.neelam028@gmail.com

S.S. Rajput
e-mail: ershyamrajput@gmail.com

A.K. Dwivedi
e-mail: amit.dwivedi@abes.ac.in

M. Shrimali
JRN Rajasthan Vidyapeeth University, Udaipur, India
e-mail: manishshrimali@gmail.com

© Springer Nature Singapore Pte Ltd. 2018
S.K. Bhatia et al. (eds.), *Advances in Computer and Computational Sciences*,
Advances in Intelligent Systems and Computing 554,
https://doi.org/10.1007/978-981-10-3773-3_62

1 Introduction

MANET [1–5] is a collection of nodes that are mobile in nature and interact without using wires. In MANET nodes that are in range can communicate directly otherwise they can take the help of other nodes to send the packet (PKT). In MANET we cannot presume the number of nodes that leads to congestion (Congn) [6, 7] onto the network.

Some challenges faced by MANET are dynamic topology, constraints on resources, bandwidth management [8, 9] and PKT broadcast overhead, which creates difficulty to design routing protocols. An application of MANET ranges from large-scale mobile networks to small, static networks limited by the power resources.

This paper is organized as follows: Sect. 2 represents the literature survey. Proposed algorithm is mentioned in Sect. 3 followed by simulation in Sect. 4. Section 5 describes the conclusion and future work.

2 Literature Survey

Simaya et al. [10] introduced an improvement of RED (IRED) algorithm for enhancement of the performance of MANET. IRED is a Priority Queue based AQM scheme. Here, the PKTs are dropped on the grounds of PKT arrival rate and the queue length that minimizes the effect of network Congn. By using IRED PKT loss rate is minimized.

Gupta and Pandey [11] proposed Congestion Control (CongnCntl) using queue based approach as Multipath. Routing under MANET. This technique minimizes the data drop. Simulation shows that proposed technique provide minimum overhead as well as minimum E2ED and increase PDR.

Choudhary and Singhal [12] proposed a different approach of CongCntl for aiding applications like streaming over MANET. Simulation shows that this approach requires lower number of acknowledgement PKTs than other CongCntl techniques. It reduces the PKT loss ratio (PLR) and increase transmission efficiency.

Soelistijanto and Howarth [13] describe transfer reliability and storage CongnCntl strategies in opportunistic networks. In these networks connection between nodes occurs unpredictably. This paper explores potential mechanism for transfer reliability service and E2E return receipt. It also explores the requisites for storage CongnCntl. Finally this paper identifies the issues in CongnCntl in opportunistic networks.

Senthilkumaran and Sankaranarayanan [14] introduced early Congn detection and adaptive routing (EDAPR) in MANET. Initially EDAPR constructs a non-congested neighbors (NHN) list and finds a route to a destination through an NHN node. All nodes calculate their queue_status periodically at node level to determine a non-congested path between two nodes so as to control Cong.

When performance of EDAPR is compared with EDAODV and EDCSCAODV simulation shows reduction of E2ED, routing overhead and improvement in PKT delivery.

Vaidyaet and Bhatnagar [15] in their paper used RED mechanism in the internet. Although RED shows better performance than its predecessor, Drop Tail, its performance is highly sensitive to parameter settings. The algorithm shows improved performance than the previously proposed algorithm that adaptively tunes a RED parameter.

3 The P-RED Algorithm

This paper discusses about the P-RED congn manipulation algorithm. In this algorithm the average queue size (AQS) is monitored for every single output queue. Congestion (Congn) is notified in this algorithm by randomized connection. The key believed behind P-RED is to present queue association established undeviatingly on PKT defeat and link utilization. P-RED estimates the AQS, whichever employing an easy EWMA (Exponentially weighted advancing average) in the forwarding trail, or employing a comparable mechanism in the background. Figure 1 displays the relation amid two thresholds and AQS of router employing P-RED.

When the AQS is less than the min_{th}, no PKTs are discarded. After the AQS is larger than the max_{th}, every single appearing PKT is discarded. After the AQS is amid the min_{th} and the max_{th}, every single appearing PKT is discarded alongside probability P_a, whereas P_a is a purpose of the PKT drop and link utilization. The finished P-RED algorithm is delineated in Fig. 2.

Hence the P-RED has two distinct algorithms. One algorithm is for computing the AQS to ascertain the degree of burstiness that will be allowed in the router queue. And the second is for computing the P_a to ascertain how oftentimes the router discards PKTs at the present level of Congn.

Fig. 1 Two thresholds and average queue size

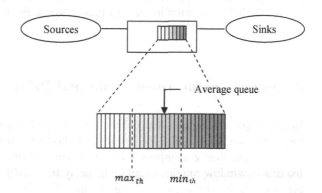

Fig. 2 Algorithm for P-RED

> For each incoming PKT
>   Calculate the AQS
>   If $min_{th} \leq AQS \leq max_{th}$
>   Calculate P_a.
>   With P_a,
>   Discard the incoming PKT
>   Else if $max_{th} \leq AQS$
> Discard the incoming PKT.
> AQS : average queue size.
> min_{th}: minimum threshold for queue.
> max_{th} : maximum threshold for queue.
> P_a: PKT drop probability

Fig. 3 Detailed algorithm for P-RED

> if $min_{th} \leq AQS \leq max_{th}$
> if AQS increments
>   if current_time-last_up_time > up_time
> $P_a + d_1$
> if AQS decrements
>   if current_time-last_up_time > up_time
> $P_a - d_2$

Pa, worth endowed by RED is a linear purpose to AQS. So difference amid two thresholds ought to be colossal plenty to become an adequate presentation of RED. But, the P-RED computes P_a worth in a pace form by employing PKT defeat and link utilization past to resolve this setback provoked by RED. In Fig. 2, the P_a is periodically notified if an AQS is increased or cut after an AQS is amid two thresholds (min_{th} and max_{th}). P_a has to be incremented by d_1, after AQS is larger than the specific value. On the supplementary hand, P_a has to be cut by d_2 after AQS is tinier than the specific worth.

RED computes probability linear to AQS plainly if AQS is amid two thresholds afterward inactive time. On the supplementary hand, P-RED can manipulate the Congn provoked by the retransmission of the discarded packet with the help of the P_a and the link utilization in the past transmission. Figure 3 displays the change of P_a benefits in P-RED.

4 Simulation for Throughputs and Delay

In this paper, throughput and delay for the three Congn algorithms are observed, i.e., Drop_tail, RED and P-RED. Figure 4 shows the network topology used.

The simulation encompasses two FTP connections, every single alongside a maximum window roughly equal to the delay-bandwidth product. Two connections are commenced at somewhat disparate time. The three algorithms are compared in

Fig. 4 Simulation network

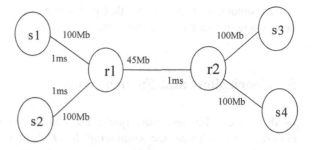

Fig. 5 Simulation result for throughput

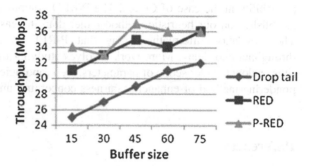

Fig. 6 Simulation result for delay

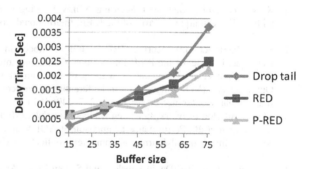

the simulation. As P-RED is an alert queue association scheme, queue size is always a vital factor on the performance. In Fig. 5 throughput of the FTP connections below disparate buffer sizes is shown. As delineated in Fig. 5, the consequence displays that the throughput of the FTP connection is enhanced as the buffer size increments. Throughput of P-RED is somewhat larger than that of RED, as shown in Fig. 5.

The stay period experienced by a PKT is a combination of the transmission delays across every single link and the processing delays generated inside every single router. Figure 6 displays stay period of the FTP connections below disparate buffer sizes. The aftermath in Fig. 6 display that after the buffer size increments, the stay period of the FTP connection increments. And we can discern that stay period of P-RED is somewhat larger than stay period of RED.

The simulation results show that performance of P-RED is better than that of other queue management algorithms.

5   Conclusions and Further Work

In this paper, the new alert queue association algorithm, shouted P-RED, is counseled to enhance the continuing RED. Congn is manipulated efficiently employing our algorithm. P-RED is an algorithm that computes a competent drop probability in the case of Congn. The P-RED computes the PKT drop probability established on our heuristic method rather than the easy method utilized in RED. The simulation aftermath display that P-RED is able to enhance fairness, throughput and delay. More work involves studying the presentation of the algorithm below a wider scope of parameters, web topologies and real traffic traces, and pondering method of enhancing fairness concerning unresponsive flow.

References

1. K. V. Arya and S. S. Rajput: Securing AODV Routing Protocol in MANET using NMAC with HBKS technique, in Proc. of the IEEE. International Conference on SPIN, pp. 281–285, 2014.
2. S. S. Rajput et al.: Securing ZRP Routing Protocol in MANET using Authentication technique, in Proc. of the IEEE International Conference on CICN, pp. 872–877, 2014.
3. S. S. Rajput et al.: Comparative analysis of Random Early Detection (RED) and Virtual Output Queue (VOQ) algorithms in Differentiated Services Network, in Proc. of the IEEE International Conference on SPIN, pp. 237–240, 2014.
4. Mobile Ad Hoc Network, https://en.wikipedia.org/wiki/Mobile_ad_hoc_network.
5. S. S. Rajput et al., Performance comparison of AODV and AODVDOR routing protocol in MANET, International Journal of Computer Applications (IJCA), vol 63, no 22, pp. 19–24, 2013.
6. Sanjeewa Athuraliya, David Lapsley, and Steven Low.: An enhanced random early marking algorithm for internet flow control. In INFOCOM 2000. Nineteenth Annual Joint Conference of the IEEE Computer and Communications Societies. Proceedings. IEEE, vol. 3, pp. 1425–1434. IEEE, 2000.
7. Network Congn, https://en.wikipedia.org/wiki/Network_Congn.
8. Jahon Koo, Byunghun Song, Kwangsue Chung, Hyukjoon Lee, and Hyunkook Kahng. "MRED: a new approach to random early detection." In Information Networking, 2001. Proceedings. 15th International Conference on, pp. 347–352. IEEE, 2001.
9. James Aweya, Michel Ouellette, and Delfin Y. Montuno.: An optimization-oriented view of random early detection. Computer Communications 24, no. 12 (2001): 1170–1187.
10. S. Simaya, A. Shrivastava, N.P. Keer.: IRED Algorithm for Improvement in Performance of MANET: Fourth International Conference on Communications System and Network Technologies (CSNT), IEEE Computer Society, pp 283–287, 2014.

11. Hitesh Gupta and Pankaj Pandey: Survey of Routing Base Congn Control Techniques under MANET: IEEE International Conference on Emerging Trends in Computing, Communication and Nanotechnology (ICECCN), pp 241–244 (2013).
12. Robin Choudhary and Niraj Singhal: A Novel Approach for Congn Control in MANET: International Journal of Engineering and Innovative Technology (IJEIT), Volume 2, pp 55–61 (2012).
13. Bambang Soelistijanto, M.P. Howarth: Transfer Reliability and Congestion Control Strategies in Opportunistic Networks - A Survey: IEEE Communications Tutorials & Surveys, pp. 538–555 (2012).
14. T. Senthilkumaran and V. Sankaranarayanan: Early Congn Detection and Adaptive Routing in Manet: Egyptian Informatics Journal, pp 165–175 (2011).
15. Rahul Vaidya, and Shalabh Bhatnagar. "Robust optimization of random early detection." Telecommunication Systems 33, no. 4 (2006): 291–316.

Analyzing Game Stickiness Using Clustering Techniques

Hycinta Andrat and Nazneen Ansari

Abstract The popularity of computer games has led to tremendous generation of gaming data. Such gaming data consists of gamer's personal information along with the game genres played and the time spent by them on a particular game. This gaming data can be utilized by the gaming industry for the purpose of extracting the knowledge needed to monitor the stickiness of the games. The raw data related to computer games can be refined, which could provide game developers the number of the gamers attracted towards a particular game. If the count of the gamers for a specific game decreases as the time passes by, then game developers need to improve the game, in order to retain the gamers. As gaming industry adds to our country's revenue to a great extent, certain technological advancements are required. Therefore, this study aims to use a data mining approach, i.e., clustering, for monitoring computer games stickiness.

Keywords Clustering · Computer games · Data mining · Gaming data · k-Means · DBSCAN

1 Introduction

Technological developments in the computers, satellites, digital storage has led to the tremendous collection of data related to various fields. As years passed by, ample of data has been gathered. So there raises a need to bifurcate and refine the data according to the requirements of a particular application; this need of the industries gave birth to the concept of data mining. Data mining is the process of

H. Andrat (✉)
Computer Department, St. Francis Institute of Technology, Mumbai, India
e-mail: hycintaandrat9@gmail.com

N. Ansari
Information Technology Department, St. Francis Institute of Technology,
Mumbai, India
e-mail: nazneenansari@sfitengg.org

© Springer Nature Singapore Pte Ltd. 2018   645
S.K. Bhatia et al. (eds.), *Advances in Computer and Computational Sciences*,
Advances in Intelligent Systems and Computing 554,
https://doi.org/10.1007/978-981-10-3773-3_63

discovering hidden knowledge and pattern from large amounts of data stored in data repositories. Data mining functionalities include association, classification and regression, the discovery of concept/class descriptions (i.e., characterization and discrimination), clustering and outlier analysis [1]. Today, the necessity for data refinement is required not only in industries like bioinformatics, agriculture, medical, social network analysis but also in digital gaming. Nowadays, digital games has its own significance; since it has led to a rise in revenue generated by the gaming industry with ~20.5 billion INR [2] and a predicted rise up to 40.6 billion INR by 2018 [3]. Amongst the other countries, India ranks 20th in terms of revenue generation from the gaming industry [2]. A computer game is a game controlled by computer where players interact with objects shown on a screen for the sake of entertainment [4]. Computer games consists of different genres namely sports, strategy, action, adventure, role playing, simulation, massively multiplayer online [5]. The gaming industry has been experiencing numerous revolutions ever since it came into existence. Therefore, data mining can be used in computer games to extract information from gaming data and utilized in taking valuable decisions.

The remainder of this paper is organized as follows. Section 2 provides an overview about a data mining approach, i.e., clustering along with its algorithms. Section 3 discusses proposed system for monitoring game stickiness. Section 4 focuses on implementation details. Section 5 provides experimental results of the proposed system. Section 6 concludes the study.

2 Literature Review

Data mining is a powerful technology with great capability to benefit company's focus on the most significant information in their data warehouses. Data mining predict future trends and behaviors, allowing businesses to make proactive, knowledge-driven decisions [6]. So, one of the data mining functions that can be utilized for refining gaming data is clustering.

Clustering involves arranging data into classes or groups. However, initially the class labels are unknown and it is up to the clustering algorithm to discover what the classes are and evaluate which data from the repositories will belong to which class. The core goal of clustering algorithms is to group or segment objects (e.g., players, asset, items, games or any observation, case or record) in such a way that the similarity between objects in one group (cluster) is high (intracluster similarity), while between groups is dissimilar (intercluster similarity) [7]. For example, group of houses are identified according to their house type, value, and geographical location. The most popular clustering algorithm is k-Means along with DBSCAN belonging to density-based clustering family, which overcome the limitations of k-Means [1, 8].

2.1 k-Means

The k-Means algorithm takes the input parameter as k, i.e., number of clusters, and divides a set of n elements into k clusters so that the resulting similarity within a cluster is high but the similarity between clusters is low. Cluster similarity is measured with reference to the mean value of the elements in a cluster, which can be considered as the cluster's centroid [1]. Figure 1 shows the flowchart of k-Means algorithm [1].

Fig. 1 k-Means flowchart

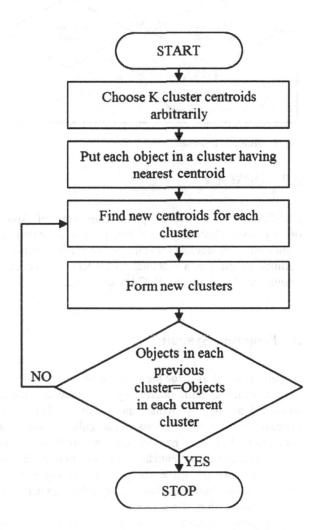

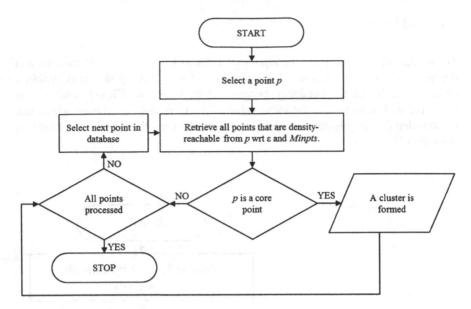

Fig. 2 DBSCAN flowchart

2.2 DBSCAN

DBSCAN (Density-Based Spatial Clustering of Applications with Noise) is a
density-based clustering algorithm. In this algorithm regions are developed with
sufficiently high density into clusters which discovers clusters of arbitrary shape. It
outlines a cluster as a maximal set of density-connected points [2]. Figure 2 represents the flowchart of DBSCAN algorithm [2].

3 Proposed System

A game is developed for the purpose of entertainment. Suppose, if an individual
finds a particular game interesting, then there are chances of that game becoming
renowned among his friends and relatives'. Thus, the popularity of that game
increases exponentially across a particular region or age group or a particular
community. But, if no new add-ons are introduced in that game, then the gamer's
interest reduces and eventually they stop playing it. To keep the track about the
attractiveness of the game, the gaming industry performs user polls. This assists
them to know about the statistics regarding gamer's retention, leading to unnecessary consumption of resources.

The gaming industry can overcome this shortcoming by developing a system
where several clustering algorithms will be applied to gaming data. Executing
clustering technique will help in formation of groups of gamers playing same game.

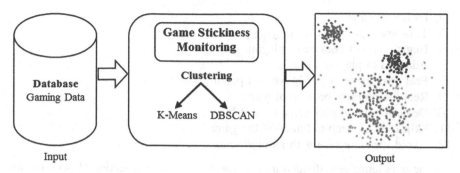

Fig. 3 Proposed system

All the identified gamers group can be examined on the basis of group size and group count. In later years, if the group size and group count decreases; then it can easily inferred that the game is in danger, and instant actions are required to be taken. Clustering algorithms like k-Means and DBSCAN can be implemented for monitoring the game stickiness. Figure 3 exhibits the proposed system for monitoring the stickiness of computer games.

The gaming data is provided as input to the system, where the two clustering algorithms, i.e., k-Means and DBSCAN will be executed and the output will in the form of charts representing different clusters. Amongst these two algorithms, the one with the better performance can be selected by the gaming industry for game stickiness monitoring.

4 Implementation

Data mining has been proved advantageous for sectors like agriculture, medical, education. According to the proposed model, it can also be beneficial to gaming sector. The gaming dataset executed using RapidMiner Studio can aid the gaming industry in monitoring the retention of gamers for a particular game. The implementation process for monitoring game attractiveness consists of following points.

4.1 Data Collection

Data gathering task is performed using a survey method. Survey form is circulated through online resources. The questionnaire consists of following questions.

1. Name
2. Age
3. Gender

4. Preferred game
5. Time spent per week for playing preferred game
6. Negative effect because of playing game
7. Addicted to playing game
8. Feel incomplete or lonely without playing game
9. Real life problem because of game
10. Dedicate time to play game even if busy
11. Skip meal or sleep to finish off the game
12. Avoid spending time with family/friends to play game

The questionnaire is divided into two sections. The first section (1–4) provides gamers details, and the second section (5–12) focuses on questions related to what extend the gamers are attracted towards a particular game. The questions from 6 to 12 are YES/NO type.

4.2 Tool Used

RapidMiner Studio incorporates technology along with the applicability to serve a user-friendly integration of the latest as well as established data mining techniques. Analysis processes with RapidMiner Studio is performed by drag and drop of operators, parameters setting and combining operators. RapidMiner Studio permits to use strong visualizations like 3D graphs, scatter matrices and self-organizing maps. It allows turning data into fully customizable, exportable charts along with support for panning, zooming, and rescaling for maximum visual impact. It also offers access to an extensive list of data sources including Microsoft Excel, Access, SQL, and Oracle. It is platform independent [9].

4.3 Procedure

The gaming dataset collected through online survey consists of 1000 records and 12 attributes. The dataset is in xlsx format, i.e., an Excel file. The following steps depict execution process of the proposed system in brief:

1. Collect a gaming dataset using survey method
2. Load the dataset in RapidMiner Studio for implementation
3. Perform data preprocessing
4. Apply k-Means and DBSCAN algorithm
5. Observe and analyze results.

(a)   **(b)**

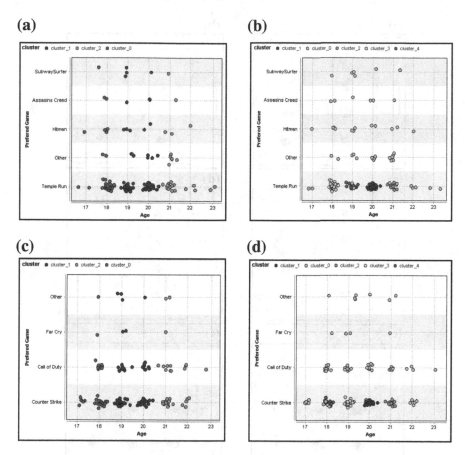

Fig. 4 Adventure games versus age. **a** k-Means. **b** DBSCAN, fighting games versus age. **c** k-Means. **d** DBSCAN

5   Experimental Results

For implementation, dataset is segmented with respect to the six genre (Adventure, Fighting, Racing, Role Playing, Sports and Strategy) options. Each genre is evaluated against age, negative effect, time spent/week; this will help in observing the changes in the size and count from the obtained gamer's clusters as the dataset will vary with time. The outcome of each genre is as follows.

Figure 4a and b depicts the outcome of k-Means and DBSCAN for adventure games against age respectively. In Fig. 4a, three clusters are obtained, where the age group 19–20 has maximum occurrence without acquiring the knowledge of particular game the gamers are interested in. Figure 4b provides five clusters, with cluster 0 representing noise points, this graph gives a precise result with individuals belonging to age group 18 are interested in Temple Run. Similarly, in Fig. 4c for

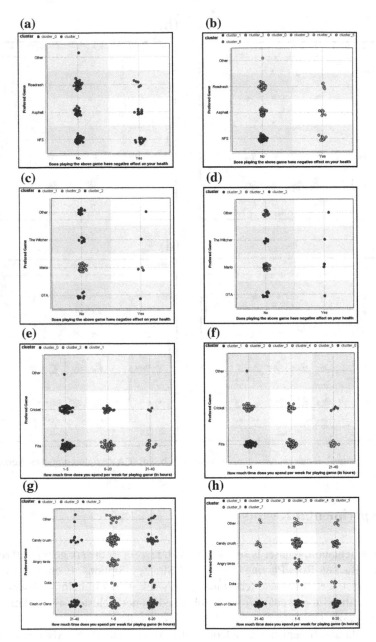

Fig. 5 Racing games versus negative effect **a** K-Means **b** DBSCAN. Role playing games versus negative effect **c** K-Means **d** DBSCAN. Sports games versus time spent/week **e** K-Means **f** DBSCAN. Strategy games versus time spent/week **g** K-Means **h** DBSCAN

fighting games, age groups of '19–20' is obtained without a precise game and in Fig. 4d individuals of age '19' are interested in 'Counter Strike'. Thus, in future if the count of the people interested in 'Counter Strike' decreases, then certain improvements are needed to retain gamers.

In Fig. 5a–d, racing games and role playing games are plotted against negative effect respectively, where Fig. 5a and c consists of both 'Yes'/'No' responses. But on the other hand, in Fig. 5b and d, option 'No' has maximum occurrence for 'Roadrash' and 'Mario', respectively. This helps gaming industry to infer that 'Roadrash' and 'Mario' does not have any adverse effect and are safe to play. Figure 5e–h represents graphs for sport and strategy games, where both are plotted against time spent/week. In Fig. 5e and g three clusters have almost equal number of elements and no particular inference can be drawn, but in Fig. 5f maximum occurrences specifies, people mostly spent around '1–5' h/week in playing 'Fifa' and Fig. 5h depicts 'Candy Crush' is played for around '1–5' h/week. Thus, the time spent in playing 'Fifa' and 'Candy Crush' is less, compared to the other options provided; therefore certain improvements are required so that gamers spent more time. From the above results, it is clear that DBSCAN performance is better than k-Means. Similar experimentation can be performed for the remaining attributes.

6 Conclusion

Data mining applications have been proven valuable to many industries till date and can also be found beneficial to the gaming industry in several ways. One of the applications among them can be monitoring game stickiness by implementing clustering algorithms such as k-Means and DBSCAN. The outcome of these algorithms has help in checking the count of the gamers interested in a particular game and keep on examining the count on regular basis. The decrease in count will help in interpreting that certain upgradations are needed to retain and to increase the number of gamers. Also execution of k-Means and DBSCAN aids in inferring that DBSCAN is far much better than k-Means, since it provides precise value or solution to the query and also helps in detecting noisy data.

References

1. Han, J. Kamber, M.: Data mining. Elsevier, Amsterdam (2011).
2. Top Countries by Game Revenues | Newzoo, http://www.newzoo.com/free/rankings/top-100-countries-by-game-revenues.
3. FICCI-KPMG.: The stage is set: FICCI-KPMG Indian Media and Entertainment Industry Report. FICCI-KPMG, New Delhi (2014).
4. Computer and video games, https://www.sciencedaily.com/terms/computer_and_video_games.htm.

5. Grace, L.: Game Type and Game Genre. Lindsay Grace (2015).
6. Tiwari, H.: Data mining, warehousing and OLAP technology. Discovery. 24, 58–62 (2014).
7. Drachen, A., Thurau, C., Togelius, J., Yannakakis, G., Bauckhage, C.: Game Data Mining. In: Game Analytics. 205–253. Springer, Heidelberg (2013).
8. Jain, A.K., Murty, M.N., Flynn, P.J.: Data Clustering: A Review. In: ACM Computing Surveys, vol. 31, No. 3, 264–323. ACM, New York (1999).
9. RapidMiner.: RapidMiner Studio Manual. RapidMiner, London (2015).

Automated Detection of Acute Leukemia Using K-mean Clustering Algorithm

Sachin Kumar, Sumita Mishra, Pallavi Asthana and Pragya

Abstract Leukemia is a hematologic cancer which develops in blood tissue and triggers rapid production of immature and abnormal-shaped White Blood Cells. Based on statistics it is found that the leukemia is one of the leading causes of death in men and women alike. Microscopic examination of blood sample or bone marrow smear is the most effective technique for diagnosis of leukemia. Pathologists analyze microscopic samples to make diagnostic assessments on the basis of characteristic cell features. Recently, computerized methods for cancer detection have been explored towards minimizing human intervention and providing accurate clinical information. This paper presents an algorithm for automated image based acute leukemia detection systems. The method implemented uses basic enhancement, morphology, filtering and segmenting technique to extract region of interest using k-means clustering algorithm. The proposed algorithm achieved an accuracy of 92.8% and is tested with Nearest Neighbor (kNN) and Naïve Bayes Classifier on the dataset of 60 samples.

Keywords Image processing · White Blood Cells · Leukemia · Clustering

S. Kumar (✉) · S. Mishra · P. Asthana
Amity University, Lucknow Campus, Lucknow, India
e-mail: skumar3@lko.amity.edu

S. Mishra
e-mail: smishra3@lko.amity.edu

P. Asthana
e-mail: pasthana@lko.amity.edu

Pragya
Department of Chemistry, MVPG College, Lucknow University, Lucknow, India
e-mail: dr.pragya2011@gmail.com

© Springer Nature Singapore Pte Ltd. 2018   655
S.K. Bhatia et al. (eds.), *Advances in Computer and Computational Sciences*,
Advances in Intelligent Systems and Computing 554,
https://doi.org/10.1007/978-981-10-3773-3_64

1   Introduction

Leukemia is a type of cancer which develops in blood tissue. Leukemia originates in the soft inner part of the bones known as Bone Marrow. The Bone Marrow comprises hematopoietic stem cells; which over an interval of period develop into various components of blood namely White Blood Cells (WBC), platelets, and Red Blood Cells (RBC), each of them has different roles to play [1]. Leukemia affects the production of White Blood Cells and interrupts normal cell activities. In addition, leukemic cells have abnormal growth and their survival time is much more than the normal cells, consequently number of abnormal blood WBC increases rapidly in the blood. Classification of leukemia is based upon how fast it becomes severe and it can be classified as chronic or acute. Acute leukemia progresses rapidly, and if not treated, becomes fatal within a few months. Chronic leukemia grows over a longer interval of time, and patients in most cases can live for many years. But chronic leukemia is generally harder to cure than acute leukemia [2–4]. It can be sub-categorized as Acute Lymphocytic Leukemia (ALL), Acute Myeloid Leukemia (AML), Chronic Lymphocytic Leukemia (CLL), and Chronic Myeloid Leukemia (CML). This paper deals with automated detection of Acute Lymphocytic Leukemia (ALL). Onset of Acute Lymphocytic Leukemia (ALL) causes production of immature lymphocytes in excessive amounts in bone marrow. These abnormal lymphocytes which are in excess will interrupt the function of normal cells.

In a conventional setup, the pathologist plays a crucial role in accurate diagnosis of ALL; since manual detection process is tedious, time-consuming, and accuracy of diagnosis also depends upon the experience of pathologist. Recent advances in digital image processing technology has led to a lot of research towards the development of automated recognition systems for identification of ALL [5–7]. CAD systems have the potential to provide valuable assistance to the pathologist in determination of the presence or the absence of the disease. In addition, it may also help in evaluation of stage of progression of disease.

Several Algorithms of identification and detection of leukemia have been implemented, S. Mohapatra et al. in 2012 [8] segregated region of interest using color-based clustering, they achieved successful classification of infected cells using SVM classifier based on Fractal dimensions, shape, and color. Later in 2012 [9], the same classification was achieved using ANN classifier, features in consideration included algorithms based on color methods. Nasir et al. in 2013 [10] did classification of acute leukemia cells using Multilayer Perceptron and simplified Fuzz ARTMAP neural networks with FNN and Bayesian classifier, features used for identification were based on shape and color of the target. In 2014, N. Chatap et al. [11] did analysis of blood samples for counting leukemia cells using SVM and KNN nearest neighbor algorithm based on shape of the cells. Later in 2015, R. Devi et al. [12] worked on the classification of Acute Myelogenous Leukemia in Blood Microscopic Images using PNN considering features based on shape, similar work was done by L. Faivdullah et al. [13] for leukemia detection from blood smears using SVM. The proposed work segregates the region of interest using K-means clustering, and then a combination of morphological, color, geometric, textural, and statistical features has been used to classify a mature lymphocyte and leukemic lymphocyte using Nearest Neighbor (kNN) and Naïve Bayes Classifier. Further devised methodology also addresses the segmentation of overlapping cells.

The paper is arranged in four sections. The next section deals with the proposed methodology and algorithm. Section 3 deals with identification and classification, and Sect. 4 provides the concluding remark.

2 Proposed Methodology

The automated detection of leukemia is performed by analyzing numerous microscopic images of the White Blood Cells or bone marrow smears obtained from the patient. First step is image acquisition; the samples for the proposed work were obtained from Dr RML Awadh hospital, Lucknow. Proposed algorithm for the automated identification of acute leukemia from microscopic image is shown in Fig. 1. Various steps involved are discussed below.

2.1 Image Preprocessing

The images of acceptable quality are subjected to preprocessing operations. The acquired microscopic digital image is usually corrupted by various kinds of noise or it may have a blurred region in the image which is important for detection. Figure 2 depicts the microscopic image of one of the blood samples.

Image preprocessing operations are performed to suppress the undesired distortions present in the image and enhance image features relevant for further analysis. The noise is removed from the image using various filtering techniques depending upon the type of noise. We have utilized Wiener filter which adequately reduced the blurriness without reducing the image sharpness. Further histogram equalization technique is applied to enhance the contrast in the image. Figure 3 depicts histogram of the image for performing adaptive histogram equalization. Gray-scale transformation brightness thresholding is chosen to modify brightness and threshold used is 192. This gray-scale transformation results in a binary image as shown in Fig. 4 [14–16].

The component mainly analyzed in the dataset is leucocyte; other than that every other component needs to be eliminated from the dataset. Further in the dataset being examined, it is possible that certain percentage of leucocytes is present on the edges of the image. In the image cleaning process, the leucocytes which are at the edge in the sample image under study and other irrelevant elements present in the image are removed in order to reduce errors in the later stages of the identification process. There are two cleaning operations which are required

Fig. 1 Proposed algorithm
for automated identification of
leukemia

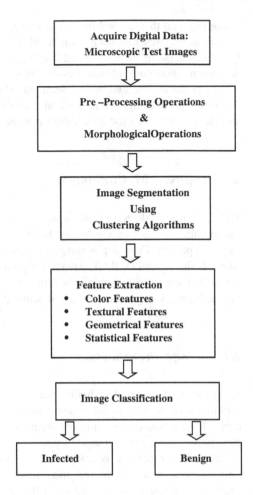

- Removal of Noise or Cleaning of Image
- Removal of abnormal components

The first one is done with help of filtering algorithms especially Wiener and Median filter. The component having a small area is usually the component located on the edges. The component with large values of area must be cells that are overlapping leucocytes. Therefore area and the convex area both need to be calculated for the removal of the unwanted components. Figure 5 depicts the morphologically cleaned image which is obtained by removing the leucocytes on the edges and irregular components.

2.2 Image Segmentation

Segmentation process partitions an image into distinct regions on the basis of features of interest. Segmentation in the present work involves segregation of White

Fig. 2 Microscopic image of blood samples

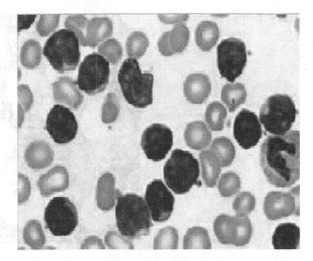

Fig. 3 Histogram for adaptive thresholding

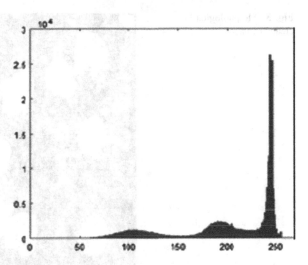

Blood Cells. Five components of White Blood Cells include: Neutrophil, Basophil, Eosinophil, Lymphocyte and Monocyte. ALL symptoms are associated only with the lymphocytes since morphological components of normal and malignant lymphocytes are significantly different; so other four components of White Blood Cells namely neutrophil, basophil, eosinophil, and myelocytes are neglected during segmentation process.

Figures 6, 7 and 8 depict the results of the k-means clustering with segregation of different nucleus and cytoplasm [17–19]. The performance of segmentation approaches such as k-means [20], texture-based segmentation [21] and color-based segmentation [22] have been compared. The brief description of the performance measures used is

Fig. 4 Binary image

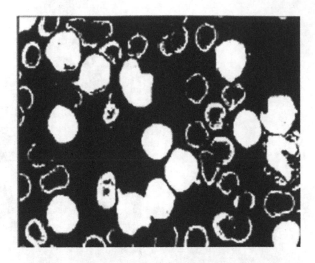

Fig. 5 Morphologically cleaned image

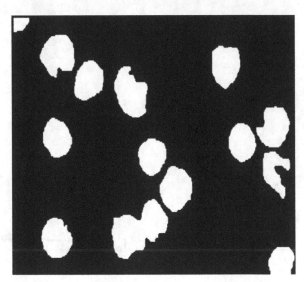

- Probability Random Index (PRI): It is a nonparametric evaluation of the goodness of segmentation. It is obtained by summing the number of pixel pairs with same label and number with different label in both S (test samples) and G (ground reality) and then dividing it by total number of pixel pairs. For a given a set of ground truth segmentations G_k, PRI is evaluated using

$$\text{PRI}(S_{\text{test}}, G_k) = \frac{1}{(N/2)} \sum_{\forall i, j \& i < j} [c_{ij} p_{ij} + (1 - c_{ij})(1 - p_{ij})], \qquad (1)$$

Fig. 6 Cluster 1: k-means clustering

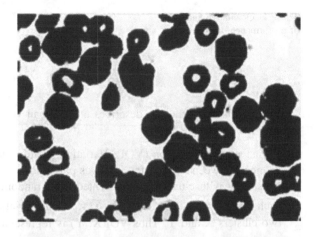

Fig. 7 Cluster 2: k-means clustering

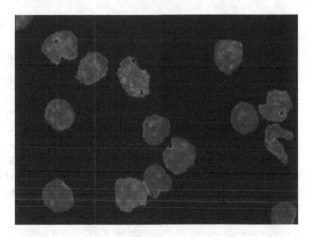

Fig. 8 Cluster 3: k-means clustering

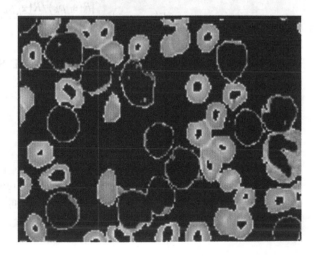

Table 1 Qualitative analysis of performance parameters

	PRI	GCE	VOI
Color k-means	0.932	0.0079	0.089
k-means	0.942	0.0091	0.092
Texture based	0.941	0.0132	0.015

where c_{ij} is an event that describes a pixel pair (i, j) having same or different label in the test image S_{test}.

- Variance of Information (VOI): Variation of Information gives the measure of distance between two clusters. It gives partition of pixels with different clusters. Clustering with clusters is represented by a random variable X, X = {1... k} such that $P_i = \frac{|X_i|}{n}$, $i \in X$, and $n = \sum_i X_i$ is the variation of information between two clusters X and Y. Thus VOI(X, Y) is represented using

$$VOI(X, Y) = H(X) = H(Y) - 2I(X, Y), \tag{2}$$

where H(X) is entropy of X and I(X, Y) is mutual information between X and Y.

- Global Consistency Error (GCE): Local refinement error is calculated using Eq. (3), where s_i and g_j contain pixel, p_k, so that $s \in S$, $g \in G$, where S segment is obtained after segmentation by the algorithm being evaluated and G denotes reference segment. The value obtained from (3) is used to evaluate global consistency errors using (4), where n denotes set of difference operation. R(x, y) represents the set of pixels corresponding to region x that includes pixel y. GCE quantifies the amount of error in segmentation. Table 1 depicts the comparative values of segmentation algorithms on the basis of VOI, PRI and GCE for the dataset evaluated [23–25].

$$E\left(s_{i,} g_{j,} p_k\right) = \frac{\left|R(s_{i,} p_k)/R\left(g_{j,} p_k\right)\right|}{\left|R(s_{i,} p_k)\right|} \tag{3}$$

$$GCE(S, G) = \frac{1}{n} \min\left\{\sum E(S, G.p_i), E(S, G.p_i)\right\} \tag{4}$$

3 Identification and Classification

Identification involves extraction of color, geometric, textural, and statistical features. Final step, i.e., the labeling of sample as malignant or benign is achieved through image classification process. Image classification analyzes various image features to arrange data into categories. Classification algorithms typically employ two processes: training and testing. In the training process, relevant properties of typical image features are isolated and a unique description of each classification category is created.

Two categories of classification algorithms namely supervised and unsupervised are generally used. In supervised classification, statistical processes are employed to extract class descriptors. Classification used in the present work relies on clustering algorithms to automatically segment the training data into various prototype classes. During the testing phase, features of sample dataset are compared with the previously calculated standard values. Depending upon the values of the input image finally classification is achieved with the help of Nearest Neighbor (kNN) and Naïve Bayes Classifier, comparison of which is also presented. The work was carried out on the dataset of 60 samples.

3.1 Identification of Grouped Leucocytes

Microscopic images of blood samples usually contain cells which are overlapping, this complicates the analysis and identification process. Segregation of Region of Interest (ROI) is achieved through k-Means clustering. In order to segregate leucocytes roundness has been used as a measure. Roundness checks whether the shape of the object is circular or not by excluding the local irregularities. Roundness can be gained by dividing the area of a circle to the area of an object by using the convex perimeter.

$$Roundness = \frac{4 X \pi X\ area}{convex\_perimeter^2} \tag{5}$$

The value of roundness is 1 if the object is circular and the value of roundness is less than 1 for the non-circular objects. Roundness as a measure is less sensitive to irregular boundaries because it excludes the local irregularities. Threshold value chosen is 0.80 to distinguish between the single leucocyte and clusters of overlapping leucocytes. The components which are having the roundness value more than the value of threshold are considered as the individual leucocyte while the components which are having value less than the threshold are considered as grouped leucocytes. Figure 9 represents roundness metrics obtained for various leucocytes. The individual leucocytes are sent next for the further study and the grouped leucocytes are sent to the separation process [26–28].

Solidity is used to find out the density of a component. It is obtained as

$$Solidity = \frac{area}{convex\ area} \tag{6}$$

Fig. 9 Metrics close to one
indicates roundness

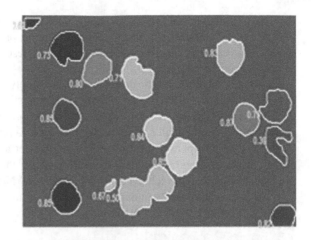

Fig. 10 Measure of solidity

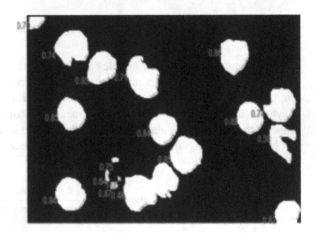

If the value is 1 then it can be identified as a solid object. If the value is less than 1 then it is a component having irregular boundaries. The threshold value for solidity which is used for identifying the abnormal components is obtained from the image which is having individual leucocytes. An optimum threshold value of 0.80 can efficiently be used to find out abnormal components from the image and this is depicted in Fig. 10. The components which are having the solidity value less than the threshold are removed [29, 30]. Figure 11 depicts standard deviation of identified leucocytes.

3.2 Nucleus and Cytoplasm Selection

The leucocytes segmented can now be used to extract the nucleus and cytoplasm. This is achieved by cropping the image with the bounding box. This step separates

Fig. 11 Standard deviation
of identified leucocytes

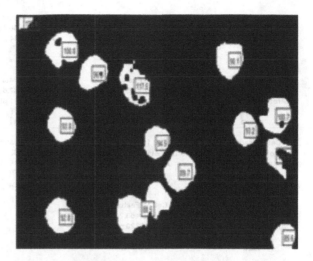

out each leucocyte. The borders of images obtained in the above step have to be cleaned in order to proceed further. Next step involves cropping out the outer portion of the leucocyte to segregate cytoplasm. This process segregates cytoplasm. From the close examination, it can be concluded that White Blood Cells nuclei are more in contrast on the green component of the RGB color space. So, we can get nucleus using the threshold [31].

3.3 Feature Extraction

Feature extraction is the process of converting the image into data so that we can check these values with the standard values and finally identify infected samples. Figure 12 depicts individual segmentation of the lymphocytes with its area and diameter. Features required to train model parameters include are

- Color Features—It includes mean color values of the gray images acquired.
- Geometric Features—It includes perimeter, radius, area, rectangularity, compactness, convexity, concavity, symmetry, elongation, eccentricity, solidity, etc.
- Texture Features—It Includes entropy, energy, homogeneity, correlation as obtained.
- Statistical Features—It includes mean, variance, and standard deviation. The values are computed are shown in Fig. 11.

$$Elongation = 1 - \frac{major\ axis}{minor\ axis} \qquad (7)$$

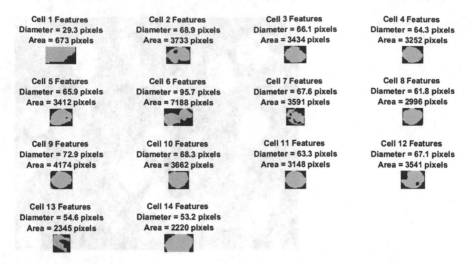

Fig. 12 Segmented leucocytes with corresponding diameter and area

$$Eccentricuty = \frac{\sqrt{(major\ axis^2 - minor\ axis^2)}}{major\ axis} \tag{8}$$

$$Rectangularity = \frac{area}{major\ axis\ X\ minor\ axis} \tag{9}$$

$$Convexity = \frac{Perimeter\_convex}{Perimeter} \tag{10}$$

$$Compactness = \frac{4 X \pi X\ area}{perimeter^2} \tag{11}$$

Elongation indicates the object elongation towards particular axis. Rectangularity depicts how well the bounding box is filled. Eccentricity is the ratio of the major axis length and the foci of the ellipse. Convexity shows the relative amount of difference of object from its convex object. Compactness is the ratio of the area of an object and area of circle having same perimeter. Figure 13 depicts the geometric characteristics of the infected cell and Fig. 14 depicts texture features of the infected cell [32, 33].

3.4 Image Classification

Proposed algorithm is tested with Nearest Neighbor (kNN) and Naïve Bayes Classifier on the dataset of 60 pretested samples, the accuracy achieved is 92.8%.

Fig. 13 Geometric features of infected cells

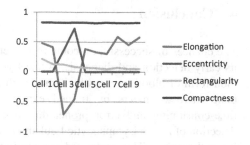

Fig. 14 Texture features of infected cells

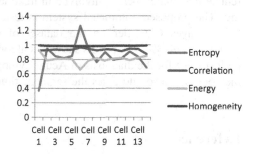

Fig. 15 Performance analysis of classifiers

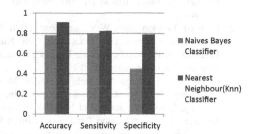

The process involved following steps:

1. Finalization of feature set
2. Selection of appropriate Algorithm
3. Mapping and Training of model parameters

Features such as Elongation, Eccentricity, Rectangularity, Convexity, Compactness, entropy, energy, homogeneity, correlation, and standard deviation are used to train model parameters to identify infected cells. Confusion Matrix has been utilized to compute performance of classifiers. Performance parameters included accuracy, sensitivity, and specificity shown in Fig. 15.

4 Conclusion

The chances of successful outcome of cancer treatment are primarily dependent on its early detection and diagnosis. Acute Lymphocytic Leukemia is the most common form of blood cancer which can identified through examination of blood and bone marrow smears by trained experts. This manual inspection process is time-consuming and error prone, thus a computer-based system for automated detection of Acute Lymphocytic Leukemia may provide an assistive diagnostic tool for pathologists. The automated segregation and identification algorithm aims to reduce the latency period involved in treatment, which is sometimes life threatening. The proposed automated system is tested with Nearest Neighbor (kNN) and Naïve Bayes Classifier on the dataset of 60 pretested samples, the accuracy achieved is 92.8%. The results show that algorithm proposed achieves an acceptable performance for the diagnosis of Acute Lymphocytic Leukemia; further the devised methodology also addresses the segmentation of overlapping cells.

References

1. Barbara JB. India: Blackwell Publishing Ltd; 2004. A Beginner's Guide to Blood Cells; pp. 64–5.
2. Bain BJ. London: John Wiley and Sons; 2010. Leukemia Diagnosis.
3. Mohapatra S, Patra D, "Automated Cell Nucleus Segmentation and Acute Leukemia Detection in Blood Microscopic Images", Systems in Medicine and Biology (ICSMB), 2010 International Conference on, IEEE.
4. Amin MM, Kermani S, Talebi A, Oghli M G., "Recognition of Acute Lymphoblastic Lukemiua Cells in Microscopic Image Using K-Means Clustering and Support Vector Machine Classifier", Journal of Medical Signals and Sensors, 2015; 5(1):49–50.
5. Subrajeet Mohapatra, Dipti Patra, Sanghamitra Satpathy, "Automated leukemia detection in blood microscopic images using statistical texture analysis", February 2011 ICCCS '11: Proceedings of the 2011 International Conference on Communication, Computing & Security sponsored by ACM, Pages: 184–187.
6. N. Z. Supardi, M. Y. Mashor et al, "Classification of Blasts in Acute Leukemia Blood Samples Using K -Nearest Neighbour", 2012, IEEE 8th International Colloquium on Signal Processing and its Applications, Pages: 461–465.
7. Mohapatra S, Dipti Patra et al, "Automated morphometric classification of acute lymphoblastic leukaemia in blood microscopic images using an ensemble of classifiers", Computer Methods in Biomechanics and Biomedical Engineering: Imaging & Visualization, Volume 4, Issue 1, January 2016, pages 3–16.
8. Mohapatra S., Patra D. and Satpathy S., "Unsupervised Blood Microscopic Image Segmentation and Leukemia Detection using Color based Clustering", International Journal of Computer Information Systems and Industrial Management Applications. ISSN 2150-7988 Vol. 4, pp. 477–485, 2012.
9. Mohapatra S., Patra D., Kumar S. and Satpathy S., "Lymphocyte Image Segmentation Using Functional Link Neural Architecture for Acute Leukemia Detection", The Korean Society of Medical & Biological Engineering and Springer, 2012.

10. Nasir A. A., Mashor M. Y., Hassan R., "Classification of Acute Leukaemia Cells using Multilayer Perception and Simplified Fuzzy ARTMAP Neural Networks", The International Arab Journal of Information Technology, Vol. 10, No. 4, pp-356–364; July 2013.

11. Chatap N., Shibu S., "Analysis of blood samples for counting leukemia cells using Support vector machine and nearest neighbour", IOSR Journal of Computer Engineering (IOSR-JCE), Vol. 16, Issue 5, Ver. III; PP 79–87, 2014.

12. Devi R., Arivu C.V., "Classification of Acute Myelogenous Leukemia in Blood Microscopic Images Using Supervised Classifier", IJESC,
DOI 10.4010/2015.314; ISSN-2321 -3361; 2015.

13. Faivdullah L., Azahar F., Htike Z. Z., Naing W. N., "Leukemia Detection from Blood Smears", Journal of Medical and Bioengineering, Vol. 4, No. 6; pp-488–491, December 2015.

14. S. Chinwaraphat, A. Sanpanich, C. Pintavirooj, M. Sangworasil, and P. Tosranon. A modified fuzzy clustering for white blood cell segmentation. In Proceedings of the Third International Symposium on Biomedical Engineering, volume 6, pages 2259–2261, 2008.

15. Scotti F. 2005 IEEE International Conference on Computational Intelligence for Measurement Systems and Applications; 2005. Automatic Morphological Analysis for Acute Leukemia Identification in Peripheral Blood Microscope Images.

16. Theera-Umpon N. Springer: Springer-Verlag Berlin Heidelberg; 2005. White Blood Cell Segmentation and Classification in Microscopic Bone Marrow Images, in Fuzzy Systems and Knowledge Discovery; pp. 787–96.

17. Scotti F. Instrumentation and Measurement Technology Conference, 2006. IMTC 2006. Proceedings of the IEEE; 2006. Robust Segmentation and Measurements Techniques of White Cells in Blood Microscope Images; pp. 43–8.

18. Reta C, Altamirano L, Gonzalez JA, Diaz R, Guichard J. In FLAIRS Conference; 2010. Segmentation of Bone Marrow Cell Images for Morphological Classification of Acute Leukemia.

19. Madhloom HT, Kareem SA, Ariffin H, Zaidan AA, Alanazi HO, Zaidan BB. Anautomated white blood cell nucleus localization and segmentation using image arithmetic and automated threshold. J Appl Sci 2010;10(11):959–66.

20. Scotti F. Robust segmentation and measurements techniques of white cells in blood microscope images. In: Daponte P, Linnenbrink T, editors. Proceedings of the IEEE instrumentation and measurement technology conference, 24–27 April. Sorrento, Italy: IEEE Publisher; 2006. p. 43–8.

21. N. T. Umpon. Patch based white blood cell nucleus segmentation using fuzzy clustering. ECTI Transaction Electrical Electronics Communications, 3(1):5–10, 2005.

22. L. Nowinski, "Medical image segmentation using k-means clustering and improved watershed algorithm," in Proceedings of the 7th IEEE Southwest Symposium on Image Analysis and Interpretation, pp. 61–65, IEEE, March 2006.

23. N. R. Pal and S. K. Pal, "A review on image segmentation techniques," Pattern Recognition, vol. 26, no. 9, pp. 1277–1294,1993.

24. M.-N. Wu, C.-C. Lin, and C.-C. Chang, "Brain tumor detection using color-based K-means clustering segmentation," in Proceedings of the 3rd International Conference on Intelligent Information Hiding and Multimedia Signal Processing (IIHMSP '07), pp. 245–248, IEEE, November 2007.

25. L.B. Dorini, R. Minetto, and N.J. Leite. White blood cell segmentation using morphological operators and scale-space analysis. In Proceedings of the Brazilian Symposium on Computer Graphics and Image Processing, pages 294–304, October 2007.

26. F. Scotti. Automatic morphological analysis for acute leukemia identification in peripheral blood microscope images. In *Proceedings of IEEE International Conference on Computational Intelligence for Measurement Systems and Applications*, pages 96–101, July 2005.

27. Kovalev VA, Grigoriev AY, Ahn H. Robust recognition of white blood cell images. In: Kavanaugh ME, Werner B, editors. Proceedings of the 13th international conference on pattern recognition, August 25–29. Vienna, Austria: IEEE Publisher; 1996. p. 371–5.

28. Mohapatra S, Patra D, Satpathy S. An ensemble classifier system for early diagnosis of acute lymphoblastic leukemia in blood microscopic images. Neural Comput Appl. 2014;24: 1887–904.

29. D. Goutam, S. Sailaja, "Classification of acute myelogenous leukemia in blood microscopic images using supervised classifier", 2015, IEEE International Conference on Engineering and Technology (ICETECH), pages 1–5.

30. Jakkrich Laosai, Kosin Chamnongthai, "Acute leukemia classification by using SVM and K-Means clustering", 2014, Proceedings of the International Electrical Engineering Congress, pages 1–4.

31. Osman Selçuk, Figen Özen, "Acute lymphoblastic leukemia diagnosis using image processing techniques", 2015 23nd Signal Processing and Communications Applications Conference (SIU), Year: 2015, Pages: 803–806.

32. M. Sarfraz; A. Ridha, "Content-based Image Retrieval using Multiple Shape Descriptors", 2007, IEEE/ACS International Conference on Computer Systems and Applications, Pages: 730–737.

33. J. Angulo, J. Serra, and G. Flandrin. Quantitative descriptors of the lymphocytes. In *Proceedings of the 7th Congress of the European Society for Analytical Cellular Pathology*, pages 69–70, France, April 2001.

Energy Data Analysis
of Green Office Building

Weiyan Li, Minnan Piao, Botao Huang and Chenfei Qu

Abstract Green office building is the future trend of office buildings, whose energy consumption data collection point is more, and so there are more data and outlier. In this paper, based on energy consumption data in a green office building of some architectural design institute, we establish the energy consumption monitoring model, which first identify the energy consumption pattern by clustering analysis of historical energy consumption data, get the decision tree of energy consumption pattern by classifying the energy consumption data, match the real-time energy consumption data with the energy consumption patterns, and then make outlier analysis with historical data of the same pattern. So we can determine whether the energy consumption data are abnormal and current energy consumption is rational or not and achieve energy saving purposes.

Keywords Green office building · Energy consumption monitoring model · Energy saving

This paper comes from the subject "the analysis and evaluation of green building operation Based on integration platform of the intelligent system" supported by the department of housing and urban–rural development, and the subject number is R12014110.

W. Li (✉) · C. Qu
Tianjin Architecture Design Institute, Tianjin, China
e-mail: lwytju1@163.com

C. Qu
e-mail: chinaqchf@163.com

M. Piao · B. Huang
Nankai University, Tianjin, China
e-mail: 2514842881@mail.nankai.edu.cn

B. Huang
e-mail: billow2008@126.com

© Springer Nature Singapore Pte Ltd. 2018
S.K. Bhatia et al. (eds.), *Advances in Computer and Computational Sciences*,
Advances in Intelligent Systems and Computing 554,
https://doi.org/10.1007/978-981-10-3773-3_65

1 Introduction

In our country annual increase of construction area is equivalent to 40% of the world's new construction area, more than 95% of which is high-energy building, and under the same climatic conditions energy consumption per unit area of which is 2 to 3 times higher than other countries, especially in high-energy office buildings. So for green office building, which is the future trend of office buildings, energy consumption data analysis and outlier detection has important significance.

Researches on building energy consumption analysis have made some achievements. In 1996, Terry Sharp established stepwise linear regression model which was used to identify high-energy consumers, and then successfully applied to office buildings and commercial buildings [1]; in 2002, Muller W and Wiede thole E studied how to apply decision tree building benchmarking the energy consumption [2]. John E. Seem studied historical energy consumption data clustering for each day of the week as the basic unit, and discovered the existence of a building energy model [3]. In 2006, Melek Yalcintas researched how to apply neural network model to evaluate energy consumption [4]. In 2007, John E. Seem used outlier detection method based on statistics GESR to build energy consumption data in history, and successfully detected the energy consumption during operation anomalies exist [5].

2 Basics

Before the analysis of energy consumption data, we need to select specific algorithm based on the characteristics of the actual data; otherwise, it will affect the accuracy of the results.

2.1 DBSCAN Clustering Algorithm

Clustering algorithm is an algorithm that the data object is grouped, based on the data relationships [6]. K-means clustering not only needs to determine the number of clusters in advance but it is high sensitivity to outliers, and can only find round cluster; the time complexity of clustering hierarchy is unfavorable to deal with a large number of data sets; while DBSCAN algorithm can automatically determine the number of clusters, discover clusters of arbitrary shape, and detect the abnormal value, so in the paper we choose the DBSCAN algorithm to cluster the data.

DBSCAN is a clustering algorithm based on density which is to find a high-density area separated from the low-density region. First, confirm all data points which they are core points, border points, or outliers. Then the two closes enough to the core point form a cluster, at the same time all border points which belong to the neighborhood of core point are also placed in the same cluster with the core point. In DBSCAN algorithm, if a point, which the number of data points within the radius of a specified distance Eps exceeds Minpts, is the core point. A point is not the core point, but in the neighborhood of a core point, then this point is the boundary point. If a point is neither a core point nor the boundary points, it is outlier. Eps and Minpts are parameters which are achieved by K-distance map of each data point or experience.

2.2 C4.5 Classification Algorithm

Classification algorithm predicts the class label for the future sample with the learning model established by the relationship between the class label and attribute value of input data set [7]. In this paper, considering the energy consumption data set has only two attributes, using the decision tree classification algorithm is not only more simple and effective, but also more easy to understand the results of classification, compared to the Bayesian method, neural network method and support vector machine (SVM) classification algorithm. So we select C4.5 decision tree classification algorithm to classify the data set.

C4.5 algorithm [7] is a classification decision tree algorithm. Model training starts from the root, and then we select the optimal attributes for data at each node, and classify data set to form new nodes, and one more time select new optimal attributes to classify data set, thereby it forms a constantly growing tree, until all data points have the same class label or the same attribute value, which is a leaf node. But the process of the data set for training may appear over-fitting, because in an unknown test data set the proportion of misclassification is very large. In order to avoid over-fitting, we use a pruning method on the resulting decision tree to prune the tree.

2.3 Dynamic LOF Algorithm

The basic idea of traditional LOF algorithm is to find the observation point from the k-neighborhood of an object p, and to use the distance from these observations point to the object p to calculate local anomaly factor of the object p, and then by the size of the LOF value to determine whether the object p is outliers.

To calculate the LOF values of an object p, first determine the k-neighborhood of the object p:

$$N_k(p) = \{q \mid d(p, q) <= k - \text{distance}(p)\}, \tag{1}$$

where $k - \text{distance}(p)$ is the k-distance between the object p and the object q.

And then get the reach distance between the object p and all points of k-neighborhood of the object p:

$$\text{reach} - \text{dist}_k(p, q) = \max(k - \text{distance}(q), d(p, q)). \tag{2}$$

On the basis, to calculate the local density of the object p:

$$lrd_k(p) = 1 / \frac{\sum_{q \in N_k(p)} \text{reach} - \text{distance}(p, q)}{N_{MinPts}(p)}. \tag{3}$$

Then calculate the LOF value of the object p:

$$LOF_k(p) = \frac{\sum_{q \in N_{KinPts}(p)} \frac{lrd(q)}{lrd(p)}}{|MinPts(p)|}, \tag{4}$$

where $|MinPts(p)|$ represents the number of collection $N_k(p)$ of the data points.

In this paper, we use advanced LOF algorithm [8], which is easier and to more clearly express the concept of outlier. The basic idea is as follows: For an object p, determine the number of points in a circular area within the circle center p and radius ε. If the number is less than M, it is determined to be abnormal; if the number of points exceeds M, p is calculated to ε-neighborhood distance from the farthest point p, and the formula is improved:

$$LOF_k(p) = \frac{\sum_{q \in \varepsilon(p)} \frac{d_{\varepsilon - \max(p)}}{d_{\varepsilon - \max(q)}}}{|\varepsilon_{MinPts}(p)|}. \tag{5}$$

3 Examples of Verification

In the paper, on the basis of data integrity and accuracy, the lighting data were analyzed using DBSCAN clustering algorithm, C4.5 classification algorithm to establish energy consumption model, and then detect outliers of the future data under the same model.

3.1 Data Sorting and Analyzing

Select 7–8 month's data to analyze; each 2 weeks is seen as a sample, and so in July and August there are four samples: Sample 1, sample 2, sample 3, and sample 4. Use the first three samples (Sample 1, Sample 2, Sample 3) to establish energy consumption model, and in turn subsequently detect three samples (Sample 2, Sample 3, Sample 4) abnormality.

According to the above, we first remove the outliers and add two properties for all data. The first is whether the property is weekday. Because there are no other legal holidays during 2 months, except Saturday and Sunday, the attribute on the weekend is set to 7, while from Monday to Friday the attribute value of data is 1. Second, 0,…, 23 represent each data point range, such as 0 represents 00: 00–1:00. Finally, all data have the above two properties.

3.2 Using DBSCAN Algorithm Clustering

According to the feature of the lighting system, select two indicators (average hourly and hourly peaks) to cluster energy consumption data.

First, we must determine the parameters Eps and MinPts. For the two-dimensional data, MinPts is usually selected a value of 4 for DBSCAN algorithm, so we also set MinPts to 4. And Eps is selected by drawing a 4-distance map method. For all of the sample data, first calculate the distance of nearly a fourth of their own data points from the 4-distance map, then the 4-distance map of all sample points will be sort and drawn in axis, and finally when the value of the image based on 4-distance map mutates, the number is suitable to Eps. At the same time, we determine the core point and outliers: if k-distance value is less than Eps, the point is core point, and k-distance value is greater than the value as outliers. It is proved that all of the distance map is the big change is after 1.5 in the distance, so choose Eps 1.5 in this algorithm.

Then all sample points are traversed to determine the core point, boundary point, and outliers; according to Eps and MinPts, make all sample points as different clusters. The below shows clustering results of sample 1, 2, 3, 4. (Fig. 1).

From the clustering results, the number of clusters is approximately 2 or 3, three clusters are formed in the sample 1, 2, 3, and the sample 4 forms two clusters. Cluster 0 and cluster 1 contain the most part of the sample points, while the sample point in cluster 2 and outliers is mainly energy consumption data in 7:00, 8:00, 17:00, 18:00, 19:00 time, because these moments are transitional moment time between work and rest, so some are in cluster 0, some are cluster 1, and some form a separate cluster. The remaining node, whose neighbors are less than 4, is marked as outliers. Since the number of sample point in cluster 2 is very small compared to other clusters, which has less influence on outlier detection, the data of the cluster 2 are classified to the cluster which is the nearest to the sample point.

3.3 Using C4.5 Algorithm to Classify the Clustering Data

To use the C4.5 classification algorithm for classifying data, we first need to determine the conditions for the termination of tree growth. This paper selects the termination of three conditions: First, it will continue until the class label of all energy values is the same; the second is that the number of energy data points,

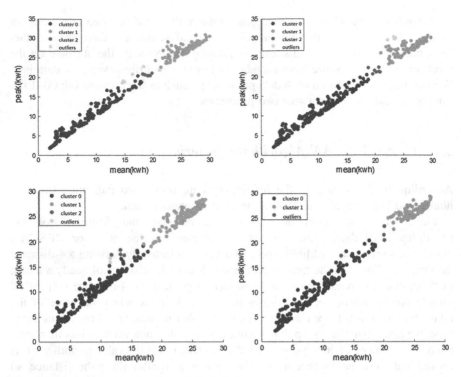

Fig. 1 The clustering results of sample 1, 2, 3, 4

which are classified, is less than a critical value, which is selected to 10; the third is that all data of node points have the same attribute values. Then, depending on two attributes which are mentioned in the second chapter, we determine the basis classification, and calculate the mean and peak value in each basis classification. After code project is run classification results are obtained as follows:

As can be seen from the classification result, classification process of each sample is different, because every attribute, which is based on the gain of the entropy maximization judgment, is not the same. But the result is as follows: Sample 1, 2, 3 classification result is the same, namely, energy consumption data 8:00–17:00 in the weekday is classified to cluster 1, and the rest of the energy consumption data to cluster 0. This can be seen that in July and August lighting energy consumption patterns have no major changes; the result of classification is in line with reality (Fig. 2).

3.4 Dynamic LOF Algorithms to Detect Outliers

According to the attributes of weekday and time range, we determine the cluster label of new energy data point, and then use the dynamic LOF algorithms to detect the outliers of new energy data by comparing to corresponding cluster.

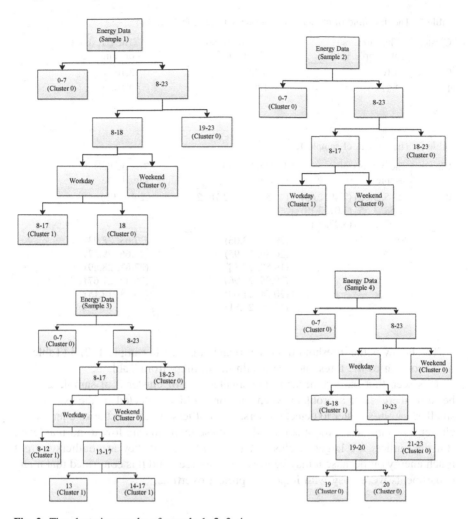

Fig. 2 The clustering results of sample 1, 2, 3, 4

The historical data for each one hour of energy consumption data, with the same Energy consumption mode together, use the dynamic LOF algorithm to detect outliers. The below shows the outliers of sample 2, 3, 4 by the energy data of sample 1, 2, 3.

Before determining whether the new energy point is abnormal data, we need to decide the threshold value of the anomaly factor, which we select to the average of the three maximum value of abnormal factor in the same energy consumption patterns of history data. Although the data are normal in the cluster 0, the local density of each data point is very large, causing that abnormal factor values around these points a little further away from the data are large, so we select the average of the three maximum values as the threshold of abnormal factor values.

Table 1 The threshold of cluster 0 and cluster 1 in sample 1, 2, 3, 4

Cluster	The threshold of sample 1	The threshold of sample 2	The threshold of sample 3
0	3.10	5.22	3.16
1	3.27	1.9	2.16

Table 2 The outliers of sample 1, 2, 3, 4

Cluster	The outliers of sample 2	The outliers of sample 3	The outliers of sample 4
0	(2.75, 3.65) (2.49, 2.89) (2.58, 3.55) (3.07, 3.13) (3.07, 3.12) (3.07, 3.13)	(2.55, 3.14) (2.76, 2.77)	(2.4, 3.93)
1	Null	(20.08, 23.06) (20.86, 21.95) (19.58, 22.38) (20.95, 23.96) (20.74, 21.09) (20.28, 24.81)	(27.63, 28.93) (27.66, 29.37) (27.87, 28.69) (26.82, 27.67) (15.88, 18.15)

The below is the threshold of cluster 0 and cluster 1 in sample 1, 2, 3 (Table 1). According to the threshold, we obtain the outliers from Table 2:

It is seen that outliers occur in the smaller part of cluster 0 of sample 2, 3, 4, because there exist a problem when the threshold is selected in the part of the smaller, and it is easier to detect outliers. The outliers are also in the smaller part of cluster 1 of sample 3. And it is proved that these moments are low-to-normal level. But the outliers are larger in cluster 1 of sample 4, and they are outliers we put much energy on because it may be waste of resource. And it is confirmed that many departments were busy with important project overtime.

4 Summary

In this paper, based on a green office building lighting data, we clustered and classified historical energy consumption data, matched dynamic data with the energy consumption patterns, and then got outlier analysis. So it is easy to determine whether energy consumption data are abnormal and current energy use is reasonable. This method is also suitable for other seasonal changes less-volatile energy systems. Therefore, if energy consumption data model is applied to existing green office buildings, it will be easy to achieve energy-efficiency goals.

References

1. T. Sharp. Energy benchmarking in commercial office buildings. Proceedings of the ACEEE1996 Summer Study on Energy Efficiency in Buildings, (4): 321–329(1996)
2. T. Sharp. Benchmarking energy use in schools. Proceedings of the ACEEE 1988 Summer Study on Energy Efficiency in Buildings, (3):305–316 (1998)
3. JOHN E. SEEM. Pattern recognition algorithm for determining days of the week with similar energy consumption profiles [J]. Energy and Buildings, 37(2):127–139 (2005)
4. M. Yalcintas. An energy benchmarking model based on artificial neural network method with a case example for tropical climates [J]. International Journal of energy research, (30):1158–1174 (2006)
5. JOHN E. SEEM. Using intelligent data analysis to detect abnormal energy consumption inBuildings [J]. Energy and Buildings, 39(1):52–58 (2007)
6. D.B. Lu, Based on Decision Tree Data Mining Algorithm and Application[D], Wuhan University of Technology (2008)
7. Department of Statistics data mining center of Renmin University of China. Data mining clustering analysis [J]. Statistics & Information Forum, 17(53):4–10 (2002)
8. G.W. Milligan, M.C. Cooper, An examination of procedures for determing the number of clusters in a data set, psychometrika, 50(2):159–179 (1985)

Location Prediction Model Based on K-means Algorithm

Yan Hu, Xiaoying Zhu and Gang Ma

Abstract Location prediction is critical to mobile service because various kinds of applications tightly combined with object's location. However, location prediction is a challenging work because the location captured is always not continuous and object's behavior is uncertain and irregular. The prediction accuracy of many models is less than 30%. But the prediction accuracy is important to location prediction. It will directly affect the mobile services. So this paper is to improve prediction accuracy to provide more efficient mobile service. This paper proposes a location prediction model based on k-means algorithm and time matching. For the mobile service always region oriented, we first cluster history location using k-means algorithm to define several regions. Then we divide every day time into several segments and calculate the maximum probability location in every time segment. A trajectory of an object in one day is formed with trajectory model and trajectory updating model which is proposed in this paper. We can predict object's location with time-matching method. At last, we do experiments with real location data which captured by APs. The prediction result with k-means is compared to the result without model based on k-means algorithm. The experiment result shows that prediction accuracy of our model is higher than the prediction without new model. So more location services can be provided to objects with this new model.

Keywords Location prediction · K-means · Prediction accuracy · Cluster

1 Introduction

Location prediction is critical to mobile service because various kinds of applications tightly combined with object's location [1]. However, location prediction is a challenging work. First, location capturing is not continuous and exists blind spot.

Y. Hu (✉) · X. Zhu · G. Ma
Network and Information Center, Beijing University of Posts and Communications,
Beijing, China
e-mail: huyan@bupt.edu.cn

© Springer Nature Singapore Pte Ltd. 2018
S.K. Bhatia et al. (eds.), *Advances in Computer and Computational Sciences*,
Advances in Intelligent Systems and Computing 554,
https://doi.org/10.1007/978-981-10-3773-3_66

So some import location source data will be lost. It leads to the prediction more complex. Second, objects' behavior is uncertain and irregular. It is difficult to predict location in a single algorithm. Third, the location data usually cannot be directly used for it is discrete and fragmented [2]. These data always need to be preprocessed through a complex approach before predicting. All the above factors make the prediction work difficult and complex.

The method of location prediction always can be divided into two ways. One way is predicting future location according to the last location. Location is predicted through calculating the transition probability. Markovian algorithm is combined with the first stage tightly. In paper [3–5], Markova algorithm and hybrid markovian model based on dynamic social ties or with time based is presented. The location prediction is only related to the last location.

The other way is collecting history location and predicting. In paper [6], algorithm of Bayesian is used to predict location and multi-factors are considered in order to improve the prediction accuracy rate. And more factors are taken into account, such as spatial, temporal, and similarity characteristics of object mobility using to construct a set of object features. In other papers, objects' pattern is calculated to predict location [5], and random Walk and Markovian algorithm are used in paper [7, 8]. The visited cell-path and time intervals spent in each cell are taken into account.

The prediction accuracy is important to location prediction. It will directly affect the effectiveness of location-based services. The prediction accuracy of most above models is less than 30%. So the model in this paper is to improve prediction accuracy to provide more efficient location-based services. Most prediction model is based on the exact place or location which can be described with longitude and latitude. Our prediction is based on region to improve prediction accuracy rate. These regions are also valuable to mobile service. Then we divide every day time into several pieces and calculate the maximum probability location in every time segment. A trajectory of an object in one day is formed according to the time and representative location. We can predict an object's location according to the comparison of the time and position in history trajectory.

The main contributions of this paper are summarized in the following aspects:

- We propose a location prediction model based on k-means algorithm to predict object location.
- The time-matching method is also used to form the prediction trajectory.
- We do experiment with prediction model on a real data and realize the prediction model. The result shows that prediction accuracy reached 60.3%.

The rest of this paper is organized as follows. The basic idea of k-means algorithm is introduced in section two. In section three, a new location prediction model is design based on k-means algorithm, and time matching is also take into account. In section four, we do experiment with this new model to predict location. A conclusion is summarized in the last section.

2 Location Prediction Model Based on K-means

2.1 Basic Idea of K-means Algorithm

K-means is a classic learning algorithm which solves the clustering problem. The procedure follows a simple and easy way to classify a given data set through a certain number of clusters (assume k clusters) fixed a priori [9]. Let $X = (x_1, x_2, \ldots x_n)$ be the set of samples and each x_i is a sample. The k-means algorithm partitions these points into K clusters which is defined as $C = (c_1, c_2, \ldots c_k)$. The main idea is to define k centroids, one for each cluster. First, choosing k original centroids from the set of X randomly. Second, the distance of sample to centroids will be calculated. Euclidean distance and cosine similarity are usually used to measure the distance. The distance of sample to centroid can be described as $\|x_i - c_j\|^2 (i \in n, j \in k)$. The sample will be assigned if it is the nearest to k centroids until no sample is pending. Second, we need to recalculate k new centroids as barycenter of the clusters resulting from the previous step. After we have these k new centroids, a new binding has to be done between the same data set points and the nearest new centroid. A loop has been generated. As a result of this loop we may notice that the k centroids change their location step by step until no more changes are done. In other words centroids do not move any more.

2.2 Location Prediction Method Based on K-means

This paper proposes a location prediction model based on k-means algorithm. There are three steps to prediction location: First, clustering the history location with k-means; second, defining different time segments in a day; finally, predicting object's location with history location and time matching. In the first step, we cluster history location using k-means algorithm to define several regions. These maybe means a building or an area in the real world. Our prediction based on these clustered locations is to improve accuracy rate of prediction. In second step, we divide every day time into several segments and calculate the maximum probability location in every time segment. Finally, a trajectory of an object in one day is formed with location and time. We can predict an object's location in the future day according the time period.

Let $L = (L1, L2, \ldots Ln)$ be a set of history location which object visited. Then we define the longitude and latitude of a location point L as $L = (x, y)$. Every history location Li is corresponding to a $L_i = (x_i, y_i)$. We use the distance of every two locations to cluster with k-means algorithm. It can be described as $D_{ij} = \|x_i - y_i\|^2$. The steps of clustering are as follows:

Input: $L_i = (x_i, y_i)$ and k
Output: result of cluster

Step 1: Choose k random samples as centroids in history locations as initial group centroids.

Step 2: Calculate the distance of history location to centroids

$$D_{ik} = \arg \min \sum_{i=1}^{k} \sum_{x_i \in L_i} \|x_i - m_k\|^2$$

The sample will cluster into k_i if k_i is the nearest k to this sample.

Step 3: When all samples have been assigned, recalculate the positions of the k centroids.

Step 4: Repeat Steps 2 and 3 until the centroids no longer move. This produces a separation of the objects into groups from which the metric to be minimized can be calculated.

The result of clustering is $Cluster = (c_1, c_2, \ldots c_k)$ and k is the input value. According to the k-means algorithm, every L belongs to a cluster:

$$L_i \in Cluster_j (0 < j < k).$$

An object always appears at a place with in a time. So the appearance time is tightly relative to the history location of an object. One day time can be divided into serval time segments:

$$T = (T[ti, tj], T[tj, tk], \ldots T[tn, ti]) (0 \le ti < tj < tn \le 24).$$

The probability of an object appearing in a place can be expressed as

$$p_i = \frac{Li}{\sum_{i=j}^{n} Lj} (1 \le i \le n).$$

The maximum probability location is defined as Mp in history in the time of $T_{ij} = [t_i, t_j]$.

$$Lp = \arg \max \frac{L_i}{\sum_{i=j}^{n} L_j} (1 \le i \le n).$$

For $L_p \in L$ and formula (3), we can derive that $L_p \in Cluster_i$. The $Cluster_i$ is our prediction location in the time of $T_{ij} = [t_i, t_j]$. So there is a sequence of history trajectory:

$$S = \{\langle T_{ij}, Cluster_a \rangle, \langle T_{jk}, Cluster_b \rangle \ldots \langle T_{ni}, Cluster_k \rangle\}$$
$$(0 \le ti < tj \ldots tn \le 24, 0 < a, b \ldots < k).$$

According to the above history trajectory sequence, we can predict object's location with time in day.

3 Experiments and Results

We test our prediction model with real data which published by University of California, San Diego. We use one-week object data as the history data to build history trajectory. Then we predict the object location at time of next week. First, we cluster the AP's location with k-means and name it. We define k = 8, so all the AP's location will be clustered into 8 clusters. Figure 1 shows the clustering result.

Second, we divide 24 h of day into six segments. They are t1 [0 am, 6 am], t2 [6 am, 9 am], t3 [9 am, 12 am], t4 [12 am, 14 pm], t5 [14 pm, 18 pm], and t6 [18 pm, 24 pm]. The maximum probability of history location in this time segments can be calculated. The prediction trajectory is formed, and we can use this trajectory of object to predict location in the future day. In our experiments, 50 objects' location data are used. Figure 2 shows the prediction correct times of comparison in different time segments. In the time of t1, the results are same because it is in sleepy time. In time t2, the prediction ratio is lower than before using our model because its activity area is large. In time t3–t6, the prediction ratio is higher than before using model. It can prove that campus objects always act in a certain region in a certain time. So our prediction model based on k-means is suitable to predict location in campus. Figure 3 shows that the prediction accuracy before using our model is 39.7% and the prediction accuracy after using our model is 60.3%. So it proves that we construct an efficient model to predict location.

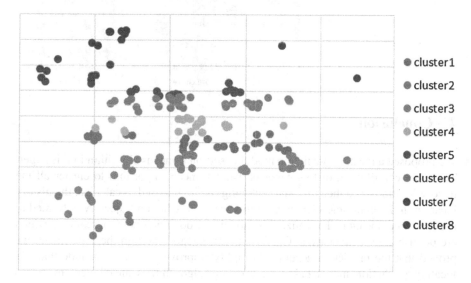

Fig. 1 Cluster result with k-means algorithm

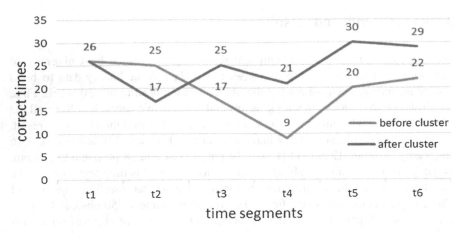

Fig. 2 Results of comparison in different time segments

Fig. 3 Prediction accuracy
before and after cluster

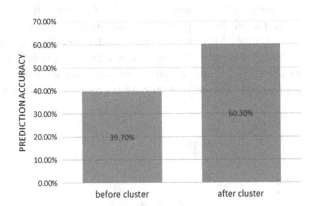

4   Conclusion

We proposed a location prediction model based on k-means algorithm in this paper. The main contribution of this paper is using k-means algorithm to cluster all the history location. And then the time-matching method is used to calculate the history location in different time segments. In this way, a prediction trajectory is formed to predict object location. To realize this model, we do experiment on real data which are published by University of California, San Diego. Through the experiment, we proved that the prediction accuracy is highly improved. So we conclude that our location prediction model based on k-means algorithm is an efficient model to predict object's location.

References

1. YU Rui-Yun, XIA Xing-You, LI Jie, ZHOU Yan, WANG Xing-Wei. Social-Aware Mobile Object Location Prediction Algorithm in Participatory Sensing Systems. Chinese Journal of Computers, vol. 38, No. 2, Feb 2015.
2. LI Wen, XIA Shi-xiong, LIU Feng, ZHANG Lei, YUAN Guan. Location prediction algorithm based on movement tendency. Journal on Communicaitons, Vol. 35, No. 2, February 2014.
3. Matthew W. Robards, Peter Sunehag. Semi-Markov kMeans Clustering And Activity Recognition From Body-Worn Sensors. 2009 Ninth IEEE International Conference on Data Mining.
4. Yi Yang, Zhiliang Wang, Qiong Zhang, Yang Yang. A Time Based Markov Model for Automatic Position-Dependent Services in Smart Home. 2010 Chinese Control and Decision Conference.
5. Wen Li, Shi-xiong XIA, Feng LIU, Lei ZHANG. Hybrid Markov Location Prediction Algorithm Based on Dynamic Social Ties. IEICE TRANS. INF. & SYST, VOL. E98-D, NO. 8 AUGUST2015.
6. LIN Shimin, TIAN Fengzhan, Lu Yuchang. Construction and applications in data mining of bayesian networks. Journal of Tsinghua University, 2001, 41(1).
7. Yucheng Zhang, Jinglong Hu, Jiantao Dong, Yao Yuan, Jihua Zhou, Jinglin Shi. IEEE GLOBAL COMMUNICATIONS CONFERENCE (IEEE GLOBECOM 2009).
8. Fatima MOURCHID, Ahmed HABBANI, Mohamed EL KOUTBI. Mining object patterns for location prediction in mobile social networks.
9. Kiran K. Rachuri and C.Siva Ram Murthy. Level Biased Random Walk for information Discovery in Wireless Sensor Networks.

Visual Tracking via Clustering-Based Patch Weighing and Masking

He Yuan, Hefeng Wu, Dapeng Feng and Yongyi Gong

Abstract A novel visual tracking method via clustering-based patch weighing and masking is presented in this paper. At initialization stage, we divide the object region defined by the given object bounding box and its surrounding background region into non-overlapping patches, and introduce a robust clustering algorithm to build patch-based object and background models. Then a structure mask is constructed from the discrimination between the object and background models. During tracking, we calculate each pixel's object probability inside the search region by patch weighing. The best object location is found by maximum a posteriori estimation with dense sampling and structure masking constraint. Experimental results demonstrate that the proposed tracking method is effective and it outperforms several state-of-the-art methods in challenging scenarios.

Keywords Object tracking · Clustering · Patch weighing · Structure mask

This work was supported in part by National Natural Science Foundation of China (Nos. 61402120 and 61370160), Natural Science Foundation of Guangdong Province (Nos. 2014A030310348 and 2015A030313578), Science and Technology Program of Guangdong Province (No. 2015B010106005), Science and Technology Program of Guangzhou (No. 2014J4100032), and Startup Program in Guangdong University of Foreign Studies (No. 299-X5122029). The corresponding author is Hefeng Wu.

H. Yuan · H. Wu (✉) · D. Feng · Y. Gong
Guangdong University of Foreign Studies, Guangzhou 510006, China
e-mail: wuhefeng@gmail.com

H. Yuan
e-mail: zxcq_123@qq.com

D. Feng
e-mail: dapengfeng@qq.com

Y. Gong
e-mail: gongyongyi@gdufs.edu.cn

© Springer Nature Singapore Pte Ltd. 2018
S.K. Bhatia et al. (eds.), *Advances in Computer and Computational Sciences*,
Advances in Intelligent Systems and Computing 554,
https://doi.org/10.1007/978-981-10-3773-3_67

1 Introduction

Object tracking is an important and fundamental task in many computer vision applications such as intelligent transportation, security monitoring, and robotics. Although much progress has been made in recent years, tracking the object robustly and stably remains challenging due to various factors including illumination variations, occlusion and background clutter, etc.

Building robust appearance models is quite critical in successful tracking. Existing methods generally fall into two categories: holistic appearance models [1, 2] and part-based appearance models [3, 4]. Holistic models can preserve the global information of the object well in some situations, however, it is difficult to adapt to deformation of non-rigid objects. On the contrary, part-based models can better utilize the partial information of the object, but background noise may easily degrade such models without proper formulation of the parts' relationship.

Therefore, in this paper, we propose a novel tracking method with patch-based models and structure masking, which utilizes part-based information by introducing holistic structure mask constraint. Experimental results demonstrate that our method outperforms several state-of-the-art tracking methods in challenging scenarios.

2 The Proposed Method

2.1 Framework Overview

The framework of the proposed tracking method is illustrated in Fig. 1. Given the object bounding box in the first frame, we divide the bounding box and its surrounding region into non-overlapping patches, and we use a clustering method to build patch-based models for the object and background, respectively. A structure mask of the object is constructed from the discrimination between the object and background models. When a new frame t comes, the search region is divided into non-overlapping patches and the object probability of each pixel is calculated by patch weighing. The object is then located by maximum a posteriori estimation with dense sampling and structure mask constraint. We detail the proposed method in the following subsections.

2.2 Appearance Modeling

The modeling of the object appearance consists of two components: clustering-based patch models and structure mask.

Clustering-based Patch Models. Assume the bounding box Ω_1 of a target is given in the first frame I_1 and the size of Ω_1 is $W \times H$, where W and H are the width and

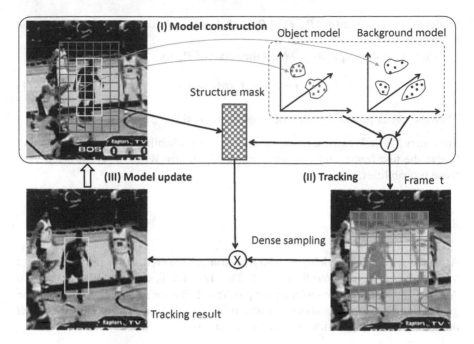

Fig. 1 The flowchart of our tracking method

height respectively. We divide the bounding box Ω_1 into non-overlapping patches of size $s \times s$ and extract a feature vector $\mathbf{u}$ from each patch. Then, we do clustering in those patches, regarding one patch as a unit. The DBSCAN clustering method [5] is applied in our method, because it has the following advantages: The number of clusters need not be predefined and it is quite robust to outliers. After DBSCAN clustering, some core points are obtained in each cluster [5]. The outliers that do not belong to any cluster will be discarded. We choose a small set of K core points randomly from each cluster. These chosen core points, which are also called the core patches, will be stored into an array and form our object model Φ_o.

Meanwhile, we define a background region Ω_1^b with an expanded size e surrounding the object region Ω_1. The size of the expanded region is $(W + 2e) \times (H + 2e)$, and the background region is the annular region outside the object bounding box. We also divide Ω_1^b into patches, and then the background model Φ_b is built in the same way as the object model.

Object Probability. For two patches P_m and P_n, we can compute their distance by $dis(P_m, P_n) = D(\mathbf{u}_m, \mathbf{u}_n)$, where $\mathbf{u}_m$ and $\mathbf{u}_n$ are feature vectors extracted from the two patches respectively.

Then for a patch P_m, we define the distance between P_m and the object model Φ_o as $d_1 = \min\{dis(P_m, P_o^i) \mid P_o^i \in \Phi_o\}$, where P_o^i denotes a core patch in the object model. Similarly, we can also define the distance between P_m and the background model Φ_b by $d_2 = \min\{dis(P_m, P_b^i) \mid P_b^i \in \Phi_b\}$, where P_b^i denotes a core patch in the background model.

After computing d_1 and d_2, we get the score of P_m by

$$\Theta(P_m) = d_2/(d_2 + d_1) \qquad (1)$$

Then the object probability $\Theta(x)$ of each pixel x inside P_m is assigned with $\Theta(P_m)$.

Structure Mask. We calculate each pixel's object probability in the object bounding box in the first frame, and a structure mask M_1 of size $W \times H$ is constructed with these probabilities.

2.3 Tracking

At frame I_t ($t \geq 2$), we define a search region R_t centered at the predicted object location $\hat{x}_{t-1}$ in the previous frame $t - 1$. The size of R_t is set to be $(W + 2e) \times (H + 2e)$. We divide R_t into non-overlapping patches of size $s \times s$ (as depicted by the blue grid in frame t in Fig. 1), and obtain a feature vector from each patch. Then the object probability of each pixel inside R_t can be calculated.

Candidate Likelihood Estimation. For each candidate object location x_t, we estimate its likelihood of being the tracked object by convolution with the structure mask M_{t-1}. Specifically, we approximate the probability $p(x_t|I_t)$ of the object being at location x_t by the following likelihood estimation:

$$p(x_t|I_t) = \frac{1}{|M_{t-1}|} \sum_v \Theta(x_t + v)M_{t-1}(v) + (1 - \Theta(x_t + v))(1 - M_{t-1}(v)) \qquad (2)$$

where $|M_{t-1}|$ is the number of pixels within the mask and v denotes the relative position in M_{t-1}.

Object Localization. We densely sample the candidate object locations in the search region horizontally and vertically with a step size of δ. We find the predicted object location $\hat{x}_t$ using maximum a posteriori estimation:

$$\hat{x}_t = \arg\max_{x_t} \ p(x_t|I_t) \qquad (3)$$

Model Update. When the object location $\hat{x}_t$ is found in frame t and it satisfies the condition $p(\hat{x}_t|I_t)/p(\hat{x}_1|I_1) > \tau$, we will update the mask M_{t-1} to M_t by

$$M_t(v) = (1 - \lambda)M_{t-1}(v) + \lambda\Theta(\hat{x}_t + v) \qquad (4)$$

where the learning rate λ is set to be 0.005. If the condition is not satisfied, we set $M_t = M_{t-1}$.

3 Experiments

3.1 Experimental Setup

The proposed tracking method is implemented in C++ without optimization. In the implementation, we set the patch size $s = 8$ and the expanded size $e = 20$. The feature vector $\mathbf{u}$ of a patch is an HSV color histogram with 48 bins, with 16 bins for each channel. Histogram intersection kernel is chosen as the distance function $D(\cdot, \cdot)$ of two feature vectors. In DBSCAN clustering, we set the minimal size of point $MinPts = 4$ and the scan radius $r_s = 0.6$. The threshold τ is set to be 0.8.

In the experiments, we evaluate our method on 6 publicly available challenging sequences, i.e., *Basketball, Doll, FaceOcc2, Fish, Liquor*, and *Woman*. These sequences cover various challenging factors that often occur in object tracking, including illumination variation, occlusion, deformation, fast motion, background clutters, etc. We compare our method with 5 state-of-the-art trackers: CT [6], CSK [2], VR [1], BSBT [7] and SMS [8].

3.2 Quantitative and Qualitative Analysis

We evaluate the compared tracking methods quantitatively in terms of Center Location Errors (CLE). CLE is defined as the Euclidean distance between the center positions of the ground truth bounding box and the predicted bounding box by a tracker.

Table 1 shows the average center location errors of each tracker on each test sequence. The first and second best results are mark in red and blue, respectively. The average in the last low of Table 1 reflects the overall performance of each tracker on all the 6 tested sequences. As can be observed, our method outperforms the conventional methods obviously in *Basketball, Liquor* and *Woman* sequences, and ranks second best in the other sequences. The overall CLE of our method is 20.4 pixels, which is superior to the other methods.

Table 1 The average center location errors (lower is better). The first and second best results are marked in red and blue

Sequences	Ours	CT	CSK	VR	BSBT	SMS
Basketball	**17.26**	89.43	59.08	55.80	289.76	88.69
Doll	24.68	**20.68**	442.82	46.14	132.84	63.35
Faceocc2	25.64	**24.65**	33.02	64.77	45.15	292.05
Fish	18.38	**14.44**	24.31	24.98	58.19	211.77
Liquor	**18.68**	114.05	27.18	102.75	29.66	57.05
Woman	**17.76**	114.70	583.60	132.15	173.24	119.86
Average	**20.40**	62.99	195.00	71.10	121.47	138.80

Fig. 2 Examples of the tracking results on Basketball (1st row) and Woman (2nd row) sequences. The predicted bounding boxes of the six trackers are presented with different colors: Ours (*red*), CT (*green*), CSK (*blue*), VR (*black*), BSBT (*purple*) and SMS (*cyan*). The CLE plots of each frame are shown in the last column. Best viewed on a high-resolution screen. (Color figure online)

Figure 2 demonstrates examples of the tracking results on two representative sequences (i.e., *Basketball* and *Woman*). In the last column of Fig. 2, the CLE plots of each frame are shown for the compared trackers. It can be observed that our tracker can track the target stably with a constantly low CLE although many challenging factors exist, while the other trackers either lose the target halfway or produce large jitters.

In these two sequences, when the moving object is partially occluded by background clutters occasionally, our method tracks the target acceptably by employing clustering-based patch models. Additionally, in the *Basketball* sequence, the target (Player No. 9) undergoes large deformation and fast motion, but our tracker deals with these situations well. Furthermore, in this sequence, there are several other players wearing the same color uniform as the target, which disturbs some conventional methods and causes them to drift away from the target to other adjacent players. But our method overcomes this challenge by convolving patches with the structure mask.

In the *Woman* sequence, our method is able to track the target well and the performance measured in average CLE is far better than that of the other compared methods. However, it should be noticed that, in this sequence, when the woman walks away from the camera and her size gets smaller and smaller, our tracker does not have the ability to adjust its size automatically. We will take further study in future work to address this problem.

4 Conclusions

In this paper, we have presented a novel visual object tracking method via clustering-based patch weighing and masking. At initialization, we introduce the DBSCAN clustering method to construct the object and background models and generate the structure mask. While tracking, we calculate the object probability of each pixel in the search region by evaluating a patch's discriminative ability between object and background. The object is located by dense sampling of candidates, which are convolved with the structure mask to restrain the influence of background noise. Experimental results demonstrate that the proposed method obtains more accurate tracking results than some conventional state-of-the-art tracking methods in challenging scenarios.

References

1. Collins, R.T., Liu, Y., Leordeanu, M.: Online selection of discriminative tracking features. IEEE Trans. Pattern Anal. Mach. Intell. 27(10), 1631–1643 (2005)
2. Henriques, J.F., Caseiro, R., Martins, P., Batista, J.: High-speed tracking with kernelized correlation filters. IEEE Trans. Pattern Anal. Mach. Intell. 37(3), 583–596 (2015)
3. Hare, S., Saffari, A., Torr, P.H.S.: Efficient online structured output learning for keypoint-based object tracking. In: CVPR, pp. 1894–1901 (2012)
4. Lee, D., Sim, J., Kim, C.: Visual tracking using pertinent patch selection and masking. In: CVPR, pp. 3486–3493 (2014)
5. Ester, M., Kriegel, H., Sander, J., Xu, X.: A density-based algorithm for discovering clusters in large spatial databases with noise. In: KDD, pp. 226–231 (1996)
6. Zhang, K., Zhang, L., Yang, M.: Fast compressive tracking. IEEE Trans. Pattern Anal. Mach. Intell. 36(10), 2002–2015 (2014)
7. Stalder, S., Grabner, H., Gool, L.V.: Beyond semi-supervised tracking: Tracking should be as simple as detection, but not simpler than recognition. In: ICCV Workshop, pp. 1409–1416 (2009)
8. Collins, R.T.: Mean-shift blob tracking through scale space. In: CVPR, pp. 234–240 (2003)

A Presenter Discovery Method Based on Analysis of Reputation Record

Jin-dong Wang, Zhi-yong Yu, Xiang Liu and Miao Sun

Abstract In order to satisfy the requirement of service presenters' credibility for recommendation system, this paper puts forward a method based on the analysis of the presenters' reputation record to find the credible presenters. First, we calculate the preference similarity between different users with their preference vectors and get the presenter initial set. Then we calculate the domain correlation, recommendation response rate and recommendation satisfaction rate to filter the presenter initial set and get presenter candidate set. According to the recommendation history we calculate the presenters' current reputation with introducing the penalty factor. Finally, we filter the presenters' reputation record with localized changes and tendentious changes and calculate the skewness coefficient and kurtosis coefficient of the reputation records. Then we get the presenters' excellent reputation values combined with the expectation and variance to choose the credible presenters who have higher reputation values. The experimental results show that it can improve the accuracy of presenter reputation calculation and the effectiveness of service recommendation.

Keywords Presenter discovery · Reputation analysis · Recommendation · Credible

J. Wang · Z. Yu (✉) · X. Liu · M. Sun
Zhengzhou Institute of Information Science and Technology,
Zhengzhou 450001, China
e-mail: yuzhiyong623@sina.com

J. Wang · Z. Yu · X. Liu · M. Sun
State Key Laboratory of Mathematical Engineering and Advanced
Computing, Zhengzhou 450001, China

© Springer Nature Singapore Pte Ltd. 2018
S.K. Bhatia et al. (eds.), *Advances in Computer and Computational Sciences*,
Advances in Intelligent Systems and Computing 554,
https://doi.org/10.1007/978-981-10-3773-3_68

1 Introduction

In the cloud environment, all kinds of resources are dynamically connected to the Internet and provide services through the Internet. But we cannot judge the quality, reliability and credibility of services before using them. So we need service presenter to recommend services. Therefore the credible service presenters are the key to guarantee service recommendation information available and effective.

For the above application requirements, this paper proposes a method finding service presenters based on the analysis of reputation records. The objective of this paper is to find reliable presenters with high reliability and good recommendation history through the analysis of the presenters' recommendation records. So that we can improve the reliability of service recommendation information and the users choose better services finally. The proposed method creatively uses the line chart of reputation and scientifically analyzes it. We first use the skewness coefficient and kurtosis coefficient in analysis of reputation records. And it takes a good use of the existing user interaction data in system and reduces the cost of information collection. Meanwhile it improves the efficiency of service trust evaluation and the accuracy of looking for service presenters and the selection of high-quality services.

2 Related Work

The service recommendation based on collaborative recommendation has made a lot of research results. Kim [1] stores and manages the feedback of service operation result with a global center and puts forward a more flexible trust evaluation method according to the factors in transaction process. Ma [2] puts forward a service evaluation collection mechanism. An absolute trusted third party manages the service interactive information and takes charge of service trust evaluation. Tang [3] defines the user's trust in the presenter as the weight of recommendation information and gets the service recommendation through correlation factor. Wang [4] transfers and shares service recommend information through a trust network according to the connection relationship and gives weights according to the length of the links. Park [5] fused the trading characteristics and service feedback to the trust evaluation model and calculated the trust of services on the basis of service interaction feedback with adding time attenuation and event impact factor and improved the ability of the algorithm to resist malicious nodes. Pan [6] proposed a presenter discovery method based on reputation. It measures the recommendation trust relationship through relevant factors and calculates the presenters' reputation values through the trust transfer and iteration. Gan [7] proposed a kind of multi-dimensional trust algorithm based on reputation. It sets a confidence factor to synthesize direct trust and recommendation trust and introduce the deal evaluation system and weight system into the multidimensional mechanism which enhance the sensitivity of the trust algorithm. Jeffrey [8] first proposed field relevant model

TBM to evaluate users' reputation in different areas and verify this algorithm combined with the real data and show more effective and robust than other algorithms.

The entity in the service system under cloud environment can be divided into two categories that users and services. Users act as service consumer and service presenter. Users use the services and evaluate services' operating result. And users form recommendation relations [9] by sharing the service using experience.

Each user stores the interaction records and personal trust information in the local. Interaction records including presenter interaction history records and history service interaction experience. Presenter interaction history records store the data of users' interaction with the service presenters including the number of interactions, the number of response, satisfactory interactions, unsatisfactory interactions, the recommended services' information, etc. History service interaction experience stores the interaction information between the users and services including the interaction time, amount, service attributes satisfaction evaluation [10] and the overall evaluation information, etc. Personal trust information includes the description of personal preference [11] to services and personal reputation record. Because each user has the personalized demand preference when he is selecting and assessing services. The users, who have high user preference similarity, have similar service judging standards and have more credibility when selecting service presenters. Personal reputation record is used to record the change of users' reputation values along with time in system and then determine the source of recommendation trust through the analysis and forecast of the user's reputation records.

Definition 1 (*Service Satisfaction*) Users give the subjective assessment of the performance of used services according to their own subjective preferences. Here users composite the each side of services' performance to give the satisfaction evaluation. And it mainly used for updating the presenters' recommendation results and the services' records of the interactions between users and services.

Definition 2 (*Presenter Reputation*) Reputation is the degree of recognition by a group to a user. So presenter reputation means the degree of recognition of user's recommendation action by a group.

Definition 3 (*Recommendation Response Rate*) The ratio of the number of presenter's responses when they receive the requests of services' recommendation information to the total number of their received requests of services' recommendation information. It is used to measure the presenters' participation in the system and filter bad nodes or deprecated nodes.

Definition 4 (*Recommendation Satisfaction Rate*) The ratio of the number of users' satisfaction after they use the recommended services by a presenter to the total number of the presenter's recommendation. It is used to show the quality of presenters' recommendation and calculate the presenters' reputation values.

3   Presenter Discovery Method

This paper proposes a method to find service presenters based on the analysis of reputation record [12] and its basic steps are shown as follows.

1. Target user issue a request of service recommendation information in system. Find the users who used this service according to the interaction of the service record to make up a presenter collection.
2. Calculate the preference similarity between target user and the users in presenter collection. Filter the presenter in collection according to the calculated results and eliminate the presenter who has a low similarity to get presenter initial set.
3. Calculate the domain correlation, recommendation response rate and recommendation satisfaction rate according to the presenters' recommendation record. Filter the presenter in presenter initial set according to the calculated results and form a presenter candidate set.
4. Calculate recommendation reputations and add them to the presenters' personal reputation records.
5. Analyze and forecast the presenters' reputation records. Filter the presenter candidate set according to the filter conditions to get trust presenters.
6. Calculate the services' recommended degree. Users select services to interact and update the presenters' reputation according to the result of users' interaction.

3.1   Presenter Initial Set

The similarity of users' evaluation weights is an important measure of select presenters. Assuming that recommendation information request U_a ask for the trust recommendation degree about service S_b. Each service contains L properties in system $\{f_1, f_2 \ldots f_l\}$. The preference vector [13] of User U_a about service properties is $W_a = (w_{a1}, w_{a2} \ldots w_{al})$ and the preference vector of presenter U_i about service properties is $W_i = (w_{i1}, w_{i2} \ldots w_{il})$ in which $\forall w_{ij} \in [0, 1]$ and $\sum_{i=1}^{l} w_{ji} = 1$. Cosine similarity [14] can calculate the similarity between the preference vectors and more accurately find the presenters with similar preferences. The preference similarity $Sim_p(U_a, U_i)$ between user U_a and U_i shows as follow.

$$Sim_p(U_a, U_i) = \frac{\sum_{j=1}^{l} w_{aj} w_{ij}}{\sqrt{\sum_{j=1}^{l} w_{aj}^2} \sqrt{\sum_{j=1}^{l} w_{ij}^2}} \tag{1}$$

The reference value of the interaction records between users and services will gradually reduce with the growth of the time. The interaction record which is too far away from the current time has no reference value and should not be considered. So we set a time window $[t_s, t_c]$ in which t_s is the current time and t_c is the cut-off time.

System looks for the users who have interactions with service S_t in the time window $[t_s, t_c]$ and make presenter collection as $U_{init} = \{U_1, U_2 \dots U_m\}$. Then it uses the formula (1) to calculate the preference similarity between target user and each presenter in presenter collection. The threshold of preference similarity *per*is set and the presenter in collection is filtered. Remove the presenters whose preference similarity are below the threshold and get the presenter initial set.

3.2 Build Presenter Candidate Set

Presenters' recommendation history can well reflect the ability and the credibility of recommendation. This paper calculates the domain correlation, recommendation response rate and recommendation satisfaction rate to reflect the performance of the presenters from different aspects.

Domain correlation [15] represents the degree of correlation between the presenter and the domain target service. The domain correlation of presenters shows the degree of presenters' understanding and awareness about certain domain. The higher the correlation degree is the more valuable the presenter's recommendation is. So when screening presenters we should choose the presenters with high domain correlation. Here we quantify the domain correlation with the interaction number of the presenter in a certain domain and noted as $\alpha = \sqrt{n/(n+1)}$ in which n is the number of interaction between presenters and certain kind of services.

Recommendation response rate refers to the presenters' response to recommendation information requests. Due to reasons such as subjective factors or network congestion recommendation information requests may not be able to get the response.

Recommendation satisfaction rate shows that the statistics of users' satisfaction with the recommendation by presenters. After a lot of interactions we can get the recommendation response rate and recommendation satisfaction rate. The scene is in accordance with Bernoulli random experiment. There are presenter initial set $U_{init} = \{U_1, U_2 \dots U_m\}$ and presenter set U_i with the recommendation record $R_i = \{rt_i, rb_i, rs_i, ru_i\}$. Among them rt_i signifies the received recommendation request collection, rb_i signifies the collection of recommendation response, rs_i signifies the satisfying recommendation feedback collection, and ru_i signifies the unsatisfying recommendation feedback collection. According to Bernoulli's laws of large numbers, recommendation response rate will converge to the probability of recommendation response with the increasing of the number of rt_i, namely $Rr_i \leftarrow Num(rb_i)/Num(rt_i)$. Recommendation satisfaction rate will converge to the probability of satisfying recommendation with the increasing of the number of rb_i, namely $Rs_i \leftarrow Num(rs_i)/Num(rb_i)$.

Set the thresholds of the domain correlation, recommendation response rate and recommendation satisfaction rate. Filter the presenters in presenter initial set $U_{init} = \{U_1, U_2 \ldots U_m\}$ according to the thresholds and remove the presenters form the initial set that is below the thresholds. The presenters who meet the requirements make up the presenter candidate set $U_{cand} = \{U_1, U_2 \ldots U_n\}$. We can size the scale of the presenter candidate set by changing the thresholds which increase the calculating flexibility.

3.3   Calculation of the Trust Presenters

Calculation of Recommendation Reputation

Presenter reputation [16] is the transmission and summary of trust relationship and it is a consensus of a group towards a presenter's recommendation ability and the degree of trust.

Definition 5 (*Penalty factor ρ*) ρ is the punishment to the presenters who give not real recommendation. According to the fragility of trust when presenters with high credibility do not give true recommendation, the reputation will sharply decrease.

In order to prevent this situation this paper introduces the transaction value impact factor $\Phi(m_{ij})$ when calculating the recommendation trust. Make the recommendation record which has a larger trade amount have the greater contribution to recommendation trust. Calculation formula is $\Phi(m_{ij}) = \exp(-1/m_{ij})$ in which m_{ij} is the trade amount of recommendation by presenter U_i in j times.

Reference value of presenters' recommendation history will decrease with the time. So we set time attenuation factor $\Psi(t_j)$. The recommendation record which is closer form the current moment has the greater contribution to the calculating of recommendation trust. Calculation formula is $\Psi(t_j) = \exp(-\lceil (t_0 - t_j)/T \rceil)$ in which t_0 is the current moment, t_j is the moment when the jth recommendation happen and T is the interval period.

There is presenter U_i recommendation record $R_i = \{rt_i, rb_i, rs_i, ru_i\}$. Among them rt_i signifies the received recommendation request collection, rb_i signifies the collection of recommendation response, rs_i signifies the satisfying recommendation feedback collection, and ru_i signifies the unsatisfying recommendation feedback collection. Thus we can get the recommendation reputation formula.

$$Rt_i = \alpha * Rr_i * \sum_{j \in rs_i} \Phi(m_{ij})\Psi(t_j) - \sum_{p \in ru_i} \rho\Phi(m_{ip})\Psi(t_p) \tag{2}$$

Among the formula, $\alpha = \sqrt{n/(n+1)}$ is the domain correlation and n is the number of presenter using this certain kind of services. α is used to adjust the different domains' influence to reputation. $Rr_i = Num(rb_i)/Num(rt_i)$ is service response rate. The higher the rate is the higher reputation is. So it has incentive effect to motivate presenters to give recommendation. $\sum_{j \in rs_i} \Phi(m_{ij}) \Psi(t_j)$ is the weighted recommendation satisfaction according to satisfaction record of recommendation feedback. The higher recommendation satisfaction of presenter means the services recommendation by the presenter are more reliable. $\sum_{k \in ru_i} \Phi(m_{ik}) \Psi(t_k)$ is the weighted recommendation unsatisfactory according to unsatisfactory record of recommendation feedback. $\sum_{p \in ru_i} \rho \Phi(m_{ip}) \Psi(t_p)$ is the penalties for presenter untrue recommendation and ρ is penalty factor which can set the size according to the concrete situation. By adding the penalty factor we can fortify the sensitivity of the presenters' reputation calculation. The reputation will sharply fall when presenter gives untrue recommendation. Algorithm 1 gives the presenter candidate set construction algorithm.

Algorithm 1 Presenter candidate set building algorithm

Input: target service S_t, all users' information

Output: Presenter candidate set, each presenter's recommendation reputation

Step 1 Look for the users who have interactions with the service S_t in the time window $[t_s, t_c]$ and make up presenter collection.

Step 2 Calculate the preference similarity between the user and the presenter according to the formula (1). Set the threshold *per* of preference similarity and filter the presenters in collection. Then get the presenter initial set.

Step 3 Calculate the domain correlation, recommendation response rate and recommendation satisfaction rate of the presenters in presenter initial set. Set the thresholds to filter the presenters and get the presenter candidate set.

Step 4 Calculate the recommendation reputation of presenters in presenter candidate set according to formula (2).

Analysis and Prediction of Reputation Record

The traditional methods select top k presenters who have highest reputation as trust presenters. However, the presenters' reputations are dynamic values change with time. This paper draws a presenter reputation record line chart [17] to analysis and predicts the presenters' reputation in order to select presenters who have excellent reputations. The reputation records of the presenters store in the local and we can set time interval to draw the reputation line chart. Here we select time interval as 12 h. The line chart is shown in Fig. 1.

If there are n points in line chart continue to rise or drop we call this situation tendentious change and parameter n is manual setting. Tendentious change represents the trend of data changing, so we can forecast the data. If the presenters' reputation records are in tendentious reduction in the current moment they may be in dishonest recommendation or other unsatisfactory condition. So we can consider removing them from the presenter candidate set. We can set the threshold Qt of tendentious change to remove the presenters who are over threshold. For example, the presenter U_3 is in tendentious change in Fig. 1.

If there are continuous m points in line chart above or below the boundary value Dt we call this situation localized changes. Parameter m and boundary value Dt are manual setting. Localized change represents that the data more than or less than the boundary value Dt for a long time. So we can forecast the data in the next phase

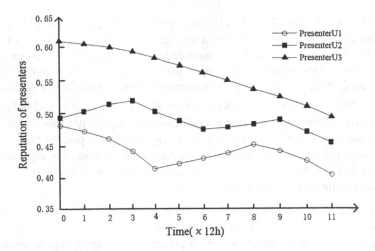

Fig. 1 Presenter reputation line chart

through the localized change. If the presenters' reputation is below the Dt in the current moment and they are in localized change we can predict their reputation will below Dt for a long time. The capability of presenter to recommend or the credibility of presenters if low may lead to its low reputation for a long time, so we can consider removing them from the presenter candidate set. For example, the presenter U_1 is in localized change in Fig. 1.

Skewness is a measurement of the direction and degree of the statistical data's distribution. It is a digital feature to represent the asymmetric degree of the data's distribution. When the data's distribution is symmetrical about the average (expectation) the skewness is 0. When skewness is more than 0, it is right skewness and the number of data which is greater than average is less than the number of data which is less than average. When skewness is less than 0, it is left skewness and the number of data which is greater than average is more than the number of data which is less than average. So we can analyze the presenter reputation records with the coefficient of skewness. When the expectation of two reputation records are the same, we can find the more excellent reputation by compare the coefficient of skewness. With the characteristics of coefficient of skewness we can know that the smaller it is the data migrate greater to the direction of more than expectation and there is more data larger than expect. It indicates that the reputation data is more stable in a value more than expectation. Therefore, under the same expectation the smaller skewness is the more excellent reputation record is. The formula of coefficient of skewness is as follows.

$$SK = \frac{\sum_{i=1}^{N} (X_i - \overline{X})^3}{N\sigma^3} \tag{3}$$

Among them, $\sigma = \sqrt{E(X - EX)^2}$, EX and $\overline{X}$ are the expectation of reputation record, N is the number of reputation data.

Kurtosis is an index used to reflect the top tip of flat of the frequency distribution curve. It can be used to measure the aggregation degree of data in center. Sometimes expectation, standard deviation and coefficient of skewness of two sets of data are the same, but their distribution curve has different top degree. Statistics use the fourth order center distance to measure the coefficient of kurtosis. In order to eliminate the influence of variable value level and different unit of measurement we usually use the ratio of fourth order center distance and variance as a measure of kurtosis index. So we can use the coefficient of kurtosis to calculate the stability of the presenter reputation record. The smaller variance and kurtosis are, the more stable reputation records are and the presenters are more outstanding. The formula of coefficient of kurtosis is as follows.

$$KU = \frac{\sum_{i=1}^{N} (X_i - \overline{X})^4}{(N-1)\sigma^4} \tag{4}$$

Among them, $\sigma = \sqrt{E(X - EX)^2}$, EX and $\overline{X}$ are the expectation of reputation record, N is the number of reputation data. Thus we give excellent reputation Rb_i formula is as follows.

$$Rb_i = \frac{\varphi(\sigma^2)Rt_i + (1 - \varphi(\sigma^2))EX \times (e^{SK} + 1)^{-1}}{In(KU)^2} \tag{5}$$

Among them, Rt_i is the reputation value in the current moment, $\varphi(\sigma^2) = \lambda^{\min(\frac{1}{\sigma^2}, 1)}$, σ^2 is the variance of reputation record, λ is the initial weights value. Algorithm 2 is the trust presenters selecting algorithm.

Algorithm 2 Trust presenters selecting algorithm

Input: The reputation records of presenters in presenter candidate set

Output: Trust presenters

Step 1 Set the time threshold Qt of the tendentious change. Remove the presenters who are in tendentious reduction in the current moment from the presenter candidate set.

Step 2 Set the parameter m and boundary value Dt of the localized change. Remove the presenters who are in localized change and their reputation is below Dt in the current moment from the presenter candidate set.

Step 3 Calculate the coefficient of skewness SK and kurtosis KU according to the formulas 3 and 4. Then get the excellent reputation Rb_i with formula 5.

Step 4 Select top k presenters who have highest excellent reputation Rb_i as the trust presenters.

3.4 Recommendation Record Update

This paper does not discuss the calculation of service recommendation degree in detail and we use the traditional collaborative recommendation mechanism. We send the recommendation information request, collect and integrate the returned recommendation information about the service. Suppose S_t is the target service, $Sat_{t,i}$ is the satisfaction of service S_t from presenter U_i, E_{set} is the collection of trust presenters. So the recommendation degree Rep_{S_t} of target service S_t can be calculated as follows.

$$Rep_{S_t} = \sum_{U_i \in E_{set}} \frac{Sat_{t,i} \times Rb_i}{\sum\limits_{U_i \in E_{set}} Rb_i} \tag{6}$$

When updating the presenters' recommendation records we should compare the service satisfactory degree feedback from the user with the service satisfactory degree from the presenter. If they are similar it indicates that the presenter's recommendation is right and the number of presenter's recommendation satisfactory record should increase. If they have large difference it indicates that the presenter's recommendation is wrong and the number of presenter's recommendation unsatisfactory record should increase.

Suppose $Sat_{t,i}$ is the satisfactory of service S_t from user U_i and its scope is [0, 5]. $Sat_{t,a}$ is the satisfactory feedback of service S_t from service's user U_a. $Sat_{t,b}$ is the recommendation satisfactory degree of service S_t from presenter U_b. Validity of recommendation $TR_{a,b}$ can be defined that $TR_{a,b} = |Sat_{t,a} - Sat_{t,b}|$ and Tr is the threshold of validity judgment. When $TR_{a,b} \leq Tr$ the presenter gives the similar service satisfactory degree with real situation and the number of presenter's recommendation satisfactory record should increase. When $TR_{a,b} > Tr$ the presenter gives the different service satisfactory degree with real situation and the number of presenter's recommendation unsatisfactory record should increase.

4 Experiment and Result Analysis

In order to verify the accuracy and effectiveness of the proposed method, we use peersim [20] simulation platform which has network nodes and the topological relationship to build a service selection application scenario. In this scenario a node represents a user and the simulation platform run according to the rounds. In every rounds, each user starts a service selection which includes that calculate the presenters' excellent reputation, determine the trust presenters, synthesize the recommendation information, select the highest recommended service to interact, document the interaction results and update the record.

4.1 The Experimental Setting

Simulation scenario settings include configurations of service model and user model. Service model contains the number of services and its related properties. Suppose there are 100 services in service selected scenario noted as $S = \{S_1, S_2, \ldots, S_{100}\}$. Each service's satisfactory performance value is initialized to the value of [0, 5]. Each service's satisfactory performance value [18] in every service's running obey normal distribution probability function $N(\mu, \sigma^2)$ in which μ is a service satisfactory performance initialized value and σ^2 is the deviation between the service operation and the μ. Service attributes are transparent for users and users can only rely on the service recommendation degree compounded by the recommendation information from presenters to determine the service's performance. User model contains the number of users and its behavioral attributes. Suppose there are 500 users in service selected scenario noted as $U = \{U_1, U_2, \ldots, U_{500}\}$. Behavioral attributes contains users' personal preference and whether it is malicious. The malicious users will give mendacious recommendation information and service satisfactory degree.

Service satisfactory rate [19] (SSR) describes the average of the service satisfactory degree after users selecting and interacting with the services. The size of SSR is closely related to the quality of the selected services. So it can intuitively reflect the accuracy of the credible presenters' selection algorithm. The more excellent selected presenter is the higher quality recommendation service is and the higher SSR is.

4.2 The Experimental Results and Analysis

We select the presenter discovery method (RRDA) based on reputation proposed by Pan [6] to compare with the algorithm this paper proposed. Compare their differences on the accuracy of calculating presenters' reputation values, finding trustworthy presenters and against the malicious recommendation.

Verify the Accuracy

The experimental values are compared with the algorithms' accuracy of the presenters' selection. We use the RRDA algorithm and this paper's algorithm to calculate the presenters' reputation. At the same time, take the reputation value Rt_i in the current moment (RT method for short) into contrast to show that the excellent reputation proposed by this paper can be more accuracy than the traditional reputation in the current moment. Then select top k presenters who are most credible through the calculation of reputation values. In every round each user selects the highest recommended service to interact and give the service satisfactory degree feedback after interaction. Set the number of selected presenters (k) is 50 and service performance deviation σ^2 is 5.

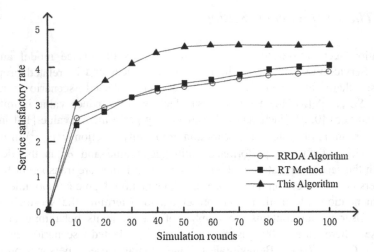

Fig. 2 Service satisfactory rate

The abscissa in Fig. 2 represents the experimental operation rounds and every round each user start a service selection. The ordinate represents the SSR of all users after the rounds run. Through the analysis of the algorithm and experiment results we can know that with the increase of simulation rounds the experience of users' interactions with services are gradually increasing and presenters' reputation are set up. The main reason of RT method slightly better than RRDA algorithm is that RRDA algorithm use adjacency graph with the PageRank algorithm to calculate the presenters' reputation and RT method compute the reputation based on time and amount weights according to the recommendation records. So its accuracy of reputation calculation is higher than PageRank method. The proposed algorithm is obviously better than RRDA algorithm and RT method. The key reason is that the excellent reputation proposed by this paper is calculated based on the analysis and prediction of the presenters' reputation records. Through mapping reputation line chart it comprehensively considers the distribution characteristics and change tendency of reputation records. So excellent reputation can reflect more presenters' information than other two methods and it can more accurately find the credible presenters.

Ability to Resist Malicious Recommendation

In order to verify the algorithms' ability to resist malicious recommendation we set 200 malicious presenters and 300 ordinary presenters. Compare the percentage of the malicious presenters in selected trust presenters calculated by these three methods.

The ordinate in Fig. 3 represents the percentage of the malicious presenters in selected trust presenters. From the analysis we can know that the interaction experience of users is very little at early experiment, so that users have difficulty to identify malicious presenters. The proposed algorithm and RT method can timely

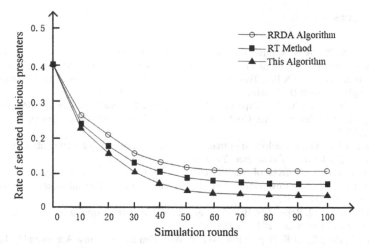

Fig. 3 Rate of selected malicious presenters

reduce the malicious presenters' reputation values by penalty factor so that malicious presenters can be quickly and effectively shielded. But RRDA algorithm which only relies on the PageRank algorithm to calculate the reputation value cannot effectively respond to malicious presenters in time. So its rate of malicious presenters declines slowly and stable value is higher than other two methods. The reason why the proposed algorithm is superior to the RT method is that the proposed algorithm can find the malicious presenters who have high reputation values in the current moment and block them based on the analysis of the reputation records.

5 Conclusions

The proposed credible presenter discovery method based on the analysis of reputation records makes some improvement and innovation. Its main advantages are as follows. First, the proposed algorithm can make presenters to actively provide recommendation information and punish malicious presenters. Thus it can improve the calculation accuracy of presenters' reputation and. Second, it puts forward presenter reputation record analysis method. It combines with the localized change, tendency change, coefficient of skewness and kurtosis to analyze and predict the presenters' reputation records through mapping reputation line chart. Not only it improves the accuracy of the trust presenters' selection but also increases the ability to resist malicious presenters.

References

1. Kim Y A, Phalak R. A trust prediction framework in rating-based experience sharing social networks without a Web of Trust [J]. Information Sciences, 2012, 191 (5):128–145.
2. Jing B X,. XuF. Tao X P.L. Trust model in open environment based on reputation records and its implementation [J]. Electronic journals, 2007,12(12):160–164.
3. Tang J, Gao H, Hu X, et al.Exploiting homophily effect for trust prediction [c]/Proceedings of the 6th ACM International Conference on Web Search and Data Mining. Rome, Italy, 2013:53–62.
4. Wang G, Gui XL. The selection of trading nodes and calculation of trust relationship in social network [J]. Journal of computer. 2013, 36(2):368–383.
5. Kim S, Park H. Effects of various characteristics of social commerce(s-commerce) on consumers' trust and trust performance [J]. International Journal of Information Management. 2013, 33 (2): 318–332.
6. Pan J, Xu F, Lv J. Reputation Based Recommender Discovery Approach for Service Selection [J]. Journal of software. 2010(2):388–400.
7. Gan ZF, Ding Q, Li K. Reputation—Based Multi-Dimensional trust Algorithm[J]. Journal of software. 2011,22(10):2401–2411.
8. Li B C, Li R H, Irwin King, Jeffrey XY. A topic-biased user reputation model in rating systems [J]. KnowlInf Syst. 2015, 44:581–607.
9. Hu J L, Zhou B, Wu Q Y. Research on incentive mechanism integrated trust management for P2P networks [J]. Journal on Communications, 2011,5(32):22–32.
10. Wang HY, Yang WB, Wang SC. A Service Recommendation Method Based on Trustworthy Community [J]. Journal of computer, 2014, 37(2):301–311.
11. Wang PY, Chen EH, Huang B. Personalized requirements oriented trustworthy services recommendation based on social network [J]. Journal of communication, 2013, 34(12):49–59.
12. Medo M, Wakeling JR. The effect of discrete vs. continuous-valued ratings on reputation and ranking systems [J]. EurophysLett, 2010, 91(4):6.
13. Shang YM, Zhang P, Cao YN. New interest-sensitive and network-sensitive method for user recommendation [J]. Journal of communication, 2015, 36(2).
14. Malik Z, Buguettaya A. Rateweb: Reputation assessment for trust establishment among web service [J]. The VLDB Journal, 2012, 18(4):885–911.
15. Vadivelou G, Ilavarasan E. Fusion of pearson similarity and slope one methods for QoS prediction for web services [C]. 2014 International Conference on Contemporary Computing and Informatics (IC3I), 2014:27–29.
16. Li R-H et al. Robust reputation-based ranking on bipartite rating networks [C]. In: Proceedings of the 2012 SIAM international conference on data mining, 2012, 612–623.
17. Abdullah A, Xining Li. An integrated-model QoS-based graph for web service recommendation [J]. 2015 IEEE International Conference on Web Services (ICWS), 2015:416–423.
18. Guo J, Yang Z. Research on PeerSim Simulation Technology [J]. Modern computer, 2010(4).
19. Tian C Q, Zou S H, Wang W D. Trust model based on reputation for peer-to-peer networks [J], Journal on Communications, 2008, 4(29):63–70.

Author Index

© Springer Nature Singapore Pte Ltd. 2018
S.K. Bhatia et al. (eds.), *Advances in Computer and Computational Sciences*,
Advances in Intelligent Systems and Computing 554,
https://doi.org/10.1007/978-981-10-3773-3